Optics: Ultrashort and Ultraintense Laser Science

Edited by **Kristie Ames**

Clanrye International

New Jersey

Published by Clanrye International,
55 Van Reypen Street,
Jersey City, NJ 07306, USA
www.clanryeinternational.com

Optics: Ultrashort and Ultraintense Laser Science
Edited by Kristie Ames

International Standard Book Number: 978-1-63240-403-9 (Hardback)

Printed in China.

Contents

Preface

Optics is generally defined as the scientific study of sight and the behavior of light, or the properties of transmission and deflection of other forms of radiation. The power of laser pulse is growing gradually every year due to development in ultra-short and ultra-intense laser technologies. Due to the easy accessibility of these new machines to researchers, experiments have witnessed total remodeling in non-linear optics. This is because an advanced physical picture of light matter correlation comes to fore whenever the laser pulse intensity approaches or exceeds intra-atomic field power. A very broad band of electromagnetic beam is ascertained when the laser rays are adeptly changed into fluxes of charged or neutral particles. The conservative phenomena of non-linear optics like harmonic generation, self-focusing, ionization, etc. exhibit widely contrasting reliance on laser pulse intensity in contrast to the established rules. This branch of research is developing at a rapid pace. The book imparts knowledge on the latest progress and testified results acquired by writers in some selected areas of this widespread domain of science. This book will provide information to readers who specialize in the subject of laser matter interactions.

The information contained in this book is the result of intensive hard work done by researchers in this field. All due efforts have been made to make this book serve as a complete guiding source for students and researchers. The topics in this book have been comprehensively explained to help readers understand the growing trends in the field.

I would like to thank the entire group of writers who made sincere efforts in this book and my family who supported me in my efforts of working on this book. I take this opportunity to thank all those who have been a guiding force throughout my life.

Editor

Part 1

Femtosecond-Time-Scale Physics

1

Magnetization Dynamic with Pulsed X Rays

Boeglin Christine
Institut de Physique et de Chimie de Strasbourg, Université de Strasbourg, France

1. Introduction

Lasers have become more and more useful and a large field of application is nowadays reached including medicine, biology but also fundamental research as physics for instance. It is also in the fundamental research area that recently a fast developing new field is growing: Ultra-short high-energy pulsed X rays. Compared with the lasers community where first technological developments were recently achieved [Spi1997, Dre2001 Schn1999, Kra2009] in order to reach higher energies (5-100 eV), the X-ray community is using high energy X rays from large facilities, for instance the synchrotron storage ring facilities were a large UV and X-ray energy range is produced but were time resolved spectroscopy is only starting since a few years [Sch2000, Scho2000, Hol2005]. It is my aim here to describe the actual state of the art in the field of X rays and especially concerning the different X-ray pulse length and intensities. In the second part I will develop the application in the field of magnetism of the time resolved X-ray spectroscopy and microscopy.

The description of the High-energy X-ray pulse section (2.) will include technical details about the energy range of the X rays, the different time resolution and density of photons produced in the facilities as Synchrotron and X-ray Free electron lasers (X-FEL). The f-slicing possibilities at BESSY (Germany) and also the X-FEL facilities in Europe and in USA will be developed. The recently launched free-electron laser at the FLASH facility in Hamburg and LCLS in Stanford are the two first free electron sources in the world.

Description and discussion of applications using the pulsed X-ray sources are given in section (3.) and will introduce some of the actual motivations in the field of ultrafast magnetization dynamics using ultrafast X-ray pulses. It is divide into two sub-sections; one concerning the spectroscopies performed using the time structures of X rays and the second the time resolved imaging techniques actually developed in the world.

2. Time resolved spectroscopy's using the temporal structure of X rays

In recent years, magnetism at ultrafast time scales has been a growing topic of interest. A thorough understanding of femtosecond magnetism will address the important questions of how fast the magnetization can be reoriented in a material and what physical processes is behind and limits to this speed. In the spatial domain, magnetism at nanometer length is a topic directly relevant to data storage, since future advances in this technology will require a further reduction in device dimensions to increase the storage density. These considerations have motivated a variety of studies using magnetooptic effects in conjunction with ultrafast light pulses to explore these fundamental limits. These studies currently make use of visible-wavelength light from ultrafast lasers, or X-rays from large-scale synchrotron x-ray facilities. Ultrafast lasers produce short pulses (~30 fs), making possible femtosecond time resolution [Beau1996, Cin2006], but with a spatial resolution that is generally limited by the wavelength of the probe light. X rays, on the other hand, allow for high spatial resolution and high contrast imaging at the elemental absorption edges of ferromagnetic materials. However, the available time resolution to date is too slow to resolve the fastest dynamics. Because of this, significant efforts have been devoted to using short or isolated electron bunches of X rays pulses at synchrotron to perform time resolved microscopy with X rays. More recently femtosecond strong laser pulses are used to slice short burst (100 fs) of X rays from synchrotron radiation [Stam2007, Boe2010] Magnetic imaging techniques as for instance X-ray PhotoEmission Electron microscopy (X-PEEM), Scanning Transmission X-Ray Microscopy (STXM) or X-ray Resonant Elastic Scattering (XRES), are currently using the short X-ray pulses in order to accede to time resolved imaging in the picosecond time range. Unfortunately, the f-slicing technique in synchrotrons produces a strongly reduced photon flux hindering the f-second magnetic imaging at facilities as synchrotrons.

2.1 Magnetic imaging using the ps time structure of the synchrotron

2.1.1 Magnetic domains and Vortices under magnetic field pulse excitations

In order to move magnetic domains one of the simplest way one can think of is to apply short magnetic field pulse perpendicular to the magnetization. In this way the field will exert a torque on the sample magnetization vector and induce a rotation of the spins. In a second step the out of equilibrium spins will start to relax in order to transfer the energy from the external field to the lattice, by characteristic precession and damping mechanism. Many experimental description of this process in soft and hard magnetic materials were performed aiming to model the dynamic of relaxation mechanisms in the pico and nanosecond time ranges. Even if the simple idea of a magnetic field pulse excitation is strike forward compared with electronic excitations, in practice this method suffers from the difficulties to produce strong and short magnetic pulses as well as sharp on and off sets (rise times) of the magnetic pulses. Several methods for the generation of magnetic field pulses have been used. Electrical pulse generators for instance (limited by the self-inductance of the electric circuit) with rise times of more than 100 ps and further lithography "stripe lines" were developed in order to reduce the rise times [Ele1996]. Further improvements of the rise-time was archived using optical switches, which can be optically controlled and which are based on lithography fabricated photoconductive " Austin" switches (based on metal-GaAs-metal junctions) [Ger2002] or alternatively "Schottky diodes" switches (based on metal-semiconductor junctions) [Acre2001)]. Beside the large ~50 ps rise times a second limitation is the low induced magnetic fields (~0.1 T) produced by the set-up at the sample location. This often limits the experiments to soft material as permalloy and soft CoFe alloy films (Fig1). Such systems where extensively studied in the past 10 years focusing on reduced dimensions in nanostructures and lithography designed vortices structures.

[Cho2004, Schne2004, Raa2005, Weg2007, Kras2005, Kuc2004, Vog2005, Vogel 2005, Fuk2006, Vog2008, Hey2010, Uhl2011]

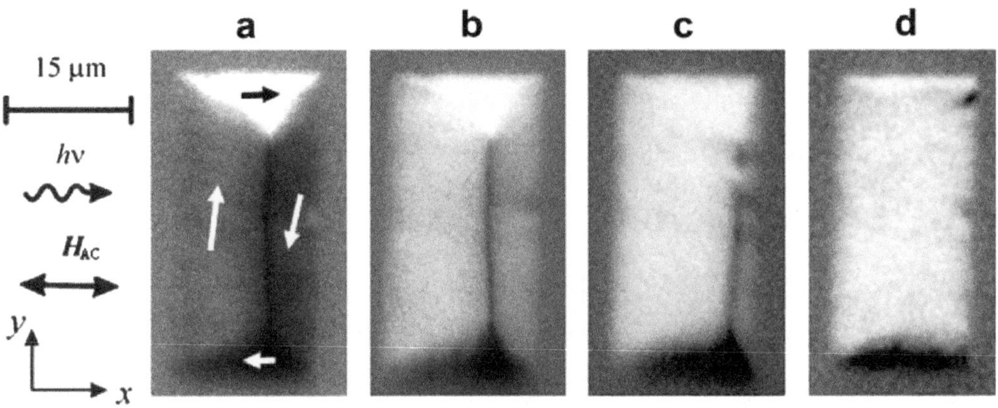

Fig. 1. Magnetic response of the x-component of the magnetization (bright areas are magnetized to the right, dark areas to the left) in a permalloy platelet of 16 · 32 µm2 size and 10 nm thickness for three different field amplitudes I (1.5 Oe), II (2.0 Oe) and III (2.5 Oe). (a) XMCD-PEEM snapshot of the domain pattern in dynamic mode at excitation amplitude I; arrows denote the local magnetization direction. (b)-(d) Snapshots of magnetic domain patterns at maximum magnetic response excited with increasing amplitudes. [Weg2007]

Furthermore, extremely large effective magnetic field pulses can be produced by femtosecond laser pulses combined with the heating of an exchange-biased system. Recently it was suggested that ultrafast switching could be induced via laser-induced reorientation of an exchange coupled antiferromagnet such as $TmFeO_3$ [Kim2005]. A strong magnetic field pulse has also be generated by a relativistic electron bunch combining short duration of 1ps and high field strength ~100 Tesla [Stam2005]. The counterpart of such experiments is that it is accompanied by a strong electric field. Up to now, no time resolved study using a pump-probe set-up has been archived using these high magnetic field pulses.

Time-resolved scanning transmission x-ray microscopy (STXM) in NiFe thin films was studied in order to define the role of domain wall pinning on the dynamic behavior of magnetic vortex structures [Van 2008]. The X-ray magnetic circular dichroism (XMCD) effect, was used as contrast mechanism for the imaging of the structures (Fig 2). In contrast with the X-PEEM, the STXM geometry is sensitive to the projection of the magnetization along the photon propagation direction; therefore, the in-plane magnetized sample was tilted over 60° with respect to the incoming photon beam in order to observe the magnetization. A full image can be constructed by scanning the sample along both in-plane directions. The lateral resolution is determined by the zone plate of the beam line and is about 30 nm. Time-resolved measurements were performed in order to investigate the dynamic behavior in magnetic vortex structures. The natural time structure in the storage ring of the synchrotron delivers photon flashes every 2 ns in the so-called multibunch mode.

This allows the experiment to follow a typical pump-and-probe scheme, with the incoming photon flashes as probe and the externally applied in-plane magnetic field pulses as pump. The magnetic structures were repeatedly excited every 82 ns by sending an electric current in the stripline underneath the structures. The current pulses induce magnetic field pulses with amplitudes of about 10 mT and a full width at half maximum of about 1 ns (500 ps of rise and falling time). The excitation was synchronized with the x-ray flashes of the synchrotron, which probe the magnetization at different times t after the pump. The analysis of the dynamic behavior of the vortex gyration frequency show that they are increased in square-shaped structures, where domain walls are present suggesting that the domain wall pinning is causing the increased frequency.

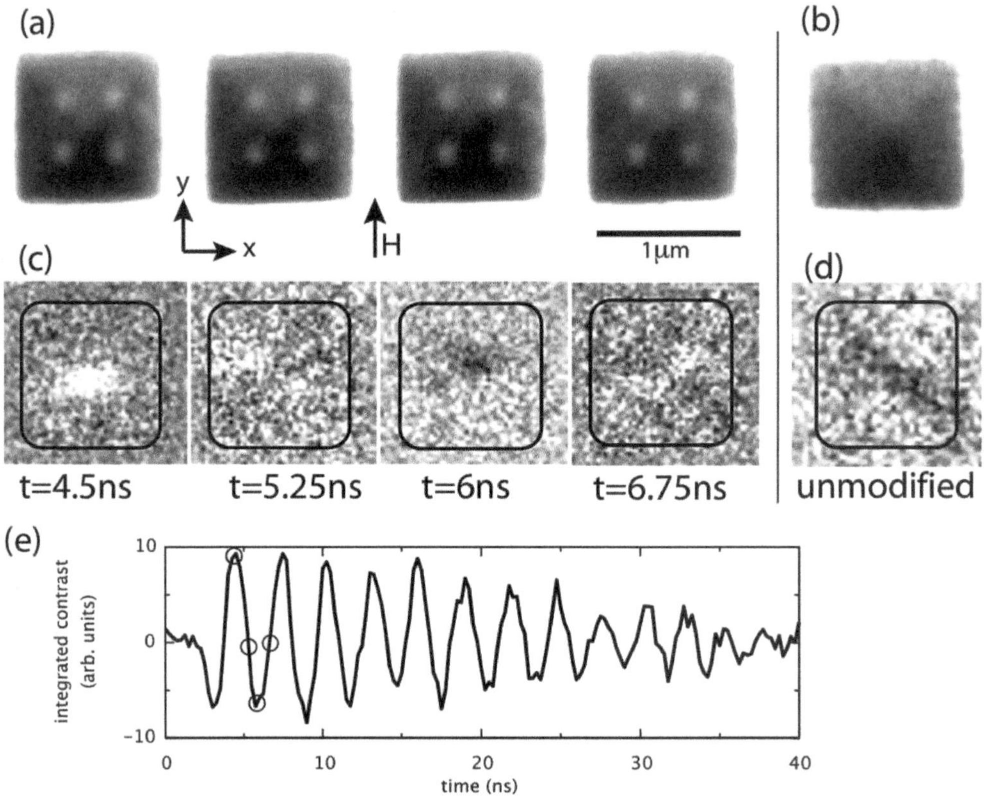

Fig. 2. (a) Sequence of STXM images for a 1 μm x 1 μm x 50 nm modified square during one period of the oscillation. (b) image of an unmodified structure and shows that the domain wall motion can span a larger area of the structure when no defects are created. The intensity is proportional to the x component of the magnetization, revealing the Landau configuration and the small gyrotropic motion of the vortex structure. The total contrast in this sequence oscillates with the resonance frequency, as shown in (e). The four STXM images in (a) and (c) correspond to the four markers in (e). The magnetic pulse H starts at t=2 ns. [Van 2008].

2.1.2 Magnetic domains under femtosecond laser excitation

In order to study the magnetization dynamics in oriented ferromagnetic domains, after a femtosecond pump laser excitation, a precise nanometer scale characterization of the magnetic domain contrast and domain configurations is of great importance. This more recent studied aspect of the space resolved dynamics aims to discribe the influence of a laser excitation on the magnetic domains in different time ranges (nanosecond, picosecond and femtosecond scales). The characterization of the dynamic of the magnetic domain configurations helps to understand the demagnetization process because it provides a description of the magnetization in space. Using X rays these studies benefit from the chemical sensitivity of the circular polarized X rays and from the high spatial resolution (30 nm) of the magnetic imaging mode of X-PEEM instruments that are nowadays currently working at synchrotron storage rings. Appropriate femtosecond pump laser can easily be implemented one such instruments in order to address thermal effects of the laser pump in the ps range. The ultrafast modifications induced by a infra-red laser pump on the magnetic domain configurations is still unknown. Questions concerning the induced changes in the magnetic contrast in a magnetic domain and the size and shape of the domains are still pending. The typical time resolution of the actuel experiment is ~60 ps using the multibunch mode and 10 ps using the low alpha operation modes currently provided in synchrotron storage rings. The time resolution limitation is strongly related with the limited X-ray flux and with the imaging technique by them self where high flux is mandatory.

One of the interesting subjects today is the study of the dynamics in the picoseconde time range of domain sizes and of the magnetic contrast provided using the X-ray circular magnetic dichroism (XMCD) as a function of the pump probe delay. This can be studied either in in-plane oriented magnetic domains or in perpendicular oriented domains.

Following the excitation of ferromagnetic materials with ultra-short laser pulses, a sequence of relaxation mechanisms takes place. The first one is related to the ultra-fast demagnetization. The second mechanism is related to electron - spin and lattice energy transfer, most important within a few picoseconds after the excitation. This mechanism depends on several parameters: the electron-phonon coupling, the material's specific heat, the magneto-crystalline anisotropies and specific interactions like the ferromagnetic or anti-ferromagnetic coupling. One of the goal is to correlate the results with the laterally averaged spectroscopic information obtained using XMCD time resolved spectroscopy.

The experimental method consists in measuring the FM domain contrast in ferromagnetic materials set into a remanent state. Magnetic imaging in the pump probe configuration is setup using the triggered imaging detection mode to obtain a XMCD contrast image at the Fe, Co, or Ni L_3 edges. Moreover the laser fluence necessary to de-magnetize the films is typically in the order of a few mJ/cm^2 and, in the best cases, this allows achieving complete demagnetization of the films. Using focalization of the laser this can easily be achieved by focusing the laser spot onto e few 10 micronmeter on the sample surface. The XMCD signals are probed in a gated mode at different time delays between the laser pump pulse and the probe pulse of circularly polarized synchrotron radiation. The time resolved magnetic signal is extracted from a time-delay sequence of XMCD images and will allow extracting the magnetic components in a semi-quantitative way as a function of time delay. Intensive research in this field is developing using the X-PEEM imaging technique and extension toward other techniques as time-resolved scanning transmission x-ray microscopy (STXM) is expected soon.

2.2 Spectroscopy using slicing techniques or X-FEL pulses

2.2.1 Pump probe with lasers using f -slicing

In order to perform experiments using ultrashort X-ray pulses of only ~100 fs in synchrotrons storage rings one had to modify the large electron bunch time structure of 60-80 ps. This can be performed by using a femtosecond laser pulse to slice the electron bunch. The first generation of fs X-ray pulses in third generation synchrotron radiation sources was proposed [Zho1996] and experimentally demonstrated at the Advanced Light Source (ALS) in Berkeley [Sch2000, Scho2000] using x-ray radiation from a bend magnet. The first undulator-based facility was constructed and successfully commissioned at BESSY [Holl2005].

Such an installation has been set up at BESSY (Berlin) and also at SLS (Villigen) and consists on a slicing of the electron bunches using a femtosecond infra-red laser [Kah2005]. The source at BESSY is based on laser-induced energy modulation ("femtoslicing") and subsequent angular separation of the short-pulse x-rays emitted by an elliptical undulator. The femtosecond X-ray source is thus delivering X-ray pulses of 100 fs (fwhm) duration with tuneable polarization.

The electronic synchronization between the laser pulse and the electron bunches is adjusted so that the electric field of the laser interacts with the bunches at the maximum of the intensity (Fig 3). A specific insertion device names Modulator hosts the laser-electron bunch interaction where the femtosecond laser pulse copropagates with an electron bunch, causing an oscillatory energy modulation of the electrons in the short overlap region. The off-energy electrons are transversely displaced by dispersive elements in order to extract the short component of radiation emitted in a subsequent device (the "radiator"). The second device (Radiator) deviates the two electron bunches with a different angle, so that the angular separation allows extracting only the short radiation component.

The THz signal is the prime diagnostics tool for optimizing the femtoslicing source, when starting an experiment. In addition to being crucial for diagnostics of the laser-electron interaction, the THz radiation itself is useful for experiments where intense ultrashort THz pulses of well-defined temporal and spectral characteristics are required [Holl2006].

The ultrashort X-ray pulses produced by slicing thus provides a strongly reduced flux of 10^4 photons s^{-1} mrad-2mm-2per 0.1% BW, compared to 10^6 photons s^{-1} mrad-2mm-2per 0.1% BW using the single electron bunch. The static measurements using all the bunches we can typically expect at 700 eV a flux of 10^{13} photons s^{-1} mrad-2mm-2per 0.1% BW. The reduction of the flux us thus extremely important when performing time resolved experiments and is in the limit of any experimental set up possibilities when using the sliced beam. This motivates to develop a Bragg-Fresnel zone plate beam line were a photon flux of more than a factor 10 is provided. The energy range of the X rays produced at the beam line at BESSY II ranges from 600 eV to 1400 eV.

The pump probe experiment using such slicing set up are done in a specific pump-probe geometry using the transmitted X-rays at the element core level threshold (ex: Fe, Co, Ni L_2 and L_3), as a probe and a femtosecond laser as a pump. They were carried out using the circularly polarized X-ray femtosecond pulses [Sta2007, Boe2010]. The pump pulse with FWHM of 60 ± 5 fs is issued from an amplified Titanium Sapphire oscillator at a central wavelength 790 nm and amplified at 1.5 kHz repetition rate. The pump pulse is synchronized with the sliced electron bunch of the storage ring. The incident X-ray beam is perpendicular to the surface or at 30 degree from the normal and transmitted through the film deposited on a Si_3N_4 membrane. This geometry allows an optimization of the pumping through the film and of the amplitude of the X-ray magnetic circular dichroic (XMCD) amplitudes when performing the time resolved measurements. The transmitted X-ray intensity is measured using a fast Si avalanche photodiode, and a gated boxcar. The measurements are made by making the difference between the transmitted signals obtained for two opposite applied magnetic fields. The spins are then aligned either parallel or anti-parallel to the incoming circularly polarized X rays.

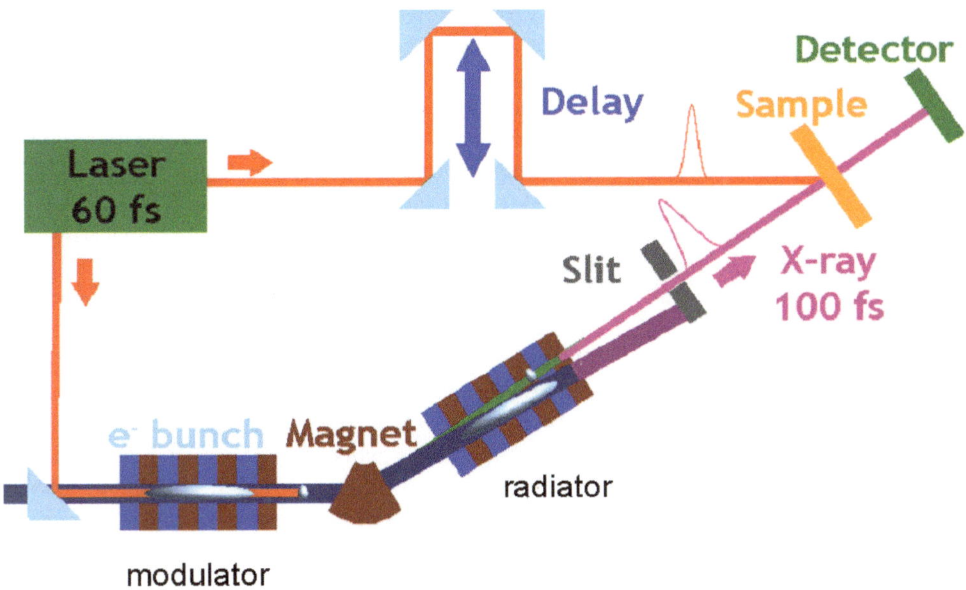

Fig. 3. Slicing experimental geometry: The femtosecond laser is divided into two branches. One is devoted to the slicing inside the ring and the second branche is used to pump the sample at the experimental end station.

The XMCD contrast is obtained by subtracting the gated signals obtained with and without pump beam. The numerical XMCD values are obtained from the normalized difference of the signals recorded near the edges, for an energy position where the static XMCD signal is maximum. The results are normalized in order to account for the degree of circular polarization of the sliced X-rays (70%) as well as the moderate energy resolution of the zone plate (5 eV). The limited energy resolution of the zone plate (5 eV) used in such experiments ensures that a "integrated signal" over 5 eV is measured and allows us to apply the sum rules and to extract the spin and orbital magnetic moments [Car2009, Boe2010].

2.2.2 Pump-probe experiments using coherent X-FEL pulses

The recent development of ultrashort soft X-ray pulses, as provided by femto-slicing in conventional synchrotron storage rings, or by X-ray free electron lasers, opens today new perspectives in the femtomagnetism field. The free-electron lasers are now operating at Stanford (LCLS- USA) and at Hamburg (FLASH- Germany) producing very short and intense coherent X-ray pulses. The energy ranges at FLASH spreads from 20 eV to 200 eV and at LCLS from 400 eV to 2000 eV. One of the new opportunities at such sources are the pump-probe single shot imaging using the coherence of the source. For this purpose intensive work has been performed in order to define the imaging techniques that will permit to reach not only the ultimate time but also the ultimate space resolution in order to progress in the understanding of ultrafast magnetism.

The ultimate X-ray microscope provides a resolution that is only limited by the wavelength of the radiation. The resolution of STXM, however, is limited by the spot size on the sample. Much simpler is the image formation process using Fourier transform holography (FTH), where the scattered radiation from the sample interferes with a reference wave and forms a hologram on the detector. Reverse Fourier transform of the measured diffraction pattern yields an unambiguous image of the object. As the phases are encoded in the hologram, several numerical contrast enhancing procedures, can be applied to the image. The spatial resolution in FTH-based methods is limited by the size of the reference aperture-today FTH masks can be routinely produced with reference holes of 30 nm size. However, the image obtained by reverse Fourier transform provides an excellent starting point for a further phase retrieval treatment. In such a way the resolution limitation of FTH can be overcome. FTH is especially attractive in the soft X-ray regime where the photon energy can be tuned to element-specific core level energies allowing for element-specific contrast in the images. This can be used for example to image magnetic domain structures, using X-ray magnetic circular dichroism or for anomalous diffraction imaging.

The recently launched free-electron lasers (FLASH and LCLS) are the first such sources covering the spectral range of relevance for magnetization studies in 3d metals. These novel x-ray sources are able to generate X-ray pulses as short as 10 fs with up to $\sim 10^{12}$ linearly polarized photons. The short pulse duration, brightness, coherence, and well-defined polarization of the x-ray radiation are the main ingredients that may allow realizing femtosecond single-shot visualization of sub 100 nm magnetic domains [Eis2004].

In particular, it has been shown that if an ultrashort, bright, and coherent X-ray pulse illuminates a sample, the resulting far-field diffraction pattern will encode the image of the sample, from which it can be reconstructed [Eis2004, Chap2006, Gun2011]. The temporal resolution of such a single X-ray pulse snapshot image is then given by the duration of the x-ray pulse (10 fs - 100 fs).
One should remember that such an approach requires not only a very short, but also a very bright x-ray pulse and the large amount of energy deposited into the sample will ultimately turn it into a plasma. Chapman *et al.* demonstrated, however, that the destruction of the sample is not an obstacle for ultrafast "flash diffractive imaging" [Chap2006] as long as the coherent diffraction pattern is 'created' before the sample is destroyed. In order to reach higher energies than the one obtained at the fundamental wavelength (at FLASH -7.97 nm) one can also operated at the fifth harmonic originating from self-amplified stimulated emission at 1.59 nm. Using this operating mode resonant magnetic scattering at FLASH has been performed recently [Gut2009] by using a Co/Pd multilayer sample that was illuminated with 20-fs-long soft X-ray pulses tuned to the Co L_3 absorption edge at 778.1 eV.
More recently, Gutt *et al.* have applied the idea of ultrafast "flash diffractive imaging" to magnetic studies [Gut2010] performing a single-pulse resonant magnetic scattering experiments (Fig 4). By tuning the wavelength to one of the magnetically dichroic absorption resonances of cobalt (the M_3 edge around 60 eV in their case), one may achieve substantially different absorption of polarized X rays in the domains with different orientation of spins. Therefore, coherent and polarized X rays will diffract from such a sample and a far-field diffraction pattern will be formed. The authors performed a simple analysis of this pattern, being able to extract information about the size distribution of the magnetic domains. Due to the linear polarization of the FEL light, however, Gutt *et al.* did not obtain a real image of the magnetic domains. Nevertheless, the technique clearly demonstrates its ability to probe sub 100 nm magnetic domains with a single fs X-ray pulse.

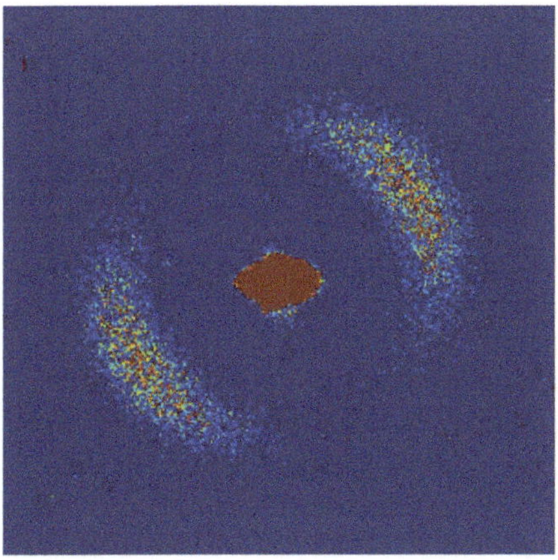

Fig. 4. CCD image of the magnetic diffraction pattern recorded with soft X-ray radiation at Co L_3 edge, using the fifth harmonic at FLASH (photon energy of 778.1 eV) [Gutt2009].

In the near future, we anticipate that such a technique, in combination with further development of 4th generation synchrotron sources will dramatically improve our understanding of ultrafast magnetization dynamics and femtosecond laser control of magnetism. The advantage of such X-ray sources for our purpose is the high X-ray peak power, the very short pulse duration (down to ~10 fs), the high coherence and the tenability of the X-ray photon energy.

3. Ultrafast magnetization dynamics on the nanoscale

3.1 Magnetization dynamics in magnetic solids

In solids the magnetization reaction upon external disturbances as for instance temperature, external magnetic field pulse or pulsed magnetic or electric fields. The induced changes in the magnetization shows different time scales, and different characteristic length scales and sizes for the magnetic structures, domains and domain walls, leading to intense work in this research field during the last decades.

Since the development of the magneto-optics using pulsed lasers has opened a new field of research named ultrafast magnetization dynamics many different experimental and theoretical work was performed. All this work concentrate on pump probe experiments were fs laser excite the ground state in ferromagnets. The development of this field was unambiguously correlated with the ability to perform time resolved spectroscopy below 1 ps which is the range of interest because they naturally corresponds to important magnetic energies, as given by the time-energy correlation t = h/E which links the cycle in time t to a characteristic energy E. For 3d elements this leads to characteristic times of a few ns for anisotropy energies in the 10^{-6} - 10^{-3} eV range, of a few ps for spin-orbit energies in the 10^{-2} - 10^{-1} eV range, and of a few fs for the inter-atomic exchange energy of ~ 5.10^{-1} eV.

From the discovery of subpicosecond demagnetization over a decade ago [Beau1996] (Fig.5) to coherent interactions between laser and spins [Zha2000, Zha 2008, Big2009] and to the recent demonstration of magnetization reversal by a single laser pulse [Stan2007], the manipulation of magnetic order by ultrashort laser pulses has become a fundamentally chanllenging topic with a potentially high impact for future spintronics, data storage and manipulation.

The signal is normalized to the signal measured in the absence of pump beam. [Beau1996]

The recent development of ultrashort soft X-ray pulses, as provided by femto-slicing in conventional synchrotron storage rings, or by X-ray free electron lasers, opens today new perspectives in the femtomagnetism field. Indeed, thanks to the use of sum rules, time resolved XMCD might be viewed as a quantitative measurement of dynamical magnetism, allowing an unambiguous assessment of the magnetization relaxation time, thus confirming previous magneto-optical measurements.

In this context a new milestone has been set by Boeglin et al. [Boe2010] who observed, using time resolved X rays, how ultrashort laser light pulses modify the orbital angular momentum of electrons before it is transfered to the spins (Fig.6). By disentangling the changes in these two components the group showed that spin-orbit coupling can be manipulated on the femtosecond time scale before any lattice or structural transformations occur. The dynamics of spin and orbit angular moments were measured separately in out of plane oriented CoPd alloys, were different orbital and spin dynamics was evidenced [Boe2010]. This work shows that the projections along the easy magnetization axis z the orbital moment is faster by around 60fs than the spin showing that an ultra-fast quenching of the magneto-crystalline anisotropy occurs. This result could be a clue for understanding the laser induced demagnetization process, since microscopic theoretical interpretations take into account spin-orbit interaction (SOI) [Zha2008, Koo2009, Kaz2009] and demonstrates that at time scales shorter than 100fs, one may enter the regime of the SOI. Since the magneto-crystalline anisotropy, which dictates the direction where the magnetization is directed, also relies on the SOI, understanding magnetization dynamics at such time scales may help finding new routes for ultrafast magnetization manipulation.

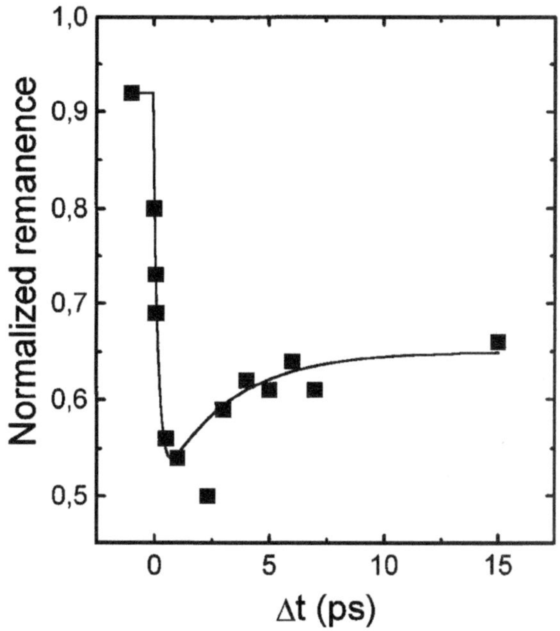

Fig. 5. Transient remanent longitudinal MOKE signal of a Ni(20 nm)/MgF2(100 nm) film for 7 mJ /cm^2 pump fluence.

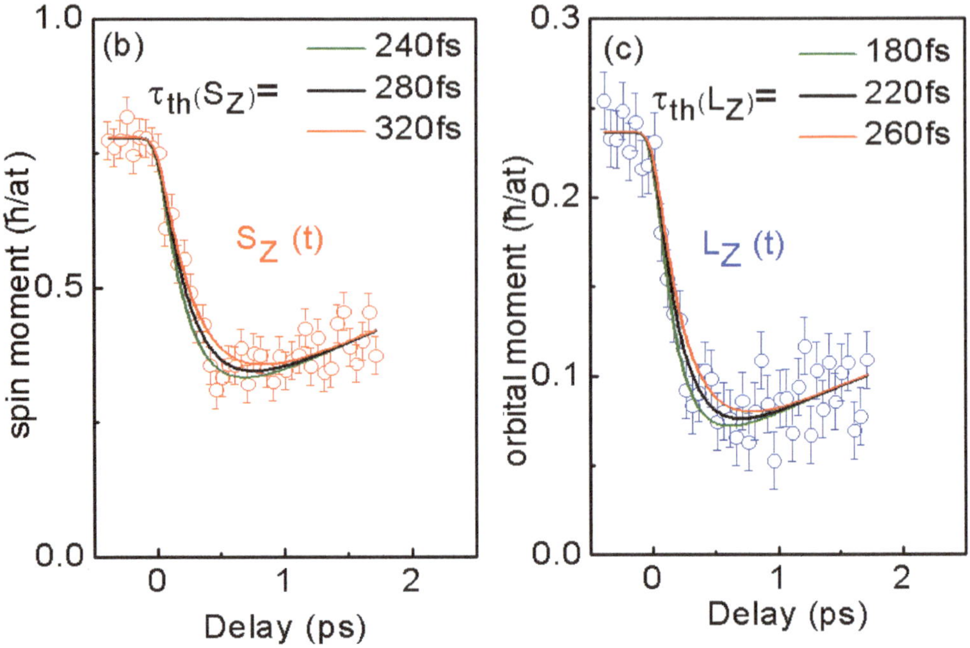

Fig. 6. Ultrafast dynamic of the spin and orbital magnetic moment measured using the pump probe set-up for a CoPd film at the $CoL_{2,3}$ edges. The X-ray probe beam are generated with the f-slicing set up at HZB-BESSY II leading to a to a time resolution of 130 fs [Boe2010]. Two different thermalization times were found for the spin and for the orbital magnetic moments. The best results of the fit procedure lead to $\tau_{th}(S_z)$ = 280 fs and $\tau_{th}(L_z)$ = 220 fs.

Developments in the field of magnetization dynamics naturally lead us to ask if there is a physical limit to the speed at which magnetic moments can be switched. Moreover, exploring this limit is complicated, partly because spin reorientation and switching from one direction to the other can occur in multiple ways and along different paths. For example, magnetic and electric fields, electric currents, and laser pulses can all stimulate magnetic reorientation and the trajectory of the magnetization vector from its initial to its final state will vary with each of these mechanisms.

So far, groups have mainly looked at ways of turning and redirecting the magnetization continuously, typically by causing it to precess with magnetic field pulses [Schu2003]. Using purely optical methods, Vahaplar *et al.* show that a faster way to switch the magnetization is to temporarily quench it [Vah2009] and restore it immediately afterwards in the opposite direction, a scheme they call a *linear* reversal.

Their experiments are an ingenious combination of the different effects by which light interacts with magnetic moments. In their setup, Vahaplar *et al.* first stimulate the magnetization of amorphous 20 nm ferromagnetic films made of $Gd_xFe_{100-x-y}Co_y$ with a short and intense circularly polarized (pump) laser pulse and then image the magnetization with a second, equally short but linearly polarized (probe) laser pulse.

The first laser pulse has two effects on the magnetization. First, it rapidly pumps energy into the film, locally heating the material and demagnetizing it. Changes in the electronic temperature affect the magnetic properties on sub-ps time scales. Most importantly, the magnitude of the magnetization M decreases as the temperature of the electronic system approaches the Curie temperature. The first laser pulse also affects the magnetization via the inverse Faraday effect [Far1846, Ziel1965]: as the circularly polarized electromagnetic field pulse traverses the sample, it acts as an effective magnetic field along the pulse's propagation direction. This effective magnetic field is proportional to the intensity of the laser pulse and to its degree of circular polarization. The inverse Faraday effect provides outstanding possibilities to control the magnetization, since it can generate locally enormously strong effective magnetic fields of up to about 20 T. It can switch the magnetization as well, since the sign of the field only depends on the pulse's chirality. This optomagnetic, nonthermal control of the magnetization was first demonstrated by the Nijmegen group in 2007 [Stan2007]. Essentially, they showed that laser pulses as short as 40 fs could induce optomagnetic switching, but it was not clear how much time the magnetization required to complete the switching process after the exposure to such a short pulse.

By carefully varying the delay between the circularly polarized pump pulse and the linearly polarized probe pulse, the authors could obtain precise information on the spatiotemporal evolution of the magnetization in the film. They found that the switching process completes within a time well below 90 ps, which is very short but still much longer than the duration of the pulse. Recently the authors showed that 30-50ps is the ultimate limite for the swiching time using the invers Faraday mecanism.

The magnetization reversal is connected with a change of angular momentum, which must be provided from somewhere. Yet, it is generally agreed that the apparently simple assumption of a direct transfer of the photon spin to the magnetic system is not the solution [Koo2000], suggesting that the atomic lattice may play an important role in angular momentum conservation. This makes the question about spin to lattice (spin-phonon coupling) an important issu for a complet theoretical understanding of femtomagnetism. Recent experiments, using femto-slicing concluded that the angular momentum transfert is not using the orbital momentum to transfert from the spins to the lattice [Boe2010]. The measurements performed on CoPd alloys (fig 6.), show that ultrashort laser light pulses modify the orbital angular momentum of electrons before it is transfered to the spins, defining the correct sequence of transfert between orbit, spin and lattice. Different from the inverse Faraday effect the ultrafast manipulation of SOI is expected to transform and redirect the spins just by a single laser pulse by modifying the electronic anisotropy of any system at speeds down to the size of the laser pulse itself, ultimately atoseconds. The ultimate time speed limitation will be the electronic response to the laser field, typically faster than 1fs.

3.2 Magnetization dynamics at short length scales

Fundamental solid-state physics and electronics have progressed enormously in the last 20 years and this progress can be characterized by the words "smaller" and "faster." In order to reach the ultimate ultrafast manipulations on the nanometer scale the challenge consists in improving our foundamental understanding of ultrafast magnetization dynamics and achieve ultrafast time resolved imaging at femtosecond time scales. In order to achive this

goal there are several aspects to consider related to ultrashort detection limits and spatial resolution capabilities (related to technical developments) which will ultimately enable a large step forward for the fundamental "ultrafast physics".
Considering the spatial resolution in femto magnetism the experimental advances are much more recent and technical improvements still in progress. When excited by a very short subpicosecond stimulus with duration much shorter than the time of thermal equilibration in the spin system (~100 ps) the magnetic medium is set into a strongly nonequilibrium state, where a conventional description of magnetic phenomena in terms of thermodynamics is no longer valid, a macrospin approximation fails and the dynamics becomes often stochastic [Stö2006], totally different from scenarios that rely on classical magnetism [Tud2004, Vaha2009, Hert2009]. Experimental studies of the ultrafast dynamics of a stochastic process in a sub-100-nm magnet are very demanding as well. Indeed, the stochastic character of the studied process excludes the possibility of averaging in the experiment. This basically means that for such a study one would need to obtain a magnetic image of a sample within a picosecond period of time and with sub-100-nm resolution. So far, there has been no method that would satisfy these requirements. The recent development of ultrashort soft X-ray pulses, provided by X-ray free electron lasers, opens new perspectives in this field. Several years after the pioneering work of S. Eisebitt et al. [Eis2004] demonstrating the possibility to image magnetic nanostructures by X-ray holography, C. Gutt and colleagues, report an experimental approach that may initiate a revolution in understanding ultrafast magnetic phenomena at the nanoscale. They show that by using one single 30 fs laser pulse it is possible to probe sub-100-nm magnetic domains in a Co/Pt multilayer sample [Gut2010]. The work was performed at FLASH free-electron laser facility at DESY in Hamburg. In fact, Gutt et al. have demonstrated an ultrafast probe of sub-100-nm magnets and thus have found a key to enter the uncharted world of femtosecond spin dynamics at the nanometer scale.
Similarly, the field of nanoscale magnetization dynamics is inseparably linked to the development of x-ray spin-sensitive methods as well as pulsed X-ray sources. Thanks to M. Faraday, who discovered the influence of a magnetic medium on the polarization of light [Far1846] magneto-optics in the visible spectral range has become one of the most popular tools for studies in magnetism. However, shorter wavelength is needed to observe down to the nanometer magnetic nanostructures. Since X rays sources have improved, since the 1990's advanced synchrotron radiation sources produce bright 50 ps pulses of polarized X rays. Using such sources, considerable progress has been achieved in understanding nanosecond magnetization dynamics at the sub-100-nm length scale [Waey2006].
A recently started photoemission electron microscopy (PEEM) study at BESSY on thin CoPd alloy films allowed us to perform time resolved domain imaging in the 50 ps time range. Using the robust stripe domain pattern induced by the large out of plane anisotropy in this system, we aim to resolve the influence of a fs laser pump on the domain pattern and spin orientation in the domains. This work performed at BESSY-UE49 is still in progress, but has yielded already a first and important nanoscale description of the magnetization dynamics occurring in the thermalized regime of the first 100 ps. Figure 7 a,b show two static X-PEEM images (time integrated) taken during IR laser excitation (fs pulses at 5 MHz repetition rate) where thermal effects, reducing the out of plane anisotropy in the film generate a ~90° rotation of the spins from out-of-plane to in-plane. The spins thereby organizing in magnetic

X-ray grazing incidence of 15°

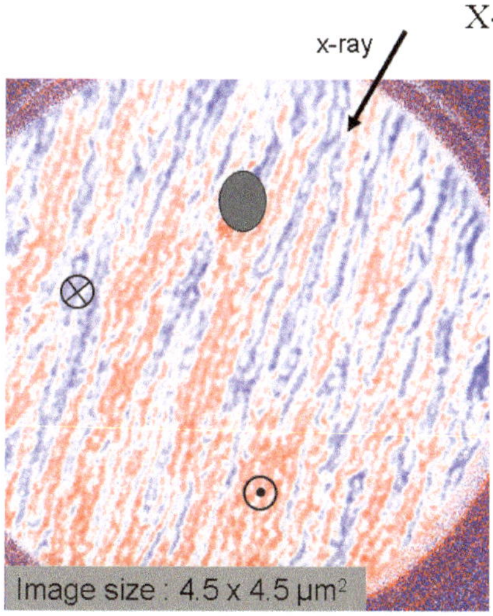

Stripe domains in CoPd film without laser. Out-of-plane spins (up = red; down = blue)

a)

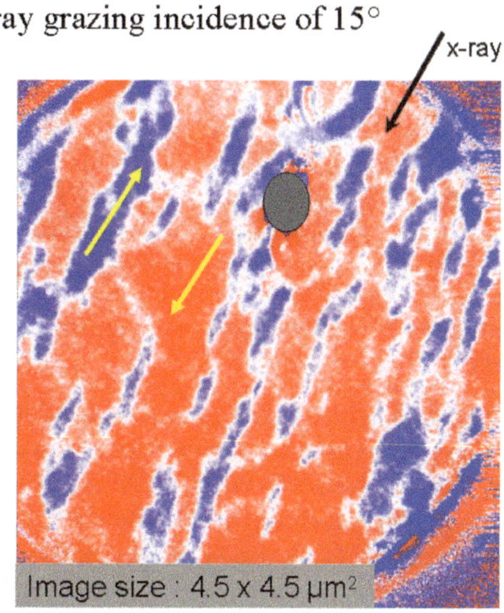

Domains in CoPd with pump laser on: In-plane spins

b)

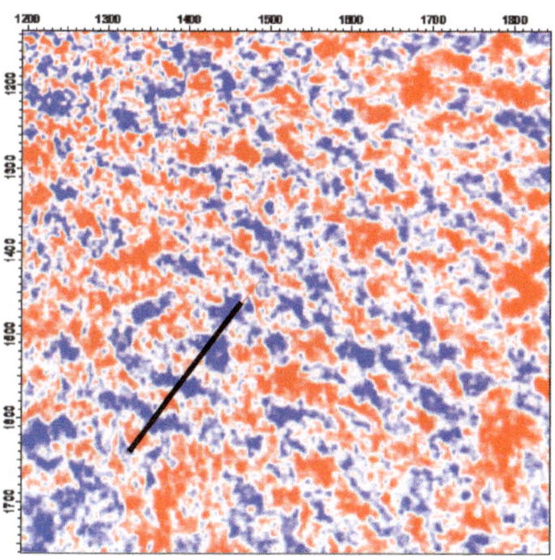

c) t< to pump-probe

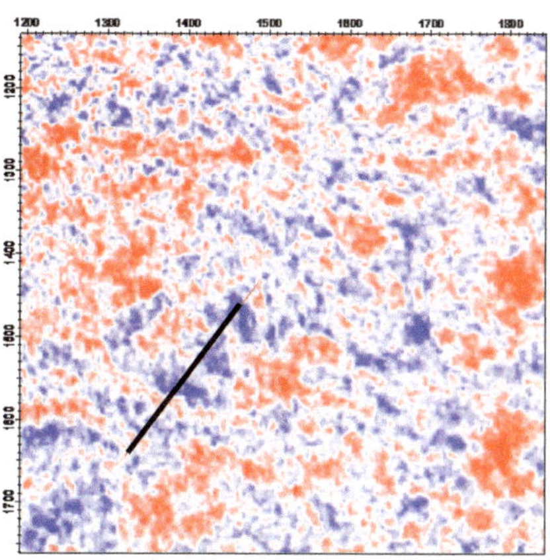

d) t > to , pump-probe

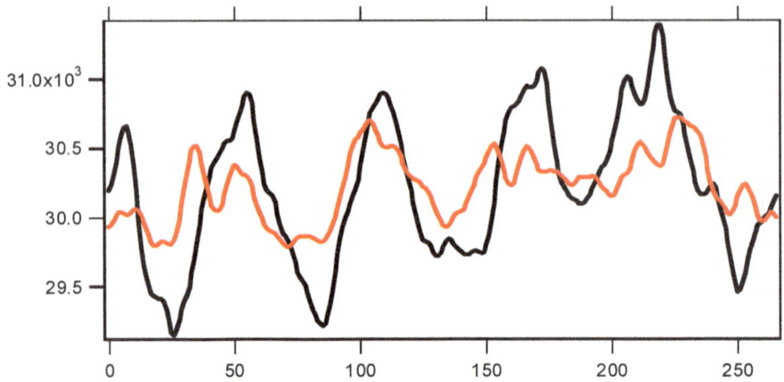

e) t < to for the black line (image c), while t > to in case of the red line (image d). Absice scale is in pixel.

Fig. 7. (a and b) X-PEEM magnetic contrast images taken in the time-integrated mode. From a to b we observe the transformations induced by thermal effects of the fs laser excitation (5MHz). The magnetic domains show a stripe domain pattern (a) whereas large "in plane domains" are revealed in (b) induced by fs laser thermal heating. The color scale is common for the two images (a,b) and is proportional to the projection of the XMCD contrast along the x-ray incidence (15° grazing / surface). This leads to high sensitivity for in-plane contributions, while out-of-plane components are strongly reduced. For example, the projected value of +45% XMCD contrast at Co L_3 along the out-of-plane leads to only +2% XMCD contrast in the upper left PEEM image, which is obtained without laser pump pulse at room temperature. The grey pattern corresponds to the defect used to align the image position. Note that the left image shows modulation superimposed to the stripes. They are coming from small components of in-plane spins (less than 1%).
(c and d). Small field of view of 1,2 x 1,2 μm² of pump probe magnetic images taken in the time resolved mode (left: t = -100 ps before pump laser; right: t = 100 ps after pump laser). The color scale is common for the two images (c, d) and enhanced relative to the one of the two top images (a, b) in order to show the XMCD contrast reduction in this case. Note that due to the azimuth alignment in respect of the incoming x rays the in-plane spins are here not observable. The statistic is in the limit to clearly show the stripes but the line scans (e) performed along the straight line in the images c and d evidence the periodicity of stripes and show that indeed the XMCD contrast of the stripes is reduced by 40 %.

domains in the plane (fig 7 b), which are large in comparison to the initial, narrow stripe domains. The origin of this rotation is the reduction of the strong out-of-plane anisotropy by thermal effects. Moreover, the symmetry of the stripes also favors a 1D orientation of spins lying in the plane (figure 7b). This anisotropy of spins in the plane has been established by a complete azimuth analysis were complete extinction of the large magnetic domains was obtained after a +90 degree rotation of the sample (stripes) in respect of the x-ray incidence (not shown - see fig 7 c and d). These first results show that fs laser induced thermal effects can switch the anisotropy from out-of-plane to in-plane. When compared with the ultrafast reduction of the SOI observed for CoPd [Boe2010] we expect that the ultrafast change in SOI is able to switch the anisotropy in the fs time scale. This expectation is supported by the fact that the ultrafast SOI effect holds on for more than 2 ps (in the regime where spins and phonons are in thermal equilibrium [Boe2010]. In the pump probe experiment using lower fluency one can observe a limited 50 % reduction of the XMCD contrast in the stripe domains (fig. 7 c, d). Note that due to the azimuthal alignment (+ 90 deg) with respect to the incoming x rays the in-plane spins (oriented along the stripe direction) are here not observable at $t > t_0$.

This specific orientation of the sample combined with reduced laser fluence allows to observe the very low XMCD contrast of 1% in the stripes at $t > t_0$. Finally, using the orientation dependent and fluence dependent information we can conclude that superposed to a contrast reduction of the stripes one can also expect a rotation of the spin into the plane and aligned along the stripe direction by fs laser pulse excitation. The results are revealing an important and new nanoscale "final state" of the fs laser induced SOI. And experimentally verifying this expectation is on of the major final goal of this work.

Pump probe PEEM imaging is still in progress, but will yield a time resolution of 100 ps only.

In order to bridge the time gap between 100 fs and 100 ps we will need to perform single shot imaging using Fourier transform Holography (FTH) with 100 fs time resolution using X-FEL sources. Compared with multi-shot imaging this single shot mode allows for imaging with higher or less adjusted laser pulses (destructive for the stripe domains at ps time scales). Alternatively, the dynamic in worm domains can be studied with multi-shot imaging, suppressing the in-plane anisotropy of the spins observed with PEEM.

For instance, it has been shown by our previous work [Beau 1998] and [Boeg2010] that for magnetically saturated $CoPt_3$ and CoPd films a 100 fs laser pulse can partially or completely demagnetizes the film.

New aspects as for instance X-ray pump and X-ray probe will also been foreseen in order to study the interaction with X-ray at core level resonances. For instance X-ray single pulse intensities could be used to change the electronic configuration at the Fermi energy and observe the induced changes on the magnetization.

Note that FLASH and LCLS generate pulses with duration of down to 10 fs. This is already comparable with the characteristic time of exchange interaction in magnetic materials. It would be extremely intriguing to employ the elemental specificity of X-ray techniques and probe spin and orbital dynamics of TM and RE sub-lattices on a time scale pertinent to the time of the exchange interaction between them.

4. Conclusion

Pulsed X rays, are nowadays a promising route toward high temporal and spatial resolution allowing for quantitative and high contrast magnetic imaging at the elemental absorption edges of ferromagnetic materials. However, the available time resolution to date is too slow to resolve the fastest dynamics of 1 fs. Because of this, significant efforts have been devoted to using short or isolated electron bunches of X rays pulses at synchrotron to perform time resolved microscopy with X rays. Nowaday we are at time resolutions of 100 fs, and better resolutions are forseen for near future using X-FEL's. Magnetic imaging techniques as for instance X-ray PhotoEmission Electron microscopy (X-PEEM), Scanning Transmission X-Ray Microscopy (STXM) or X-ray Resonant Elastic

Scattering (XRES), are currently using the short X ray pulses in order to accede to time resolved imaging in the picosecond time range. Unfortunately, the f-slicing technique in synchrotrons produces a strongly reduced photon flux hindering the f-second magnetic imaging at facilities as synchrotrons. Recent projects using other synchrotron techniques are planed in order to increase the photons per pulse in the picosecond range allowing to image magnetic domains at 1 ps time resolution using synchrotron light. In parallel a large variety of new physics develops at X-FEL's where intense X-ray pulses modifies strongly the electronic and magnetic structures of mater. The development of such instruments will also allow new scientific approaches fare from the actual quasi-static physics.

5. Acknowledgments

We are indebted to E. Beaurepaire, V. Halté, V. Lopez-Flores, C. Stamm, N. Pontius, F. Kronast, T. Quast and T. Kachel, N. Jaouen, J. Lüning, S. Eisebitt, Vincent Cros, Richard Mattana, Franck Fortuna, Y. Acremann, A. Schertz, J-Y. Bigot and H. Dürr for the help and support during the pompe-probe femtoslicing, LCLS and X-PEEM experiments and to J. Arabski and V. Da Costa for sample elaboration and characterization.
This work was supported by the CNRS - PICS, by Université de Strasbourg and the E.U. Contract Integrated Infrastructure Initiative I3 in FP6-Project No. R II 3 CT-2004-5060008, BESSY IA-SFS Access Program.

6. References

Spi1997 Ch. Spielmann et al. Science 278,661 (1997),

Dre2001 M. Drescher et al. Science 291, 1923 (2001),

Sch1999 Schnurer M et al.PRL 83, 722-725 (1999),

Kra2009 Krausz F, M. Ivanov, "Attosecond Physics" Rev. Mod. Phys. 81, 163 (2009)

Sch2000 R.W. Schoenlein *et al.*, Applied Physics (New York) 71, 1 (2000),

Scho2000 R.W. Schoenlein *et al. Science* 287, 2237-2247 (2000),

K. Holldack, S. Khan, R. Mitzner, and T. Quast, Phys. Rev. ST Accel. Beams 8, 040704 (2005)

Bea1996 E. Beaurepaire, J. C. Merle, A. Daunois, and J. Y. Bigot, Phys Rev Lett 76, 4250 (1996).

M. Cinchetti et al., Phys. Rev. Lett. 97, 177201 (2006).

Stam2007 C. Stamm, T. Kachel, N. Pontius, R. Mitzner, T. Quast, K. Holldack, S. Khan, C. Lupulescu, E. F. Aziz,M. Wietstruk, H. A. Dürr, W. Eberhardt, Nature Materials 6, 740 (2007).

Boe2010 C. Boeglin, E. Beaurepaire, V. Halte, V. Lopez-Flores, C. Stamm, N. Pontius, H. A. Durr, and J. Y. Bigot, Nature 465, 458 (2010).

Ele1996 A.Y. Elezzabi et al. Phys. Rev. Lett. 77, 3220 (1996)

Gerr2002 T. Gerrits et al. Nature 418, 509 (2002)

Acr2001 Y. Acremann et al. Nature 414, 51 (2001)

Cho2004 S.B. Choe Y. Acremann, A. Scholl, A. Bauer, A. Doran, J. Stöhr, H. Padmore, Science 304, 420 (2004),

Sch2004 C. Schneider, A. Kuksov, A. Krasyuk, A. Oelsner, D. Neeb, S. Nepijko, G. Schonhense, I. Monch, R. Kaltofen, J. Morais, C. de Nadai, N. Brookes, Appl. Phys. Lett. 85, 2562 (2004)

Raa2005 J. Raabe, C. Quitmann, C. Back, F. Nolting, S. Johnson, C. Bühler, Phys. Rev. Lett. 94, 217204 (2005)

Weg2005 F. Wegelin et al. Surface Science 601 (2007) 4694 - 4699

Kra2005 A. Krasyuk, F. Wegelin, S. Nepijko, H. Elmers, G. Schonhense, M. Bolte, C. Schneider, Phys. Rev. Lett. 95, 207201 (2005)

Kuc2004 W. Kuch, J. Vogel, J. Camarero, K. Fukumoto, Y. Pennec, S. Pizzini, M. Bonfim J. Kirschner, Appl. Phys. Lett. 85, 440 (2004)

Vog2005 J. Vogel, W. Kuch, J. Camarero, K. Fukumoto, Y. Pennec, S. Pizzini, M. Bonfim, F. Petroff, A. Fontaine, J. Kirschner, Phys. Rev. B 71, 060404 (2005)

Vog2005 J. Vogel,W. Kuch, R. Hertel, J. Camarero, K. Fukumoto, F. Romanens, S. Pizzini, M. Bonfim, F. Petroff, A. Fontaine, J. Kirschner, Phys. Rev. B 72, 220402 (2005)

Fuk2006 K. Fukumoto, W. Kuch, J. Vogel, F. Romanens, S. Pizzini, J. Camarero, M. Bonfim, J. Kirschner, Phys. Rev. Lett. 96, 097204 (2006)

Vog2011 J.Vogel et al. Appl. Phys. A (2008) 92: 505-510

Hey2010 L. Heyne et al. Phys. Rev. Lett. 105, 187203 (2010)

Uhl2011 V. Uhlir et al Phys. Rev. B 83, 020406(R) (2011)

Kim2005 A.V. Kimel et al. Nature 429, 850 (2005)

Sta2005 C. Stamm et al. Phys. Rev. Lett. 94, 197603 (2005)

Van2008 A. Vansteenkiste et al. PRB 77, 144420 (2008)

Zho1996 A. A. Zholents and M. S. Zolotorev, Phys. Rev. Lett. 76, 912 (1996)

Sch2000 R.W. Schoenlein *et al.*, Applied Physics (New York) 71, 1 (2000),

R.W. Schoenlein *et al. Science* 287, 2237-2247 (2000).

Hol2005 K. Holldack, S. Khan, R. Mitzner, and T. Quast, Phys. Rev. ST Accel. Beams 8, 040704 (2005).

Kah2006 S. Kahn et al. Proceedings of 2005 Particle Accelerator Conference, Knoxville, Tennessee page 2309 and A. Steun et al. Proceedings of EPAC 2006, Edinburgh, Scotland, page 3427.

Hol2006 Holldack, K et al. PRL 96, 054801 (2006)

Car2009 K. Carva(a), D. Legutand P. M. Oppeneer, EPL, 86 (2009) 57002

Eis2004 S. Eisebitt, J. Lüning, W. F. Schlotter, M. Lörgen, O. Hellwig, W. Eberhardt, and J. Stöhr, *Lensless imaging of magnetic nanostructures by X-ray spectro-holography*, Nature 432, 885 (2004).

Chap2006 H. C. Chapman *et al.*, Nature Phys. 2, 839 (2006)

Gun2011 C. Günther, B. Pfau, R. Mitzner, B. Siemer, S. Roling, H. Zacharias, O. Kutz, I. Rudolph, D. Schondelmaier, R. Treusch, and S. Eisebitt, « Sequential femtosecond x-ray imaging, Nature Photonics, 5, 99(2011).

Gut2009 C. Gutt, et al. Phys Re B 79, 212406 (2009)

Gut2010 C. Gutt *et al.*, Phys. Rev. B 81, 100401 (2010).

Zha2000 G. P. Zhang and W. Hubner, Phys. Rev Lett 85, 3025 (2000).

Zha2008 G. P. Zhang, and T. F. George, Phys. Rev. B 78, 052407 (2008).

Big2009 J. Y. Bigot, M. Vomir, and E. Beaurepaire, Nature Physics 5, 515 (2009).

Stan2007 C. D. Stanciu, F. Hansteen, A. V. Kimel, A. Tsukamoto, A. Itoh, A. Kiriklyuk, and T. Rasing, Phys. Rev. Lett. 94, 237601 (2007).

Far1846 M. Faraday, Phil. Trans. R. Soc. London 136, 104 (1846).

Ziel1965 J. P. van der Ziel, P. S. Pershan, and L. Malmstrom, Phys. Rev. Lett. 15, 190 (1965).

Waey2006 B. Van Waeyenberge *et al.*, Nature 444, 461 (2006).

2

Electron Acceleration Using an Ultrashort Ultraintense Laser Pulse

Eisuke Miura
National Institute of Advanced Industrial Science and Technology (AIST)
Tsukuba central 2, 1-1-4 Umezono, Tsukuba, Ibaraki
Japan

1. Introduction

With recent progress in ultrashort ultraintense laser technologies such as chirped pulse amplification (CPA) (Strickland & Mourou, 1985), the peak power of a laser pulse is increasing year by year, and the focused intensity of 10^{21} W/cm^2 has been achieved (Aoyama et al., 2003; Perry et al., 1999). When the focused intensity of a laser pulse is higher than 10^{18} W/cm^2, quiver velocity of an electron is close to the speed of light in such a high electromagnetic field. Various nonlinear phenomena are caused by the relativistic effect of the electron motion. Self-focusing, higher harmonic generation, and so on, which are well-known phenomena in nonlinear optics, have been observed in laser-plasma interactions.

An ultrashort ultraintense laser pulse propagating in a plasma can excite a plasma wave by the nonlinear force of a high electromagnetic field, called the ponderomotive force. A longitudinal electric field is formed by the plasma wave, and electrons trapped in the potential of the plasma wave can be accelerated. This is the concept of laser-driven plasma-based electron acceleration (LPA) (Tajima & Dawson, 1979). The longitudinal accelerating electric field of the plasma wave is higher than 100 GV/m, which is a thousand times higher than that of present radio-frequency (rf) accelerators. Such a high accelerating field enables us to realize compact electron accelerators and/or obtain extremely high energy electrons. Furthermore, the electron pulse duration is extremely short, of the order of tens of femtoseconds, because the wavelength of the accelerating field, that is the plasma wave, is of the order of tens of micrometers. Next-generation electron accelerators with such unique characteristics will be realized using LPA.

Since the concept was proposed, various experimental and theoretical studies have been conducted (Esarey et al., 2009; 1996). Pioneering works of the proof-of-principle such as generation of a high accelerating field and energetic electron beams have been so far presented (Joshi et al., 1984; Kitagawa et al., 1992; Malka et al., 2002; Modena et al., 1995; Nakajima et al., 1995). However, the energy spectra of the electron beams were Maxwell-like distributions, and the beam qualities were far from those required for various applications. In 2004, a major breakthrough was brought about with the generation of well-collimated electron beams with a narrow energy spread, that is quasi-monoenergetic electron (QME) beams (Faure et al., 2004; Geddes et al., 2004; Mangles et al., 2004; Miura et al., 2005). This result is a significant step toward the realization of a laser electron accelerator.

In this chapter, we provide the overview of the present status of research on LPA. First, we briefly describe the principle of LPA. Second, we present recent results of works conducted at the National Institute of Advanced Industrial Science and Technology (AIST). Generation of QME beams is mainly presented. We also present particle-in-cell simulations to discuss the mechanism and the conditions of QME beam generation. Using a femtosecond electron pulse obtained by LPA, a compact, all-optical, ultrashort X-ray source can be realized on the basis of laser Compton scattering. Third, we present X-ray generation by laser Compton scattering using a laser-accelerated electron beam. Finally, we briefly review recent progress toward a next step and future prospects.

2. Principle of laser-driven plasma-based electron acceleration

2.1 Electron motion in a electromagnetic field

Let us consider an electron motion in a electromagnetic field. The equation of motion for a free electron of charge e in an electromagnetic (laser) field is given by the Lorentz equation

$$\frac{d\mathbf{p}}{dt} = -e(\mathbf{E} + \mathbf{v} \times \mathbf{B}) \,, \tag{1}$$

where $\mathbf{p}$=$\gamma m_e \mathbf{v}$, $\mathbf{v}$, $\mathbf{E}$, and $\mathbf{B}$ are the momentum and velocity of an electron, and the electric and magnetic field, respectively. Here, m_e and γ are the electron mass and the relativistic Lorentz factor. In the case of a weak electromagnetic field, $\mathbf{v} \times \mathbf{B}$ component is negligible, and Eq. 1 is simplified to $m_e \frac{d\mathbf{v}}{dt} = -e\mathbf{E}$. For a linearly polarized laser field, the electron quiver velocity v_q is given by $v_q = eE_L/m_e\omega_L$, where E_L and ω_L are the amplitude and frequency of the laser field. The ratio of the electron quiver velocity to the speed of light c is defined by

$$a_0 = \frac{eE_L}{m_e \omega_L c} = 8.5 \times 10^{-10} \lambda_L[\mu\mathrm{m}] \sqrt{(I_L[\mathrm{W/cm^2}])} \,, \tag{2}$$

where λ_L and I_L are the laser wavelength and intensity. This gives an expression for the normalized vector potential a_0. Relativistic effects are brought about in an electron motion in a laser field yielding $a_0 \geq 1$. This region is called relativistic region. For example, the intensity of 2.2×10^{18} W/cm^2 gives $a_0 = 1$ for 800-nm laser light.

By averaging $\frac{1}{2} m_e {v_q}^2$ over one oscillation period of a field, the electron quiver energy is defined by

$$U_P[\mathrm{eV}] = \frac{1}{2} m_e \langle {v_q}^2 \rangle = \frac{e^2 {E_L}^2}{4 m_e {\omega_L}^2} = 9.3 \times 10^{-14} (\lambda_L[\mu\mathrm{m}])^2 I_L[\mathrm{W/cm^2}] \,. \tag{3}$$

This is an expression for the ponderomotive potential U_P. The ponderomotive potential results in a force $\mathbf{F_P}$=$-\nabla U_P$, that is the ponderomotive force. The ponderomotive force is directed along the intensity gradient of a laser pulse envelope, and perpendicular to the laser propagation direction. The force pushes electrons out of the region of the laser pulse, and becomes the driving force for exciting a plasma wave in a plasma.

2.2 Excitation of plasma wave

Let us consider the propagation of an ultrashort ultraintense laser pulse in a low density plasma. The electron density of the plasma is much lower than the critical density n_c given by $n_c = \varepsilon_0 m_e {\omega_L}^2/e^2$, where ε_0 is the vacuum permittivity. Electrons in the plasma are pushed out of the region of the laser pulse and separated from ions by the ponderomotive force. A local charge separation is formed, because it is regarded that ions are at rest in the short time close to the laser pulse duration. After some time, the laser pulse overtakes electrons pushed out and the space charge force starts to pull the electrons back. In turn, the charge excess is formed by the electrons pulled back near the laser propagation axis. The electrons are pushed out again. By the repetition of this process, plasma oscillation is driven. The plasma frequency ω_p for the electron density n_e is given by

$$\omega_p = \sqrt{\frac{e^2 n_e}{\varepsilon_0 m_e}} . \quad (4)$$

Because the plasma oscillation is driven together with the propagation of the laser pulse, the phase velocity of the plasma wave v_p is equal to the group velocity of the laser pulse given by

$$v_p = c\sqrt{1 - \frac{n_e}{n_c}} . \quad (5)$$

Electrons trapped in the plasma wave can be accelerated almost to the speed of light, because the phase velocity of the plasma wave is close to the speed of light in an underdense plasma ($n_e \ll n_c$).
In the linear region, a longitudinal electric field of a plasma wave E_x is give by

$$E_x = \delta n \frac{m_e \omega_p c}{e} , \quad (6)$$

where δn is the modulation of the electron density. For example, assuming $n_e = 10^{19}$ cm^{-3} and $\delta n = 0.3$, the electric field is 100 GV/m, corresponding to more than a thousand times as accelerating fields of present rf accelerators.
To explain the principle of LPA, typical particle-in-cell (PIC) simulation results are shown in Fig. 1. The initial electron density of a plasma is 1.7×10^{19} cm^{-3}, and the laser intensity of 800-nm light corresponds to the normalized vector potential of 5. A laser pulse propagates along the $x-$axis and is polarized along the $y-$axis. The snapshots of electron density distribution (a) and (b), the distributions of the electric field along the $y-$axis corresponding to the laser field (c) and (d), and the distributions of the electron density (solid curve) and the electric field (dashed curve) along the $x-$axis (e) and (f) are shown for different laser propagation lengths. As shown in Figs. 1(a) and (e), the period of the low-density and high-density parts is formed behind the laser pulse along the propagation axis, and a plasma wave is excited. As shown in Fig. 1(e), the peak longitudinal field along the $x-$axis, that is the accelerating field in LPA, reaches 700 GV/m.
The situation, in which a laser pulse drives a plasma wave behind itself, is similar to that in which a boat drives a wake on a sea. Then, LPA is also called laser wakefield acceleration (LFWA).

2.3 Trapping and acceleration of electrons

As described above, because the phase velocity of the plasma wave is close to the speed of light, the velocity of electrons should be also close to the speed of light for trapping into a plasma wave. To feed accelerated electrons, an electron gun was used as an external electron source (Amiranoff et al., 1995; Clayton et al., 1993; Ebrahim, 1994). In some experiments, hot electrons produced from a laser-irradiated solid target were used (Mori, Sentoku, Kondo, Tsuji, Nakanii, Fukumochi, Kashihara, Kimura, Takeda, Tanaka, Norimatsu, Tanimoto, Nakamura, Tampo, Kodama, Miura, Mima & Kitagawa, 2009; Nakajima et al., 1995). However, it has been demonstrated that electrons in a plasma can be almost

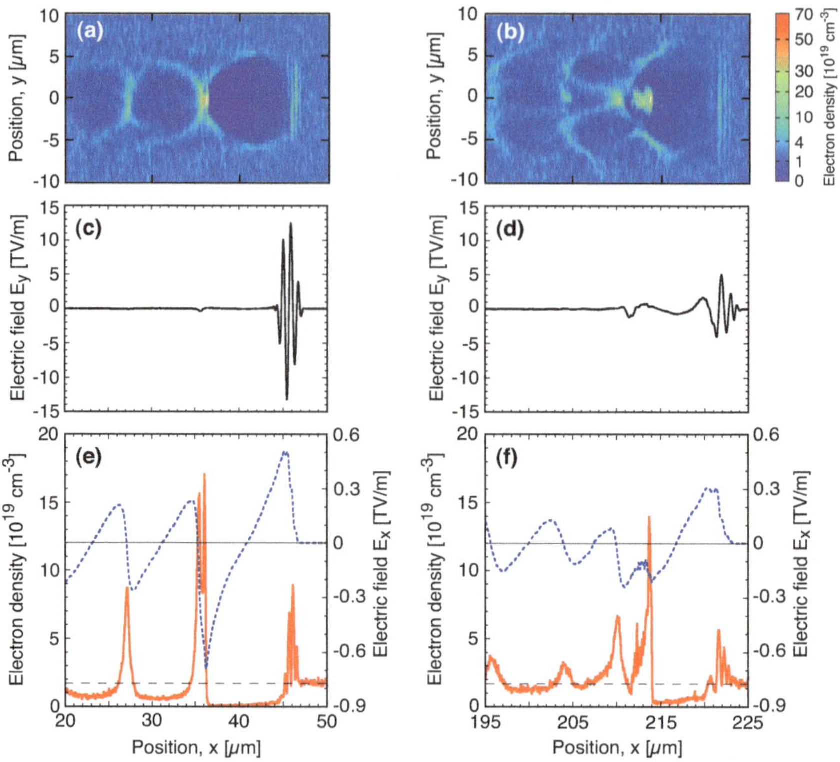

Fig. 1. Typical simulation results to explain the principle of LPA. Snapshots of electron density distribution on the $x - y$ plain (a) and (b), the distributions of the electric field along the $y-$axis corresponding to the laser field (c) and (d), and the distributions of the electron density (solid curve) and the electric field (dashed curve) corresponding to the accelerating field along the $x-$axis (e) and (f) are shown for different laser propagation lengths.

automatically injected and trapped into a plasma wave. This scheme is called self-injection scheme or self-trapping scheme. In the self-injection scheme, wave-breaking plays an important role (Bulanov et al., 1997; Decker et al., 1994). As shown in Fig. 1(e), a plasma wave with a large amplitude is excited, and the steeping of the wave occurs. As the laser pulse propagates further, the amplitude of the plasma wave becomes larger, and the wave finally breaks. This phenomenon is seen as whitecaps at the ocean. Electrons slip from the wave and are trapped into the plasma wave. As shown in Figs. 1(b) and (f), electrons are trapped and accelerated in the first period of the plasma wave and form an electron bunch.

The plasma wavelength $\lambda_p = 2\pi c/\omega_p$, which is the wavelength of the accelerating field, is the order of tens of micrometers. In the case shown in Fig. 1, the plasma wavelength is $\sim 10\ \mu$m. Electrons are trapped in the narrow region of the acceleration phase and the length of the electron bunch is a few micrometers as shown in Fig. 1(f). This electron bunch is regarded as an ultrashort electron pulse with a duration of tens of femtoseconds. Thus, a femtosecond electron pulse can be obtained by LPA.

The energy gain W_{max} is simply given by the product of the accelerating field and the acceleration length. In the linear region, the accelerating field is given by Eq. 6. The acceleration length is mainly limited by dephasing (phase slippage) of electrons trapped in a plasma wave. The dephasing is brought about by the difference between the phase velocity of a plasma wave and the velocity of accelerated electrons. Accelerated electrons in a plasma wave outrun the acceleration phase of the plasma wave, and enter the deceleration phase. In the linear region, the dephasing length L_d is given by

$$L_d = \frac{c\lambda_p/2}{c - v_p} \simeq \frac{n_c}{n_e}\lambda_p \,. \tag{7}$$

Then, the maximum energy gain is given by

$$W_{max} = 4m_e c^2 \frac{n_c}{n_e} \,. \tag{8}$$

For example, when the laser wavelength is 800 nm and the electron density is 10^{19} cm^{-3}, the dephasing length and the maximum energy gain are 1.8 mm and 350 MeV. In other words, electrons with an energy of 350 MeV can be accelerated in only 1.8-mm length.

The acceleration length is also limited by the diffraction of a laser beam and/or the energy depletion of a laser pulse for driving a plasma wave. The limit of the laser propagation length by the diffraction is not serious, because it is possible to achieve a longer propagation length by guiding a laser pulse using relativistic self-focusing, a capillary discharge plasma and so on. If the pump depletion length is defined as the length in consuming half of the initial laser pulse energy for driving a plasma wave, the pump depletion length is comparable to the dephasing length (Esarey et al., 1996). Then, the limit of the acceleration length by dephasing is dominant.

The description in this section is concentrated in the liner region. However, as shown in Fig. 1, most of phenomena in LPA are in the nonlinear region, and the treatment is important. The overview of analytical description in the nonlinear region has been provided in detailed review reports (Esarey et al., 2009; 1996).

2.4 Generation of quasi-monoenergetic electron beam with a narrow energy spread

Since the middle of 1990's, the generation of energetic electron beams based on a self-injection scheme has been demonstrated (Modena et al., 1995; Nakajima et al., 1995; Umstadter, Chen, Maksimchuk, Mourou & Wagner, 1996). Although collimated electron beams were generated, the energy spectra of the electron beams were Maxwell-like distributions and the energy spreads were large. These experiments were based on self-modulated laser wakefield acceleration (SM-LWFA) using a laser pulse with a picosecond duration and a high-density plasma with an electron density close to 10^{20} cm^{-3}. In such cases, the laser pulse length is longer than the plasma wavelength. The plasma wave with a large amplitude is excited via Raman scattering and/or the self-modulation instability and can accelerate trapped electrons. Heating the plasma occurs, when the plasma wave grows. In addition, the trapped electrons in the plasma wave interact with the laser field and can be also accelerated by the laser field directly, that is so-called direct laser acceleration (DLA) (Gahn et al., 1999). The combination of these electron acceleration mechanisms can lead to the broad electron energy spectrum.

However, in 2004, the generation of QME beams with a narrow energy spread was demonstrated by four groups (Faure et al., 2004; Geddes et al., 2004; Mangles et al., 2004; Miura et al., 2005). After that, many groups have so far reported the generation of QME beams (Hidding et al., 2006; Hosokai, Kinoshita, Ohkubo, Maekawa, Uesaka, Zhidkov, Yamazaki, Kotaki, Kando, Nakajima, Bulanov, Tomassini, Giulietti & Giulietti, 2006; Hsieh et al., 2006; Maksimchuk et al., 2007; Mori et al., 2006; Yamazaki et al., 2005). Most experimental results on QME beam generation have been explained by the acceleration in the highly nonlinear broken-wave regime, that is so-called bubble acceleration regime (Pukhov & Meyer-ter Vehn, 2002). In simulation results shown in Fig. 1, the conditions are close to those of the bubble acceleration regime. When an ultraintense laser pulse propagates in a plasma, it undergoes the self-focusing and the longitudinal pulse compression. Such laser pulse drives a highly nonlinear plasma wave and expels a large amount of electrons. Then, the electron cavitated region (bubble) is formed behind the laser pulse. The radially expelled electrons move along the boundary of the bubble and collide at the rear vertex of the bubble. Transverse wave-breaking (Bulanov et al., 1997) occurs and a large amount of electrons is injected into the bubble at the rear vertex. The field produced by the injected electrons terminates further electron injection. Electrons are trapped at the fixed narrow phase of the accelerating field. Thus, the QME beam with a narrow energy spread can be obtained. Because a large amount of electrons is instantaneously injected, the QME beam with high charge can be obtained. The electrons trapped near the rear vertex of the bubble are located behind the laser pulse. Because the trapped electrons are free from the laser pulse, an electron beam with low emittance can be obtained. These phenomena have been thoroughly investigated by many theoretical and numerical studies (Geissler et al., 2006; Gordienko & Pukhov, 2005; Kostyukov et al., 2004; Lu et al., 2007). For such bubble formation, one of the required conditions is that the spatially transverse and longitudinal sizes of the laser pulse match the plasma wavelength. To satisfy this condition, the experiments have been conducted using a focusing mirror with the long focal length yielding the spot size of $\sim$ 10 μm and an ultraintense laser pulse with a few tens of femtoseconds yielding the pulse length of $\sim$ 10 μm for a low density plasma around or less than 10^{19} cm^{-3} yielding the plasma wavelength of $\sim$ 10 μm.

3. Experimental and numerical studies on laser-driven plasma-based electron acceleration

In this section, recent results of works conducted in the National Institute of Advanced Industrial Science and Technology (AIST) (Masuda & Miura, 2008; Miura & Masuda, 2009) are presented as an example of experimental and numerical studies on LPA.

3.1 Experimental conditions

Laser system

A Ti:sapphire laser system with a repetition rate of 10 Hz based on CPA method was used for experiments. In our laser system, two intense laser pulses were available. A 20-fs laser pulse from a Kerr-lens mode-locked oscillator was stretched to 400 ps through an Öffner type pulse stretcher, which was an aberration-free type stretcher(Cheriaux et al., 1996). To control a laser pulse shape, an acousto-optic programmable dispersive filter (Dazzler: Fastlite)(Verluise et al., 2000) was installed after the pulse stretcher. The laser pulse was amplified by a regenerative amplifier and multi-pass amplifiers. To suppress and control a prepulse, a pulse cleaner composed of a Pockels cell and a thin film polarizer was set between the regenerative and the first muti-pass amplifiers. After the first multi-pass amplifier, the laser pulse was split into three pulses. The first pulse referred as main laser pulse was amplified by a multi-pass amplifier, and a laser pulse with an energy of 750 mJ, a pulse duration of 35-50 fs, and a center wavelength of 800 nm was obtained after a vacuum pulse compressor. The main laser pulse was used for electron acceleration. The second pulse referred as colliding laser pulse was amplified by two multi-pass amplifiers, and a laser pulse with an energy of 150 mJ, a pulse duration of 60 fs, and a center wavelength of 800 nm was obtained after a vacuum pulse compressor. Due to the priority to optimize a pulse shape of a main laser pulse, the spectrum control of a colliding laser pulse was not sufficient. Then, the pulse duration of the colliding laser pulse was little bit longer than that of the main laser pulse. The colliding laser pulse was used for X-ray generation by laser Compton scattering shown in Sec. 4. The third pulse referred as probe laser pulse with an energy of 0.5 mJ, a pulse duration of 60 fs, and a center wavelength of 800 nm was used for plasma diagnostics such as shadowgraph and interferogram.

Experiment setup

The experimental setup for electron acceleration is shown in Fig. 2. A p-polarized main laser pulse (400 mJ, 50 fs, 800 nm) was focused onto the edge of a helium gas jet using an $f/6$ off-axis parabolic mirror with a focal length of 300 mm. The laser spot size in vacuum was 9 μm at the full-width-at-half-maximum (FWHM), and the Rayleigh length was 230 μm. The energy concentration within the e^{-2} spot was 58% of the total laser energy. The peak intensity was 5.8×10^{18} W/cm^2, corresponding to the normalized vector potential of 1.6. The helium gas jet was ejected from a supersonic nozzle with a conical shape driven by a pulsed valve. The focal position was set at 1 mm above the nozzle exit. The diameter of the nozzle exit was 0.7 mm and the Mach number was 5.3. The density profile of the gas jet was measured with a Jamin interferometer. The density at the center was $\sim 5 \times 10^{18}$ cm^{-3}, corresponding to the electron density of $\sim 10^{19}$ cm^{-3}.

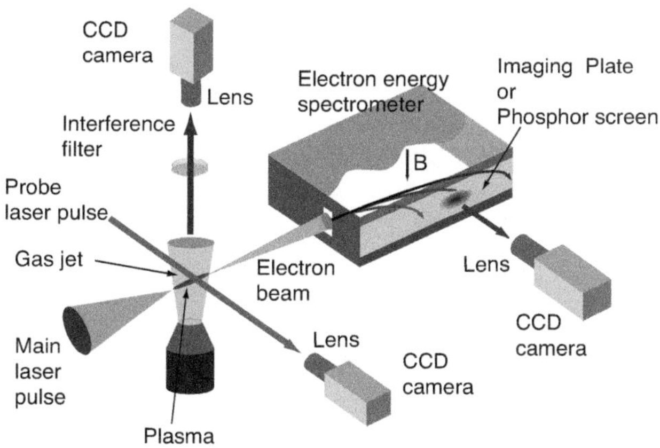

Fig. 2. Experimental setup for electron acceleration.

The energy spectrum of an electron beam was measured by an electron energy spectrometer with a dipole magnet. The measured energy range was from 1 to 80 MeV. An energy-resolved electron image was recorded on an imaging plate [(IP), Fujifilm: BAS-SR]. To estimate the absolute charge of electrons , the sensitivity curve reported in Ref.(Tanaka et al., 2005) was used. In addition, we have developed an absolutely-calibrated in-situ observation system for the electron energy spectrum. A Gd_2O_2S : Tb phosphor screen (Mitsubishi chemical: DRZ-HIGH) coupled with a CCD camera was used as an electron detector. The sensitivity of the detection system was calibrated using an IP as the reference detector (Masuda et al., 2008). The absolute electron energy spectrum was in-situ observed with a single shot at a repetition rate of 1 Hz.

To observe the propagation of a laser pulse and the formation of a preformed plasma, a side-scattered light image and a shadowgraph image were observed simultaneously with the measurement of an electron energy spectrum. The side-scattered light image was observed through an interference filter of 800 nm from the top of the gas jet. The shadowgraph image was observed with a 60-fs, 800-nm probe laser pulse. The spatial resolutions in both the measurements were approximately 40 μm.

Contrast ratios of femtosecond prepulses, generated in a regenerative amplifier, preceding the main pulse by more than 4 ns were always kept to be less than 10^{-6} using a pulse cleaner. Our another work has demonstrated that the number of accelerated electrons was dramatically decreased for low contrast ratios of the femtosecond prepulses, although the laser power and the electron density of the plasma were different from those for the present experiment (Masuda & Miura, 2010). Then, the femtosecond prepulses were suppressed.

3.2 Quasi-monoenergetic electron beam generation

Figure 3(a) shows an energy-resolved electron image of a typical QME beam from a plasma with an electron density of 1.6×10^{19} cm^{-3} produced by an 8.3-TW laser pulse. The image was recorded on an IP with a single shot. The small spot indicates the generation of the monoenergetic beam with a narrow energy spread and a small divergence angle. The divergence angle in the vertical direction was estimated to be 7 mrad at the FWHM from the vertical size of the spot. The electron energy spectrum is shown in Fig. 3(b). A monoenergetic peak was observed at the energy of 38 MeV with the relative energy spread of 19%. Here, the relative energy spread is defined as the ratio of the energy spread at the FWHM to the peak energy. The observed energy spread was little bit larger than the instrumental resolution of several percent. The total number of electrons in the monoenergetic peak was 3.1×10^{8}, corresponding to the charge of 50 pC.

Figure 4 shows the electron energy spectra obtained by sequential 14 shots under the same conditions as those giving the result shown in Fig. 3. All the spectra were obtained with a single shot. To evaluate the stability of QME beam generation, the criterion values of principal

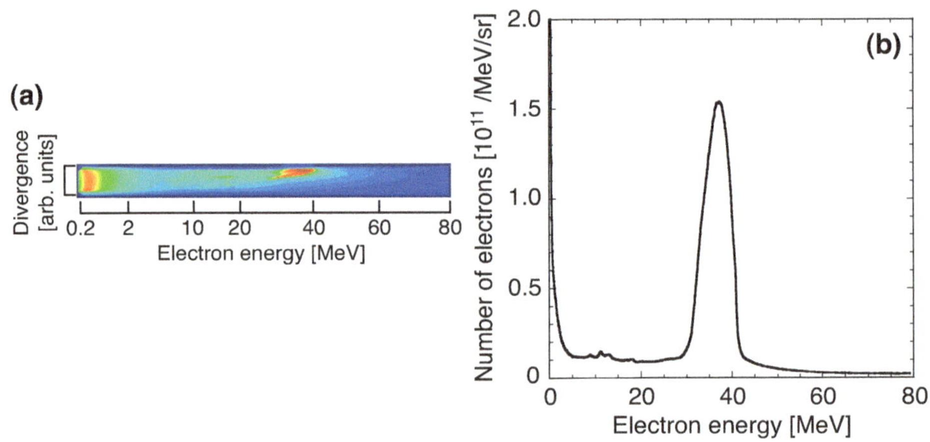

Fig. 3. (a) Typical energy-resolved electron image recorded on an IP with a single shot and (b) electron energy spectrum of a QME beam obtained from a plasma with an electron density of 1.6×10^{19} cm^{-3} produced by an 8.3-TW laser pulse.

beam parameters should be defined for judging QME beam generation. We set the following four criterion values. First, the peak energy is higher than 10 MeV. Second, the number of electrons in the electron energy spectrum is higher than 10^{10} /MeV/sr at the monoenergetic peak. Third, the ratio of the peak to the background at the monoenergetic peak is higher than 2. Forth, the relative energy spread is less than 50%. When all of these conditions are fulfilled, we consider that a QME beam has generated. According to the criteria, QME beams are obtained in 10 shots among 14 shots in the case of Fig. 4. The probability of QME beam generation is 70%, if the shot-to-shot fluctuations of the peak energy, the number of electrons, and so on are ignored. In our previous experiment (Masuda et al., 2007), the probability of

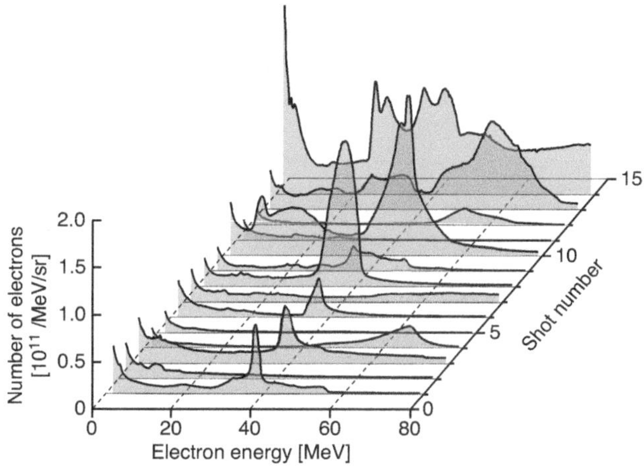

Fig. 4. Electron energy spectra obtained by sequential 14 shots under the same experimental conditions as those giving the result shown in Fig. 3. The probability of QME beam generation is 70%.

QME beam generation was a few tens of percent. The probability in the present experiment increased by a factor of 3, as compared with our previous experiment. It has been pointed out that there is a threshold energy of a laser pulse for stable generation of QME beams, which depends on the electron density of the plasma and the laser pulse duration (Mangles et al., 2007). Our present experimental conditions are in the range for stable generation predicted by scaling in Ref. (Mangles et al., 2007).

Figure 5 shows the distribution of the peak energy and the charge in the monoenergetic peak of QME beams for about 50 shots obtained under the same conditions as those giving the result shown in Fig. 3. A QME beam with a peak energy of up to 75 MeV was produced. A QME beam containing up to 88 pC in the peak at the energy of 48 MeV was also produced. In this case, the total energy of electrons in the peak was 4.2 mJ. This means that the efficiency of energy conversion from the laser pulse to the electron beam was 1%. Table 1 lists mean values and standard deviations of the parameters of QME beams obtained in about 50 shots shown in Fig. 5. The beam pointing means the angular deviation of the beam center from the laser propagation axis in the vertical direction. Although the probability of QME beam generation increases, shot-to-shot fluctuations in the charge and the energy spread are still large. The suppression of the fluctuations in the beam parameters is a key issue toward a next step.

Figure 6 shows the dependence of electron energy spectra on the electron density of the plasma, where the laser power is fixed at 8.3 TW. The electron density was varied by

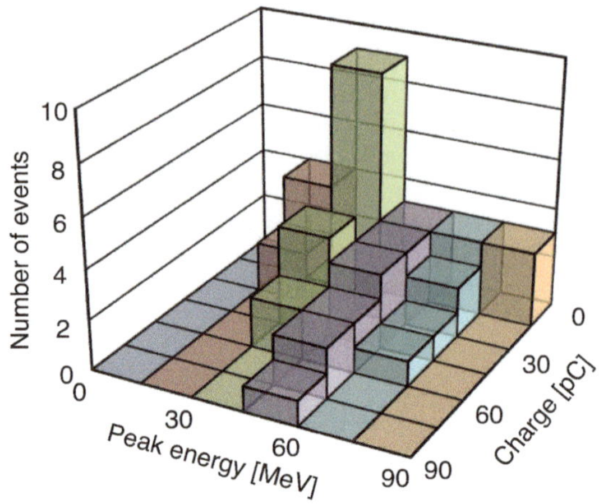

Fig. 5. Histogram showing the distribution of the peak energy and the charge in the monoenergetic peak of QME beams for about 50 shots obtained under the same conditions as those giving the result shown in Fig. 3.

Parameters	Mean $\pm$ Standard deviation
Peak energy	49 ± 15 MeV
Relative energy spread	22 ± 12 %
Charge in the monoenergetic peak	24 ± 20 pC
Divergence angle in the vertical direction (FWHM)	7.1 ± 4.0 mrad
Beam pointing in the vertical direction	± 7.8 mrad

Table 1. Statistics of the parameters of QME beams obtained in about 50 shots shown in Fig. 5.

controlling the gas jet density. As described above, QME beams were generated at an electron density of 1.6×10^{19} cm^{-3}, as shown by the solid curve. At a lower electron density of 1.3×10^{19} cm^{-3}, high-energy electrons were not observed, as shown by the dotted curve. On the other hand, at a higher density of 1.9×10^{19} cm^{-3}, the electron energy spectrum was a Maxwell-like distribution, and no clear monoenergetic peaks were seen, as shown by the dash-dotted curve. QME beams can be generated only in the narrow electron density region around 1.6×10^{19} cm^{-3}. This result shows that plasma density control is crucially important for the generation of QME beams.

3.3 Effect of a nanosecond prepulse on quasi-monoenergetic electron beam generation

There are many issues that must be controlled for stable generation of QME beams in laser-plasma interactions. Among them, the control of a prepulse and a preformed plasma is one key issue. In this subsection, we show the effects of a nanosecond prepulse on QME beam generation and propagation of an intense laser pulse.

The experimental conditions have been already shown in Sec. 3.1. The identical off-axis parabolic mirror and super sonic nozzle were used. As described in Sec. 3.1, the contrast ratios of femtosecond prepulses to a main pulse were kept to be less than 10^{-6}. The length of a nanosecond prepulse caused by amplified spontaneous emission (ASE) was controlled by varying the extraction time of the main pulse at the pulse cleaner.

Dependence of electron beam characteristics on length of nanosecond prepulse

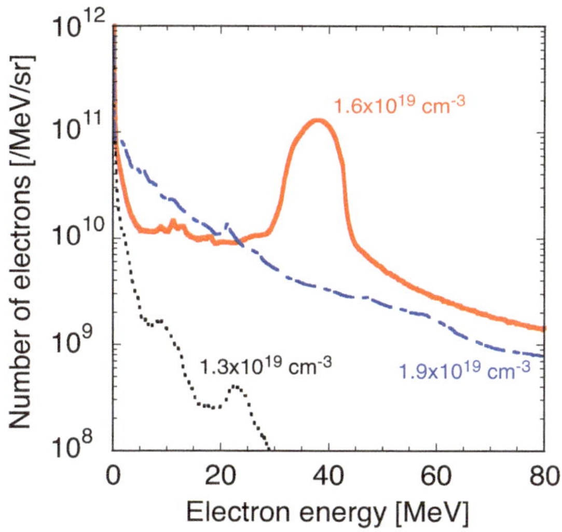

Fig. 6. The electron energy spectra observed for different electron densities of 1.3×10^{19} (dotted curve), 1.6×10^{19} (solid curve), and 1.9×10^{19} cm^{-3} (dash-dotted curve), respectively. To obtain QME beams, plasma density control is important.

Figure 7(a) shows a series of electron energy spectra obtained by 21 consecutive shots at a repetition rate of 1 Hz from a plasma with an electron density of 1.9×10^{19} cm^{-3} produced by an 8.5-TW laser pulse. The electron energy spectra were obtained by optimizing the extraction time of the main pulse at the pulse cleaner in addition to the optimization of the electron density. Here, this time is defined as the extraction time of the main pulse, $t_{ext} = 0$ ns. Figure 7(b) shows a series of electron energy spectra obtained by 17 consecutive shots at a repetition rate of 1 Hz for the different extraction time $t_{ext} = -2.5$ ns at the same laser power and electron density. The length of the nanosecond prepulse is longer, as the extraction time becomes earlier. As shown in Fig. 7(a), in most of shots, a monoenergetic peak was observed. In contrast, when the length of the nanosecond prepulse was long, monoenergetic peaks in the low energy range were observed only in a few shots among 17 consecutive shots as shown in Fig. 7(b). No monoenergetic peaks in the high energy range were observed. Furthermore,

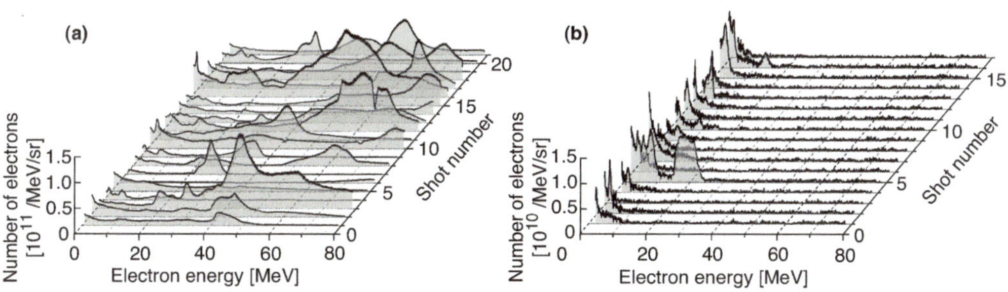

Fig. 7. Series of electron energy spectra obtained by about 20 consecutive shots at a repetition rate of 1 Hz for the different extraction times of the main pulse at the pulse cleaner: (a) $t_{ext} = 0$ ns and (b) $t_{ext} = -2.5$ ns.

the scale of the number of electrons in Fig. 7(b) is one-tenth of that in Fig. 7(a). The probability of the generation of QME beams was estimated according to the criteria described in Sec. 3.2. QME beams are obtained in 17 shots among 21 shots in the case of Fig. 7(a). The probability of QME beam generation is 81%, if the shot-to-shot fluctuations of the peak energy, the number of electrons, and so on are ignored. The probability of QME beam generation is 0 in the case of Fig. 7(b) according to the criteria described in Sec. 3.2.
Figure 8 shows the dependence of the probability of the QME beam generation (closed circles) and the laser energy (open triangles) on the extraction time of the main pulse at the pulse cleaner. Each data point of the probability was obtained with more than 10 consecutive shots at a repetition rate of 1 Hz. The electron density of the plasma was 1.9×10^{19} cm^{-3} for all data points. Notably, a few data points overlap at $t_{ext} = -2.5, -1.5$, and $+1.5$ ns, when the probability is 0. The data points for the probability of QME beam generation are divided into three groups. When the length of the nanosecond prepulse is long ($t_{ext} \sim -2$ ns), the probability is quite low. In contrast, when the nanosecond prepulse is suppressed ($t_{ext} \sim 0$ ns), the probability dramatically increases and is close to 90%. This result shows that it is necessary to suppress a nanosecond prepulse for generation of QME beams. When the extraction time is too late ($t_{ext} = +1.5$ ns), the probability becomes quite low again. This is due to the reduction in the laser energy as shown in Fig. 8.

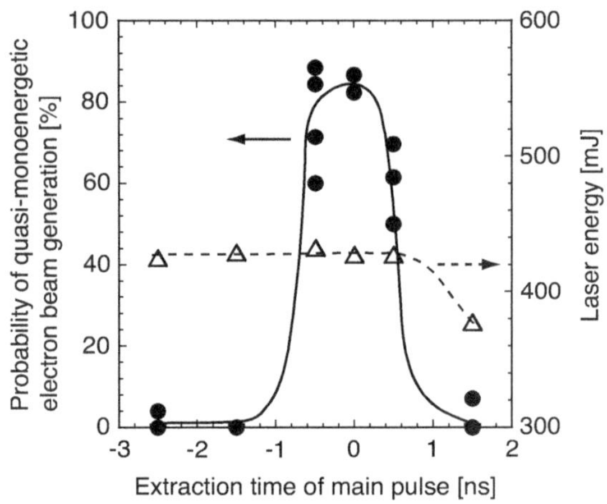

Fig. 8. Dependence of the probability of QME beam generation (closed circles) and the laser energy (open triangles) on the extraction time of the main pulse at the pulse cleaner. With the earlier extraction time, the length of a nanosecond prepulse is longer. For generation of QME beams, the suppression of the nanosecond prepulse is necessary.

The optimum electron density for QME beam generation in the experiment presented in this subsection was slightly higher than that of the results in shown in Sec. 3.2, although the experimental conditions were very close. The small difference of the laser conditions may cause the difference of the optimum electron density.

Plasma diagnostics

Figures 9(a) and (b) show a shadowgraph image and a horizontal profile of the image along the laser propagation axis for $t_{ext} = -2.5$ ns, respectively. The laser pulse propagates from left to right in the image. The black area is the shadow of the nozzle and the nozzle exit position is shown by the white area. The image was taken at 1 ps before the arrival of the main pulse at the focal position. When the length of a nanosecond prepulse is long, a preformed plasma is observed around the vacuum focal position as denoted by arrows, although the image in Fig. 9(a) is not clear. As shown in Fig. 9(b), the length of the preformed plasma is approximately 200 μm, which is close to the Rayleigh length. Because the time delay of the probe pulse is -1 ps, the preformed plasma may be produced by a picosecond prepulse. In another experiment using an 8-TW laser pulse, formation of a preformed plasma was observed in a shadowgraph image taken at several tens of picoseconds before the arrival of the main pulse for $t_{ext} = -2.5$ ns. Then, we think that the preformed plasma shown in Fig. 9 was produced by a nanosecond prepulse.

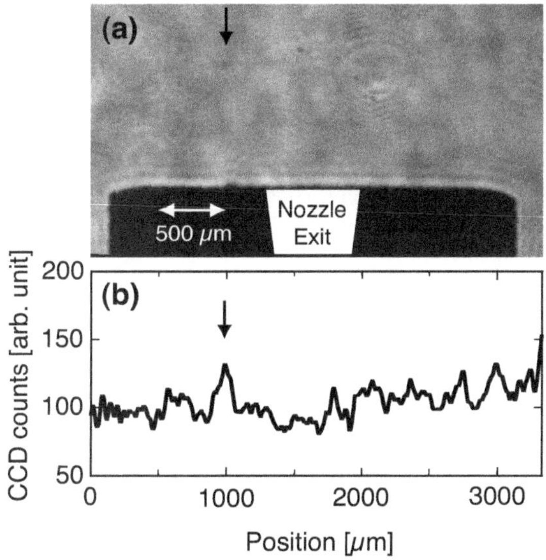

Fig. 9. (a) Typical shadowgraph image and (b) horizontal profile of the image along the laser propagation axis for the extraction time, $t_{ext} = -2.5$ ns. The image is taken at 1 ps before the arrival of the main pulse at the focal position. A preformed plasma is observed as denoted by the arrows.

Figure 10 shows typical side-scattered light images observed for the different extraction times: (a) $t_{ext} = 0$ ns and (b) $t_{ext} = -2.5$ ns, respectively. The laser pulse propagates from left to right in the images. The dotted circle indicates the nozzle exit position. Both the images were observed through filters with the same attenuation and the laser energies were almost same shown in Fig. 8. When the nanosecond prepulse was suppressed (Fig. 10(a)), the image was similar to that suggesting the formation of a plasma channel reported in Refs. (Masuda et al., 2007; Miura et al., 2005). In contrast, when the length of the nanosecond prepulse was long, strong side-scattered light was observed as shown in Fig. 10(b). The image was quite different from that reported in Refs. (Masuda et al., 2007; Miura et al., 2005). The main pulse is strongly scattered, because the density profile of the preformed plasma is unsuitable for guiding the laser pulse. The remnant energy of the main pulse is insufficient to excite the plasma wave with a large amplitude. Then, no energetic electrons are accelerated. The side-scattered light image shown in Fig. 10(b) suggests that a certain fraction of the energy of the main pulse is scattered by the preformed plasma.

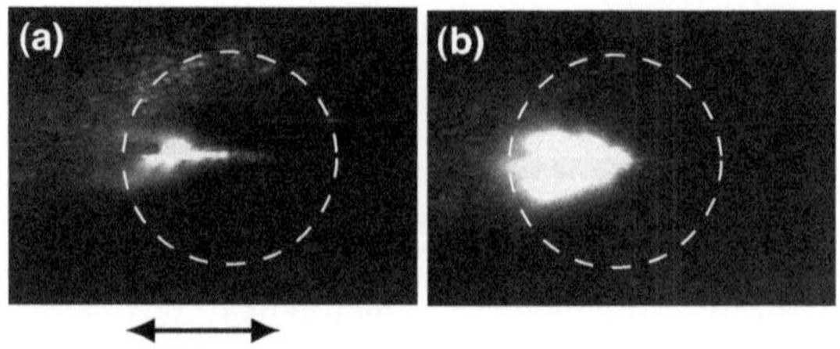

Fig. 10. Typical side-scattered light images observed for the different extraction times: (a) $t_{ext} = 0$ ns and (b) $t_{ext} = -2.5$ ns. The both images are observed through filters with the same attenuation. When the length of the nanosecond prepulse is long, the strong side-scattered light is observed.

3.4 Analysis on electron acceleration using two-dimensional particle-in-cell simulations

3.4.1 Analysis on quasi-monoenergetic electron beam generation

Conditions of two-dimensional particle-in-cell simulations

To investigate in detail the generation of QME beams shown in Sec. 3.2, we have developed a two-dimensional particle-in-cell (2D-PIC) simulation code using a moving window technique. The moving window had $127 \times 127 \mu m^2$ size with 2000×1000 cells containing 5 particles per cell. Simulations were conducted on the x-y plane. The laser propagation direction and the polarization direction were set in the x- and the y-axes. The laser pulse had a transverse profile with a TEM_{00} Hermite-Gaussian mode and a cosine temporal envelope. The laser pulse energy, the spot size, and the pulse duration were 400 mJ, 9 μm at the FWHM and 50 fs at the FWHM, respectively. The electron density profile along the x-axis was set to be similar to the measured density profile of a gas jet. The center of the gas jet in the experiment was set at $x = 0$. The initial electron density raised at $x = -960$ μm, and smoothly increased to the maximum of 1.6×10^{19} cm^{-3} at $x = -350$ μm. To investigate the interaction for a long propagation length, it was assumed that the electron density was constant from the position where the density reached the maximum value. The focal position in vacuum was set at $x = -300$ μm. The density profile was uniform along the y-axis.

Simulation results

Figure 11 shows snapshots of the electron distribution on the x-y plane (a)-(c) and the number of electrons as functions of the energy and the angle of the momentum vector (d)-(f) for three different positions. In Figs. 11(a)-(c), the leading edge of the laser pulse is near the right edge and each electron is colored by its energy. At $x = 170$ μm, electrons are not trapped in a plasma wave, because the amplitude of the plasma wave is small. The high-energy electrons are not observed in Fig. 11(d). After some time, trapping of electrons occurs due to sufficient growth of the plasma wave, and a monoenergetic electron bunch is trapped and accelerated in the first period of the plasma wave at $x = 570$ μm, as seen in Figs. 11(b) and (e). The fine structure in the monoenergetic bunch seen in Fig. 11(b) is due to the interaction of electrons with the laser field. At $x = 1050$ μm, the trapped electrons enter the deceleration phase and dissipate the energy. Then, the monoenergetic component disappears, as seen in Fig. 11(f). This result shows that the extraction position of electrons from the plasma is important for QME beam generation. The predicted optimum laser propagation length from the focal position is 870 μm, which is close to the gas jet length of approximately 1 mm in the experiment. This suggests that the generation of a QME beam is brought about by matching the laser propagation length with the gas jet length.

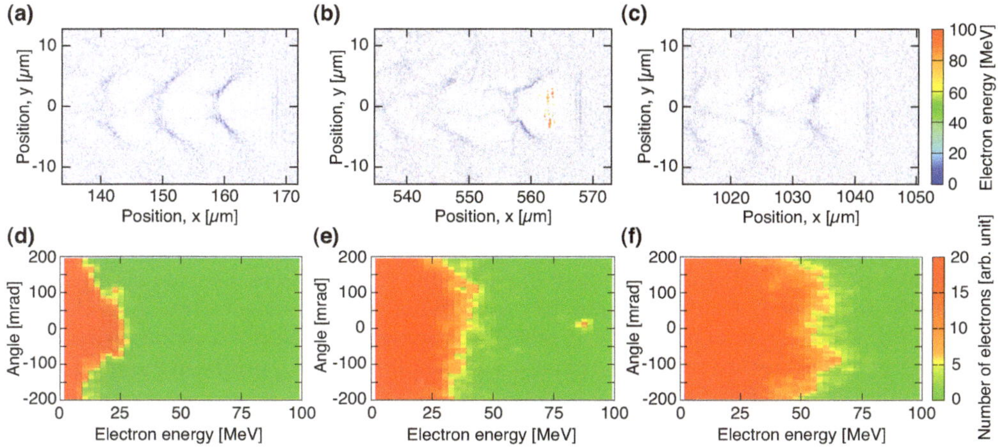

Fig. 11. 2D-PIC simulation results for an electron density of 1.6×10^{19} cm^{-3}. Snapshots of the electron distribution on the x-y plane (a)-(c) and the number of electrons as functions of the energy and the angle of the momentum vector (d)-(f) are shown for three different positions.

The generation of a monoenergetic bunch with a peak energy of 90 MeV, a relative energy spread of 6% at the FWHM, and a divergence angle of 28 mrad at the FWHM is predicted. The predicted values of the monoenergetic component are fairly close to the experimental results. As seen in Fig. 11(e), only the monoenergetic component has a small divergence angle. In contrast, the low-energy component has a large divergence. The simulation result explains the difference in the beam divergence between the monoenergetic and the low-energy components, as seen in Fig. 3(a).

Simulations were conducted for electron densities of 1.3 and 1.9×10^{19} cm^{-3}. For all the densities, including 1.6×10^{19} cm^{-3}, electron injection occurs and a monoenergetic electron bunch is formed in the first period of the plasma wave, immediately after the amplitude of the plasma wave reaches the maximum. However, the growth rate of the plasma wave depends on the electron density, and the laser propagation lengths required for electron injection are different for each electron density. The growth rate is larger as the electron density is higher. At an electron density of 1.3×10^{19} cm^{-3}, the laser propagation length required for electron injection is longer than the gas jet length of 1 mm in the experiment. Because the electron injection does not occur inside the gas jet, high-energy electrons are not observed. In contrast, at an electron density of 1.9×10^{19} cm^{-3}, the electron injection occurs too early, and the trapped electrons enter the deceleration phase, so the energy dissipates at the end of the gas jet. Then, a monoenergetic peak is not observed. The simulation results also explain the density dependence of the electron energy spectra shown in Fig. 6 (Masuda & Miura, 2009).

3.4.2 Analysis on effect of a nanosecond prepulse on quasi-monoenergetic electron beam generation

Preformed plasma model in simulation

To investigate effects of a nanosecond prepulse on QME beam generation shown in Sec. 3.3, 2D-PIC simulations were also conducted. The dynamics both of a preformed plasma produced by a nanosecond prepulse and the propagation of a femtosecond laser pulse in a millimeter-scale plasma can not be treated simultaneously in the PIC simulation, because their time scales are significantly different. In the present simulation, the preformed plasma is formed by giving a density modification of the initial electron and ion distributions.

It is assumed that a preformed plasma with a transversely hollow density profile is formed due to the expansion during a few nanoseconds. Singly ionized helium ions are distributed inside a preformed plasma region, and neutral helium atoms are distributed in the calculation region except for the preformed plasma region. The initial electron density is zero in the entire calculation region. The PIC simulation code includes the optical field ionization process. The leading edge of a laser pulse ionizes neutral atoms and singly ionized ions up to doubly ionized state. The electron density inside the preformed plasma region, where singly ionized helium ions are initially distributed, is half of that in the region, where neutral helium atoms are initially distributed. As a result, a preformed plasma with a transversely hollow density profile is formed. The main body of a laser pulse propagates in the preformed plasma. The details in a model to create a preformed plasma with a hollow density profile in the calculation have been described in Ref. (Masuda & Miura, 2010).

It is found that the longitudinal size of the preformed plasma is approximately 200 μm from the shadowgraph image as shown in Fig. 9(b). Then, the longitudinal size of the preformed plasma region is assumed to be 230 μm, corresponding to the Rayleigh length. Although the observed transverse size of the preformed plasma is limited by the spatial resolution, the transverse size is assumed to be 9 μm, corresponding to the laser spot diameter. The preformed plasma region is set at the vacuum focal position of the laser pulse, because the preformed plasma is observed around the vacuum focal position as shown in Fig. 9(a). Except for the preformed plasma conditions, simulation conditions are the same as those shown in Sec. 3.4.1. The case above described is referred as preformed plasma case. In no preformed plasma case, neutral helium atoms are initially distributed in the entire calculation region and the initial electron density is zero.

Simulation results

Figures 12(a) and (b) show spatial evolutions of transverse envelopes of the laser electric field represented by the normalized vector potential on the x-y plane in no preformed plasma case and preformed plasma case. The spatial evolutions are obtained from the transverse envelopes of the peak amplitude of the laser electric field for every 255 time steps. In no preformed plasma case, the laser pulse focused at $x = -300$ μm propagates by forming a narrow channel over the Rayleigh length due to the relativistic self-focusing effect. A part of the laser energy is scattered away from the channel. In preformed plasma case, the scattering of the laser pulse increases as shown in Fig. 12(b). This result explains the experimental observation of the strong side-scattered light image shown in Fig. 10(b). Due to the increase in the laser scattering, the peak intensity of the laser pulse in the channel decreases in preformed plasma case.

Figures 13(a) and (b) show the electron energy spectra as a function of the position of the laser pulse in no preformed plasma case and preformed plasma case. The spatial evolutions are obtained from the electron energy spectra calculated for the electrons with the angle of the momentum vector with respect to the laser propagation direction in the range of ± 50 mrad in the moving window for every 255 time steps. In no preformed plasma case, a clear monoenergetic peak is formed at $x \sim 600$ μm and the maximum peak energy reaches ~ 80 MeV as shown in Fig. 13(a). This position ($x \sim 600$ μm), which is the optimum extraction position of a monoenergetic electron bunch, is close to the end of the gas jet in the experiment. In no preformed plasma case, a QME beam is produced, because matching the length of the laser propagation with the gas jet length is achieved. This result is quite similar to that shown in Sec. 3.4.1.

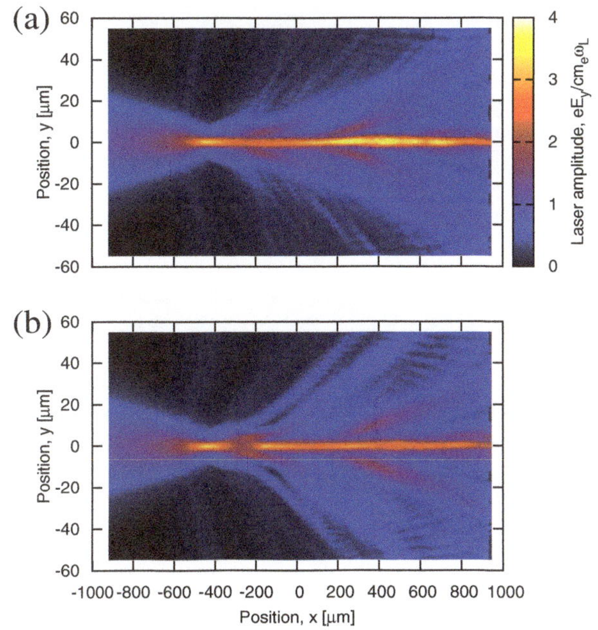

Fig. 12. Spatial evolution of transverse envelopes of the laser electric field on the x-y plane in (a) no preformed plasma case and (b) preformed plasma case. Scattering of the laser pulse increases and the laser intensity on the propagation axis decreases in preformed plasma case.

In contrast, in preformed plasma case, the growth rate of the plasma wave is smaller than that in no preformed plasma case due to the low peak amplitude of the laser field as shown in Fig. 12(b). The time, when the amplitude of the plasma wave reaches the maximum and the electron injection occurs, is delayed and the required laser propagation length becomes longer than that in no preformed plasma case. A monoenergetic peak is formed at $x \sim 700\,\mu$m as shown in Fig. 13(b). However, the monoenergetic peak is not clear, because the amount of injected electrons is small. The optimum extraction position of a monoenergetic electron bunch ($x \sim 700\ \mu$m) is outside the gas jet region in the experiment. This means that the acceleration length is extremely short in the experiment. In preformed plasma case, a QME beam is not obtained, because the required condition, that is matching the laser propagation length with the gas jet length, is not satisfied due to the low growth rate of the plasma wave. The simulation results using a simple model for a preformed plasma qualitatively explain the experimental results of effects of a nanosecond prepulse on QME beam generation.

4. X-ray generation by laser Compton scattering using laser-accelerated quasi-monoenergetic electron beam

4.1 Ultrashort X-ray source based on laser Compton scattering

An ultrashort electron pulse with a duration of the order of a few tens of femtoseconds can be obtained in LPA. In addition, there is the potential for realizing compact electron accelerators.

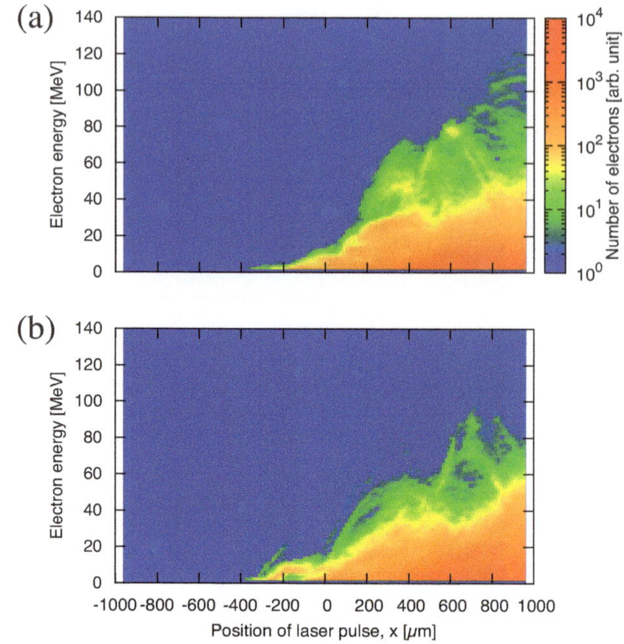

Fig. 13. Electron energy spectra as a function of the position of the laser pulse in (a) no preformed plasma case and (b) preformed plasma case. In no preformed plasma case, a clear monoenergetic peak with an energy of 80 MeV is formed at the position, $x \sim 600\,\mu$m, which corresponds to the end of the gas jet in the experiment. At the same position, a clear monoenergetic peak is not observed in preformed plasma case.

The set of such unique characteristics enables us to realize compact, all-optical, ultrashort radiation sources in the wavelength range from extreme ultraviolet to X-ray. One of them is an X-ray source based on laser Compton scattering, that is scattering of photons by energetic electrons. Laser Compton scattering X-ray sources have attracted much attention for their potential of applications to medical imaging (Ikeura-Sekiguchi et al., 2008), because they produce a well-collimated, quasi-monochromatic X-ray beam, the photon energy of which is tunable, in a small facility, as compared with synchrotron radiation sources.

So far, generation of a femtosecond X-ray pulse by laser Compton scattering has been demonstrated by using a femtosecond laser pulse and a picosecond electron pulse from rf accelerators (Schoenlein et al., 1996; Yorozu et al., 2002). The X-ray pulse duration is determined by the interaction time between the laser and electron pulses. To obtain a femtosecond X-ray pulse, 90° scattering geometry should be adopted for a picosecond electron pulse. In contrast, 180° scattering (head-on collision) geometry can be adopted for a femtosecond electron pulse obtained by LPA. There are some advantages of using 180° scattering geometry. The photon energy of X-rays is higher than that for 90° scattering geometry, even though the electron energy and the wavelength of a laser pulse are the same. In addition, the X-ray yield is higher than that for 90° scattering geometry, even though the charge of an electron beam and the energy of a laser pulse are the same. Generation of X-rays with energies around 1 keV by laser Compton scattering has been demonstrated using a laser-accelerated electron beam with a Maxwell-like energy distribution. (Schwoerer et al., 2006). In this section, X-ray generation by laser Compton scattering using a laser-accelerated quasi-monoenergetic electron beam is described.

4.2 Experimental conditions

Figure 14 shows the experimental setup. Laser pulses for electron acceleration and laser Compton scattering are referred as main laser pulse and colliding laser pulse. A p-polarized main laser pulse (700 mJ, 40 fs, 800 nm) was focused onto the edge of a helium gas jet using an $f/14$ off-axis parabolic mirror with a focal length of 720 mm. The laser spot diameter in vacuum was 13 μm at the FWHM. The energy concentration within the e^{-2} spot was 50% of the total laser energy. The peak intensity was 4.7×10^{18} W/cm^2, corresponding to the normalized vector potential of 1.4. The gas jet was ejected from a supersonic nozzle with a conical shape. The diameter of the nozzle exit was 1.6 mm and the Mach number was 5. The focal position was set at 1 mm above the nozzle exit.

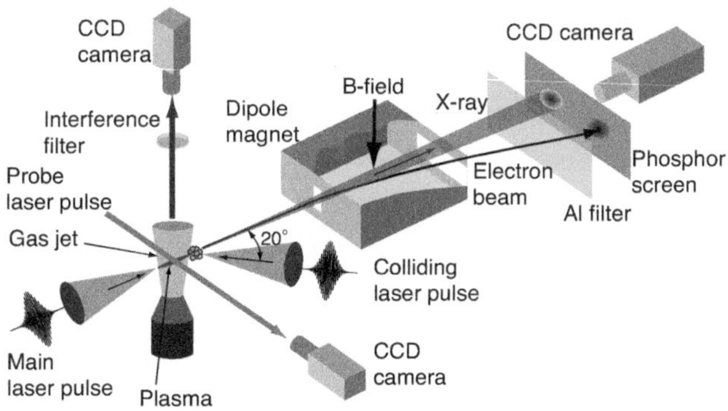

Fig. 14. Experimental setup for X-ray generation by laser Compton scattering using laser-accelerated electron beam.

A p-polarized colliding laser pulse (140 mJ, 100 fs, 800 nm) was focused around the exit of the main laser pulse from the gas jet using an $f/6$ off-axis parabolic mirror with a focal length of 300 mm. The laser spot diameter in vacuum was 9 μm at the FWHM. The incident angle of the colliding laser pulse was 20° to the propagation axis of the main laser pulse. X-rays produced by laser Compton scattering were emitted on the coaxial direction of an electron beam. The electron beam was bended by a magnetic field and spatially separated from the X-rays. Both X-rays and electrons were simultaneously incident on a $Gd_2O_2S:Tb$ phosphor screen (DRZ-HGH) through a 115-μm-thick Al filter. The images of X-rays and energy-resolved electrons were observed with a CCD camera. Electrons with an energy higher than 35 MeV were detected. The charge can be estimated, because the sensitivity of the detection system was calibrated for electrons (Masuda et al., 2008). The photon energy range of detected X-rays was higher than 10 keV from the calculated sensitivity of the phosphor screen.

For synchronized collision of a colliding laser pulse with a laser-accelerated electron beam, a time-resolved shadowgraph image along the propagation axis of the main laser pulse was observed using a 60-fs probe laser pulse. A side-scattered light image through an interference filter of 800 nm, that is Thomson-scattered light image, was also observed from the top of the gas jet.

The conditions for electron acceleration were different from those shown in Sec. 3.2. In prior, we confirmed that QME beams with a peak energy from 50 to 100 MeV and a charge in the monoenergetic peak of several tens of picocoulombs were obtained from a plasma with an electron density of 1.7×10^{19} cm^{-3} (Miura et al., 2011).

4.3 Synchronized collision of femtosecond laser pulse with electron beam

For X-ray generation by laser Compton scattering, synchronized collision of a colliding laser pulse with a laser-accelerated electron beam is required. A laser-accelerated electron beam is emitted in the coaxial direction of a main laser pulse and temporally close to a main laser pulse. Then, achieving the synchronized collision of the colliding laser pulse with the laser-accelerated electron beam is equivalent to achieving the synchronized collision of the main and colliding laser pulses. Figures 15(a)-(c) show typical shadowgraph images observed for different delay times of the probe laser pulse to the main laser pulse: (a) -1.33 ps, (b) -0.67 ps, and (c) 0 ps, respectively. The main laser pulse propagated from right to left, and the

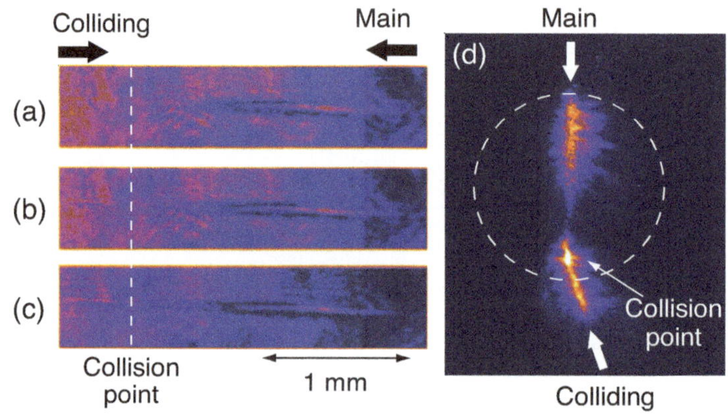

Fig. 15. (a)-(c) Shadowgraph images observed for different delay times of a probe laser pulse to a main laser pulse: (a) -1.33 ps, (b) -0.67 ps, and (c) 0 ps. (d) Thomson-scattered light image when the synchronized collision of the main and colliding laser pulses is achieved. The bright spot indicates the collision point of the two laser pulses.

colliding laser pulse propagated from left to right. The ionization fronts of the two laser pulses approached each other with varying the delay of the probe laser pulse. As seen in Fig. 15(c), the ionization fronts of the two laser pulses overlapped, and two laser pulses collided. The optical path length and the focal position in the vertical direction of the colliding laser pulse were adjusted using the shadowgraph images.

The collision point on the horizontal plane was set using a Thomson-scattered light image observed from the top of the gas jet as shown in Fig. 15(d). The dotted circle shows the position of the nozzle exit with 1.6-mm diameter. In Fig. 15(d), the main laser pulse propagated from top to bottom and the colliding laser pulse propagated from lower right to upper left in the direction of 20° to the main laser propagation axis. As shown by the thin arrow, a bright spot was observed, only the synchronized collision of the two laser pulses was achieved. It is supposed that the bright spot indicates the collision point of the two laser pulses. The collision point was set at the edge of the nozzle exit, which was near the extraction position of an electron beam from a plasma.

4.4 X-ray generation

X-rays produced by laser Compton scattering were observed, when a QME beam with a considerably high charge was obtained. Figure 16(a) shows an image of X-rays. The image was obtained with a single shot. From the energy-resolved electron image simultaneously observed with the image shown in Fig. 16(a), the peak energy and the charge in the monoenergetic peak of the QME beam were estimated to be 50 MeV and 30 pC, respectively. Figure 16(b) shows the intensity profile of the image in the vertical direction. The divergence angle in vertical direction was 5 mrad at the FWHM. The divergence angle in horizontal direction was 7 mrad at the FWHM. In laser Compton scattering, a collimated X-ray beam can be obtained. The divergence angle of the X-ray beam is given by $\sim 1/\gamma$. Here, γ is the Lorentz factor of an electron energy. The divergence of an X-ray beam is estimated to be approximately 10 mrad from the observed peak energy of 50 MeV. The observed divergence angle was close to the predicted value from the electron energy. The maximum photon energy

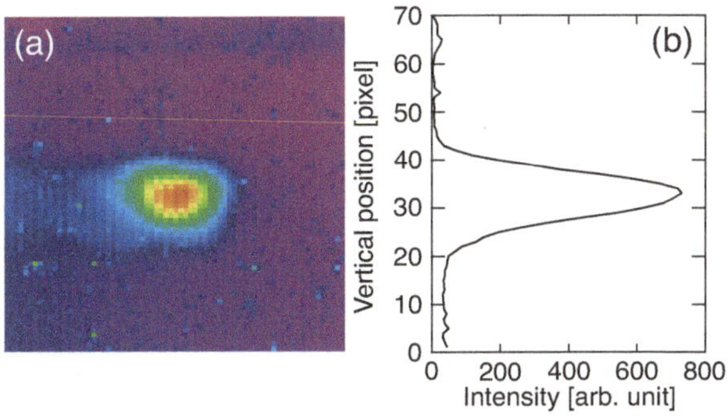

Fig. 16. (a) Image of X-rays produced by laser Compton scattering and (b) the vertical profile of the image. A well-collimated X-ray beam with a divergence angle of 5 mrad at the FWHM in the vertical direction is observed.

of the X-rays was estimated to be 60 keV from the peak energy of the QME beam and the interaction angle between the electron and laser pulses. The X-ray yield was also estimated to be approximately 10^5 photons/pulse from the charge of the QME beam and the irradiation conditions of the colliding laser pulse by including the dependence of the differential cross section of scattering on the scattered angle of X-rays.

The allowance range of the delay between the main and colliding laser pulses for X-ray generation was investigated. The allowance range was approximately 100 fs, which was close to the duration of the colliding laser pulse. This result suggests that a pulse duration of a QME beam is nearly equal to or less than 100 fs. The generation of an ultrashort electron pulse has been demonstrated at the same time.

5. Recent progress toward next step

The present status of LPA research has been presented as an example of works conducted at the AIST. In addition to our works, various works on improvement of the performance of electron beams have been so far conducted, and characterization and applications of electron beams have been also demonstrated. In this section, these woks are briefly reviewed.

5.1 Toward more energetic electron acceleration

LPA is expected as technologies obtaining extremely high energy particles using the extremely high accelerating field. A longer plasma and a longer acceleration length are required for obtaining more energetic electrons. One of the approaches is to form a plasma waveguide for guiding an intense laser pulse using a preformed plasma produced by a discharge, which has a transversely hollow density profile. A 1-GeV QME beam has been produced from a 3-cm-long capillary discharge plasma (Leemans et al., 2006). After that, several groups have reported the generation of GeV-class QME beams from a centimeter-scale capillary discharge plasma (Kameshima et al., 2008; Karsch et al., 2007; Rowlands-Rees et al., 2008). On the other hand, GeV-class electron beams have been also produced using a centimeter-scale gas jet based on a self-guiding of an intense laser pulse (Clayton et al., 2010; Hafz et al., 2008).

Experiments to accelerate extremely high energy electrons have been also conducted using a PW-class laser system of single shot operation. Energetic electron beams have been produced from a hollow glass capillary attached with a gold cone irradiated by an intense laser pulse (Kitagawa et al., 2004; Mori, Sentoku, Kondo, Tsuji, Nakanii, Fukumochi, Kashihara, Kimura, Takeda, Tanaka, Norimatsu, Tanimoto, Nakamura, Tampo, Kodama, Miura, Mima & Kitagawa, 2009). In this experiment, the gold cone attached at the entrance of a laser pulse plays an important role for the electron injection into a plasma wave driven inside the hollow capillary. Energetic electrons have been produced by the interaction of an intense laser pulse with a plasma preformed from a hollow plastic cylinder via laser-driven implosion (Nakanii et al., 2008). Electrons with energies of more than 600 MeV are observed from a 3-mm-long plasma tube.

5.2 Improvement, stabilization, and control of electron beam quality

Optical injection scheme

In a self-injection scheme, the injection of electrons into a plasma wave is based on the wave-breaking. The wave-breaking is a nonlinear phenomenon which is substantially unstable, and it is difficult to actively control it. Characteristics of electron beams are strongly affected by the shot-to-shot fluctuations of conditions of a gas jet, a laser pulse, and so on. Stabilization of electron beam qualities will be achieved by controlling the injection of electrons into a plasma wave. It is possible to control the injection of electrons into a plasma wave using multiple laser pulses. One laser pulse drives a plasma wave with an amplitude, in which wave-breaking does not occur, and other laser pulses control the injection of electrons into a plasma wave. This scheme is called optical injection scheme, or colliding pulse scheme. Although there have been several theoretical proposals (Esarey et al., 1997; Kotaki et al., 2004; Umstadter, Kim & Dodd, 1996), it took time for the experimental demonstration due to the requirement for using multiple laser pulses. Recently, experimental demonstrations of an optical injection scheme have been reported (Faure et al., 2006; Kotaki et al., 2009). When two counter-propagating laser pulses collide, a beat wave is formed and preaccelerates electrons in a plasma. The preaccelerated electrons are trapped into a plasma wave and accelerated. The stability of QME beam qualities has been improved, as compared with a self-injection scheme.

Plasma control in self-injection scheme

In a self-injection scheme, stabilization and improvement of electron beam qualities have been achieved by controlling the characteristics of a plasma.

In most experiments using a gas jet, a helium gas jet has been used. The pointing stability and divergence of QME beams have been improved by using an argon gas jet (Mori, Kondo, Mizuta, Kando, Kotaki, Nishiuchi, Kado, Pirozhkov, Ogura, Sugiyama, Bulanov, Tanaka,

Nishimura & Daido, 2009). For an argon gas jet, a preformed plasma suitable for guiding a main pulse is produced by ASE accompanying a main pulse due to the lower ionization potential of argon than that of helium.

The increase in a charge and the decrease in a beam divergence have been achieved using a gas jet composed helium and controlled amounts of various high-Z gases (McGuffey et al., 2010; Pak et al., 2010). Optical field ionization of inner shell electrons of the high-Z gas plays an important role in increasing the number of electrons injected into a plasma wave. This is referred as ionization induced trapping.

As described in Sec. 5.1, a capillary discharge plasma has been used for producing a long plasma and guiding an intense laser pulse. With a steady-state-flow gas cell using a hollow capillary without discharge, stable generation of QME beams has been achieved (Osterhoff et al., 2008). The steady-state gas flow forms a reproducible, homogeneous gas distribution along the laser propagation direction, which brings about the stable QME beam generation.

An increase in a charge and a decrease in a divergence of electron beams have been observed by applying a static external magnetic field along the laser pulse propagation axis, although the electron energy spectra are Maxwell-like distributions (Hosokai et al., 2006). The shape of a preformed plasma suitable for guiding a main pulse is formed by applying a static magnetic field.

5.3 Demonstration of femtosecond electron pulse generation

A femtosecond electron pulse can be produced in LPA. To prove the generation of a femtosecond electron pulse, a temporal characterization of an electron pulse has been conducted. In the temporal characterization, coherent transition radiation (CTR) emitted at the plasma-vacuum boundary or through a thin metallic foil is used. An electron pulse duration is measured with an electro-optical sampling technique using CTR in the THz region, and the temporal resolution of several tens of femtoseconds has been achieved. (Debus et al., 2010; van Tilborg et al., 2006) An electron pulse duration is also estimated from the spectrum of CTR (Glinec et al., 2007; Ohkubo et al., 2007). Recently, generation of a few femtosecond electron pulse has been demonstrated using an optical injection scheme (Lundh et al., 2011). Such ultrashort electron pulse is a useful tool to investigate physical and chemical kinetics of ultrafast phenomena initiated by ionization radiation referred as pulsed radiolysis. Applications of a laser-accelerated electron beam to pulsed radiolysis have been demonstrated (Brozek-Pluska et al., 2005).

5.4 Ultrashort X-ray radiation

Using a femtosecond electron pulse obtained in LPA, ultrashort short-wavelength radiation sources from extreme ultraviolet to X-ray can be obtained as the secondary source. Such ultrashort short-wavelength sources are attractive for applications to observe ultrafast phenomena such as time-resolved X-ray diffraction. As shown in Sec. 4, X-ray generation by laser Compton scattering using a laser-accelerated electron beam has been demonstrated (Miura et al., 2011; Schwoerer et al., 2006).

The generation of synchrotron radiation has been demonstrated by using a laser-accelerated QME beam through an undulator (Fuchs et al., 2009; Schlenvoigt et al., 2008). The wavelength of the radiation still remains visible to extreme ultraviolet due to the low electron energy. In the near future, the wavelength range will be extended to X-rays, as the electron energy increases.

Generation of keV X-rays by betatron radiation in a plasma has been demonstrated (Rousse et al., 2004). Laser-accelerated electrons undergo a transverse electric field in an ion channel,

which is formed inside a low electron density region of a plasma wave. They undergo betatron oscillation and emitted a collimated X-ray beam. Generation of a short X-ray pulse of less than 1 ps has been proven by measuring the temporal variation of X-ray reflectivity of a crystal pumped by a femtosecond laser pulse (Phuoc et al., 2007).
Extreme ultraviolet radiations have been demonstrated by the laser light reflection from a plasma wave driven by an intense laser pulse (Kando et al., 2007; 2009). The plasma wave moving almost at the speed of light acts as a relativistic flying mirror, and brings about frequency upshift and compression of the incident laser pulse due to the double Doppler effect. This regime is one of notable methods for generation of ultrashort short-wavelength radiations using an intense laser pulse, although laser-accelerated electron beams are not used.

6. Summary and future prospects

In this chapter, we provide the overview of the present status of research on LPA. Remarkable progress such as QME beam generation has been made especially in the last several years. Although the quality and stabilization of electron beams have been improved, there is still room for improvement of beam qualities for practical applications. For further improvement of electron beam qualities, theoretical and experimental studies should be continued to answer various questions in physics of laser-plasma interactions and to explore new acceleration regime.
Compactness of a plasma accelerating electrons, corresponding to an electron gun and a series of rf cavities in rf accelerators, has been proven. However, an ultrashort ultraintense laser system for producing the plasma is still large, although the laser system is called "table-top" system. More compact laser system is necessary. In addition, the repetition rate of present laser systems is still 10 Hz. For increasing the average flux and luminosity, the development of an efficient laser system with much higher repetition rate is also dispensable. It is essential to conduct the research on plasma physics and the development of laser technologies in parallel.
There are several prominent features in LPA such as femtosecond electron pulse generation. It is important to prove promising applications that make the best use of the features of LPA. Several promising applications have been already demonstrated. It becomes a stage when the design of laser electron accelerators for practical applications should be conducted by including a laser system. The recent progress of ultrashort ultraintense laser technologies is rapid and remarkable. In addition, the recent progress of high performance computer system is also rapid and remarkable. It will play a major role to investigate physics in laser-plasma interactions by numerical simulations. The set of rapid progress of laser technologies and investigation of physics will dramatically improve the performance of laser-accelerated electron beams. In the near future, laser electron accelerators useful for fundamental physics, and industrial and societal applications will be realized.

7. Acknowledgments

I thank all the coworkers for their supports and fruitful discussion, especially S. Masuda, S. Ishii, K. Tanaka, T. Ooyama, R. Kuroda, and H. Toyokawa, who have contributed to the works presented here. A part of our works presented here was financially supported by the Budget for Nuclear Research of the Ministry of Education, Culture, Sports, Science and Technology (MEXT), Japan, based on the screening and counseling by the Atomic Energy Commission, the Advanced Compact Accelerator Development of the MEXT, and the Matsuo Foundation. I also wish to acknowledge the support of the Advanced Compact Accelerator Development Office of the National Institute of Radiological Sciences, Japan.

8. References

Amiranoff, F., Bernard, D., Cros, B., Jacquet, F., Matthieussent, G., Miné, P., Mora, P., Morillo, J., Moulin, F., Specka, A. E. & Stenz, C. (1995). Electron acceleration in Nd-laser plasma beat-wave experiments, *Phys. Rev. Lett.* 74: 5220–5223.

Aoyama, M., Yamakawa, K., Akahane, Y., Ma, J., Inoue, N., Ueda, H. & Kiriyama, H. (2003). 0.85-PW, 33-fs Ti:sapphire laser, *Opt. Lett.* 28: 1594–1596.

Brozek-Pluska, B., Gliger, D., Hallou, A., Malka, V. & Gauduel, Y. A. (2005). Direct observation of elementary radical events: low- and high-energy radiation femtochemistry in solutions, *Rad. Phys. Chem.* 72: 149 – 157.

Bulanov, S. V., Pegoraro, F., Pukhov, A. M. & Sakharov, A. S. (1997). Transverse-wake wave breaking, *Phys. Rev. Lett.* 78: 4205–4208.

Cheriaux, G., Rousseau, P., Salin, F., Chambaret, J. P., Walker, B. & Dimauro, L. F. (1996). Aberration-free stretcher design for ultrashort-pulse amplification, *Opt. Lett.* 21: 414–416.

Clayton, C. E., Marsh, K. A., Dyson, A., Everett, M., Lal, A., Leemans, W. P., Williams, R. & Joshi, C. (1993). Ultrahigh-gradient acceleration of injected electrons by laser-excited relativistic electron plasma waves, *Phys. Rev. Lett.* 70: 37–40.

Clayton, C. E., Ralph, J. E., Albert, F., Fonseca, R. A., Glenzer, S. H., Joshi, C., Lu, W., Marsh, K. A., Martins, S. F., Mori, W. B., Pak, A., Tsung, F. S., Pollock, B. B., Ross, J. S., Silva, L. O. & Froula, D. H. (2010). Self-guided laser wakefield acceleration beyond 1 GeV using ionization-induced injection, *Phys. Rev. Lett.* 105: 105003.

Debus, A. D., Bussmann, M., Schramm, U., Sauerbrey, R., Murphy, C. D., Major, Z., Hörlein, R., Veisz, L., Schmid, K., Schreiber, J., Witte, K., Jamison, S. P., Gallacher, J. G., Jaroszynski, D. A., Kaluza, M. C., Hidding, B., Kiselev, S., Heathcote, R., Foster, P. S., Neely, D., Divall, E. J., Hooker, C. J., Smith, J. M., Ertel, K., Langley, A. J., Norreys, P., Collier, J. L. & Karsch, S. (2010). Electron bunch length measurements from laser-accelerated electrons using single-shot THz time-domain interferometry, *Phys. Rev. Lett.* 104: 084802.

Decker, C. D., Mori, W. B. & Katsouleas, T. (1994). Particle-in-cell simulations of Raman forward scattering from short-pulse high-intensity lasers, *Phys. Rev. E* 50: R3338–R3341.

Ebrahim, N. A. (1994). Optical mixing of laser light in a plasma and electron acceleration by relativistic electron plasma waves, *J. Appl. Phys.* 76: 7645–7647.

Esarey, E., Hubbard, R. F., Leemans, W. P., Ting, A. & Sprangle, P. (1997). Electron injection into plasma wakefields by colliding laser pulses, *Phys. Rev. Lett.* 79: 2682–2685.

Esarey, E., Schroeder, C. B. & Leemans, W. P. (2009). Physics of laser-driven plasma-based electron accelerators, *Rev. Mod. Phys.* 81: 1229–1285.

Esarey, E., Sprangle, P., Krall, J. & Ting, A. (1996). Overview of plasma-based accelerator concepts, *IEEE Trans. Plasma Sci.* 24: 252–288.

Faure, J., Glinec, Y., Pukhov, A., Kiselev, S., Gordienko, S., Lefebvre, E., Rousseau, J. P., Burgy, F. & Malka, V. (2004). A laser-plasma accelerator producing monoenergetic electron beams, *Nature* 431: 541–544.

Faure, J., Rechatin, C., Norlin, A., Lifschitz, A., Glinec, Y. & Malka, V. (2006). Controlled injection and acceleration of electrons in plasma wakefields by colliding laser pulses, *Nature* 444: 737–739.

Fuchs, M., Weingartner, R., Popp, A., Major, Z., Becker, S., Osterhoff, J., Cortrie, I., Zeitler, B., Hörlein, R., Tsakiris, G. D., Schramm, U., Rowlands-Rees, T. P., Hooker, S. M., Habs,

D., Krausz, F., Karsch, S. & Gruener, F. (2009). Laser-driven soft-x-ray undulator source, *Nature Phys.* 5: 826–829.

Gahn, C., Tsakiris, G. D., Pukhov, A., Meyer-ter Vehn, J., Pretzler, G., Thirolf, P., Habs, D. & Witte, K. J. (1999). Multi-MeV electron beam generation by direct laser acceleration in high-density plasma channels, *Phys. Rev. Lett.* 83: 4772–4775.

Geddes, C. G. R., Tóth, C., van Tilborg, J., Esarey, E., Schroeder, C. B., Bruhwiler, D., Nieter, C., Cary, J. & Leemans, W. P. (2004). High-quality electron beams from a laser wakefield accelerator using plasma-channel guiding, *Nature* 431: 538–541.

Geissler, M., Schreiber, J. & ter Vehn, J. M. (2006). Bubble acceleration of electrons with few-cycle laser pulses, *New J. Phys.* 8: 186.

Glinec, Y., Faure, J., Norlin, A., Pukhov, A. & Malka, V. (2007). Observation of fine structures in laser-driven electron beams using coherent transition radiation, *Phys. Rev. Lett.* 98: 194801.

Gordienko, S. & Pukhov, A. (2005). Scalings for ultrarelativistic laser plasmas and quasimonoenergetic electrons, *Phys. Plasmas* 12: 043109.

Hafz, N. A. M., Jeong, T. M., Choi, I. W., Lee, S. K., Pae, K. H., Kulagin, V. V., Sung, J. H., Yu, T. J., Hong, K.-H., Hosokai, T., Cary, J. R., Ko, D.-K. & Lee, J. (2008). Stable generation of GeV-class electron beams from self-guided laser-plasma channels, *Nature Photon.* 2: 571–577.

Hidding, B., Amthor, K.-U., Liesfeld, B., Schwoerer, H., Karsch, S., Geissler, M., Veisz, L., Schmid, K., Gallacher, J. G., Jamison, S. P., Jaroszynski, D., Pretzler, G. & Sauerbrey, R. (2006). Generation of quasimonoenergetic electron bunches with 80-fs laser pulses, *Phys. Rev. Lett.* 96: 105004.

Hosokai, T., Kinoshita, K., Ohkubo, T., Maekawa, A., Uesaka, M., Zhidkov, A., Yamazaki, A., Kotaki, H., Kando, M., Nakajima, K., Bulanov, S. V., Tomassini, P., Giulietti, A. & Giulietti, D. (2006). Observation of strong correlation between quasimonoenergetic electron beam generation by laser wakefield and laser guiding inside a preplasma cavity, *Phys. Rev. E* 73: 036407.

Hosokai, T., Kinoshita, K., Zhidkov, A., Maekawa, A., Yamazaki, A. & Uesaka, M. (2006). Effect of external static magnetic field on the emittance and total charge of electron beams generated by laser-wakefield acceleration, *Phys. Rev. Lett.* 97: 075004.

Hsieh, C.-T., Huang, C.-M., Chang, C.-L., Ho, Y.-C., Chen, Y.-S., Lin, J.-Y., Wang, J. & Chen, S.-Y. (2006). Tomography of injection and acceleration of monoenergetic electrons in a laser-wakefield accelerator, *Phys. Rev. Lett.* 96: 095001.

Ikeura-Sekiguchi, H., Kuroda, R., Yasumoto, M., Toyokawa, H., Koike, M., Yamada, K., Sakai, F., Mori, K., Maruyama, K., Oka, H. & Kimata, T. (2008). In-line phase-contrast imaging of a biological specimen using a compact laser-Compton scattering-based x-ray source, *Appl. Phys. Lett.* 92: 131107.

Joshi, C., Mori, W. B., Katsouleas, T., Dawson, J. M., Kindel, J. M. & Forslund, D. W. (1984). Ultrahigh gradient particle-acceleration by intense laser-driven plasma-density waves, *Nature* 311: 525–529.

Kameshima, T., Hong, W., Sugiyama, K., Wen, X., Wu, Y., Tang, C., Zhu, Q., Gu, Y., Zhang, B., Peng, H., Kurokawa, S., Chen, L., Tajima, T., Kumita, T. & Nakajima, K. (2008). 0.56 GeV laser electron acceleration in ablative-capillary-discharge plasma channel, *Applied Physics Express* 1: 066001.

Kando, M., Fukuda, Y., Pirozhkov, A. S., Ma, J., Daito, I., Chen, L.-M., Esirkepov, T. Z., Ogura, K., Homma, T., Hayashi, Y., Kotaki, H., Sagisaka, A., Mori, M., Koga, J. K., Daido, H., Bulanov, S. V., Kimura, T., Kato, Y. & Tajima, T. (2007). Demonstration of laser-frequency upshift by electron-density modulations in a plasma wakefield, *Phys. Rev. Lett.* 99: 135001.

Kando, M., Pirozhkov, A. S., Kawase, K., Esirkepov, T. Z., Fukuda, Y., Kiriyama, H., Okada, H., Daito, I., Kameshima, T., Hayashi, Y., Kotaki, H., Mori, M., Koga, J. K., Daido, H., Faenov, A. Y., Pikuz, T., Ma, J., Chen, L.-M., Ragozin, E. N., Kawachi, T., Kato, Y., Tajima, T. & Bulanov, S. V. (2009). Enhancement of photon number reflected by the relativistic flying mirror, *Phys. Rev. Lett.* 103: 235003.

Karsch, S., Osterhoff, J., Popp, A., Rowlands-Rees, T. P., Major, Z., Fuchs, M., Marx, B., Hörlein, R., Schmid, K., Veisz, L., Becker, S., Schramm, U., Hidding, B., Pretzler, G., Habs, D., Gruner, F., Krausz, F. & Hooker, S. M. (2007). GeV-scale electron acceleration in a gas-filled capillary discharge waveguide, *New J. Phys.* 9: 415.

Kitagawa, Y., Matsumoto, T., Minamihata, T., Sawai, K., Matsuo, K., Mima, K., Nishihara, K., Azechi, H., Tanaka, K. A., Takabe, H. & Nakai, S. (1992). Beat-wave excitation of plasma wave and observation of accelerated electrons, *Phys. Rev. Lett.* 68: 48–51.

Kitagawa, Y., Sentoku, Y., Akamatsu, S., Sakamoto, W., Kodama, R., Tanaka, K. A., Azumi, K., Norimatsu, T., Matsuoka, T., Fujita, H. & Yoshida, H. (2004). Electron acceleration in an ultraintense-laser-illuminated capillary, *Phys. Rev. Lett.* 92: 205002.

Kostyukov, I., Pukhov, A. & Kiselev, S. (2004). Phenomenological theory of laser-plasma interaction in "bubble" regime, *Phys. Plasmas* 11: 5256–5264.

Kotaki, H., Daito, I., Kando, M., Hayashi, Y., Kawase, K., Kameshima, T., Fukuda, Y., Homma, T., Ma, J., Chen, L.-M., Esirkepov, T. Z., Pirozhkov, A. S., Koga, J. K., Faenov, A., Pikuz, T., Kiriyama, H., Okada, H., Shimomura, T., Nakai, Y., Tanoue, M., Sasao, H., Wakai, D., Matsuura, H., Kondo, S., Kanazawa, S., Sugiyama, A., Daido, H. & Bulanov, S. V. (2009). Electron optical injection with head-on and countercrossing colliding laser pulses, *Phys. Rev. Lett.* 103: 194803.

Kotaki, H., Masuda, S., Kando, M., Koga, J. K. & Nakajima, K. (2004). Head-on injection of a high quality electron beam by the interaction of two laser pulses, *Phys. Plasmas* 11: 3296–3302.

Leemans, W. P., Nagler, B., Gonsalves, A. J., Tóth, C., Nakamura, K., Geddes, C. G. R., Esarey, E., Schroeder, C. B. & Hooker, S. M. (2006). Gev electron beams from a centimetre-scale accelerator, *Nature Phys.* 2: 696–699.

Lu, W., Tzoufras, M., Joshi, C., Tsung, F. S., Mori, W. B., Vieira, J., Fonseca, R. A. & Silva, L. O. (2007). Generating multi-gev electron bunches using single stage laser wakefield acceleration in a 3D nonlinear regime, *Phys. Rev. ST Accel. Beams* 10: 061301.

Lundh, O., Lim, J., Rechatin, C., Ammoura, L., Ben-Ismail, A., Davoine, X., Gallot, G., Goddet, J.-P., Lefebvre, E., Malka, V. & Faure, J. (2011). Few femtosecond, few kiloampere electron bunch produced by a laser-plasma accelerator, *Nature Phys.* 7: 219–222.

Maksimchuk, A., Reed, S., Naumova, N., Chvykov, V., Hou, B., Kalintchenko, G., Matsuoka, T., Nees, J., Rousseau, P., Mourou, G. & Yanovsky, V. (2007). Energy scaling of quasi-monoenergetic electron beams from laser wakefields driven by 40-TW ultra-short pulses, *Appl. Phys. B* 89: 201–207.

Malka, V., Fritzler, S., Lefebvre, E., Aleonard, M.-M., Burgy, F., Chambaret, J.-P., Chemin, J.-F., Krushelnick, K., Malka, G., Mangles, S. P. D., Najmudin, Z., Pittman, M., Rousseau, J.-P., Scheurer, J.-N., Walton, B. & Dangor, A. E. (2002). Electron acceleration by a wake field forced by an intense ultrashort laser pulse, *Science* 298: 1596–1600.

Mangles, S. P. D., Murphy, C. D., Najmudin, Z., Thomas, A. G. R., Collier, J. L., Dangor, A. E., Divall, E. J., Foster, P. S., Gallacher, J. G., Hooker, C. J., Jaroszynski, D. A., Langley, A. J., Mori, W. B., Norreys, P. A., Tsung, F. S., Viskup, R., Walton, B. R. & Krushelnick,

K. (2004). Monoenergetic beams of relativistic electrons from intense laser-plasma interactions, *Nature* 431: 535–538.

Mangles, S. P. D., Thomas, A. G. R., Lundh, O., Lindau, F., Kaluza, M. C., Persson, A., Wahlström, C.-G., Krushelnick, K. & Najmudin, Z. (2007). On the stability of laser wakefield electron accelerators in the monoenergetic regime, *Phys. Plasmas* 14: 056702.

Masuda, S. & Miura, E. (2008). Generation of quasi-monoenergetic electron beams with high charge by laser-driven plasma-based acceleration, *Appl. Phys. Express* 1: 086002.

Masuda, S. & Miura, E. (2009). Generation and analysis of quasimonoenergetic electron beams by laser-plasma interaction in transitional region from the self-modulated laser wakefield to bubble acceleration regime, *Phys. Plasmas* 16: 093105.

Masuda, S. & Miura, E. (2010). Effect of femtosecond prepulse with threshold intensity for optical field ionization on electron acceleration and propagation of intense laser pulse in plasma, *Jpn. J. Appl. Phys.* 49: 096401.

Masuda, S., Miura, E., Koyama, K. & Kato, S. (2008). Absolute calibration of an electron spectrometer using high energy electrons produced by the laser-plasma interaction, *Rev. Sci. Instrum.* 79: 083301.

Masuda, S., Miura, E., Koyama, K., Kato, S., Adachi, M., Watanabe, T., Torii, K. & Tanimoto, M. (2007). Energy scaling of monoenergetic electron beams generated by the laser-driven plasma based accelerator, *Phys. Plasmas* 14: 023103.

McGuffey, C., Thomas, A. G. R., Schumaker, W., Matsuoka, T., Chvykov, V., Dollar, F. J., Kalintchenko, G., Yanovsky, V., Maksimchuk, A., Krushelnick, K., Bychenkov, V. Y., Glazyrin, I. V. & Karpeev, A. V. (2010). Ionization induced trapping in a laser wakefield accelerator, *Phys. Rev. Lett.* 104: 025004.

Miura, E., Ishii, S., Tanaka, K., Kuroda, R. & Toyokawa, H. (2011). X-ray generation via laser compton scattering by laser-accelerated electron beam, *Proceedings of International Conference on Physics in Intense Fields PIF2010*, pp. 159–161.
URL: *http://ccdb4fs.kek.jp/tiff/2010/1025/1025013.pdf*

Miura, E., Koyama, K., Kato, S., Saito, N., Adachi, M., Kawada, Y., Nakamura, T. & Tanimoto, M. (2005). Demonstration of quasi-monoenergetic electron-beam generation in laser-driven plasma acceleration, *Appl. Phys. Lett.* 86: 251501.

Miura, E. & Masuda, S. (2009). Stable generation of quasi-monoenergetic electron beams with laser-driven plasma-based acceleration by suppressing nanosecond prepulse, *Appl. Phys. Express* 2: 126003.

Modena, A., Najmudin, Z., Dangor, A. E., Clayton, C. E., Marsh, K. A., Joshi, C., Malka, V., Darrow, C. B., Danson, C., Neely, D. & Walsh, F. N. (1995). Electron acceleration from the breaking of relativistic plasma-waves, *Nature* 377: 606–608.

Mori, M., Kando, M., Daito, I., Kotaki, H., Hayashi, Y., Yamazaki, A., Ogura, K., Sagisaka, A., Koga, J., Nakajima, K., Daido, H., Bulanov, S. & Kimura, T. (2006). Transverse dynamics and energy tuning of fast electrons generated in sub-relativistic intensity laser pulse interaction with plasmas, *Phys. Lett. A* 356: 146 – 151.

Mori, M., Kondo, K., Mizuta, Y., Kando, M., Kotaki, H., Nishiuchi, M., Kado, M., Pirozhkov, A. S., Ogura, K., Sugiyama, H., Bulanov, S. V., Tanaka, K. A., Nishimura, H. & Daido, H. (2009). Generation of stable and low-divergence 10-MeV quasimonoenergetic electron bunch using argon gas jet, *Phys. Rev. ST Accel. Beams* 12: 082801.

Mori, Y., Sentoku, Y., Kondo, K., Tsuji, K., Nakanii, N., Fukumochi, S., Kashihara, M., Kimura, K., Takeda, K., Tanaka, K. A., Norimatsu, T., Tanimoto, T., Nakamura, H., Tampo, M.,

Kodama, R., Miura, E., Mima, K. & Kitagawa, Y. (2009). Autoinjection of electrons into a wake field using a capillary with attached cone, *Phys. Plasmas* 16: 123103.

Nakajima, K., Fisher, D., Kawakubo, T., Nakanishi, H., Ogata, A., Kato, Y., Kitagawa, Y., Kodama, R., Mima, K., Shiraga, H., Suzuki, K., Yamakawa, K., Zhang, T., Sakawa, Y., Shoji, T., Nishida, Y., Yugami, N., Downer, M. & Tajima, T. (1995). Observation of ultrahigh gradient electron acceleration by a self-modulated intense short laser pulse, *Phys. Rev. Lett.* 74: 4428–4431.

Nakanii, N., Kondo, K., Kuramitsu, Y., Mori, Y., Miura, E., Tsuji, K., Kimura, K., Fukumochi, S., Kashihara, M., Tanimoto, T., Nakamura, H., Ishikura, T., Takeda, K., Tampo, M., Takabe, H., Kodama, R., Kitagawa, Y., Mima, K. & Tanaka, K. A. (2008). Spectrum modulation of relativistic electrons by laser wakefield, *Appl. Phys. Lett.* 93: 081501.

Ohkubo, T., Maekawa, A., Tsujii, R., Hosokai, T., Kinoshita, K., Kobayashi, K., Uesaka, M., Zhidkov, A., Nemoto, K., Kondo, Y. & Shibata, Y. (2007). Temporal characteristics of monoenergetic electron beams generated by the laser wakefield acceleration, *Phys. Rev. ST Accel. Beams* 10: 031301.

Osterhoff, J., Popp, A., Major, Z., Marx, B., Rowlands-Rees, T. P., Fuchs, M., Geissler, M., Hörlein, R., Hidding, B., Becker, S., Peralta, E. A., Schramm, U., Grüner, F., Habs, D., Krausz, F., Hooker, S. M. & Karsch, S. (2008). Generation of stable, low-divergence electron beams by laser-wakefield acceleration in a steady-state-flow gas cell, *Phys. Rev. Lett.* 101: 085002.

Pak, A., Marsh, K. A., Martins, S. F., Lu, W., Mori, W. B. & Joshi, C. (2010). Injection and trapping of tunnel-ionized electrons into laser-produced wakes, *Phys. Rev. Lett.* 104: 025003.

Perry, M. D., Pennington, D., Stuart, B. C., Tietbohl, G., Britten, J. A., Brown, C., Herman, S., Golick, B., Kartz, M., Miller, J., Powell, H. T., Vergino, M. & Yanovsky, V. (1999). Petawatt laser pulses, *Opt. Lett.* 24: 160–162.

Phuoc, K. T., Fitour, R., Tafzi, A., Garl, T., Artemiev, N., Shah, R., Albert, F., Boschetto, D., Rousse, A., Kim, D.-E., Pukhov, A., Seredov, V. & Kostyukov, I. (2007). Demonstration of the ultrafast nature of laser produced betatron radiation, *Phys. Plasmas* 14: 080701.

Pukhov, A. & Meyer-ter Vehn, J. (2002). Laser wake field acceleration: the highly non-linear broken-wave regime, *Appl. Phys. B* 74: 355–361.

Rousse, A., Phuoc, K. T., Shah, R., Pukhov, A., Lefebvre, E., Malka, V., Kiselev, S., Burgy, F., Rousseau, J.-P., Umstadter, D. & Hulin, D. (2004). Production of a keV x-ray beam from synchrotron radiation in relativistic laser-plasma interaction, *Phys. Rev. Lett.* 93: 135005.

Rowlands-Rees, T. P., Kamperidis, C., Kneip, S., Gonsalves, A. J., Mangles, S. P. D., Gallacher, J. G., Brunetti, E., Ibbotson, T., Murphy, C. D., Foster, P. S., Streeter, M. J. V., Budde, F., Norreys, P. A., Jaroszynski, D. A., Krushelnick, K., Najmudin, Z. & Hooker, S. M. (2008). Laser-driven acceleration of electrons in a partially ionized plasma channel, *Phys. Rev. Lett.* 100: 105005.

Schlenvoigt, H. P., Haupt, K., Debus, A., Budde, F., Jäckel, O., Pfotenhauer, S., Schwoerer, H., Rohwer, E., Gallacher, J. G., Brunetti, E., Shanks, R. P., Wiggins, S. M. & Jaroszynski, D. A. (2008). A compact synchrotron radiation source driven by a laser-plasma wakefield accelerator, *Nature Phys.* 4: 130–133.

Schoenlein, R. W., Leemans, W. P., Chin, A. H., Volfbeyn, P., Glover, T. E., Balling, P., Zolotorev, M., Kim, K.-J., Chattopadhyay, S. & Shank, C. V. (1996). Femtosecond x-ray pulses at 0.4 Å generated by 90° Thomson scattering: A tool for probing the structural dynamics of materials, *Science* 274: 236–238.

Schwoerer, H., Liesfeld, B., Schlenvoigt, H.-P., Amthor, K.-U. & Sauerbrey, R. (2006). Thomson-backscattered x rays from laser-accelerated electrons, *Phys. Rev. Lett.* 96: 014802.

Strickland, D. & Mourou, G. (1985). Compression of amplified chirped optical pulses, *Opt. Commun.* 56: 219–221.

Tajima, T. & Dawson, J. M. (1979). Laser electron accelerator, *Phys. Rev. Lett.* 43: 267–270.

Tanaka, K. A., Yabuuchi, T., Sato, T., Kodama, R., Kitagawa, Y., Takahashi, T., Ikeda, T., Honda, Y. & Okuda, S. (2005). Calibration of imaging plate for high energy electron spectrometer, *Rev. Sci. Instrum.* 76: 013507.

Umstadter, D., Chen, S.-Y., Maksimchuk, A., Mourou, G. & Wagner, R. (1996). Nonlinear optics in relativistic plasmas and laser wake field acceleration of electrons, *Science* 273: 472–475.

Umstadter, D., Kim, J. K. & Dodd, E. (1996). Laser injection of ultrashort electron pulses into wakefield plasma waves, *Phys. Rev. Lett.* 76: 2073–2076.

van Tilborg, J., Schroeder, C. B., Filip, C. V., Tóth, C., Geddes, C. G. R., Fubiani, G., Huber, R., Kaindl, R. A., Esarey, E. & Leemans, W. P. (2006). Temporal characterization of femtosecond laser-plasma-accelerated electron bunches using terahertz radiation, *Phys. Rev. Lett.* 96: 014801.

Verluise, F., Laude, V., Cheng, Z., Spielmann, C. & Tournois, P. (2000). Amplitude and phase control of ultrashort pulses by use of an acousto-optic programmable dispersive filter: pulse compression and shaping, *Opt. Lett.* 25: 575–577.

Yamazaki, A., Kotaki, H., Daito, I., Kando, M., Bulanov, S. V., Esirkepov, T. Z., Kondo, S., Kanazawa, S., Homma, T., Nakajima, K., Oishi, Y., Nayuki, T., Fujii, T. & Nemoto, K. (2005). Quasi-monoenergetic electron beam generation during laser pulse interaction with very low density plasmas, *Phys. Plasmas* 12: 093101.

Yorozu, M., Yang, J., Okada, Y., Yanagida, T., Sakai, F., Takasago, K., Ito, S. & Endo, A. (2002). Fluctuation of femtosecond x-ray pulses generated by a laser-compton scheme, *Appl. Phys. B* 74: 327–331.

3

Fast Charged Particles and Super-Strong Magnetic Fields Generated by Intense Laser Target Interaction

Vadim Belyaev and Anatoly Matafonov
Central Research Institute of Machine Building
Russian Federation

1. Introduction

The development of a new generation of solid-state lasers has resulted in unique conditions for irradiating laser targets by light pulses, with radiation intensity ranging from 10^{17} to 10^{21} W/cm^2 and a duration of 20 - 1000 fs.

At such intensities, the laser pulse produces superstrong electric fields which could not be obtained earlier and considerably exceed the atomic electric field of strength $E_a = 5.14\times10^9$ V/cm. In these conditions, there arises a new physical picture of laser pulse interaction with plasma produced when the pulse leading edge or a pre-pulse affects solid targets. Laser radiation is rather efficiently transformed into fluxes of fast charged particles such as electrons and atomic ions. The latter interact with the ambient material of the target, which leads to the generation of hard X-rays, when inner atomic shells are ionized, and to various nuclear and photonuclear reactions.

One important area in investigating the interaction of sub-picosecond laser pulses with solid targets is related to the important role which arising superstrong quasistatic magnetic fields and electronic structures play in laser plasma dynamics. This area of research became most attractive after carrying out the direct measurements of quasistatic magnetic fields on the Vulcan laser system (Great Britain) (Tatarakis et al., 2002), in particular, after the pinch effect has been found experimentally in laser plasma (Beg et al., 2004).

The relativistic character of laser radiation with intensity I is realized at the magnitude of a dimensionless parameter $a > 1$. This parameter represents the dimensionless momentum of the electron oscillating in the electric field of linearly polarized laser radiation and can be expressed as

$$a = \frac{eE}{mc\omega} = 0.85\left(\frac{\lambda}{\mu\text{m}}\right)\left(\frac{I}{10^{18}\,\text{W/cm}^2}\right)^{\frac{1}{2}} \quad (1)$$

$$\frac{E}{\text{V/cm}} = 27.7\left(\frac{I}{\text{W/cm}^2}\right)^{\frac{1}{2}} \quad (2)$$

where e and m are the charge and mass of the electron, respectively, E is the amplitude of electric field strength (in units of V/cm) of laser radiation, λ is the radiation wavelength (in μm), ω is the frequency of laser radiation, c is the speed of light, and I is the radiation intensity (in W/cm^2).

Terawatt-power laser systems of moderate size can fulfill the condition a > 1, which corresponds to the electric field strength above 10^{10} V/cm. In such intense fields, the overbarrier ionization of atoms occurs in atomic time on the order of 10^{-17} s, and the electrons produced are accelerated and reach MeV-range relativistic energies during the laser pulse.

The acceleration of atomic ions in femto- and picoseconds laser plasmas constitutes a secondary process. It is caused by the strong quasistatic electric fields arising due to spatial charge separation. Such separation is related to the motion of a bunch of fast electrons. For laser radiation intensities exceeding $I \geq 10^{18}$ W/cm^2, it is possible to obtain directed beams of high-energy ions with the energies $\varepsilon_i > 1$ MeV.

The generation of high-energy proton and ion beams in laser plasma under the action of ultrashort pulses is a quickly developing field of investigations. This is explained, in particular, by their important applications in such fields as proton accelerators, the study of material structure, proton radiography, the production of short-living radioisotopes for medical purposes, and laser controlled fusion (Umstadter, 2003; Mourou et al., 2006). For a laser radiation intensity of $I \geq 10^{18}$ W/cm^2, a number of nuclear reactions can be initiated that have only been realized in elementary particle accelerators (Andreev et al., 2001).

Later on, we will consider the principal mechanisms for generating fast charged particles and quasistatic magnetic fields in laser plasmas, as well as experimental results obtained both abroad and on the native laser setup NEODIM in the Central Research Institute of Machine Building (Russ. abbr. TsNIIMash) (Korolev, Moscow reg.) (Belyaev et al., 2004; Belyaev et al., 2005).

2. Generation of fast electrons in laser plasma

In irradiating a target by a high-intensity ultrashort laser pulse, the radiation energy is rather efficiently converted into the energy of fast electrons which later partially transfer their energy to the atomic ions of the target. Presently, several mechanisms are being discussed concerning the generation of fast electrons when a laser pulse affects plasma with a density well above the critical value. If the laser pulse is not accompanied by a pre-pulse (the case of high contrast), then the laser radiation interacts with plasma of a solid-state density, possessing a sharp boundary. In this case, the mechanism of `vacuum heating' is realized (Brunel, 1987), as is the so-called $\mathbf{v}\times\mathbf{B}$ mechanism (Wilks et al., 1992) (here, $\mathbf{B}$ is the magnetic field induction of the laser field) caused by a longitudinal ponderomotive force acting along the propagation direction of the laser pulse). This $\mathbf{v}\times\mathbf{B}$ mechanism becomes substantial at relativistic intensities where the energy of electron oscillations is comparable with or exceeds the electron rest energy $mc^2 = 511$ keV - that is, for the parameter $a > 1$ [see formula (1)]. In addition, fast electrons can be generated on the critical surface of plasma at a plasma resonance (Gus'kov et al., 2001; Demchenko et al., 2001) if the vector of the laser radiation electric field has a projection along the density gradient (usually at an inclined incidence of laser radiation to target) and the laser frequency coincides with the plasma frequency. In contrast to the ponderomotive **v**×**B** mechanism, vacuum heating and resonance absorption arise at nonrelativistic (substantially lower, with $a < 1$) intensities as well. In the case of the ponderomotive mechanism, the average energy of fast electrons can be estimated as the maximum energy of transverse electron oscillations in an electromagnetic field, which in the general case takes a relativistic value. In a underdense part of the laser plasma, we have

$$\varepsilon_e = mc^2\left[\left(1+\frac{Q}{Q_0}\right)^{\frac{1}{2}}-1\right],\ Q = I\lambda^2,\ Q_0 = 1.37\times10^{18}\frac{\mathrm{W}}{\mathrm{cm}^2}\mu\mathrm{m}^2 \qquad (3)$$

In the ultrarelativistic limit $Q \gg Q_0$, we hence obtain

$$\varepsilon_e = mc^2\left(\frac{Q}{Q_0}\right)^{\frac{1}{2}}.$$

By contrast, in the nonrelativistic limit $Q \ll Q_0$, we derive from formula (3) that

$$\varepsilon_e = mc^2\frac{Q}{2Q_0}.$$

In the overdense part of the plasma, the ponderomotive heating of electrons is noticeably weaker due to a difficult penetration of the laser field into this region.

In the case of vacuum heating, the maximum energy of an electron flying into the depths of a dense target is given by the formula similar to equation (3), however, with a different numerical factor.

There is one more mechanism for generating fast electrons in the underdense part of plasma in front of a target due to the betatron resonance in the arising magnetic field (Pukhov, 2003). In this regime, electrons are accelerated by the transverse ultrarelativistic electric field of the laser wave in the direction of wave polarization, and the azimuthal magnetic field produced by the current of fast electrons is responsible for the magnetic part of the Lorentz force. This force turns electrons in such a way that they gradually change to the opposite direction of motion. In the case of an exact betatron resonance, the reflection occurs at the instant when the transverse electric field changes its direction, so the electrons are accelerated at all times. This mechanism yields an energy of fast electrons three times greater than formula (3) does:

$$\varepsilon_e = 3mc^2\left(\frac{Q}{Q_0}\right)^{\frac{1}{2}}. \qquad (4)$$

There are also further mechanisms of electron acceleration that require special experimental conditions, for example, the wake field acceleration (Esarey, 1996; Amiranoff, 1998). In the case of resonance absorption, the electric field near the plasma critical surface is much stronger than that of incident laser radiation. The result is that the heating of electrons upon their impact with atomic ions is greater than follows from formulae (3) and (4).

Electrons are also accelerated by a transverse ponderomotive force (acting in the radial direction) due to a focal distribution of laser intensity. This acceleration leads to the maximum electron energy also expressed by formula (3) (in the underdense part of plasma) if electrons succeed in acquiring this energy moving from the focus to the periphery during the laser pulse. Thus, the duration τ of a laser pulse should meet the inequality $\tau \rangle\rangle\ m\omega R/eE$ (in the nonrelativistic case). Here, R is the radius of the focal spot of a laser beam. This inequality holds for picosecond- and longer-duration laser light pulses with an intensity on the order of 10^{16} W/cm^2. In fields with an intensity of 10^{18} W/cm^2, the right-hand side of this inequality reaches dozens of femtoseconds, whereas in the overdense part of plasma this ponderomotive force is noticeably weaker.

We have discussed the above mentioned mechanisms in more detail in our article (Belyaev et al., 2008).

We suggested and investigated the new mechanism of high-energy electrons formation in ultra-high intensity laser pulse interaction with solid targets (Belyaev, 2004). This investigation is an attempt to reveal and describe, based on the model suggested, the high-energy electron formation mechanism in laser plasmas so as to derive theoretical dependences which would represent specific relations between the parameters of fast electrons, laser radiation and target substance.

Any theory can be accepted only after reliable experimental verification. The degree of reliability is determined not only by the sufficient diversity of independent experimental data, but also by the ability to choose out of these data those best representative of the overall pattern. Analysis of numerous experiments to measure energy of fast electrons formed in laser plasmas shows that with a particular laser facility, given its available radiation intensity, fast electron maximum energy can be determined most closely. Generally, it is electron maximum energy values that are most widely presented in experimental investigations. This is motivated not only by experimenters' striving to get extreme record-breaking output parameters, but also by the possibility to most closely determine the electron maximum energy around their spectrum extrapolation at specified intensity of laser radiation incident on a target. On this basis we will establish our theoretical model of the maximum-energy electron formation process for a given laser radiation intensity.

Without going into details of magnetic field generation mechanisms, it can be noted that a vortical electron structure develops eventually in plasma. Given the applied electric field (constituent of the incident laser radiation) and the dominance of tunnel ionization, a great number of electrons (practically determined by solid density) are accelerated. This current of electrons generates a magnetic field which bends their trajectory. Under certain conditions these trajectories can close at skin-layer depth within larmor-radius circle. The high electron density and, correspondingly, the circular current strength cause super-strong magnetic fields generation.

Condition for such fields generation can be written as a condition for electron movement around such a circle in the form of a balance between the centrifugal force and the Lorentz force:

$$\frac{mV^2}{r} = \frac{eVB}{c}, \tag{5}$$

where $r = \delta/2$, δ - skin-layer thickness, e, m, V – charge, mass, electron velocity, c – velocity of light, B – magnetic induction in the electron orbit.
Taking electromagnetic field penetration depth δ to be equal to incident radiation wave length λ, we have $r = \lambda / 2$.
Given the relationship between mass and velocity, the kinetic energy change due to the action of the forces applied is always equal to

$$E_{KIN} = (m_V - m_0)c^2 = m_0 c^2 (\gamma - 1), \tag{6}$$

where m_V - relativistic mass, m_0 – electron rest mass, $\gamma = 1/\sqrt{1 - V^2/c^2}$ - relativistic factor.
Considering:

- relativistic expression of electron momentum

$$\mathbf{P} = \mathbf{p} - \frac{e}{c}\mathbf{A} \tag{7}$$

where V – electron velocity;

- use of generalized momentum

$$\mathbf{P} = \mathbf{p} - \frac{e}{c}\mathbf{A}, \tag{8}$$

where $\mathbf{p}$ – ordinary momentum (8), $\mathbf{A}$ – vector-potential;

- magnetic field B cylindrical symmetry: $B_X = 0$; $B_Y = 0$; $B_Z = -\frac{\partial A_X}{\partial y} = B$, which allows to put $A = Br$ and

$$P = m_0 V = \frac{e}{c} Br \tag{9}$$

So from (6) – (9) we have

$$E_{KIN} = m_0 c^2 \left[1 + \left(\frac{eBr}{m_0 c^2} \right)^2 \right]^{\frac{1}{2}} - m_0 c^2, \tag{10}$$

To find the electron maximum kinetic energy at specified intensity of laser radiation incident on the target we need the maximum value of B – magnetic field induced in laser plasma. This value can be estimated using the energy conservation law.
Omitting calculations we can use the following formula easy to keep in mind:

$$B_{MAX}\ [\mathrm{Gs}] = 10^{-1}\sqrt{J[\mathrm{W/cm^2}]}, \tag{11}$$

Substitution this formulae in expression (10) for kinetic electron energy gives:

$$E_{KIN} = 0{,}5 \left[1 + \frac{9J\lambda^2}{10^{18}} \right]^{\frac{1}{2}} - 0{,}5, \tag{12}$$

where intensity J expressed in W/cm², λ - in micrometer, kinetic energy - MeV. Graph of this dependence show Fig. 1 by curve 1.

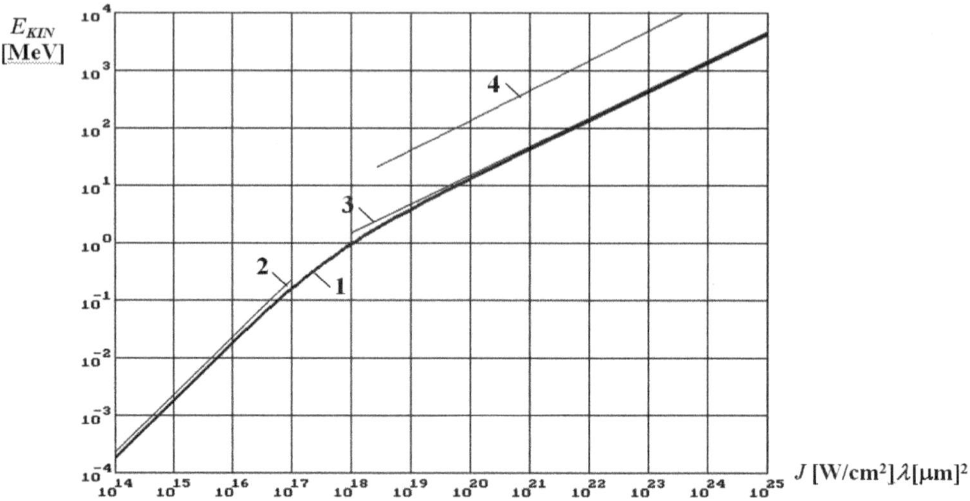

Fig. 1. Dependence of electron kinetic energy on laser radiation intensity.

Consider limiting cases.

1. $E_{KIN} = m_0c^2 = 0{,}5$ MeV.

Expression (12) gives this value at intensity

$$J_R = \frac{1}{3}10^{18}\left[\frac{1}{\lambda\ [\mu\text{m}]}\right]\frac{\text{W}}{\text{cm}^2}, \qquad (13)$$

The intensity J_R can be called as relativistic intensity.

2. $E_{KIN} \langle\langle\ m_0c^2;\ J < J_R$.

In this case

$$E_{KIN} = 2{,}25\frac{J\lambda^2}{10^{18}}, \qquad (14)$$

Graph of this dependence show at Fig. 1 by curve 2.

3. $E_{KIN} \rangle\rangle\ m_0c^2;\ J > J_R$.

For this case

$$E_{KIN} = 1{,}5\left(\frac{J\lambda^2}{10^{18}}\right)^{\frac{1}{2}}, \qquad (15)$$

and graph of this dependence show on Fig. 1 by curve 3.

Equations obtained for small (< m_0c^2) and large (> m_0c^2) values of kinetic energy agree with those in use for calculations of particle energy in a cyclotron and in a betatron, correspondingly. In both cases electrons are accelerated under the action of an electric field. In a cyclotron, this is a periodically changing electric field applied externally. In a betatron, this is a vortex electric field occurring with axisymmetric magnetic field rise in time. In laser plasmas a magnetic field is generated giving rise to a vortex electric field accelerating electrons. Thus the laser-plasma electron acceleration mechanism resembles the betatron case.

Equation (10) for electron kinetic energy was derived on the assumption that the electron acceleration is governed only by the laser radiation incident on the target without considering the processes going within the target substance, specifically, ionization process. Formally, it is reflected in the fact that the skin-layer size is determined by the laser radiation frequency

$$\delta = c/\nu = 2\pi c/\omega = \lambda \,, \tag{16}$$

meaning that the laser radiation frequency ω is an effective frequency. This assumption is true only at the first stage of interaction with the substance when a vortical electron structure develops on skin-layer scales, its characteristic size being in accordance with (16). This structure is unstable and there is a possibility of its transformation to smaller-scale structures. This process is known as a dynamic pinch.

It is demonstrated in (Belyaev & Mikhailov, 2001) that in case of laser plasmas produced by the action of high-intensity ($J > 10^{16}$ W/cm^2) laser radiation of ultrashort duration ($\tau < 10^{-12}$ sec) on a solid target this process is of quantum nature and can be described by the diffusion equation. Without going into the process nature, note that under tunnel ionization the vortical electron structure generated on skin-layer scales (16) transforms to another one, its characteristic size now being determined by the ionization frequency as an effective frequency at the next stage of laser radiation interaction with the substance, i.e. at the stage of tunnel ionization development:

$$l_i = 2\pi c/\omega_i \,, \tag{17}$$

Assuming that the vortical electron structure transformation process goes with the magnetic flow kept unchanged, we have

$$B_0 \lambda^2 = B_i l_i^2 \,, \tag{18}$$

where B_i - magnetic field within the vortical structure, its characteristic size l_i, being determined by (17). Such a vortical structure provides the following kinetic energy to the electrons:

$$E_{кин} = eB_i l_i = eB_0 \frac{\lambda^2}{l_i} \,, \tag{19}$$

Equation (19) determines the maximum energy of the small group (tail) of high-energy electrons. This dependence can be represented via the energy or ionization potential of the target substance atoms:

$$E_{KIN} = eB_0 \lambda \frac{\omega_i}{\omega} = 1.5 \left(\frac{J\lambda^2}{10^{18}} \right)^{\frac{1}{2}} \frac{\omega_i}{\omega} = 1{,}5 \left(\frac{J\lambda^2}{10^{18}} \right)^{\frac{1}{2}} \frac{I}{\hbar\omega} \text{ [MeV]}, \tag{20}$$

Here J is in W/cm^2, λ - in μm, I and $\hbar\omega$ - in eV. This dependence is plotted in Fig. 1 (curve 4).
The equation obtained demonstrates the proportionality between the electron energy and ionization frequency, hich determines physical nature of the electron acceleration process. The physics of the electron acceleration processes as a result of high-intensity laser radiation action on a substance is closely related to the physics of the ionization processes in superatomic intensity fields.
The ionization frequency is generally one or two orders higher than the laser one. This results in the high acceleration rate and electron energy.
The process of dynamic pinch development give rises formation of high-energy tail (20) and has threshold nature. Our estimations give value of threshold 0.31×10^{18} - 3.2×10^{19} W/cm^2. Threshold smearing evidences for stochastic character of the process.
The good agreement between theory and experiment (Matafonov & Belyaev, 2001; Malka & Miquel, 1996; Borodin et al., 2000; Nickles et al., 1999; Ledingham & Norreys, 1999; Cowan et al., 1999; He et al., 2004; Mangles et al., 2005) suggests the realizability of the proposed high-energy electron formation mechanism in laser plasmas.
In our article (Belyaev, Kostenko et al., 2003) we also have investigated cyclotron mechanism of electron acceleration.
The magnetic activity of picosecond laser plasma offers new mechanisms for the generation of fast electrons due to the presence of such strong quasistatic magnetic fields regardless of the mechanisms of their origin. Such a possibility is related to the emergence of cyclotron resonances when the laser frequency ω coincides with the Larmor gyration frequency $\Omega = eB_0/m_rc$ of an electron in an external constant magnetic field with the induction B_0 (here, e and m_r are the charge and relativistic mass of the electron, respectively; c is the speed of
light). Indeed, the typical laser frequency ω (in the Hartree atomic system of units) is on the order of 0.05 a.u. and coincides with the cyclotron frequency at an induction of B_0 = 7 a.u. ~ 100 MG. This value may become much greater with allowance made for the relativistic increase in electron mass, which is typical at laser radiation intensities on the order of 10^{19} - 10^{20} W/cm^2.Hence, the generation of a constant magnetic field results in stronger interaction of laser radiation with plasma. The situation is to a certain extent similar to the radiation self-focusing effect, in which case the variations in the refraction index of the medium in the field of a laser wave influence wave propagation through the medium.
In the general relativistic case, the interaction of electrons with the field of a laser wave and with the constant magnetic field $\mathbf{B}_0$ is written out in the form

$$\frac{d\mathbf{p}}{dt} = e\left\{\mathbf{E} + \frac{1}{c}\mathbf{v}\times\left(\mathbf{B}+\mathbf{B}_0\right)\right\} \tag{21}$$

for electrons possessing a momentum **p** and a velocity **v**.
For circular polarization, the problem is solved analytically, whereas in the general case of linear polarization the problem reduces to a system of nonlinear equations, which can only be solved numerically. The solution to these equations is specific in that there are resonances between the periodicmotion of electrons in the magnetic field and electron oscillations in the field of the laser wave. This fact leads to drastic changes in electron trajectory and energy at certain instants of time.

Figure 2 depicts the variations in the kinetic energy of an electron (a) and its trajectory (b) for motion with the zero initial velocity in a field with a radiation intensity of 10^{20} W/cm^2 and a frequency that is at resonance with the cyclotron frequency (Belyaev & Kostenko, 2003). The constant magnetic field is normal to the polarization of laser radiation. One can see that an electron acquires an energy of approximately 100MeV in a time on the order of hundreds of femtoseconds.

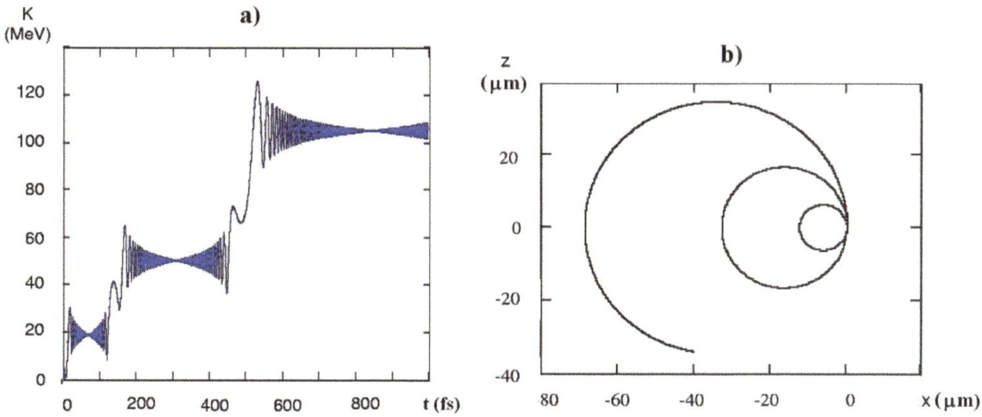

Fig. 2. Electron kinetic energy (a) and trajectory (b) in a linearly polarized laser wave with an intensity of 10^{20} W/cm^2 and a constant transverse magnetic field in the resonance case of $\Omega = \omega$.

Electron acceleration in the field of a circularly polarized laser wave propagating along a strong magnetic field was theoretically investigated in lectures (Pavlenko, 2002). It was shown that the relativistic factor of the electron may increase by an order of magnitude under a high intensity of laser radiation.

3. Generation of fast protons and ions in the interaction of ultrashort high-intensity laser pulses with solid targets

On the basis of the results of experimental and theoretical investigations performed in recent years, one can determine the following ranges for product plasma parameters: the electron temperature is about 1 to 10 keV; the mean energy of "fast" electrons is about 0.1 to 10 MeV (the maximum energy is as high as 300 MeV); the mean energy of fast ions ranges from several hundred keV units to a few MeV units (the maximum energy is 430 MeV); the relativistic longitudinal ponderomotive pressure of laser light is 1 to 50 Gbar; and the amplitudes of the electric field and spontaneous-magnetic-field strength range, respectively, between about 10^9 and 10^{12} V/cm and between about 1 and 500 MG (Belyaev et al., 2008; Salamin et al., 2006). Product terawatt-pulse picosecond laser plasmas appear to be some kind of a "table" pulsed "microaccelerator" and a nuclear "microreactor," which is relatively compact and cheap and on which one must not impose special radiation-safety requirements. Such a source admits a relatively simple possibility for controlling energy and other parameters of corpuscular and electromagnetic radiations.

At the present time, the production of high-energy proton beams in laser plasmas under the effect of ultrashort pulses is a rapidly developing field of investigations (Carrier et al., 2009; Fucuda et al., 2009; Yan et al., 2009; Willingale et al., 2009; Gonoskov et al., 2009; Psikal et al., 2010; Huang et al., 2010).

Several models that claim for explaining observed results that concern the production of directed beams of high-energy protons were proposed in theoretical investigations. One of them is based on the mechanism of proton acceleration at the front surface of the target owing to the ponderomotive pressure of a laser pulse (Sentoku et al., 2003; Maksimchuk et al., 2000). According to a different model (MacKinnon et al., 2001; Wilks et al., 2001), relativistic hot electrons produced by a laser field in a solid-state target penetrate through the target, and some electrons escape from the rear surface of the target to a distance of about the Debye radius. These electrons generate an electrostatic field at the rear surface of the target. This field, which may exceed 10^{12} V/cm, accelerates protons.

However, the efficiency of the proposed proton acceleration mechanisms has so far been debated (Salamin et al., 2006). In view of this, the our experimental studies were aimed at exploring various mechanisms of the acceleration of fast protons in laser plasmas under identical conditions of the irradiation of a solid-state target at a laser-radiation intensity of about 2×10^{18} W/cm^2.

The experiments in question were performed at the 10-TW picosecond laser facility Neodymium (Belyaev, Vinogradov et al., 2006). This laser facility has the following laser-pulse parameters: a pulse energy of up to 10 J, the wavelength of 1.055 μm, and the pulse duration of 1.5 ps. Its focusing system, which is based on an off-axis parabolic mirror whose focal length is 20 cm, ensures a concentration of not less than 40% of the laser beam energy within a spot D = 15 μm in diameter and, accordingly, an average intensity of 10^{18} W/cm^2 at the target surface and a peak intensity of 2×10^{18} W/cm^2.

Laser radiation generated by the Neodymium facility is characterized by the presence of two prepulses—one of picosecond and the other of nanosecond duration. The first prepulse appears 14 ns before the main laser pulse; it has a duration of 1.5 ps and an intensity below 10^{-8} with respect to the main pulse. The second prepulse results from amplified spontaneous emission. Its FWHM duration is 4 ns, while its intensity with respect to the main pulse is below 10^{-8}.

The layout of the experiment is shown in Fig. 1. A beam of linearly polarized laser radiation of p-type polarization is focused by an off-axis parabolic mirror onto the surface of a solid-state target (T) at an angle of 40° with respect to the normal to the target surface. For targets, we employed slabs from LiF and Cu 1 to 30 mm in thickness and the Al, Cu, and Ti foils 1 to 100 μm in thickness. The targets were arranged in a vacuum chamber 30 cm in diameter and 50 cm in height. The pressure of the residual gas in the chamber was not more than 10^{-3} torr. Detectors D1 based on CR-39 track detectors of size 24 to 20 mm and equipped with aluminum filters of different thickness, from 11 to 100 μm, which make it possible to cut off the energy interval 0.8–3.5 MeV for protons, were used to detect protons and to measure their energy spectrum. The detectors D1 were arranged upstream and downstream of the target at a distance of 20 mm from it along the normal.

The secondary activated targets D2, which were manufactured from LiF, Cu, and Ti and which are characterized by different threshold energies for (p, n) reactions (from 1.88 MeV for ^{7}Li to 5 MeV for ^{48}Ti), were also used to detect protons and to determine their number and maximum energy. The secondary activated targets D2 were slabs 30×30 mm^2 in cross-sectional dimensions and 1 to 6 mm in thickness and were installed at the same positions as the track detectors D1. Thus, either the track detectors D1 or the secondary activated targets D2 were used in our experiment.

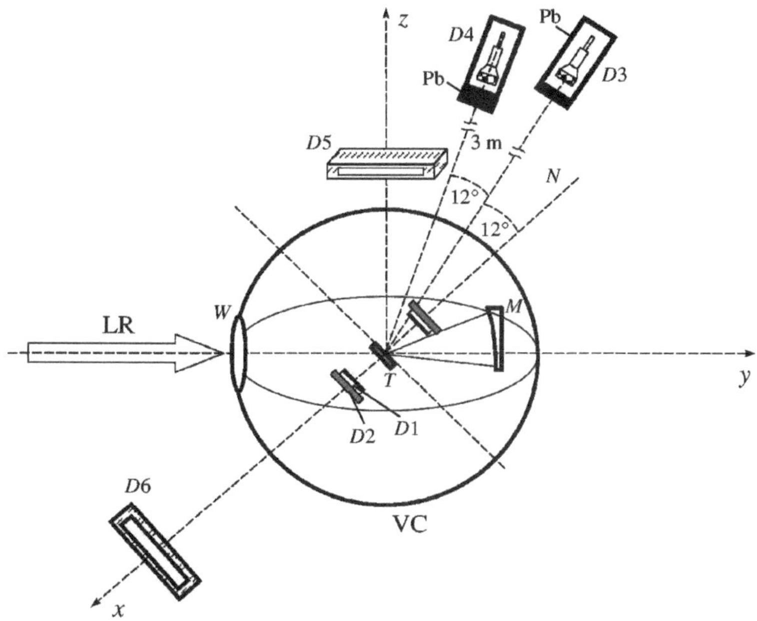

Fig. 3. Layout of the experiment: (T) target, (VC) vacuum chamber, (W) vacuum-chamber window; (M) off-axis parabolic mirror, (LR) laser radiation, (*N*) normal to the target, (D1) CR-39 track detectors equipped with aluminum filters, (D2) secondary activated targets from LiF, Cu, and Ti, (D3, D4) scintillation detectors for gamma radiation, and (D5, D6) neutron detectors on the basis of helium counters. Detectors D1–D4 and D6 lie in the xy plane.

Two scintillation detectors D3 and D4 positioned at distances of 4.3 and 3.0 m from the target, respectively, were used to record hard x-ray radiation. Lead filters 8 cm thick for D3 and 13.5 cm thick for D4 were installed in front of the detectors. The detectors D3 and D4 are scintillation detectors on the basis of plastic scintillators ∅5×10 cm in dimension. The detectors D3 and D4 were used to record hard x-ray photons of energy 0.5 to 10 MeV.

The detectors D5 and D6, which are based on helium counters, were used to determine the yield of neutrons generated in (p, n) reactions. The detector D5 was arranged along the tangent to the target surface at a distance of 25 cm, while the D6 detector was positioned behind the target at a distance of 60 cm. The detectors D5 and D6 consisted of the following units: a block of neutron counters on the basis of three CNM-18 helium counters, a voltage transducer, a signal-selection device, and a power amplifier. The side surfaces of the detectors D5 and D6 were surrounded by polyethylene 2 cm thick.

The layout of the experimental facility used to study various mechanisms of fast-proton production is displayed in Fig. 4. As targets, we employed metallic foils from titanium 30

μm thick (see Figs. 4*a* and 4*b*) and a LiF plate 6 mm thick (see Fig. 4*c*). As the secondary, activated, target, we took a LiF plate 6 mm thick. In the case presented in Fig. 4*c*, the primary target serves as the activated target as well.

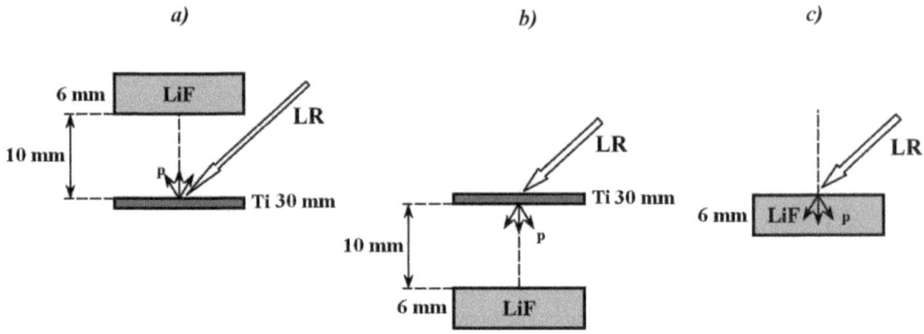

Fig. 4. Layout of experiments aimed at studying various mechanisms of the acceleration of fast protons: (*a*) acceleration of protons from the front surface of the target toward the laser pulse, (*b*) acceleration of protons from the rear surface of the target in the outward direction, and (*c*) acceleration of protons from the front surface of the target in the inward direction.

Taking into account the solid angle covered by the detector D5 and its efficiency, we found that the number of neutrons generated on average in the LiF secondary, activated, target over 4π sr per laser pulse is 50 in the first case (see Fig. 4*a*), about 2×10^3 in the second case (see Fig. 4*b*), and about 2×10^2 in the third case (see Fig. 4*c*).

The number of fast protons can be estimated by the formula $N_p \cong Y_n/(n\sigma l_i)$, where Y_n is the yield of neutrons from the reaction ^{7}Li(p, n)^{7}Be, n is the concentration of ^{7}Li atoms in the target, $\sigma \approx 60$ mb (Youssef et al., 2006) is the cross section for the reaction ^{7}Li(p, n)^{7}Be at proton energies around 1.9MeV, and l_i is the proton braking length in the target. At distances longer than l_i, protons have have energies below the threshold energy of 1.88 MeV, so that the reaction ^{7}Li(p, n)^{7}Be, which leads to neutron production, cannot proceed. Under the conditions of our experiments, $l_i \approx 10$ μm. Taking into account the yield of neutrons for the three cases considered here, we ultimately find that the number of accelerated protons from the front surface toward the laser pulse that have an energy in excess of 1.88 MeV is 10^7; the number of protons accelerated from the rear surface of the target in the outward direction is 4×10^8, while the number of protons accelerated from the front surface in the inward direction is 4×10^7. Thus, the results of our experiments have revealed that the proton-acceleration process occurs most efficiently in the case of proton acceleration from the rear surface of the target in the outward direction.

This conclusion is also confirmed by the results obtained by measuring the spectra of fast protons for various mechanisms of their acceleration.

Figures 5 shows the measured spectra of protons for various proton-acceleration mechanisms. These spectra were obtained both by using track detectors CR-39 equipped with aluminum filters of various thickness and by using the activation procedure. From these spectra, it follows that the energy distribution of fast protons corresponds to the Boltzmann distribution at a temperature of 180 keV for protons accelerated from the front surface of the target toward the laser pulse, a temperature of 500 keV for protons accelerated from the rare surface in the outward direction, and a temperature of 250 keV for protons accelerated from the front surface of the target in the inward direction.

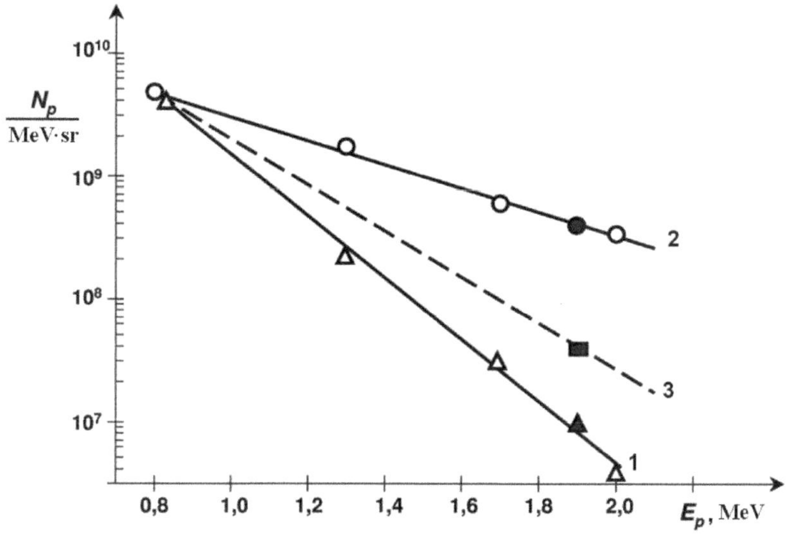

Fig. 5. Spectra of protons for various mechanisms of acceleration: (1) acceleration of protons from the front surface of the target toward a laser ray (Ti, 30 μm; $T \approx 180$ keV); (2) acceleration of protons from the rear surface of the target in the outward direction (Ti, 30 μm; $T \approx 500$ keV), and (3) acceleration of protons from the front surface of the target in the inward direction (LiF, 6 μm; $T \approx 250$ keV). The open circles and triangles represent data from track detectors, while the closed circles, triangles, and boxes stand for data obtained on the basis of the activation procedure [$p(^7Li, {}^7Be)n$, $E_{thr} = 1.88$ MeV].

Figure 6 shows the results of our experiments aimed at determining the maximum energy of protons for aluminum targets of various thickness—from 2.5 to 100 μm. These results were obtained both by using the CR-39 track detectors equipped with aluminum filters of various thickness and by using the activation procedure. From Fig. 6, one can see that there is an optimum aluminum-target thickness of 10 μm, at which protons of maximum energy 5 MeV are produced.

We will now compare the experimentally measured maximum energies of protons accelerated from the front surface of the target in the inward direction ($E \approx 2$ MeV) and protons accelerated from the rear surface of the target in the outward direction ($E \approx 5$ MeV) with the estimates of the maximum energy of protons subjected to the effect of the ponderomotive force (Pukhov, 2001),

$$E_{iMAX} \approx 2\sqrt{2} \cdot aZmc^2, \tag{22}$$

and in the case of plasma expansion into a vacuum (Zepf et al., 2003; Cowan et al., 2004; Robson et al., 2007),

$$E_{iMAX} \approx 2ZT_e \ln^2\left(\omega_{pi}\tau\right). \tag{23}$$

At $I = 2\times10^{18}$ W/cm^2, expression (22) yields the energy value of 1.73 MeV, while expression (23) leads to 5.1 MeV. These energy values agree reasonably well with experimental data.
We will now discuss the results obtained experimentally for the target-thickness dependence of the maximum energy of protons accelerated from the rare surface of the target in the outward direction. From our analysis, it follows that the reduction of the maximum proton energy at the foil thickness smaller than 10 μm is due to the effect of the nanosecond prepulse because of the pulse of enhanced spontaneous emission. The nanosecond prepulse generates shock waves in the foil, which deform the rear surface of the target, and this leads to a increase in the size of the plasma inhomogeneity at the rear surface of the target and to a decrease in the energy of protons produced at the rare surface of the target. The decrease in the maximum proton energy for target thicknesses in excess of 10 μm is due to the decrease in the energy of electrons as they pass through the target and to the increase of their angular spread. This in turn leads to a less efficient acceleration of protons from the rear surface of the target.

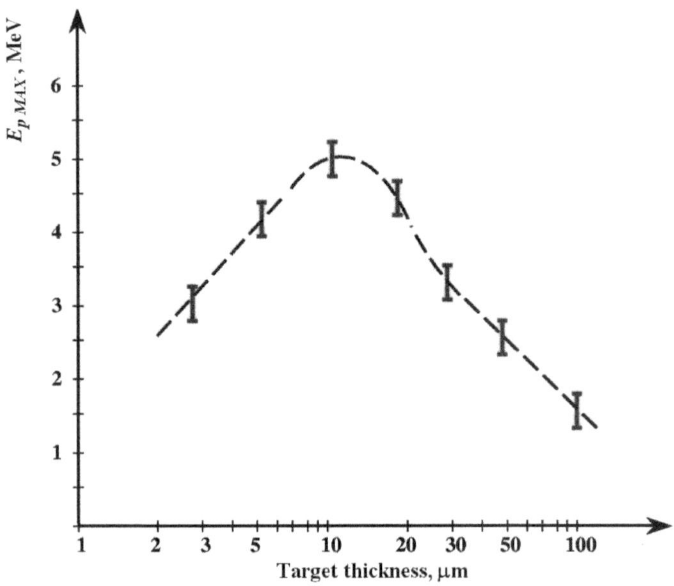

Fig. 6. Maximum proton energy E_{pMAX} as a function of the aluminum-foil thickness.

In our paper (Belyaev et al., 2005), experimental data are presented on the generation of fast ions in a laser picosecond plasma at a laser radiation intensity of 2×10^{18} W/cm^2. The results were obtained from Doppler spectra of hydrogen-like fluoride ions. An important peculiarity of the energy distribution of fast fluoride ions is the slow fall in ion energy to 1.4 MeV. In Fig. 7, the energy distribution of fast fluoride ions is plotted based on the results of measurements of L_{ya} line profile for F IX ion. The solid curves are calculated by the formula

$$\frac{dN}{dE} \sim \exp\left[-\frac{M(v-v_0)^2}{2T_{fast}}\right] \tag{24}$$

where v is the ion velocity in the observation direction, $Mv^2{}_0 /2$ = 25 keV, and the temperature of the fast fluoride ions is T_{fast} = 350 keV.
In addition, using the red shift of the Doppler profile for the L_{ya} line it was found that fast ions move inwards from a target surface. In (Belyaev et al., 2005), the parameters of the fluoride ion energy distribution were also estimated theoretically.

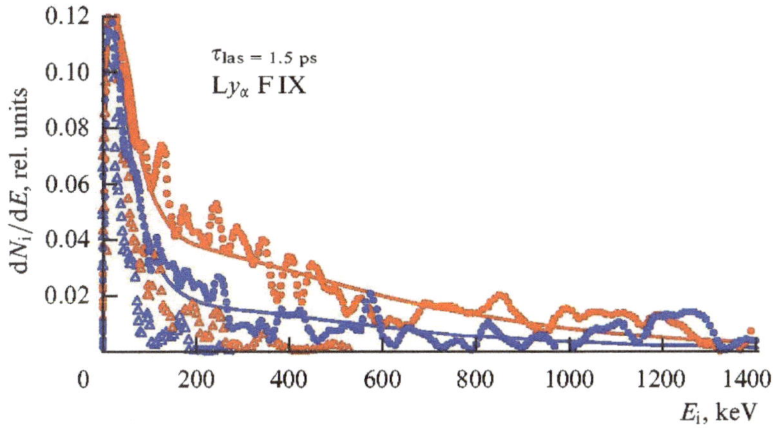

Fig. 7. Energy distribution of fast fluoride ions derived from the profile measurements of Lya line emitted by F IX ion. Top curve corresponds to ions moving towards the target, while bottom curve refers to ions moving outward from the target.

In our paper (Belyaev et al., 2009) the results of experiments devoted to studying the excitation of the promising nuclear fusion reactions $^6Li(d, \alpha)^4He$, $^3He(d, p)^4He$, $^{11}B(p, 3\alpha)$, and $^7Li(p, \alpha)^4He$, along with the standard reaction $D(d, n)^3He$, in picosecond laser plasmas are presented. For the first time, it was shown that these reactions may proceed at a moderate laser-radiation intensity of 2×10^{18} W/cm^2, the respective yield being 2×10^3 to 10^5 per laser pulse. A brief survey of the main processes responsible for the generation of fast electrons and fast ions (protons) at the front surface of the target and for the excitation of nuclear fusion reactions is given. The calculated and experimental results on the yield from nuclear fusion reactions in picosecond laser plasmas are compared. The possibilities for optimizing the yield from the promising fusion reactions excited in femto- and picosecond laser plasmas are discussed.

4. Relativistic magneto-active laser plasmas

Principal results of investigations of relativistic laser plasmas are presented here. We found parameters of magnetic fields generated in laser plasma - the amplitude of the magnetic field, its lifetime, and the increment, the spatial structure. Mechanisms of acceleration of charged particles have been investigated which are related with considered magnetic fields.
Main peculiarities that determine properties of relativistic laser plasmas are:

1. Electrons interacting with a field of electromagnetic wave can be considered as free particles.
2. Free electrons in relativistic laser plasmas interact only with an electromagnetic wave.

3. The conservation laws and motion integrals are valid also in the range of relativistic laser intensities.

Equations describing quasi-stationary magnetic fields which are generated in laser plasmas can be derived from the conservation law for generalized momentum:

$$\mathbf{P} = m\mathbf{v}\gamma - \frac{e}{c}\mathbf{A} \tag{25}$$

Here $\mathbf{A}$ is the vector-potential of an electromagnetic wave. The relativistic equation of motion is of the form

$$\frac{d}{dt}m\mathbf{v}\gamma = e\mathbf{E} + \frac{e}{c}[\mathbf{V}\times\mathbf{B}] \tag{26}$$

Deleting the intermediate derivations, we present final equations for vortex electron structures producing magnetic field in laser plasmas:

$$\boldsymbol{\omega} = rot\mathbf{v} \tag{27}$$

$$div\mathbf{V} = 0 \tag{28}$$

$$\frac{d\boldsymbol{\omega}}{dt} + rot[\boldsymbol{\omega}\times\mathbf{V}] = 0 \tag{29}$$

Here $\boldsymbol{\omega} = \dfrac{e\mathbf{B}}{mc\gamma}$ is a cyclotron frequency for electron rotation in the magnetic field $\mathbf{B}$, and $\gamma = 1\Big/\sqrt{1 - v^2/c^2}$ is the relativistic factor.

These equations mean conservation laws for vortex electron structure: Eq. (27) is the conservation law for a generalized momentum (25); Eq. (28) is the conservation law for a number of particles, and Eq. (29) is the conservation law for a magnetic flow, or for an angular momentum.

It should be noted that these equations allow undamped solutions. In general case solution of these equations taking into account losses is a difficult mathematical problem knowing as a problem of magnetic field generation. In particular, explanation of Earth magnetism is a part of this problem.

Equations (27) - (29) coincide with equations for real potential vortexes in mechanics of continuum matter which correspond to three Helmholtz theorems (Sedov, 1983).

The potential vortex presents good description of the observed vortex. Uniform rotation is unfit for description of the observed vortex. The velocity inside the observed vortex is high and outside of it is small, while the inverse statement is valid for the case of the uniform rotation. Coincidence of equations for a magnetic field in laser plasmas and for a potential vortex results in identity of their spatial structures (see Fig. 8).

An electron vortex producing a quasi-stationary magnetic field and their analogous classical potential vortex can exist only in motion. In general case the transformation of rotational energy into a translational motion is a relativistic effect. This fact follows from requirement of relativistic invariance for motion of charged particles; it takes place also at small nonrelativistic velocities. The expression for the motion integral follows from the equation (26), taking into account also the Maxwell equations for an electron in the field of an electromagnetic wave which propagates along the direction **n**:

$$\frac{1-\mathbf{n}\cdot\mathbf{v}/c}{\sqrt{1-V^2/c^2}} = Const \tag{30}$$

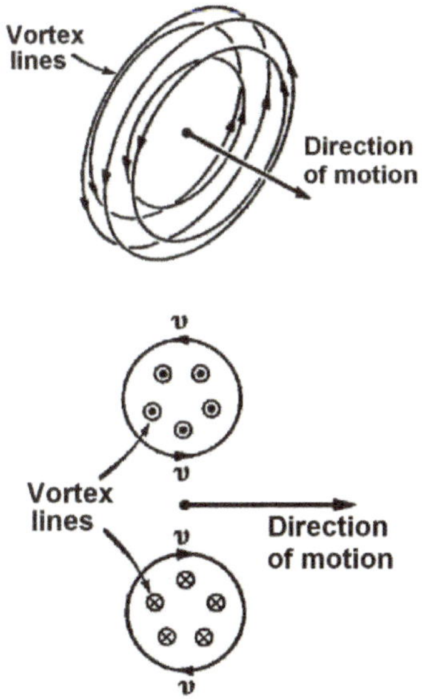

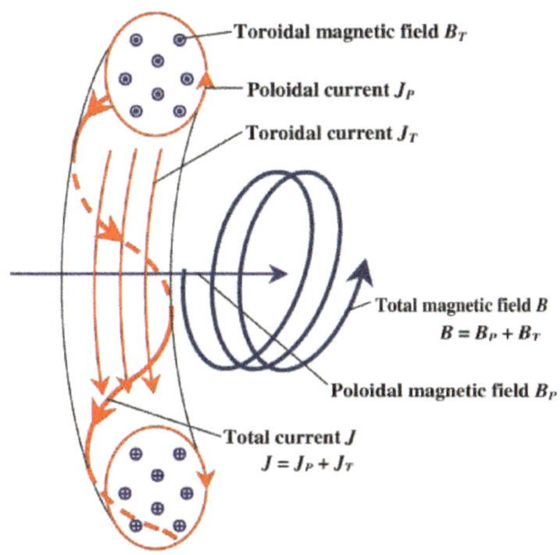

Fig. 8a. Vortex lines of moving potential vortex and its cross section

Fig. 8b. Structure of magnetic field produced in laser plasmas

This expression is useful at the consideration of dynamics of relativistic particles in a field of an electromagnetic wave. For example, if a charged particle (for example, an electron) rotates with the velocity V in a circularly polarized field of an electromagnetic wave, then this particle acquires obligatory some velocity along the direction n of the wave propagation. When $\mathbf{V}/c = 0$, the expression (30) is equal to unity. This value does not change also for other velocities. Hence, one obtains the next expression for the particle velocity along the direction of propagation of electromagnetic wave:

$$\frac{V}{c} = \frac{\gamma - 1}{\gamma} = \frac{\sqrt{1+a^2}-1}{\sqrt{1+a^2}} \tag{31}$$

Here the quantity a is determined by the electromagnetic wave intensity J: $a = 0{,}85\sqrt{\frac{J}{10^{18}}}$, J [W/cm^2].

Positively charged atomic ions prevent from motion of the considered electron vortex in a target because of the forces of the Coulomb attraction.

The requirement of quasi-neutrality results in motion of positively charged atomic ions. Omitting details of derivations and taking into account the Vlasov equations for a quasi-neutral two-component plasma and conservation law of the generalized momentum both for ions and for electrons, we present the final result:
Electrons and ions in relativistic laser plasmas form the one vortex structure - a potential vortex. This structure moves together with produced electromagnetic fields having the velocity of an electric drift (at **E** < **B**):

$$\mathbf{v} = c\frac{[\mathbf{EB}]}{\mathbf{B}^2} \tag{32}$$

Let us remark one peculiarity. The ion velocity and the ion free path are small in the process of ion motion. Ions are decelerated in a target; then new ions take their place, and finally the whole vortex structure occurs on the rear side of the target. If l_i is the depth for ion deceleration, the last ions propagate together with electrons producing quasi-neutral potential plasma vortex.
The drift motion does not produce the electric current and charge separation, since particles with positive and negative charge drift in the same direction with the same velocity. Thus, drift produces motion of neutral plasma.
Plasma magnetization results in small divergence of these flows. It is explained by a stability of vortex quasi-neutral structures as quasi-particles
Some publications report about experimental confirmation of generation of magnetized toroidal plasma structures. Ring-shaped proton flows with small divergence were observed (Nakamura & Mima, 2008; Clark et al., 2000). The magnetic field of about 100 MG has been measured by direct spectral method on large distance (several hundreds of microns) from the target surface (Belyaev et al., 2004).
Our experiments at the peak laser intensity of 2×10^{18} W/cm^2 allows us to observe on the rear side of thin (30 μm) titanium target clear ring-shaped structures by the proton detector CR-39 placed on a distance of 20 mm. Photo of ring-shaped proton structure is presented in Fig. 9a, and proton distributions with the energy of 2.5 MeV are presented in Fig. 9b. The divergence of the proton beam is $\varphi_{1/2} \approx 14°$. Protons with the energy higher than 2.5 MeV present narrow collimated beam with the divergence angle of $\varphi_{1/2} = 3°$. Inside this narrow collimated beam with the divergence angle $\varphi_{1/2} = 3°$ we observed well collimated proton beams with the divergence angle of $\varphi_{1/2} = 0.1°\div0.3°$.
Note, that drift velocity can increase significantly under condition of development of pinch-effect up to relativistic values. Respectively, not only electron velocity, but also the velocity of heavy positively charged atomic ions can increase up to relativistic values (Belyaev, Faenov et al., 2006).
Deleting the intermediate derivations, we present expressions for lifetime considered magnetic field:

$$T = 2\frac{\varepsilon}{\Delta\varepsilon}\Delta t \tag{33}$$

where ε - laser pulse energy, $\Delta\varepsilon$ - losses of an energy for electron vortex structure, $\Delta t = \lambda^2/D$, D – coefficient of Bohm's diffusion. This lifetime does not depend on duration of

laser action and can exceed it on one-two order. For this reason the superstrong magnetic fields generated in laser plasma, term quasistationary.
Increment of the considered magnetic field is equal to the ionization rate ω_i, which is larger than the plasma frequency.

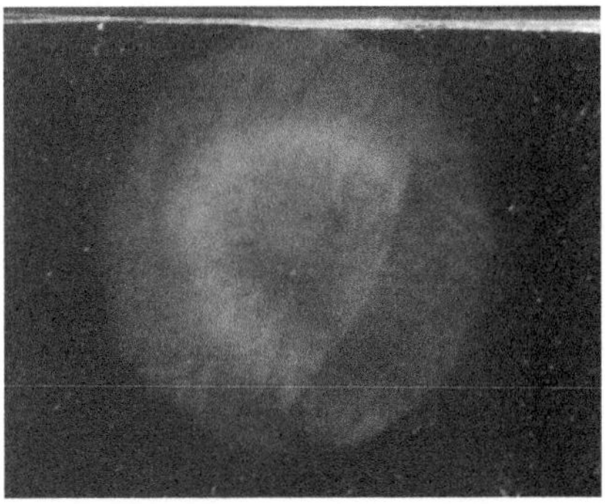

Fig. 9a. The photo of the track detector CR-39 covered by 11 mm Al filter. Detector CR-39 shows the tracks of protons with energies $E_p > 0.8$ MeV; $\varphi_{1/2}$ » 14° (cone half angle)

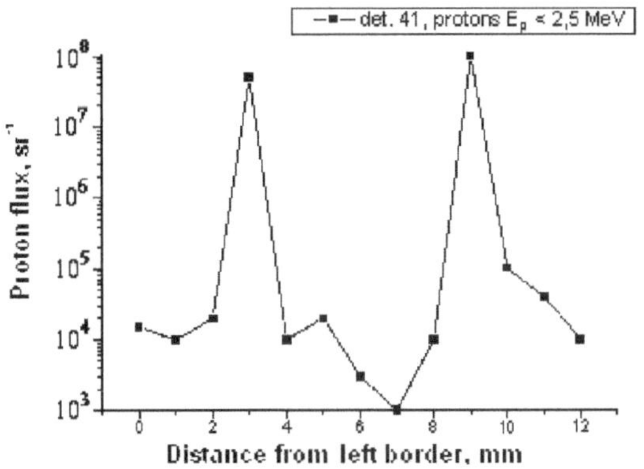

Fig. 9b. The proton distribution inside the spot for detector with 11 μm Al ($E_p > 0.8$ MeV). Target Cu 25 μm. Protons with energy $E < 2{,}5$ MeV

Mechanisms of generating magnetic fields are a subject of numerous investigations performed in recent years (Tatarakis et al., 2002; Beg et al., 2004; Borghesi et al., 1998). Various mechanisms for generating a magnetic field in the interaction of intensive laser radiation with solid targets are described in a number of theoretical works (Stamper, 1991; Wilks et al., 1992; Bell et al., 1993; Buchenkov et al., 1993; Sudan, 1993; Haines, 1997; Mason & Tabak, 1998; Krainov, 2003). In particular, they predicted the origin of magnetic fields with induction of up to 1GG in the dense plasma produced during the interaction process. These fields are localized near the critical surface, where the laser energy is mainly absorbed. The arising magnetic fields noticeably affect the dynamics of laser plasma. The principal mechanisms of generating quasistatic magnetic fields were considered: (1) different directions of the temperature and plasma density gradients; (2) the flux of fast electrons accelerated by ponderomotive forces in the longitudinal and transversal directions with respect to the direction of laser pulse propagation, and (3) the collisionless Weibel instability (Weibel, 1959).
Measurements of superstrong quasistatic magnetic fields in laser plasma and their theoretical interpretation have been discussed in more detail in our article (Belyaev et al., 2008).
Measurements of magnetic fields in plasma by various independent methods are very important for both proving the existence of such fields and determining their spatial structure (topology). For this purpose, we measured the profiles of X-ray spectral lines of hydrogen-like fluoride ions in laser plasma with a radiation intensity of 10^{17} W/cm^2 and

pulse duration of 1 ps (Belyaev et al., 2004). The structure observed is characterized by distinct dips and peaks on the spectral line profiles (see Fig. 10). These features can be explained by invoking a conception of the strong turbulent noise that develops in the superstrong magnetic field generated in laser plasma.

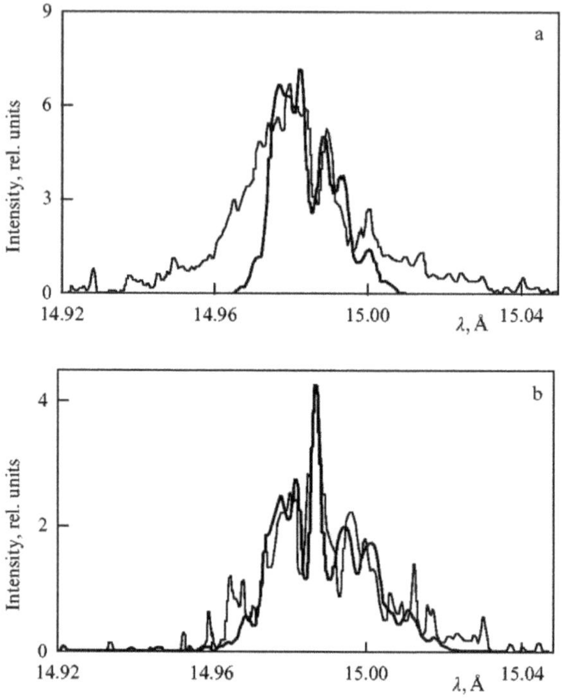

Fig. 10. Comparison of experimental (thin curves) and theoretical (thick curves) profiles for the $L_{y\alpha}$ line of F IX ion: (a) the experiment was performed at I_{las} = 2×10^{17} W/cm^2, and calculation was made at T_i = 100 eV, n_e = 10^{20} cm^{-3}, ω = 7×10^{14} s^{-1}, E_0 = 4×10^8 V/cm; (b) the experimental I_{las} = 3×10^{17} W/cm^2, and the calculation was done at T_i = 100 eV, n_e = 2×10^{20} cm^{-3}, ω = 10^{15} s^{-1}, and E_0 = 6×10^8 V/cm.

5. Conclusion

The above-described mechanisms for accelerating electrons and ions to a greater or lesser degree comply with up-to-date concepts on the generation of fast particles in laser plasma. According to these concepts, the energy of an initial laser pulse is converted to the energy of electron motion. The mechanisms for such energy conversion are mainly related to (1) a ponderomotive potential; (2) a phase interruption of electron oscillations in the laser wave due to various mechanisms, among which the main one is electron ejection beyond the sharp boundary of a target (vacuum heating), and (3) various resonance mechanisms where the electron motion is at resonance with plasma waves (wake-field resonance absorption or acceleration) or the cyclotron or betatron oscillation of an electron in the channel produced by laser radiation in the presence of a magnetic field. Ion acceleration in this case is a secondary effect mainly caused by the electric fields of the spatial charge produced when fast accelerated electrons are separated from ions. The detailed distribution of such fields substantially depends on the target thickness, which makes a difference in the ion acceleration at the front and rear target surfaces.

As a whole, particle acceleration is characterized by the multifactor character of the parameters involved. Such parameters are the intensity, frequency, and duration of the laser pulse; the contrast, which determines pre-plasma parameters; the thickness and structure of the target; the presence of magnetic fields, and some other factors. By combining these parameters, one can reach the optimal (in certain limits) conditions of particle acceleration. There is a wide range of various applications of such laser-driven accelerators, starting from fundamental investigations concerning nuclear processes for isotope production, to the initiation of thermonuclear reactions using laser setups that are quite small in size compared to standard accelerators, and ending by particular applications such as sources of proton radiation for medical purposes.

Nevertheless, there are a sufficiently large number of problems to be solved related to particle acceleration. These are, for example, ion beam focusing and annular structures arising in the beam. In electron acceleration, the problem of forming a monoenergetic beam of fast electrons with a maximum energy has not yet been solved.

As far as the generation of super-strong magnetic fields is concerned, the main problems are determination of their lifetime and topology. Experimental results definitely indicate that the lifetimes of magnetic fields are considerably longer (by orders of magnitude) than the laser pulse duration.

From our point of view, this is direct evidence that long-living magnetic configurations exist in laser plasma. This is also confirmed by investigations into the dynamics of pinch structures in irradiating wire targets by laser pulses. The topology and dynamics of such structures are, as was noted above, in surprisingly good agreement with those obtained under the pulse action of mega-ampere currents.

It is clear that the presence of high-intensity fast particles and magnetic fields in plasma, in addition to the specific features of particle acceleration mentioned above, should result in numerous instabilities arising in plasma. This is directly illustrated by the results mentioned above on measuring the profiles of spectral lines for multiply charged ions. Profile irregularity is indicative of the existence of intense electrostatic oscillations possessing definite frequencies and intensities. Thus, in view of all the specific features mentioned above, one can conclude that in the case of ultra-short laser pulses we are dealing with magneto-active turbulent plasma, numerous properties of which are not clear presently. Nevertheless, it is possible to choose sufficiently optimal conditions for generating high-energy charged particles in such plasmas.

6. Acknowledgment

This research has been supported by the Russian Foundation for Basic Research, projects NN 09-02-00041, 10-02-01095, 10-08-00752, 11-02-12026.

7. References

Andreev, A.V.; Gordienko, V.M. & Savel'ev, A.B. (2001). Nuclear processes in a high-temperature plasma produced by an ultrashort laser pulse. *Quantum Electronics*, Vol.31, No.11, (November 2001), pp. 941-956, ISSN 0018-9197

Amiranoff, F.; Baton, S.; Bernard, D.; Cros, B.; Descamps, D.; Dorchies, D.; Jacquet, F.; Malka, V.; Marques, J.R.; Matthieussent, G.; Mine, P.; Modena, A.; Mora, P.; Morillo, J. & Najmudin, Z. (1998). Observation of Laser Wakefield Acceleration of Electrons. *Physical Review Letters*, Vol.81, No.5, (August 1998), pp.995-998, ISSN 0031-3727

Beg, F.N.; Clark, E.L.; Wei, M.S.; Dangor, A.E.; Evans, R.G.; Gopal, A.; Lancaster, K.L.; Ledingham, K.W.D.; McKenna, P.; Norreys, P.A.; Tatarakis, M.; Zepf, M. & Krushelnick, K. (2004). High-Intensity-Laser-Driven *Z* Pinches. *Physical Review Letters*, Vol.92, No.9, (March 2004), pp. 095001, ISSN 0031-9007

Bell, A.R.; Beg, F.N.; Chang, Z.; Dangor, A.E.; Danson, C.N.; Edwards, C.B.; Fews, A.P.; Hutchinson, M.H.R.; Luan, S.; Lee, P.; Norreys, P.A.; Smith, R.A.; Taday, P.F. & Zhou, F. (1993). Observation of plasma confinement in picosecond laser-plasma interactions. *Physical Review E*, Vol.48, No.3, (September 1993), pp.2087-2093, ISSN 1539-3755

Belyaev, V.S. & Mikhailov, V.N. (2001). Quantum Plasmas: Criteria, Anomalous Diffusion, and Bleaching in a Magnetic Field. *Laser Physics*, Vol.11, No.8, (December 2000), pp.957-973, ISSN 1054-660X

Belyaev, V. S.; Kostenko, O.F. & Lisitsa, V.S. (2003). Cyclotron mechanism of electron acceleration in subpicosecond laser plasma. *Letters to Jounal of Experimental and Theoretical Physics*, Vol.77, No.12, (June 2003), pp.653-656, ISSN 0021-3640

Belyaev, V.S. (2004). Mechanism of high-energy electron production in a laser plasma. *Quantum Electronics*, Vol.34, No.1, (January 2004), pp.41-46, ISSN 0018-9197

Belyaev, V.S.; Vinogradov, V.I.; Kurilov, A.S.; Matafonov, A.P.; Lisitsa, V.S.; Gavrilenko, V.P.; Faenov, A.Ya.; Pikuz, T.A.; Skobelev, I.Yu.; Magunov, A.I. & Pikuz, S.A. Jr. (2004). Plasma Satellites of X-ray Lines of Ions in a Picosecond Laser Plasma. *Journal of Experimental and Theoretical Physics*, Vol.99, No.4, (April 2004), pp. 708-719, ISSN 1090-6509

Belyaev, V.S.; Vinogradov, V.I.; Matafonov, A.P.; Krainov, V.P.; Lisitsa, V.S.; Faenov, A.Ya.; Pikuz, T.A.; Skobelev, I.Yu.; Magunov, A.I.; Pikuz, S.A. Jr.; Andrianov, V.P.; Ignat'ev, G.N.; Kozhunov, Yu.I.; Kozlov, O.B. & Chekmarev, A.M. (2005). Effective Temperature and the Directional Motion of Fast Ions in a Picosecond Laser Plasma. *Letters to Jounal of Experimental and Theoretical Physics*, Vol.81, No.12, (May 2005), pp.616-620, ISSN 0021-3640

Belyaev, V.S.; Faenov, A.Ya.; Magunov, A.I.; Matafonov, A.P.; Pikuz, T.A.; Pikuz S.A.Jr., Skobelev, I.Yu. & Vinogradov, V.I. (2006). Directed inward beams of ions in laser produced plasma. *Journal De Physique IV France*, Vol.133, No., (June 2006), pp. 525-527, ISBN 2-86883-925-8

Belyaev, V.S.; Vinogradov, V.I.; Matafonov, A.P.; Krainov, V.P.; Lisitsa, V.S.; Andrianov, V.P. & Ignatyev, G.N. (2006). Effect of Prepulses with Various Durations on the Neutron Yield in Laser Picosecond Plasma. *Laser Physics*, Vol.16, No.12, (December 2006), pp.1647-1657, ISSN 1054-660X

Belyaev, V.S.; Krainov, V.P.; Lisitsa, V.S. & Matafonov, A.P. (2008). Generation of fast charged particles and superstrong magnetic fields in the interaction of ultrashort high-intensity laser pulses with solid targets. *Physics – Uspekhi*, Vol.51, No.8, (August 2008), pp. 793-814, ISSN 1063-7869

Belyaev, V.S.; Vinogradov, V.I.; Matafonov, A.P.; Rybakov, S.M.; Krainov, V.P.; Lisitsa, V.S.; Andrianov, V.P.; Ignatiev, G.N.; Bushuev, V.S.; Gromov, A.I.; Rusetsky, A.S. &

Dravin, V.A. (2009). Excitation of Promising Nuclear Fusion Reactions in Picosecond Laser Plasmas. *Physics of Atomic Nuclei*, Vol.72, No.7, (September 2009) pp.1077–1098, ISSN 1063-7788

Borghesi, M.; MacKinnon, A.J.; Bell, A.R.; Gaillard, R. & Willi, O. (1998). Megagauss Magnetic Field Generation and Plasma Jet Formation on Solid Targets Irradiated by an Ultraintense Picosecond Laser Pulse. *Physical Review Letters*, Vol.81, No.1, (July 1998), pp.112-115, ISSN 0031-3727

Borodin, V.G.; Gilev, O.N.; Zapysov, A.L.; Komarov, V.M.; Lykov, V.A.; Malinov, V.A.; Migel', V.M.; Nikitin, N.V.; Pokrovskii, V.G.; Pronin, V.A.; Saprykin, V.N.; Charukhchev, A.V. & and Chernov, V.N. (2000). Hard X Radiation and Fast Particles in Laser Plasma Experiments at Laser Intensities of up to 5×10^{18} W/cm^2 on the Target Surface. *Letters to Jounal of Experimental and Theoretical Physics*, Vol.71, No.6, (April 1999), pp.246-250, ISSN 0021-3640

Brunel, F. (1987). Not-so-resonant, resonant absorption. *Physical Review Letters*, Vol.59, No.1, (July 1987), pp.52-55, ISSN 0031-3727

Bychenkov V.Yu., Kasyanov, Yu.S.; Sarkisov, G.S. & Tikhonchuk, V.T. (1993). Mechanism of generation of magnetic fields in laser plasma study. *Letters to Jounal of Experimental and Theoretical Physics*, Vol.58, No., (1993), pp.184-, ISSN 0021-3640

Carrier, M.; Lefebvre, E.; Flacco, A. & Malka, V. (2009). Influence of subpicosecond laser pulse duration on proton acceleration. *Physics of Plasmas*, Vol.16, No.5, (May 2009), pp.053105, ISSN 1070-664X

Clark, E.L.; Krushelnick, K.; Zepf, M.; Beg, F.N.; Tatarakis, M.; Machacek, A.; Santala, M.I.K.; Watts, I.; Norreys, P.A. & Dangor, A.E. (2000). Energetic Heavy-Ion and Proton Generation from Ultraintense Laser-Plasma Interactions with Solids. *Physical Review Letters*, Vol.85, No.8, (August 2000), pp.1654-1657, ISSN 0031-3727

Cowan, T.E.; M.D. Perry, M.D.; Key, M.H.; Ditmire, T.R.; Hatchett, S.P.; Henry, E.A.; Moody, J.D.; Moran, M.J.; Pennington, D.M.; Phillips, T.W.; Sangster, T.C.; Sefcik, J.A.; Singh, M.S.; Snavely, R.A.; Stoyer, M.A.; Wilks, S.C.; Young, P.E.; Takahashi, Y.; Dong, B.; Fountain, W.; T. Parnell, T.; Johnson, J.; Hunt, A.W. & and Kuhl, T. (1999). High energy electrons, nuclear phenomena and heating in petawatt laser-solid experiments. *Laser and Particle Beams*, Vol.17, No.4, (October 1999), pp.773-783, ISSN 0263-0346

Cowan, T.E.; Fuchs, J.; Ruhl, H.; Kemp, A.; Audebert, P.; Roth, M.; Stephens, R.; Barton, I.; Blazevic, A.; Brambrink, E.; Cobble, J.; Fernandez, J.; Gauthier, J.-C.; Geissel, M.; Hegelich, M.; Kaae, J.; Karsch, S.; Le Sage, G.P.; Letzring, S.; Manclossi, M.; Meyroneinc, S.; Newkirk, A.; Pepin, H. & Renard-LeGalloudec, N. (2004). Ultralow Emittance, Multi-MeV Proton Beams from a Laser Virtual-Cathode Plasma Accelerator. *Physical Review Letters*, Vol.92, No.20, (May 2004), pp.204801, ISSN 0031-3727

Demchenko, N.N. & Rozanov, V.B. (2001). A Hydrodynamic Model of the Interaction of Picosecond Laser Pulses with Condensed Targets. *Journal of Russian Laser Research*, Vol.22, No.3, (May 2001), pp.228-242, ISSN 1071-2836

Esarey, E.; Sprangle, P.; Krall, J. & Ting, A. (1996). Overview of plasma-based accelerator concepts. *IEEE Transaction on Plasma Science*, Vol.24, No.2, (April 1996), pp.252-288, ISSN 0093-3813

Fukuda, Y.; Faenov, A.Ya.; Tampo, M.; Pikuz, T.A.; Nakamura, T.; Kando, M.; Hayashi, Y.; Yogo, A.; Sakaki, H.; Kameshima, T.; Pirozhkov, A.S.; Ogura, K.; Mori, M.; Esirkepov, T.Zh.; Koga, J.; Boldarev, A.S.; Gasilov, V.A.; Magunov, A.I.; Yamauchi, T.; Kodama, R.; Bolton, P.R.; Kato, Y.; Tajima, T.; Daido, H. & Bulanov, S.V. (2009). Energy Increase in Multi-MeV Ion Acceleration in the Interaction of a Short Pulse Laser with a Cluster-Gas Target. *Physical Review Letters*, Vol.103, No.16, (October 2009), pp.165002, ISSN 0031-3727

Haines, M.G. (1997). Saturation Mechanisms for the Generated Magnetic Field in Nonuniform Laser-Matter Irradiation. *Physical Review Letters*, Vol.78, No.2, (January 1997), pp.254-257, ISSN 0031-3727

He, F.; Yu, W.; Lu, P.; Xu, H.; Qian, L.; Shen, B.; Yuan, X.; Li, R. & Xu, Z. (2004). Ponderomotive acceleration of electrons by a tightly focused intense laser beam. *Physical Review E*, Vol.68, No.4, (October 2004), pp.046407, ISSN 1539-3755

Huang, L.G.; Lei, A.L.; Bin J.N.; Bai, Y.; Wei Yu; Yu, M.Y. & Cowan, T.E. (2010). Improving proton acceleration with circularly polarized intense laser pulse by radial confinement with heavy ions. *Physics of Plasmas*, Vol.17, No.1, (January 2010), pp.013106, ISSN 1070-664X

Gonoskov, A.A.; Korzhimanov, A.V.; Eremin, V.I.; Kim, A.V. & Sergeev, A.M. (2009). Multicascade Proton Acceleration by a Superintense Laser Pulse in the Regime of Relativistically Induced Slab Transparency. *Physical Review Letters*, Vol.102, No.18, (May 2009), pp.184801, ISSN 0031-3727

Gus'kov, S.Yu.; Demchenko, N.N.; Makarov, K.N.; Nishchuk, S.G.; Pikuz, T.A.; Satov, Yu.A.; Skobelev, I.Yu.; Smakovski, Yu.B.; Faenov, A.Ya.; Khomenko, S.V. & Sharkov, B.Yu. (2001). Influence of Angles of Incidence of Laser Radiation on the Generation of Fast Ions. *Letters to Jounal of Experimental and Theoretical Physics*, Vol.73, No.12, (May 2001), pp.665-661, ISSN 0021-3640

Krainov, V. P. (2003). Generation of high magnetic fields in an atomic plasma by irradiation with a super-intense femtosecond laser pulse. *Journal of Physics B: Atomic, Molecular and Optical Physics*, Vol.36, No.15, (August 2003), pp.3187-3202, ISSN 0953-4075

Ledingham, K.W.D. & Norreys, P.A. (1999). Nuclear physics merely using a light source. *Contemporary Physics*, Vol.40, No.6, (December 1999), pp.367-383, ISSN 0010-7514

Mackinnon, A.J.; Borghesi, M.; Hatchett, S.; Key, M.H.; Patel, P.K.; Campbell, H.; Schiavi, A.; Snavely, R.; Wilks, S.C. & Willi, O. (2001). Effect of Plasma Scale Length on Multi-MeV Proton Production by Intense Laser Pulses. *Physical Review Letters*, Vol.86, No.9, (February 2001), pp.1769-1772, ISSN 0031-3727

Maksimchuk, A.; Gu, S.; Flippo, K.; Umstadter, D. & Bychenkov, V.Yu. (2000). Forward Ion Acceleration in Thin Films Driven by a High-Intensity Laser. *Physical Review Letters*, Vol.84, No.18, (May 2000), pp.4108-4111, ISSN 0031-3727

Malka, G. & Miquel, J.L. (1996). Experimental Confirmation of Ponderomotive-Force Electrons Produced by an Ultrarelativistic Laser Pulse on a Solid Target. *Physical Review Letters*, Vol.77, No.1, (July 1996), pp.75-78, ISSN 0031-3727

Mangles, S.P.D.; Walton, B.R.; Tzoufras, M.; Najmudin, Z.; Clarke, R.J.; Dangor, A.E.; Evans, R.G.; Fritzler, S.; Gopal, A.; Hernandez-Gomez, C.; Mori, W.B.; Rozmus, W.; Tatarakis, M.; Thomas, A.G.R.; Tsung, F.S.; Wei, M.S. & Krushelnick, K. (2005). Electron Acceleration in Cavitated Channels Formed by a Petawatt Laser in Low-

Density Plasma. *Physical Review Letters*, Vol.94, No.24, (June 2005), pp.245001, ISSN 0031-3727

Mason, R. J. & Tabak, M. (1998). Magnetic Field Generation in High-Intensity-Laser–Matter Interactions. *Physical Review Letters*, Vol.80, No.3, (January 1998), pp.524-527, ISSN 0031-3727

Matafonov, A. & Belyaev, V. (2001). Experimental Investigation of Gamma and Neutron Radiation from Laser Produced Plasma, *Proceedings of 28th EPS Conference on Controlled Fusion and Plasma Physics-2001*, Funchal, Madeira, Portugal, 18 - 22 June, 2001, http://www.cfn.ist.utl.pt/EPS2001/

Mourow, G.A.; Tajima, T. & Bulanov, S.V. (2006). Optics in the relativistic regime. *Review of Modern Physics*, Vol.78, No.2, (April 2006), pp. 309-371, ISSN 0034-6861

Nakamura, T. & Mima, K. (2008) . Magnetic-Dipole Vortex Generation by Propagation of Ultraintense and Ultrashort Laser Pulses in Moderate-Density Plasmas. *Physical Review Letters*, Vol.100, No.20, (May 2008), pp.205006, ISSN 0031-3727

Nickles, P.V.; Kalachnikov, M.P.; Warwick, P.J.; Janulewicz, K.A.; Sandner, W.; Jahnke, U.; Hilscher, D.; Schnurer, M.; Nolte, R. & Rousse, A. (1999). Energetic electrons and x-ray photons from multiterawatt Ti:sapphire lasers. *Quantum Electronics*, Vol.29, No.5, (May 1999), pp.444-448, ISSN 0018-9197

Pavlenko, Yu.G. (2002). *Lectures on theoretical mechanics*, FIZMATLIT, ISBN 5-9221-0241-9, Moscow (in russian)

Pukhov, A. (2001). Three-Dimensional Simulations of Ion Acceleration from a Foil Irradiated by a Short-Pulse Laser. *Physical Review Letters*, Vol.86, No.16, (April 2001), pp.3562-3565, ISSN 0031-3727

Psikal, J.; Tikhonchuk, V.T.; Limpouch, J. & Klimo, O. Lateral. (2010). Lateral hot electron transport and ion acceleration in femtosecond laser pulse interaction with thin foils. *Physics of Plasmas*, Vol.17, No.1, (January 2010), pp.013102, ISSN 1070-664X

Pukhov, A.; Sheng, Z-M. & Meyer-ter-Vehn, J. (1999). Particle acceleration in relativistic laser channels. *Physics of Plasmas*, Vol.6, No.7, (April 1999), pp.2847-2855, ISSN 1070-664X

Robson, L.; Simpson, P.T.; Clarke, R.J.; Ledingham, K.W.D.; Lindau, F.; Lundh, O.; McCanny, T.; Mora, P.; Neely, D.; Wahlstrom, C.-G.; Zepf, M. & McKenna, P. (2007). Scaling of proton acceleration driven by petawatt-laser–plasma interactions. *Nature Physics*, Vol.3, No., (December 2007), pp.58-62, ISSN 1745-2473

Salamin, Y.I.; Hu, S.X.; Hatsagortsyan, K.Z. & Keitel, C.H. (2006). Relativistic high-power laser–matter interactions. *Physical Reports*, Vol.427, No.2-3, (April 2006), pp.41-155, ISSN 0370-1573

Sedov, L.I. (1983). *Mechanics of continuum matter* (Vol. 1), ISBN 978-3-540-88467-5, Nauka, Moscow

Sentoku, T.; Cowan, T.E.; Kemp, A. & Ruhl, H. (2003). High energy proton acceleration in interaction of short laser pulse with dense plasma target. *Physics of Plasmas*, Vol.10, No.5, (December 2003), pp.2009-2016, ISSN 1070-664X

Stamper, J.A. (1991). Review on spontaneous magnetic fields in laser-produced plasmas: Phenomena and measurements. *Laser and Particle Beams*, Vol.9, No.4, (March 1991), pp.841-862, ISSN 0263-0346

Sudan, R.N. (1993). Mechanism for the generation of 10^9 G magnetic fields in the interaction of ultraintense short laser pulse with an overdense plasma target. *Physical Review Letters*, Vol.70, No.20, (May 1993), pp.3075-3078, ISSN 0031-3727

Tatarakis, M. ; Gopal, A.; Watts, I.; Beg F.N.; Dangor, A.E.; Krushelnick, K.; Wagner, U.; Norreys, P.A.; Clark, E.L.; Zepf, M. & Evans, R.G. (2002). Measurements of ultrastrong magnetic fields during relativistic laser–plasma interactions. *Physics of Plasmas*, Vol.9, No.5, (May 2002), pp. 2244-2250, ISSN 1070-664X

Umstadter, D. (2003). Relativistic laser–plasma interactions. *Journal of Physics D: Applied Physics*, Vol.36, No.8, (April 2003), pp. R151-R165, ISSN 0022-3727

Weibel, E.S. (1959). Spontaneously Growing Transverse Waves in a Plasma Due to an Anisotropic Velocity Distribution. *Physical Review Letters*, Vol.2, No.3, (February 1959), pp.83-84, ISSN 0031-3727

Wilks, S.C.; Kruer, W.L.; Tabak, M. & Langdon, A.B. (1992). Absorption of Ultra-Intense Laser Pulses. *Physical Review Letters*, Vol.69, No.9, (August 1992), pp.1383-1386, ISSN 0031-3727

Wilks, S.C.; Langdon, A.B.; Cowan, T.E.; Roth, M.; Singh, M.; Hatchett, S.; Key, M.H.; Pennington, D.; MacKinnon, A. & Snavely, R.A. (2001). Energetic proton generation in ultra-intense laser–solid interactions. *Physics of Plasmas*, Vol.8, No.2, (August 2001), pp.542-550, ISSN 1070-664X

Willingale, L.; Nagel, S.R.; Thomas, A.G.R.; Bellei, C.; Clarke, R.J.; Dangor, A.E.; Heathcote, R.; Kaluza, M.C.; Kamperidis, C.; Kneip, S.; Krushelnick, K.; Lopes, N.; Mangles, S.P.D.; Nazarov, W.; Nilson, P.M. & Najmudin, Z. (2009). Characterization of High-Intensity Laser Propagation in the Relativistic Transparent Regime through Measurements of Energetic Proton Beams. *Physical Review Letters*, Vol.102, No.12, (March 2009), pp.125002, ISSN 0031-3727

Yan, X.Q.; Wu, H.C.; Sheng, Z.M.; Chen, J.E. & Meyer-ter-Vehn, J. (2009). Self-Organizing GeV, Nanocoulomb, Collimated Proton Beam from Laser Foil Interaction at 7×10^{21} W/cm^2. *Physical Review Letters*, Vol.103, No.13, (September 2009), pp.135001 (2009), ISSN 0031-3727

Youssef, A.; R. Kodama, R. & M. Tampo, M. (2006). Study of proton acceleration at the target front surface in laser-solid interactions by neutron spectroscopy. *Physics of Plasmas*, Vol.13, No.3, (March 2006), pp.030702, ISSN 1070-664X

Zepf, M.; Clark, E.L.; Beg, F.N.; Clarke, R.J.; Dangor, A.E.; Gopal, A.; Krushelnick, K.; Norreys, P.A.; Tatarakis, M.; Wagner, U. & and Wei, M.S. (2003). Proton Acceleration from High-Intensity Laser Interactions with Thin Foil Targets. *Physical Review Letters*, Vol.90, No.6, (February 2003), pp.064801, ISSN 0031-3727

4

Physics of Quasi-Monoenergetic Laser-Plasma Acceleration of Electrons in the Blowout Regime

Serguei Y. Kalmykov[1], Bradley A. Shadwick[1],
Arnaud Beck[2] and Erik Lefebvre[2]
[1]*Department of Physics and Astronomy, University of Nebraska – Lincoln, Lincoln*
[2]*CEA, DAM, DIF, Arpajon F-91297*
[1]*USA*
[2]*France*

1. Introduction

Progress in the technology of optical pulse amplification (Herrmann et al., 2009; Ross et al., 2000; Spence et al., 1991; Strickland & Mourou, 1985) has made sub-50 fs pulse length, 0.1–10 Hz repetition rate, multi-terawatt (TW) lasers available to university-scale laboratories. These new instruments, accessible to a large community of researchers, revolutionized experiments in relativistic nonlinear optics (Mourou et al., 2006), and enabled the compact design of plasma-based particle accelerators (Esarey et al., 2009; Tajima & Dawson, 1979). Owing to continuous improvements in laser systems and gas target technology (Semushin & Malka, 2001; Spence & Hooker, 2001), stable generation of well-collimated, quasi-monoenergetic, hundred-megaelectronvolt (MeV)-scale electron beams from millimeter to centimeter-length plasmas has become experimentally routine (Brunetti et al., 2010; Faure et al., 2006; Hafz et al., 2008; Leemans et al., 2006; Maksimchuk et al., 2007; Malka et al., 2009; Mangles et al., 2007; Osterhoff et al., 2008). These beams have been used for a broad range of technical and medical physics applications – γ-ray radiography for material science (Glinec et al., 2005; Ramanathan et al., 2010), testing of radiation resistivity of electronic components used in harsh radiation environments (Hidding et al., 2011), efficient on-site production of radioisotopes (Leemans et al., 2001; Reed et al., 2007), and radiotherapy with tunable, high-energy electrons (DesRosiers et al., 2000; Glinec et al., 2006; Kainz et al., 2004). Their unique properties – femtosecond (fs)-scale duration and multi-kiloampere current (Buck et al., 2011; Lundh et al., 2011) – are clearly favorable for ultrafast science applications, such as high-energy radiation femtochemistry (Brozek-Pluska et al., 2005), spatio-temporal radiation biology and radiotherapy (Malka et al., 2010), and compact x-ray sources (Fuchs et al., 2009; Grüner et al., 2007; Hartemann et al., 2007; Kneip et al., 2010; Pukhov et al., 2010; Rousse et al., 2007; Schlenvoigt et al., 2008). The current record of accelerated electron energy is close to one gigaelectronvolt (GeV) (Clayton et al., 2010; Froula et al., 2009; Kneip et al., 2009; Leemans et al., 2006; Liu et al., 2011). Furthermore, ongoing introduction of sub-150 fs, compact, high repetition rate petawatt (PW) lasers (Aoyama et al., 2003; Gaul et al., 2010; Hein et al., 2006; Korzhimanov et al., 2011; Sung et al., 2010) opens possibilities beyond the GeV energy frontier (Gorbunov et al., 2005; Kalmykov et al., 2010a; Lu et al., 2007; Martins et al., 2010), enabling further steps towards practical designs of high-brightness x-

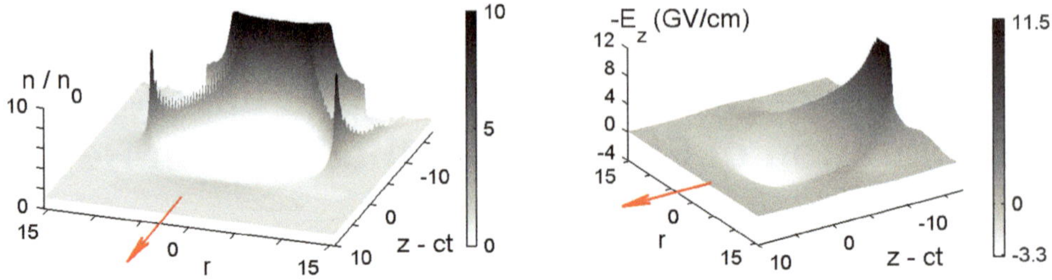

(a) Normalized electron density (b) Accelerating gradient, $-E_z$ (in GV/cm)

Fig. 1. Electron density bubble driven by the 70 TW laser pulse in a plasma of density $n_0 = 6.5 \times 10^{18}$ cm^{-3} (cf. Fig. 3(d.1)). The laser pulse (not shown) is centered at $z \approx ct$ and propagates in the direction indicated by the arrow. The wake bucket (plot (a)) is devoid of electrons; the peak of electron density at its rear is well above the cutoff value $10n_0$. The accelerating gradient (plot (b)) reaches 11.5 GV/cm at the rear of the bucket.

and γ-ray sources and compact high energy physics particle colliders (Schroeder et al., 2010). Success of these applications critically depends on high collimation and low energy spread of the multi-GeV beams. Presently, however, laser-plasma accelerators (LPAs) produce GeV-scale electrons with polychromatic energy distributions spanning from a few MeV to the maximum energy observed; sometimes with a quasi-monoenergetic feature at the high-energy end (Clayton et al., 2010; Froula et al., 2009; Kneip et al., 2009; Liu et al., 2011). In this Chapter, we explore the relationship between the electron beam quality and the nonlinear evolution of the accelerating structure – a three-dimensional (3-D) bucket of a strongly nonlinear, fully electromagnetic (EM) plasma wake – and propose dynamical scenarios to help reduce electron energy spread and suppress the poorly collimated polychromatic background.

In a modern LPA experiment, the ponderomotive force of a focused pulse produces a full cavitation of the surrounding electron fluid. All plasma electrons facing the pulse are expelled by the radiation pressure (whereas fully stripped ions remain immobile). Fields due to this charge separation attract bulk electrons to the axis, and their trajectories overshoot. The resulting closed cavity of electron density (the "bubble" (Pukhov & Meyer-ter-Vehn, 2002)) surrounded by a dense shell (sheath) of relativistic electrons encompasses the pulse and guides it over many Rayleigh lengths until depletion (Lu et al., 2007; Mora & Antonsen, 1996). Figure 1 presents one example of such fully cavitated bucket.

The bubble is a high-quality 3-D EM accelerating structure. Its longitudinally uniform but radially linear focusing gradient implies strict conservation of normalized transverse emittance. In addition, the accelerating field is radially uniform, which helps mitigate longitudinal emittance dilution (Lu et al., 2006; Rosenzweig et al., 1991). The bubble propagates with the group velocity of laser pulse in the plasma, which in a linear approximation can be expressed as $v_g = c(1 - \gamma_g^{-2})^{1/2}$, where c is a speed of light in vacuum, and $\gamma_g = \omega_0/\omega_{pe} \gg 1$ (here, ω_0 is the laser frequency, $\omega_{pe} = (4\pi e^2 n_0/m_e)^{1/2}$ is the Langmuir electron frequency, m_e is the electron rest mass, n_0 is the background electron density, and e is the electron charge). Therefore, even with the Lorentz factor γ_g approaching 100, the bubble remains a "slow" accelerating structure capable of trapping initially quiescent electrons from the ambient plasma (Kalmykov et al., 2010a; Lu et al., 2007; Martins et al., 2010). Single-shot optical diagnostics, such as second harmonic generation from the sheath (Gordon et al., 2010; Helle et al., 2010) and frequency-domain holography/shadowgraphy (Dong et al., 2010a;b), show a direct correlation between the generation of collimated electron beams and bubble formation. Also, the moment of injection can be precisely identified (Thomas et al., 2007a). Electron self-injection greatly simplifies the experimental design enabling a single-stage acceleration with considerable flexibility in beam parameters.

The bubble shape and potentials are determined by the electron flow surrounding the cavitated area. The majority of electrons slip behind the bubble within a time interval close to the period of plasma oscillations, $\tau_p = 2\pi/\omega_{pe}$. This means that the bubble, on the whole, is a quasistatic structure (Lu et al., 2006; Mora & Antonsen, 1996). At the same time, the sheath electrons, exposed to the highest fields and preaccelerated to relativistic energies, stay with the bubble much longer (Kalmykov et al., 2011b). If the quasistatic bubble expands during their slippage time (satisfying semi-empirical condition discussed in section 2.1), the sheath electrons can penetrate into the bubble near its rear, synchronize with it (i.e. obtain the longitudinal momentum $p_{\|} \approx \gamma_g m_e c$), and then travel inside the cavity, continuously gaining energy (Kalmykov et al., 2011b). Diffraction of the driving laser pulse is usually sufficient to cause this kind of dynamic behavior (Kalmykov et al., 2009; Xu et al., 2005). Therefore, the quality of accelerated beam, sensitive to the details of self-injection process, appears to be tied to the self-consistent optical evolution of the driver.

Accelerated electrons eventually outrun the slow bubble. They exit the accelerating phase within a time interval $\tau_d = L_d/c$, where $L_d = (2/3)\gamma_g^2 R_b$ is the dephasing length, and R_b is the bubble radius. Acceleration until dephasing maximizes the energy gain, yielding $E_{\max} \approx 0.085\gamma_g^2(k_p R_b)^2$ MeV, where $k_p = \omega_{pe}/c$ (Lu et al., 2007). Numerical simulations presented in this Chapter correspond to $\gamma_g \approx 16$, $k_p R_b \approx 5$, yielding $L_d \approx 1.7$ mm and $E_{\max} \approx 550$ MeV. In strongly rarefied plasmas, such as $\gamma_g \gg 3k_p R_b/4$, dephasing takes many Rayleigh lengths (this estimate implies that the laser pulse spot size approximately equals to the bubble radius). To propagate the pulse over this distance in a uniform background plasma, majority of modern experiments rely on a combination of relativistic and ponderomotive self-guiding (Hafizi et al., 2000; Ralph et al., 2009). The following dynamical scenario is usually the case. Upon entering the plasma, the pulse with $P/P_{\rm cr} \gg 1$ and $\tau_L < \tau_p$ self-focuses until full electron cavitation is achieved, and the charge-separation force of an electron density channel (bubble) balances the radial ponderomotive force; the pulse is then guided until depletion (here, $P_{\rm cr} = 16.2\gamma_g^2$ GW is the critical power for relativistic self-focusing (Sun et al., 1987)). The condition of the force balance matches the peak vector potential of the pulse (normalized to $m_e c^2/|e|$) to the bubble radius as $a_{\rm sg} \approx (k_p R_b/2)^2 \gg 1$ (Lu et al., 2007). Even though the bubble is a natural attractor for the relativistic laser-plasma dynamics (Gordienko & Pukhov, 2005), both transient dynamics *before* the onset of self-guiding (Kalmykov et al., 2010a) and the laser evolution *during* the self-guiding (Froula et al., 2009; Kalmykov et al., 2010a; 2011b; Kneip et al., 2009; Oguchi et al., 2008) may cause unwanted additional electron injection (dark current), degrading the beam quality. The lack of balance between the light pressure and charge-separation force makes the pulse spot size oscillate (Kalmykov et al., 2010a; Oguchi et al., 2008); self-steepening (Vieira et al., 2010) and depletion (Decker et al., 1996) gradually turn the pulse into a relativistically intense piston (Kalmykov et al., 2011b); and relativistic filamentation (Andreev et al., 2007; Kalmykov et al., 2010a; Thomas et al., 2007b; 2009) distorts the transverse profile of the pulse. Resulting deformations of the bubble, to which self-injection is extremely sensitive (Kalmykov et al., 2010b; 2011b), lead to the rapid degradation of electron beam quality (Froula et al., 2009; Kalmykov et al., 2010a; 2011b; Kneip et al., 2009; Martins et al., 2010). Novel optical diagnostics, such as frequency-domain streak cameras (Dong et al., 2010b; Li et al., 2010; 2011), allow capturing the details of bubble evolution and associating them with the final quality of electron beam in a given experimental shot, providing necessary feedback for the real-time optimization. The combined effort of theory and experiment should thus be aimed at suppressing dark current through a proper optimization and control of laser pulse evolution without complicating the experimental design (ideally, involving a single laser pulse and a single gas target).

It has been understood that to reduce the nonlinear effects and stabilize the bubble during the pulse self-guiding, it is necessary to work with matched PW-class pulses in low-density plasmas, $\gamma_g > 35$ (Lu et al., 2007). At even lower densities, $\gamma_g > 100$, initial mismatching of the pulse (e.g. over-focusing) helps enforce and control the injection (Kalmykov et al., 2010b; 2011a). With $P \sim 10P_{cr}$, and $\tau_L < \tau_p/2$, the initially over-focused laser diffracts, the bubble expands, and electrons are injected continuously. When self-guiding sets in, the bubble stabilizes, and self-injection terminates. Secondary injection into the same bucket remains suppressed, and low-energy tails do not develop in the electron spectra. Electrons injected during the period of expansion travel deep inside the bucket and are continuously accelerated. At the same time, electrons injected late are located in the region of the highest accelerating gradient. They rapidly equalize in energy with earlier injected particles. Thus, a nanocoulomb (nC), quasi-monoenergetic bunch forms long before dephasing (Kalmykov et al., 2009; 2010b; 2011a). It can be further accelerated without emittance dilution, reaching multi-GeV energy near dephasing, keeping a few percent energy spread and a few mrad divergence. By varying plasma density and/or laser focusing geometry, one can control initial mismatching, changing the system behavior during the brief initial transient stage, thus precisely controlling the charge and emittance (Kalmykov et al., 2011a). High optical quality of the PW pulse is critically important for the realization of this scheme (Kalmykov et al., 2010a; Wang et al., 2011).

With a higher power ratio, $P/P_{cr} > 10$, and/or longer pulse, $\tau_L > \tau_p/2$, the described above scheme becomes compromised (Kalmykov et al., 2010a). The length of the transient stage increases, giving rise to multiple oscillations of the laser spot size and periodic injection (Kalmykov et al., 2010a), resulting in polychromatic electron energy distributions (Martins et al., 2010). At higher plasma densities, $\gamma_g < 20$, longitudinal deformation of the pulse becomes another source of the dark current (Kalmykov et al., 2011b). The leading edge of the pulse that pushes away plasma electrons rapidly accumulates frequency red-shift, and group velocity dispersion concurrently compresses the pulse (Fang et al., 2009; Faure et al., 2005; 2006; Pai et al., 2010; Vieira et al., 2010). Concomitant depletion of the leading edge further enhances the self-steepening effect (Decker et al., 1996; Lu et al., 2007). Initially smooth driver thus turns into a relativistically intense "piston", or a "snow-plow" that pre-accelerates and compresses the initially quiescent electron fluid. A large charge separation immediately behind the piston results in sheath electrons receiving strong longitudinal kick, increasing their inertia, and delaying their return to the axis. As a result, the bubble elongates, and massive, uninterrupted self-injection follows (Froula et al., 2009; Kalmykov et al., 2011b; Kneip et al., 2009). In spite of high injected charge, this scenario remains the same in both quasistatic and fully explicit, 3-D EM particle-in-cell (PIC) simulations (Kalmykov et al., 2011b). Beam loading (Rechatin et al., 2010; Tzoufras et al., 2009) becomes important only in the final stage of this process. Notably, in this situation, transverse matching of the pulse precludes neither periodic nor continuous injection (Martins et al., 2010), and thus does not help improve the beam characteristics. However, the dark current-free acceleration can be achieved even in these unfavorable regimes. Giving up acceleration until dephasing, and limiting the plasma length to one cycle of laser waist oscillation results in a quasi-monoenergetic beam formation (Hafz et al., 2011; Kalmykov et al., 2009). This approach does not maximize electron energy and thus does not optimize the accelerator performance. However, high-quality, a few-hundred MeV electron beams with tunable parameters can be produced, which is valuable for applications. In addition, as we show in section 3, using a broad bandwidth, negatively chirped pulse may help compensate for the nonlinear red-shift and delay formation of the relativistic piston, thus reducing the amount of dark current.

In this Chapter, we elucidate the intrinsic connection of electron injection with the laser pulse optical evolution and demonstrate the mechanism of monoenergetic electron beam formation. We also discuss adverse scenarios of the pulse evolution leading to the continuous injection, and propose ways to mitigate them. To examine electron injection during various stages of laser pulse evolution in a single numerical experiment, we use two complementary simulation approaches. In section 2, we explain the physics, and develop the conceptual framework of the problem using the quasistatic, cylindrically symmetric, fully relativistic PIC code WAKE (Mora & Antonsen, 1997). A fully 3-D, non-averaged, dynamic test electron tracking module incorporated in WAKE (Kalmykov et al., 2009; Malka et al., 2001) emulates the non-quasistatic response of initially quiescent electrons to a high-frequency quasi-paraxial laser field and slowly varying EM plasma wakes. In section 2.5 we validate the test-particle results in a full 3-D PIC simulation using the code CALDER-Circ (Lifschitz et al., 2009).

The formation of a quasi-monoenergetic electron bunch during one period of laser spot oscillation is the subject of sections 2.1, 2.2, and 2.3. By analyzing quasistatic trajectories and using the results of test electron tracking in section 2.1, we identify precisely the injection candidates, collection volume, and evaluate the minimal bubble expansion rate for the initiation of self-injection. Formation of the quasi-monoenergetic beam during one period of the bubble size oscillation is described in section 2.2. Results of section 2.3 show that self-injection and subsequent acceleration of an electron require initial reduction of its moving-frame Hamiltonian. Laser pulse self-compression and resulting continuous injection are considered in section 2.4. In section 2.5, we validate the self-injection scenarios discussed in sections 2.1 – 2.4 in a CALDER-Circ simulation. We find that the test-particle modeling correctly identifies physical processes responsible for the initiation and termination of self-injection. Using our fast simulation toolkit (WAKE with test particles) and CALDER-Circ code, we explain in section 3 that laser self-compression and concomitant continuous injection can be suppressed by using a negatively chirped, broad bandwidth driver pulse. In section 4 we summarize the results and point out directions of future work.

2. Self-injection scenarios in the blowout regime

We study self-injection and acceleration of electrons until dephasing in a standard regime of LPA experiments at the University of Nebraska (Ramanathan et al., 2010). We examine various scenarios of injection and relate them to nonlinear dynamical processes involving the laser pulse. The quasistatic behavior of the bulk plasma makes it possible to elucidate the physics of self-injection using a conceptually simple and computationally efficient toolkit: a fully relativistic 3-D particle tracking module built into the cylindrically symmetric time-averaged (over ω_0^{-1}) quasistatic PIC code WAKE (Mora & Antonsen, 1997). WAKE uses an extended paraxial solver for the slowly varying laser pulse envelope, which preserves the group velocity dispersion in the vicinity of the carrier frequency and calculates precisely radiation absorption due to the wake excitation. Electron response to the time-averaged ponderomotive force is calculated assuming that *all* plasma electrons (macroparticles) eventually fall behind the bubble. This approach, termed the quasistatic approximation (QSA), tremendously speeds-up the simulation, enabling serial runs and extensive parameter scans using small workstations. On the other hand, neglecting the long-term contribution of near-luminous-speed macroparticles traveling with the bubble prohibits self-consistent modeling of electron self-injection and trapping. To simulate self-injecton without compromising computational efficiency, we use test particles. The test-particle module is fully dynamic, making no assumption of cylindrical symmetry, and is not time-averaged. In particular, it takes into account the interaction of test electrons with the non-averaged, linearly polarized laser field with non-paraxial corrections (Quesnel & Mora, 1998). To capture the laser pulse interaction with non-quasistatic background electrons (and thus to model self-injection into non-stationary quasistatic wakefields), a group of quiescent test electrons is placed before the laser pulse at each time step. In this way, electron self-injection associated with bubble and driver evolution is separated from the effects brought about by the collective fields of the trapped electron bunch, i.e. from effects due to beam loading. This simulation approach allows us to fully characterize the details of self-injection process and relate them to the evolution of the laser and the bubble using a non-stationary Hamiltonian formalism (Kalmykov et al., 2010b; 2011b).

In the simulation, a transform-limited, linearly polarized Gaussian laser pulse with a full width at half-maximum in intensity $\tau_L = 30$ fs and central wavelength $\lambda_0 = 0.805$ μm is focused at the plasma border ($z = 0$) to a spot size $r_0 = 13.6$ μm, and propagates towards positive z. The plasma has a 0.5 mm linear entrance ramp followed by a 1.7 mm plateau. The length of the plateau is equal to the dephasing length. Plasma density, $n_0 = 6.5 \times 10^{18}$ cm^{-3}, corresponds to $\gamma_g = \omega_0/\omega_{pe} \approx 16.3$. The laser power is 70 TW, which yields $P/P_{cr} = 16.25$, a peak intensity at the focus 2.3×10^{19} W/cm^2, and normalized vector potential $a_0 = 3.27$. The WAKE simulation uses a grid $dr \approx 0.1k_p^{-1} \approx 0.21$ μm, with 30 macroparticles per radial cell, $d\xi = dr/3$ (where $\xi = z - ct$), and time step $dt = dz/c \approx 1.325\omega_0^{-1}$.

2.1 Injection candidates, collection volume, and minimal expansion rate to initiate injection

Upon entering the plasma, the laser pulse self-focuses and reaches the highest intensity at $z \approx 0.8$ mm. Full blowout is maintained over the entire propagation distance. Bubble expansion *and* electron injection begin soon after the laser pulse enters the density plateau. The wakefield potential, time-averaged (over ω_0^{-1}) Lorentz forces, electron density, and sample trajectories of macroparticles $r_i(\xi)$ are shown in Fig. 2 for the fully expanded bubble after 1 mm of propagation (cf. position (2) of Fig. 3).

The quasistatic electron density and number density of non-quasistatic test particles in (r, ξ) space are strikingly similar in Fig. 2(c). Thus, even though the macroparticles cannot be trapped, analysis of their trajectories helps identify the injection candidates, and specify the scenario of bubble evolution favorable for injection. This analysis also provides precise estimates of the collection volume and the bubble expansion rate necessary to initiate the injection. Each macroparticle can be put into one of three clearly defined groups [color coded in Figs. 2(d) – 2(f)]. The majority of electrons, viz. those expelled by the radiation pressure (black) and those attracted from periphery to the axis (green), are passing. They fall behind the bubble roughly within a time interval $\tau_b = L_b/c$ (where $L_b \approx 2R_b$ is the bubble length). The bulk plasma electrons thus obey the QSA restrictions exceptionally well, which enables precise WAKE modeling. Sheath electrons (red) are different; they may travel with the bubble over a long distance. Figure 2(e) shows that their slippage time,

$$T_{\text{slip}} = \int_0^{L_b} \frac{d\zeta}{c - v_z}, \tag{1}$$

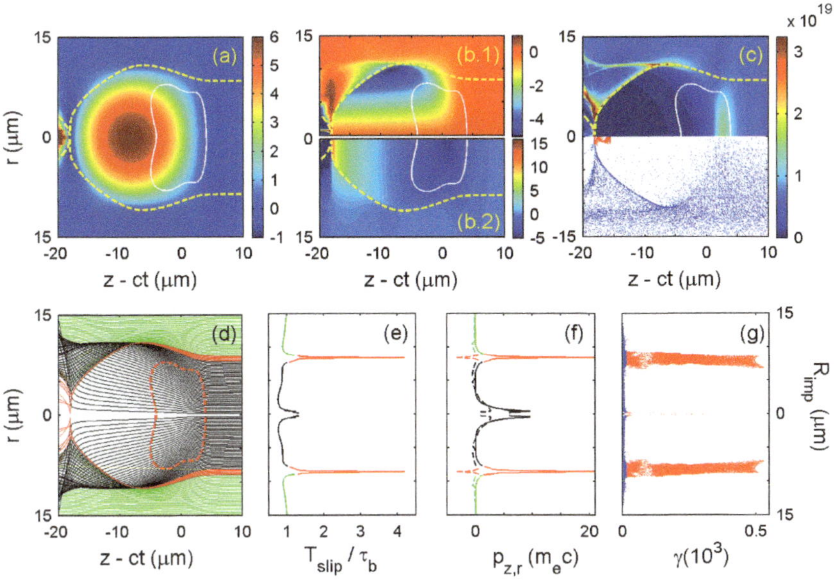

Fig. 2. Fully expanded bubble from the WAKE simulation (cf. position (2) of Fig. 3). Solid white and dashed red contours in panels (a–d) are the iso-contours of laser intensity at $\exp(-2)$ of the peak (the pulse propagates to the right). (a) Wake potential $\Phi = \phi - A_z$ in units $m_e c^2/|e|$. (b) Focusing ($-E_r + B_\theta$, top) and accelerating ($-E_z$, bottom) Lorentz forces (in GV/cm). (c) Top: the quasistatic electron density (in cm^{-3}). Bottom: radial positions of non-quasistatic test electrons. Red markers are the test particles with $\gamma > \gamma_g = 16.3$. (d) Trajectories $r(\xi)$ of the quasistatic macroparticles. Green and black trajectories correspond to passing electrons. Red trajectories correspond to sheath electrons – injection candidates. (e) Normalized slippage time as a function of the impact parameter, $R_{\text{imp}} = r(\xi \gg c\tau_L)$. (f) Longitudinal ($p_z$, solid line) and transverse (p_r, dashed line) momenta of macroparticles at the rear of the bubble (the point of trajectory crossing). Sheath electrons have the largest slippage time, and become relativistic before crossing the axis. The yellow dashed curve in panels (a–c) is a trajectory of the macroparticle with the greatest slippage time ("the innermost electron"). (g) Impact parameters of test electrons from panel (c) vs energy.

significantly exceeds τ_b (here, $\zeta = -\xi$, and $\zeta = L_b \approx 18\ \mu$m is the coordinate of the rear of the bubble). A yellow broken line in Figs. 2(a) – 2(c) shows the trajectory of the macroparticle with the largest slippage time, $T_{\text{slip}} \approx 4.2\tau_b$. Figure 2(d) indicates that the sheath electrons originate from a hollow cylinder with a radius close to the laser pulse spot size. While slipping through the structure, they are exposed to the highest wakefields. At the rear of the bubble, they are strongly pre-accelerated in *both* longitudinal *and* transverse directions: Fig. 2(f) gives $p_{z\,\text{max}} \approx 21 m_e c > \gamma_g m_e c$, and $|p_{r\,\text{max}}| \approx 4 m_e c$. The large longitudinal momentum of these electrons makes them the best injection candidates; their promotion to fully dynamic macroparticles may result in their self-injection and acceleration (Morshed et al., 2010).

Longitudinal synchronization of sheath electrons, $p_z \geq \gamma_g m_e c$, is necessary and, in high-density plasmas (such as $\gamma_g \approx k_p R_b$), also sufficient for their injection (Kostyukov et al., 2009). In the opposite limit of strongly rarefied plasmas, $\gamma_g > 10 k_p R_b$, the sheath electrons do not

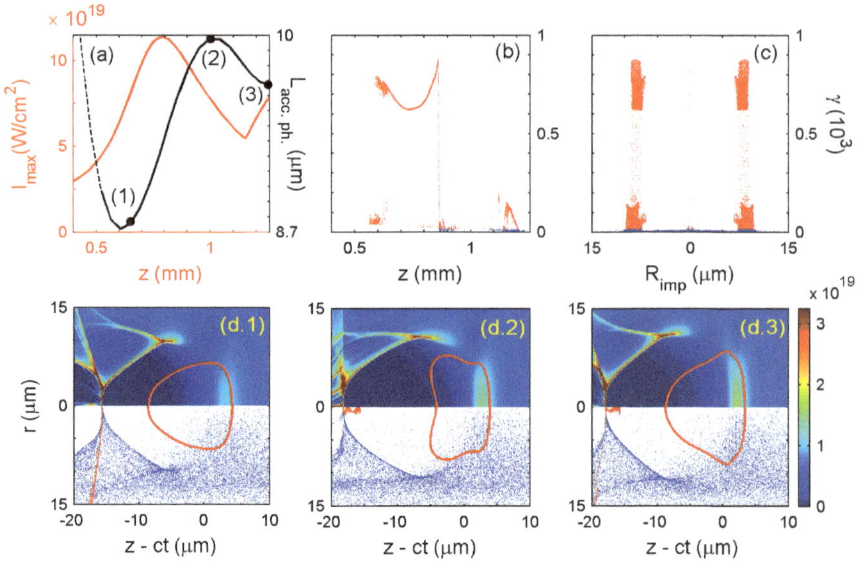

Fig. 3. Electron injection into the oscillating bubble. (a) Peak laser intensity (red) and length of the accelerating phase $L_{\text{acc.ph.}}$ (black) vs propagation distance; $L_{\text{acc.ph.}}$ is the distance between the positions of zero and peak accelerating gradient on axis – see Fig. 4(b). Panels (b) and (c): energy of test electrons vs their initial longitudinal (b) and radial (c) positions after one period of bubble size oscillation (cf. position (3) of panel (a)). (d.1) – (d.3) Quasistatic electron density in cm^{-3} (color map) and number density of non-quasistatic test particles (dots) corresponding to the positions (1) – (3) of the plot (a) (labeled accordingly). Red contour is an iso-contour of laser intensity at $\exp(-2)$ of the peak. Data from positions (1) – (3) are used to describe the process of monoenergetic electron bunch formation in Fig. 4.

synchronize with (and thus cannot be injected into) a non-evolving bubble (i.e. depending on variables r and ξ only); in this case, evolution of the bucket is vital for self-injection (Kalmykov et al., 2009; 2010a;b; 2011a). In the intermediate regime with $\gamma_g \approx 3.5k_pR_b$ (which is the focus of this Chapter), longitudinal synchronization appears to be insufficient for injection. The sheath electrons crossing the axis have relativistic transverse momenta, and thus tend to exit the cavity in the radial direction. Such electrons have been earlier observed in the laboratory in the absence of any noticeable trapping (Helle et al., 2010; Kaganovich et al., 2008). To be injected (i.e. return to axis), the injection candidates must be confined into a focusing cavity *after* crossing the axis. To this effect, the bubble must continuously *expand*, changing its size by an appreciable fraction during the electron transit time T_{slip}. Energetic sheath electrons can then outrun the boundary of the bubble and stay inside long enough to *both* synchronize longitudinally *and* make a U-turn transversely. To separate the most energetic electrons from the sheath, elongation of the bubble over the slippage time has to exceed the sheath thickness Δ_{sh} at the rear of the bucket (Kalmykov et al., 2010b; 2011b),

$$\Delta L_b = L_b(t + T_{\text{slip}}) - L_b(t) \geq \Delta_{\text{sh}}, \tag{2}$$

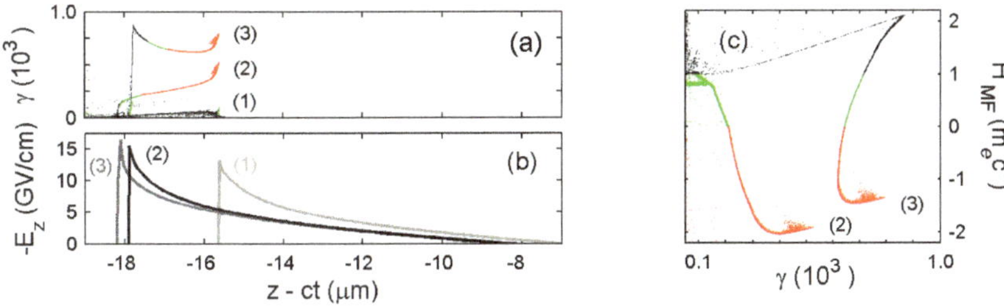

Fig. 4. Phase space rotation and formation of a quasi-monoenergetic bunch (Kalmykov et al., 2011b). (a) Phase space rotation of injected test electrons. Longitudinal phase space is shown at the positions (1)–(3) of Fig. 3(a). (1) Injection begins. (2) The bubble is fully expanded, injection stops, and phase space rotation begins. (3) The bucket slightly contracts. Electrons injected lately equalize in energy with those injected earlier. Quasi-monoenergetic bunch forms. Test electrons are color coded according to $H_{MF} < 0$ (red), $0 < H_{MF} < 1$ (green), $H_{MF} > 1$ (black). (b) Axial line-outs of the accelerating gradient (in GV/cm). (c) H_{MF} vs energy gain for the fully expanded (2) and contracted (3) bubble.

where the parameter Δ_{sh} has to be found empirically from simulations. For the fully expanded bubble of Fig. 2, $\Delta L_b \approx 3\Delta_{sh} \approx 0.5\ \mu m$ (the estimate obtained using Fig. 3(a)), which is sufficient to maintain the injection. Indeed, a group of test particles accelerated to $\gamma > \gamma_g$ (red markers) can be seen at the base of the bubble in Fig. 2(c). Figure 2(g) indicates that, in agreement with earlier studies (Pukhov et al., 2010; Tsung et al., 2006; Wu et al., 2009), only electrons with impact parameters such that they enter the sheath are collected and accelerated. And, the condition (2) always holds, sometimes rather closely, when self-injection occurs in low-density plasmas, $\gamma_g > 65$ (Kalmykov et al., 2009; 2010a;b; 2011a).

All accelerated electrons in Figs. 2(c,g) are collected from a cylindrical shell with thickness $\Delta R_{coll} \approx 2\ \mu m$ and radius close to the laser spot size, $R_{coll} \approx 8\ \mu m$. The length of this hollow cylinder equals to the interval of bubble expansion from Fig. 3(a), $\Delta z_{exp} \approx 400\ \mu m$. This collection volume contains 2.6×10^{11} electrons – injection candidates (42 nC charge). To calculate the injected charge correctly, one has to take into account the self-fields of both sheath and recently trapped electrons near the base of the bubble, making resort to the fully kinetic simulations (Kalmykov et al., 2011b; Morshed et al., 2010). 3-D PIC simulation of section 2.5 shows that only 0.5% of particles from the collection volume are actually injected. Injection from a very narrow range of impact parameters, together with low collection efficiency in realistic PIC modeling, makes massive self-injection during the slippage time very unlikely. Injected particles thus remain the minority, and their contribution to the bubble evolution is insignificant. This justifies the quasistatic treatment of the plasma electrons making up the accelerating structure, and validates the test-particle model of self-injection process.

The background plasma is never perfectly uniform in the laboratory experiment. Localized density depressions, which may naturally occur in gas jet targets, can also cause electron self-injection (Hemker et al., 2002; Suk et al., 2001). Even if the driver evolution is negligible (as in the beam-driven case (Lu et al., 2006; Rosenzweig et al., 1991)), the bubble crossing a density down-ramp necessarily expands; if the ramp is longer than a slippage distance, $L_{ramp} \gg cT_{slip}$, then the bubble evolution is slow, and Eq. (2) applies. The bubble length in the non-uniform plasma is $L_b(z) = \kappa(2\pi/k_{p0})/\sqrt{\tilde{n}_e(z)}$, where the parameter $\kappa \sim 1$ is determined empirically from simulations. Assuming a power-law density profile, $\tilde{n}_e(z) \equiv n_e(z)/n_0 = 1 - A[(z - z_{in})/L_{ramp}]^\alpha$, where $\tilde{n}_e(z < z_{in}) = 1$, and $\tilde{n}_e(z > z_{in} + L_{ramp}) = 1 - A < 1$, and substituting $z - z_{in} = cT_{slip}$ into Eq. (2), we find the relation between the length of the ramp L_{ramp} and density depression A necessary to incur injection: $L_{ramp} < cT_{slip}[A\kappa\pi/(k_{p0}\Delta_{sh})]^{1/\alpha}$. Using parameters of the bubble from Fig. 2, we find that a linear down-ramp with a 10% density depression may produce self-injection if $L_{ramp} < 0.47$ mm. If the plasma density is more homogeneous than that, the self-injection into the bubble can be enforced by the pulse evolution only.

2.2 Self-injection into an oscillating bubble: formation of quasi-monoenergetic collimated electron beam

The first period of bubble size oscillations is displayed in Fig. 3. Figure 3(d.1) shows the bubble and the laser pulse at the beginning of bubble expansion (cf. position (1) of Fig. 3(a)); Fig. 3(d.2) at the end of expansion (cf. position (2) of Fig. 3(a)); and Fig. 3(d.3) at the end of oscillation period (cf. position (3) of Fig. 3(a)). The laser pulse entering the density plateau at $z = 0.5$ mm is longer than one-half of the electron plasma period, $\tau_L \approx 0.7\tau_p$. Its head, residing in an incompletely evacuated nonlinear channel, remains guided with an almost invariant spot size, $r(\xi \approx 0) \approx 6.5$ μm, whereas the tail, confined within an evacuated bubble, is strongly mismatched. Beating of the mismatched tail, causing alternating expansion and contraction of the bubble, are clearly seen in progression from Fig. 3(d.1) to 3(d.3).

Figure 3(a) shows that bubble expansion starts near the edge of the density plateau and continues until $z \approx 1$ mm. The bubble expands by 14% of its size over a 400 μm distance ($\sim$ 20 bubble lengths). Injection of non-quasistatic test electrons continues uninterrupted during this stage, and, as is clear from Fig. 4(a), their momentum distribution is continuous. Contraction of the bubble between $z = 1$ and 1.25 mm extinguishes injection and truncates the bunch: electrons injected at the very end of expansion are expelled. Particles remaining in the bucket are further accelerated. At this stage, the bunch becomes quasi-monoenergetic. According to Fig. 4(b), the longitudinally nonuniform, co-moving accelerating gradient changes insignificantly during the contraction. The tail of the bunch, constantly exposed to the highest gradient, equalizes in energy with earlier injected electrons (cf. position (3) of Fig. 4(a)). This rotation of longitudinal phase space, responsible for the formation of a quasi-monoenergetic bunch long before dephasing, is clearly different from that discussed in literature (Tsung et al., 2006). According to Figs. 3(b) and (c), electrons remaining in the bucket at the end of bubble contraction are collected during the interval of bubble expansion from a cylindrical shell with the radius close to the laser spot size. They form the bunch with the energy $E = 360^{+40}_{-20}$ MeV and 4.3 mrad divergence. *Therefore, limiting the plasma length to a single period of the bubble size oscillations gives a quasi-monoenergetic, collimated electron beam* (Hafz et al., 2011; Kalmykov et al., 2009).

2.3 Evolution of test electron Hamiltonian during injection

WAKE calculates all potentials directly, which makes the Hamiltonian analysis of test particle tracking straightforward. Using the definitions of normalized momentum $\mathbf{p} \equiv \mathbf{p}/(m_e c)$, wake potential $\Phi = |e|(\varphi - A_z)/(m_e c^2)$ (where ϕ is a scalar potential, and A_z is the longitudinal component of vector potential), envelope of the laser vector potential $\mathbf{a} \equiv |e|\mathbf{a}/(m_e c^2)$, and $\gamma_e = (1 + \mathbf{p}^2 + \mathbf{a}^2/2)^{1/2}$, we introduce the normalized time-averaged moving-frame (MF) Hamiltonian $H_{MF}(r, z, \xi) = \gamma_e + \Phi - p_z$. For the quasistatic macroparticles, $H_{MF} \equiv 1$ (Mora & Antonsen, 1997). Test electrons (which are not assumed to be quasistatic) move in explicitly time-dependent potentials; hence, H_{MF} changes in the course of propagation according to $dH_{MF}/dt = \partial H_{MF}/\partial t$. For a test electron moving away from the bubble, $H_{MF} = \gamma_e + \Phi - p_z \rightarrow \sqrt{1+\mathbf{p}^2} - p_z > 0$. Hence, the electron is confined inside the bucket at all times (trapped) if the H_{MF} remains *negative* in the course of interaction. As soon as the bubble stabilizes, H_{MF} is conserved. All test electrons can be then divided into 3 groups: (1) $H_{MF} < 0$ — trapped; (2) $0 < H_{MF} < 1$ — injected (accelerated); and (3) $H_{MF} > 1$. All the three groups are represented in Fig. 4(a), where the phase space of test electrons is shown at the stationary points of full expansion (labeled (2)) and full contraction of the bubble (labeled (3)). Electron phase space for the fully expanded bubble shows that the bubble expansion causes a reduction in H_{MF} (Kalmykov et al., 2009; 2010b). The condition $H_{MF} < 1$ is thus *necessary* for injection and initial acceleration. For instance, it can be used for promotion of test electrons into the non-quasistatic electron beam particles in order to self-consistently incorporate beam loading into the model. Conversely, even minimal bubble contraction may raise H_{MF} significantly. Figure 4(c) shows that electrons with $0 < H_{MF} < 2$ are accelerated as effectively as those which are formally trapped. Hence, the natural evolution of the structure may result in violation of the sufficient trapping condition; this, however, does not disrupt acceleration with good collimation and low energy spread (Kalmykov et al., 2011b).

2.4 Continuous self-injection caused by self-compression of the driving pulse

Although a monoenergetic electron bunch forms early, the general experimental trend is to push the accelerator efficiency to the limit and use the entire dephasing length. Electron beam quality, however, can be compromised in this pursuit. The driver pulse evolves continuously, which may cause uninterrupted electron injection and emittance growth. Understanding the physical mechanism of continuous injection will help control the beam quality by limiting injection via a judicious choice of laser-plasma interaction geometry and target design (Kalmykov et al., 2011a) or by manipulating the phase and envelope of the incident pulse.

Running the simulation until the nonlinear dephasing limit, we find two distinct stages of the system evolution. Stage I, discussed above, corresponds to a single oscillation of the laser spot size and produces a monoenergetic electron bunch. Stage II is characterized by a gradual increase of laser intensity (up to 2×10^{20} W/cm^2) and a steady elongation of the bubble. Figure 5 shows that bubble elongation is accompanied by continuous injection and growth of the energy spread. At the end of the run, the number of continuously injected test electrons ($\gamma > \gamma_g$) is factor of 7.5 larger than the number of electrons in the leading quasi-monoenergetic bunch. Similar development of self-injection process has been reported elsewhere (Froula et al., 2009; Kneip et al., 2009). Figures 5(c.1) – 5(d.3) show that the asymmetric growth (elongation) of the bubble is accompanied by self-compression of the driver pulse from 25 to roughly 5.5 fs, and simultaneous 5-fold increase in intensity. The compressed pulse (relativistic piston) acts as a snow-plow. The ponderomotive push of its front pre-accelerates plasma electrons to $\gamma > \gamma_g$ and, as is seen in Figs. 5(c.2) and 5(c.3), creates a strongly compressed electron slab an order of magnitude denser than the ambient plasma. As a result of this strong charge separation, plasma electrons entering the sheath (as can be seen in Figs. 6(b) and 6(d)) are exposed to the positive electric field a factor 2.25 higher than in the case of smooth driver (Figs. 6(a) and 6(c)). Hence, upon passing the piston, sheath electrons receive a large kick in the backward direction, and quickly become relativistic, $p_z \approx -1.65 m_e c$ (in contrast to $-0.55 m_e c$ in the smooth driver case). Therefore, in the piston case, it takes nearly twice as long for the sheath electron to reach the point of return, $p_z = 0$, and to start getting accelerated; reaching the axis also takes longer time, which explains the bubble elongation.

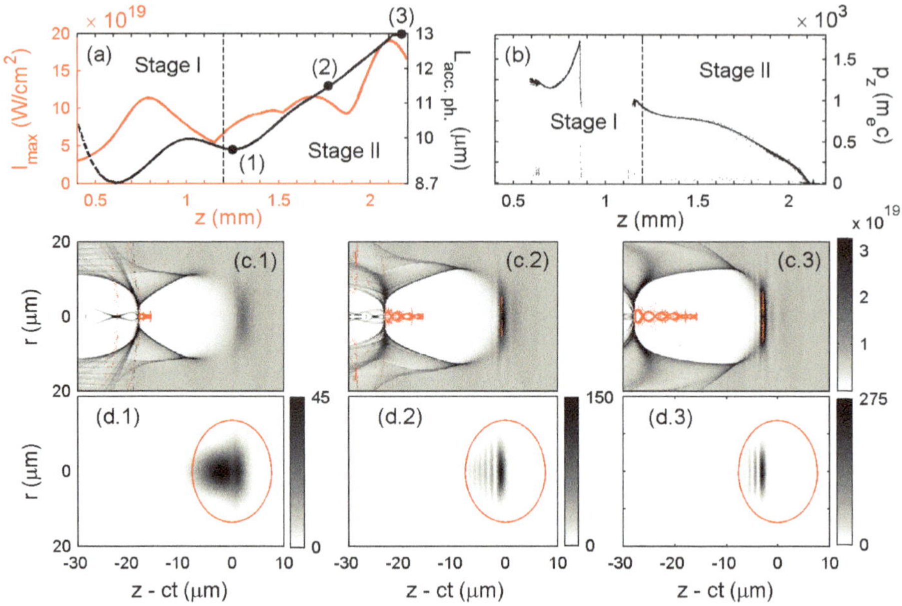

Fig. 5. Pulse self-compression and continuous injection (Kalmykov et al., 2011b). (a) Peak laser intensity (red) and the length of the accelerating phase vs propagation length (black). (b) Energy of test electrons vs their initial positions at $z = 2.2$ mm. The leading quasi-monoenergetic bunch forms during Stage I (one period of bubble size oscillation). Bubble expansion during Stage II causes continuous injection with broad energy spectrum. The bubble and injected test electrons at the positions (1) – (3) of panel (a) are shown in panels (c.1) – (c.3). Grayscale: electron density (in cm^{-3}); red dots: test electrons with $\gamma > \gamma_g$. (d.1) – (d.3) Normalized laser intensity, $|a|^2$, at the positions (1) – (3) of panel (a); red contour: iso-contour of an incident pulse intensity at $\exp(-2)$ of the peak. Self-steepening of the pulse (formation of a relativistic piston) causes elongation of the bubble and continuous injection.

This physical interpretation, deduced from the analysis of quasistatic electron trajectories, is validated in section 2.5 in fully explicit 3-D PIC simulations. It suggests that the root cause of continuous injection is the pulse self-steepening. The steepening is partly caused by depletion due to the wake excitation ($\sim$ 33% at $z = 2.2$ mm) (Decker et al., 1996; Fang et al., 2009; Lu et al., 2007), and is partly a nonlinear optical effect (Faure et al., 2005; Pai et al., 2010; Vieira et al., 2010). Figures 7(a) and 7(b) show axial lineouts of normalized intensity and of the nonlinear index of refraction. The pulse leading edge witnesses the *index down-ramp* at all times. Hence, the laser frequency *red-shifts* in the region of index gradient. At the same time, the tail traveling inside the bubble remains unshifted. With lower frequencies temporally leading higher frequencies, the pulse acquires a *positive chirp*. The anomalous group velocity dispersion of plasma compresses the positively chirped pulse: the red-shifted leading edge slows down with respect to the non-shifted tail, building up the field amplitude first in the pulse head (Fig. 7(a)), and later near the pulse center (Fig. 7(b)). The chirp, as shown in Fig. 7(c), broadens the laser spectrum towards $\omega = \omega_{pe}$; envelope oscillations in Fig. 7(b) result from the strong reduction of the pulse central frequency. The

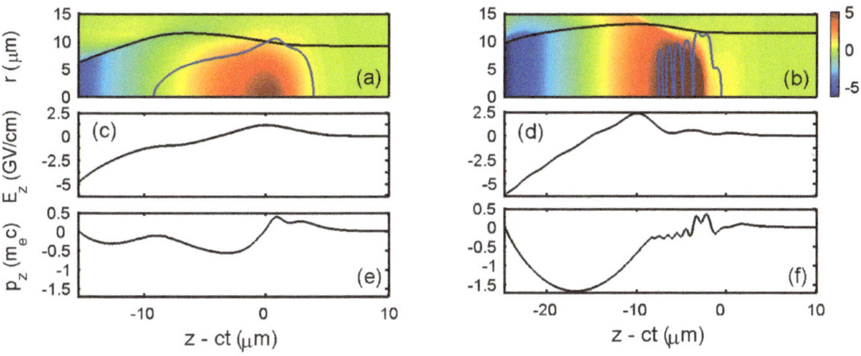

Fig. 6. Bubble elongation due to formation of a relativistic piston: quasistatic analysis. Physical quantities are shown before (left column) and after (right column) the piston formation. The left column corresponds to position (1), and the right column – to the position (3) of Fig. 5(a). (a), (b) Longitudinal electric field E_z (in GV/cm). Black lines are the innermost electron trajectories, $r_{\text{in}}(\xi)$, plotted until the point of return, $p_z = 0$. Blue contours are iso-contours of laser intensity at 5×10^{18} W/cm^{-2}. (c), (d) Longitudinal electric field in the points of the innermost electron trajectory, $E_z(r_{\text{in}}(\xi))$, from plots (a) and (b), respectively. (e), (f) Longitudinal momentum of the innermost electron along its trajectory, $p_z(r_{\text{in}}(\xi))$.

large bandwidth explains pulse compression to roughly two cycles. The red shift of central frequency together with the continuous front etching additionally slows down the pulse and the bubble and provides another reason for the occurrence of continuous injection (Fang et al., 2009). We show in section 3 that *negative chirp* of the incident pulse may compensate for the gradually accumulating red-shift, thus delaying the pulse contraction and partly suppressing continuous injection.

2.5 Validation of self-injection scenarios in full 3-D PIC simulation: role of beam loading

Collective fields of the electron beam, neglected in the test-particle treatment, are known to change the shape of the sheath and thus reduce accelerating gradient, eventually terminating self-injection (Tzoufras et al., 2009). In this section, we verify the test-particle results by running a fully explicit 3-D PIC simulation with the identical set of initial conditions. We use the quasi-cylindrical code CALDER-Circ (Lifschitz et al., 2009), which preserves realistic geometry of interaction, and accounts for the axial asymmetry by decomposing EM fields (laser and wake) into a set of poloidal modes (whereas the particles remain in full 3-D). Well preserved cylindrical symmetry during the interaction enables us to use just the two lowest order modes and thus reduce a 3-D problem to an essentially 2-D one. We suppress sampling noise by using a large number of macroparticles (45 per cell) and high resolution in the direction of propagation, $dz = 0.125c/\omega_0$. The aspect ratio $dr/dz = 15.6$, and the time step $dt = 0.1244\omega_0^{-1}$. Figure 8 shows that despite a much coarser grid, larger time step and underlying approximations, the WAKE simulation correctly captures all relevant physics of plasma wake evolution and dynamics of electron self-injection. In addition, CALDER-Circ having fully self-consistent macroparticle dynamics yields the complete electron phase space, and thus calculates precisely injected charge and beam emittance.

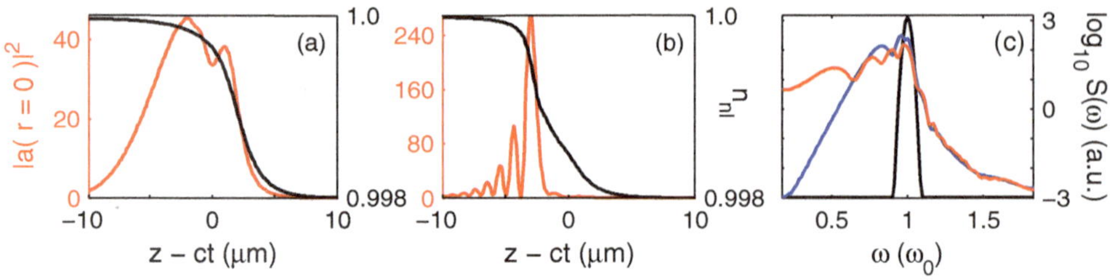

Fig. 7. Pulse red-shifting and formation of the relativistic piston (Kalmykov et al., 2011b). (a) Axial lineouts of normalized intensity (red) and nonlinear refractive index (black) at the position (1) of Fig. 5(a). (b) Same for the position (3) of Fig. 5(a). (c) Laser frequency spectra (radially integrated). Blue corresponds to panel (a), red – to panel (b), black – to the incident pulse. The co-moving index gradient causes frequency red shifting and spectral broadening. The spectrally broadened pulse in panel (b) is compressed to approximately two cycles.

In spite of the great difference in the algorithms and physics content, both codes demonstrate the same correlation between the laser and bubble evolution. Self-injection begins, terminates, and resumes at exactly the same positions along the propagation axis in both runs. Figures 8(a), 8(d), and 8(g) show the result of Stage I — formation of quasi-monoenergetic electron bunch before dephasing. Self-fields of the bunch are unable to prevent the bucket contraction and partial de-trapping of electrons. The bunch phase space has a characteristic "U"-shape produced by the phase space rotation. The bunch has 8% energy spread around 245 MeV and a charge $Q_{\rm mono} \approx 230$ pC (in addition, electrons from the second bucket produce a separate, rather diffuse peak around 150 MeV in Fig. 8(d)). The bunch duration, $t_b = 10$ fs, is the same as in the test-particle simulation; whereas divergence, 13 mrad, is three times higher. The normalized transverse emittance in the plane of polarization is $\varepsilon_{N,x} = (m_e c)^{-1} \left(\langle \Delta x^2 \rangle \langle \Delta p_x^2 \rangle - \langle \Delta x \Delta p_x \rangle^2\right)^{1/2} \approx 8.74\pi$ mm mrad. It was understood that the absence of beam self-fields in the test particle model leads to strong underestimation of emittance, which validates the importance of 3-D PIC simulations for precise calculation of the phase space volume of self-injected electrons (Kalmykov et al., 2011b).

The difference between the phase spaces of WAKE test electrons and CALDER-Circ macroparticles, clearly seen in Fig. 8(g), can be attributed to the effect of beam loading which reduces the accelerating gradient along the bunch and slows down phase space rotation, ultimately reducing the bunch energy by 30%. Using the formalism of (Tzoufras et al., 2009), we find that the repulsive EM fields of electron bunch are not high enough to prevent sheath electrons crossing the axis and are thus unable to prevent further injection (Kalmykov et al., 2011b).

Figures 8(a)–8(c) show that continuous injection develops in both CALDER-Circ and WAKE runs in exactly the same fashion, which validates the physical origin and continuous injection scenario inferred from the analysis of quasi-static electron trajectories in section 2.4. Near dephasing, continuously injected charge in CALDER-Circ simulation reaches $Q_{\rm cont} \approx 1.06$ nC; the beam divergence is 36 mrad. The ratio $Q_{\rm cont}/Q_{\rm mono} \approx 4.6$ is lower than the test particle result of section 2.4 on account of beam loading. Figure 8(c) shows that, in spite of the high injected charge, the bubble shape at the dephasing point is almost unaffected by the presence of the electron beam; this observation rules out beam loading as a cause of continuous injection. The bubble is still not fully loaded and injection in the CALDER-Circ simulation continues beyond the dephasing point. Therefore, apart from slight reduction of the accelerating gradient, beam loading brings no new physical features into the scenario of

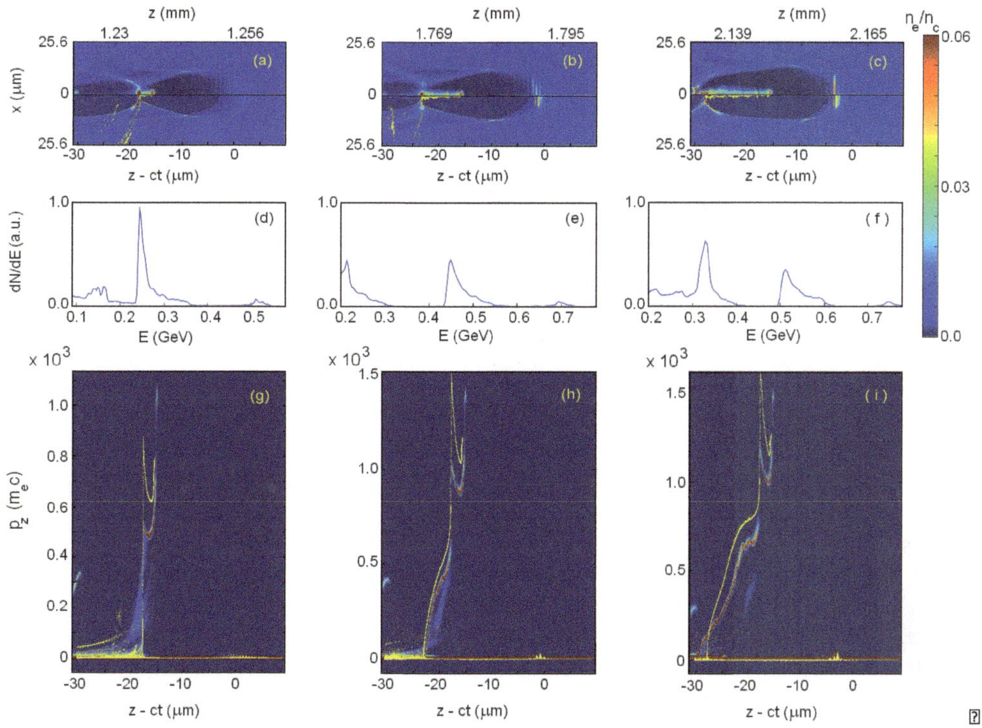

Fig. 8. Continuous injection in quasistatic (WAKE with test particles) and full 3-D PIC (CALDER-Circ) simulations (Kalmykov et al., 2011b). (a)–(c) Electron density from CALDER-Circ (top half) and WAKE (bottom) runs. Yellow dots are the test electrons with $\gamma > \gamma_g$. (d)–(f) Electron energy spectrum (CALDER-Circ). (g)–(i) Longitudinal phase space (colormap – CALDER-Circ; test electrons – yellow dots). Panels (a), (b), (c) are counterparts of Figs. 5(c.1), 5(c.2) and 5(c.3).

continuous injection discussed in section 2.4. Continuous injection can be thus associated solely with frequency red-shift and self-compression of the driver pulse, causing the growth of the quasistatic bubble.

3. Suppression of continuous injection using negatively chirped driving pulse

We have learned in the last section that ploughing through the electron fluid significantly reduces the frequency of the pulse leading edge. Subsequent self-steepening and compression of the driver to a few cycles causes gradual expansion of the bubble, bringing about massive continuous injection, degrading the beam quality. This physical scenario prevents effective use of the entire dephasing length for high-energy quasi-monoenergetic electron acceleration. Using a *negatively chirped* pulse may naturally alleviate this issue. If the pulse has a sufficiently large frequency bandwidth (corresponding to a few-cycle transform-limited duration), the blue shift of the leading edge would largely compensate for a gradually accumulating nonlinear red-shift. This flattening of the phase would then reduce the pulse self-steepening

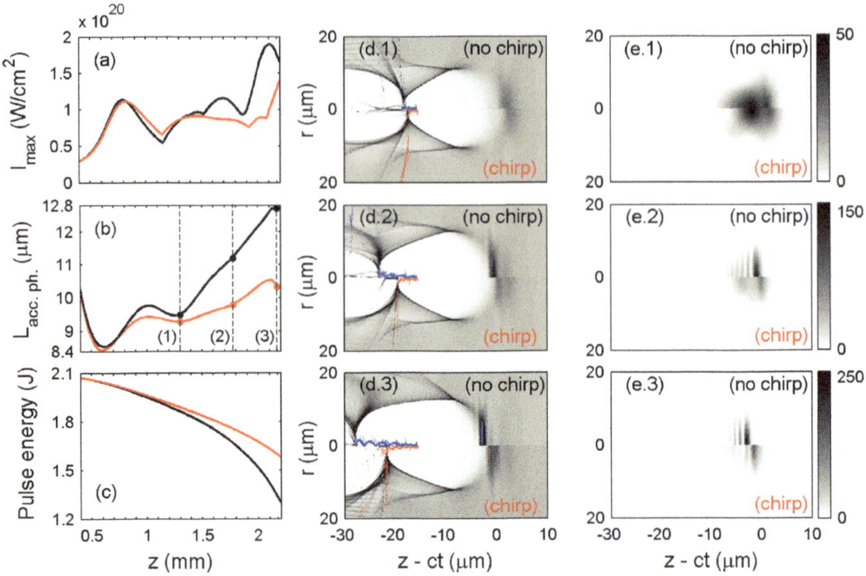

Fig. 9. Effect of the negative chirp on the evolution of driver pulse and plasma bubble. Red lines in panels (a) – (c) correspond to the chirped pulse; black lines to the non-chirped pulse. (a) Peak intensity, (b) length of the accelerating phase, and (c) pulse energy vs propagation distance. Positions (1) – (3) are the same as in Fig. 5(a). Panels (d.1) – (d.3) show electron density and test electrons (red and blue dots) with $\gamma > \gamma_g$ at the positions (1) – (3) (labeled accordingly). Grayscale is linear with a cutoff at $n_e = 3.25 \times 10^{19}$ cm^{-3}. (e.1) – (e.3) Normalized laser intensity, $|a|^2$, at the positions (1) – (3). Negative chirp prevents rapid self-compression of the pulse, slowing down bubble expansion and reducing the number of injected test electrons.

and delay the relativistic piston formation. Elongation of the bubble, vividly demonstrated in Fig. 8, would be thus reduced, and concomitant continuous injection partly suppressed.
We first verify this scenario in a WAKE simulation with test particles. We take the 30 fs, 70 TW Gaussian pulse with the parameters specified at the beginning of section 2 and introduce a linear chirp, temporally advancing higher frequencies. The center of this negatively chirped pulse corresponds to the carrier frequency, and the bandwidth corresponds to a transform-limited 5 fs duration (viz. the relativistic piston of the last section). Multi-Joule amplification systems delivering such broad-bandwidth pulses are not available yet; their development, however, is being actively pursued (Herrmann et al., 2009). As in section 2.4, we run the WAKE simulation until $z = 2.2$ mm. Figures 9 and 10 show that the negative chirp profoundly changes the pulse evolution, and, hence, the dynamics of self-injection.
Figure 9(a) indicates that the initial stage of laser evolution, corresponding to the Stage I of Fig. 5(a), remains almost unaltered. Relativistic self-focusing appears to be quite insensitive to the frequency chirp. Figure 9(b) shows that the bubble experiences one pulsation between $z = 0.6$ and 1.3 mm. The maximal bubble expansion around $z = 1$ mm appears to be 25% smaller than in the non-chirped driver case, with a proportionally smaller number of injected test particles.

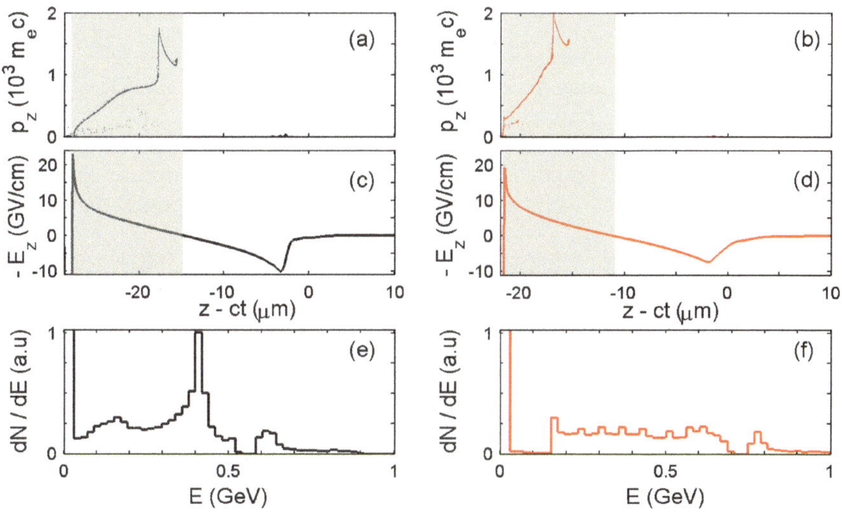

Fig. 10. Reduction of dark current using the negatively chirped driver pulse. Panels on the left correspond to the non-chirped driver, and on the right to the chirped pulse. (a,b) Phase space of test electrons showing signatures of continuous injection. (c,d) Accelerating gradient. Shaded area covers the accelerating phase ($E_z < 0$). (e,f) Energy spectrum of test electrons (number of test particles per spectrometer energy bin). The data are extracted from WAKE simulations at $z = 2.2$ mm (cf. position (3) of Fig. 9(b)). Chirp of the driver reduces expansion of the bubble (cf. panel (d) vs (c)); early dephasing is thus prevented (cf. (b) vs (a)). Injected electrons, exposed to higher accelerating gradient (cf. (d) vs (c)), gain higher energy. The amount of continuously injected charge is significantly reduced (cf. (f) vs (e)).

It is during the Stage II ($z > 1.2$ mm) that the chirp changes the pulse behavior most drastically. From Fig. 9(a), the chirped pulse stabilizes after one spot size oscillation and propagates until $z = 2.1$ mm with almost invariable intensity. As expected, the front steepening remains almost unnoticeable till the end of the run: the chirped pulse envelopes in Figs. 9(e.1) and 9(e.2) show just minimal longitudinal compression. As a result, according to Fig. 9(b), the bubble driven by the chirped pulse expands by roughly 12% between positions (1) and (3), in contrast to 35% in the non-chirped driver case. This stabilizing effect of the pulse chirp is the most vividly seen in progression from Fig. 9(d.1) to 9(d.3).

Figure 9(c) shows 22% depletion of the chirped pulse, in contrast to 33% in the non-chirped driver case. Therefore, the chirped pulse energy is transferred to the plasma wake less effectively. The chirp not only compensates for the nonlinear frequency shift and concomitant pulse self-compression (nonlinear optical effects unrelated to the pump depletion), but also reduces the pulse front etching due to the local pump depletion (Decker et al., 1996). As a result, the chirped pulse propagates with a higher group velocity. The peak of the chirped pulse in Figs. 9(e.1) – 9(e.3) is temporally advanced with respect to its non-chirped counterpart. Larger velocity of the structure extends the dephasing length and, as we shall see below, results in higher final electron energy.

Despite all promising tendencies, continuous injection is not fully shut down. As shown in Figs. 10(a) and 10(b), longitudinal phase space at the end of simulation ($z = 2.2$ mm) consists

	Q_{mono}	E_{mono}	ΔE_{mono}	$\Delta E/E_{mono}$	$\varepsilon_{N,x}$	$\varepsilon_{N,y}$	$\langle\Delta\alpha\rangle_{mono}$
Chirp	0.181	637	30	0.047	8.1π	7.4π	10.6
No chirp	0.230	515	44	0.085	8.74π	8.5π	12.9

Table 1. Parameters of the quasi-monoenergetic electron bunch (CALDER-Circ simulation, $z = 2.2$ mm). Q_{mono} is the charge in nC; E_{mono} is the central energy in MeV; ΔE_{mono} is the absolute energy spread (FWHM) in MeV; $\Delta E/E_{mono}$ is the normalized energy spread; $\varepsilon_{N,x}$ and $\varepsilon_{N,y}$ are the root-mean-square (RMS) normalized transverse emittances (in mm mrad) in and out of the laser polarization plane; $\langle\Delta\alpha\rangle_{mono}$ is the RMS divergence in mrad.

	Q_{cont}	E_{max}	$\langle\Delta\alpha\rangle_{cont}$
Chirp	0.46	500	33
No chirp	1.06	400	36

Table 2. Parameters of the low-energy background with continuous spectrum, 50 MeV $< E < E_{max}$ (CALDER-Circ simulation, $z = 2.2$ mm). Q_{cont} is the charge in nC; E_{max} is the high-energy cutoff (in MeV); $\langle\Delta\alpha\rangle_{cont}$ is the RMS divergence in mrad.

of two distinct components: the leading (quasi-monoenergetic) bunch and long continuous tail. In the non-chirped driver case (cf. Fig. 10(a)), the leading bunch has reached dephasing. In the chirped driver case, however, the electrons are still located deeply inside the accelerating phase and continue gaining energy (cf. Fig. 10(b)). Owing to very slow expansion of the bubble, these particles have been exposed to a higher gradient, gaining additional 150 MeV energy over the same acceleration distance. Comparison of Figs. 10(e) and 10(f) also shows that the dark current is much lower in the chirped driver case, on account of much slower bubble expansion during the second half of the simulation.

We quantify the effect of dark current suppression running a CALDER-Circ simulation with the same initial conditions. We find that the electron beam components – leading quasi-monoenegretic bunch and a polychromatic background – are affected by the pulse chirp differently. Data presented in Tables 1 and 2 show that negative chirp reduces the charge of the background more than twice, and noticeably improves the leading bunch quality, increasing its energy by 20% and reducing the energy spread from 8.5% to 4.7%.

Poor collimation of the continuously injected electrons, $\langle\Delta\alpha\rangle_{cont} \approx 3\langle\Delta\alpha\rangle_{mono}$, together with their distribution over the energy range 10-15 times broader than the absolute energy spread of the leading bunch, reduce dramatically the brightness of the energy tail. These poorly collimated beam components can be dispersed in vacuum using miniature magnetic quadrupole lenses, further improving the beam collimation and reducing the energy spread (Weingartner et al., 2011).

More details on electron acceleration in the wake of negatively chirped pulse are shown in Fig. 11. First, comparison of Figs. 11(a.1) and 11(a.2) indicates that the chirped driver can accelerate the leading bunch to a given energy (515 MeV in our case) in a shorter plasma, and with much weaker background. Figures 11(b) also demonstrate that using the negatively chirped driver improves the LPA efficiency, increasing the dephasing length and final electron energy without compromising beam quality. Indeed, Figs. 11(b.2) and 11(b.3) show that, in the chirped driver case, the leading bunch with 181 pC charge, energy 660 MeV and 5.8% relative energy spread remains the dominating spectral feature in the energy range $E > 100$ MeV. In contrast, according to Fig. 11(b.1), electrons accelerated in the wake of non-chirped driver are completely dominated by the dark current by this point.

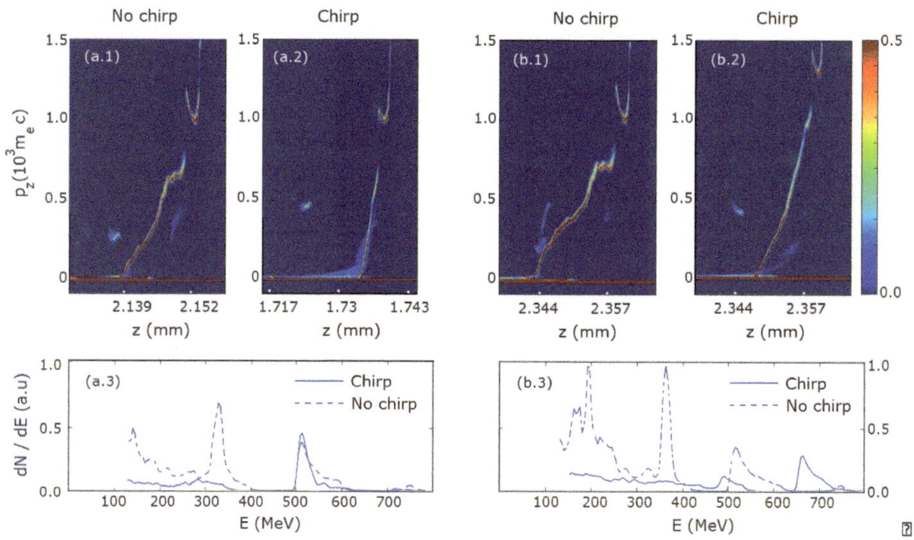

Fig. 11. Acceleration of electrons in the wake of negatively chirped pulse: suppression of dark current (CALDER-Circ simulation). (a.1), (a.2), (b.1), (b.2) Longitudinal phase space; (a.3), (b.3) electron energy spectra. Panel (a.1) – dephasing point of electrons accelerated with the *non-chirped* driver – is the same as Fig. 8(i). The leading bunch with the energy 515 MeV and relative spread 8.5% is followed by a continuous component carrying 4.6 times higher charge. (a.2), (a.3) Electrons accelerated with the chirped driver achieve 515 MeV energy earlier, and with much weaker energy tail. (b.2) Dephasing point of electrons accelerated with the *chirped* driver. Leading bunch with the energy 660 MeV and 5.8% relative spread is the dominating spectral feature in the energy range $E > 100$ MeV (cf. panels (b.2) and (b.3)). At this point, electrons accelerated with the non-chirped pulse are completely dominated by the dark current (cf. panels (b.1) and (b.3)).

In conclusion, 3-D PIC simulations successfully support the idea of suppressing the dark current using negative chirp of the driving pulse. Even though complete elimination of dark current is hard to achieve in high-density plasmas ($\gamma_g = 10 - 15$), strong reduction of the charge in the poorly collimated, continuous low-energy tail is useful for applications. Subsequent manipulations with the beam using permanent magnets may further improve its quality (Weingartner et al., 2011).

4. Conclusion

A time-varying electron density bubble created by the radiation pressure of a tightly focused laser pulse guides the pulse through a uniform, rarefied plasma, traps ambient plasma electrons and accelerates them to GeV-level energy. Natural pulse evolution (nonlinear focusing and self-compression) *is in most cases sufficient* to initiate and terminate self-injection. Bubble dynamics and the self-injection process are governed primarily by the driver evolution. Expansion of the bubble facilitates injection, whereas stabilization and contraction extinguishes injection and suppresses the low-energy background. Simultaneously, longitudinal non-uniformity of the accelerating gradient causes rapid phase space rotation.

Although beam loading reduces the accelerating gradient and slows down phase space rotation, a quasi-monoenergetic, well collimated electron bunch forms long before dephasing. At the same time, extending the acceleration length to the dephasing limit without compromising electron beam quality is not straightforward. In modern experiments, phase self-modulation and frequency red-shift due to the wake excitation cause gradual compression of the driver pulse, turning it into a relativistic piston. This process causes bubble elongation and massive continuous secondary injection (dark current). Compensation of the nonlinear frequency shift by negatively chirping the pulse is one way to delay the piston formation. As a result, nearly 60% reduction of the dark current is observed in our 3-D PIC simulations. The same set of simulations also shows that higher stability of the broad bandwidth negatively chirped pulse in plasma leads to a 20% increase in the central energy, a 50% reduction of relative energy spread, and a 20% emittance reduction of the quasi-monoenergetic, high-energy component of electron beam.
The reported results highlight the importance of reduced physics models. Reduced models not only lower the the computational cost of simulations (sometimes by many orders of magnitude), but also allow for the identification of the underlying physical processes responsible for the observed phenomena. The self-injection dynamics and its relation to the nonlinear optical evolution of the driver was understood using especially simple simulation tools (cylindrical quasistatic PIC code with fully 3-D dynamic test particle module). In practical terms, this means that the system performance (electron beam duration, mean energy, energy spread, and, very roughly, divergence) can be approximately assessed without recourse to computationally intensive 3-D PIC simulations. It appears, however, that calculation of the beam charge and transverse emittance still needs a 3-D fully kinetic simulation. Clarifying the nature of self-consistent effects affecting the phase space volume of self-injected electrons in various numerical models, and establishing the true physical origin of these effects is the subject of ongoing work.

5. Acknowledgements

S. Y. K. is grateful to S. A. Yi, V. Khudik, and M. C. Downer for many stimulating discussions and keen interest in the work presented in this Chapter. The work is partly supported by the U. S. Department of Energy grant DE-FG02-08ER55000. A. B. and E. L. acknowledge the support of LASERLAB-EUROPE/LAPTECH through EC FP7 contract no. 228334. The authors acknowledge CCRT at CEA (www-ccrt.cea.fr) for providing computing resources.

6. References

Andreev, N. E., Gorbunov, L. M., Mora,. P & Ramazashvili, R. R. (2007). Filamentation of ultrashort laser pulses propagating in tenuous plasmas, *Phys. Plasmas* 14(8): 083104.

Aoyama, M., Yamakawa, K., Akahane, Y., Ma, J., Inoue, N., Ueda, H. & Kiriyama, H. (2003). 0.85-PW, 33-fs Ti:sapphire laser, *Opt. Lett.* 28(17): 1594–1596.

Brozek-Pluska, B., Gliger, D., Hallou, A., Malka, V. & Gauduel, Y. A. (2005). Direct observation of elementary radical events: low- and high-energy radiation femtochemistry in solutions, *Radiat. Phys. Chem.* 72: 149–157.

Brunetti, E., Shanks, R. P., Manahan, G. G., Islam, M. R., Ersfeld, B., Anania, M. P., Cipiccia, S., Issac, R. C., Raj, G., Vieux, G., Welsh, G., H., Wiggins, S. M. & Jaroszynski, D. A. (2010). Low emittance, high brilliance relativistic electron beams from a laser-plasma accelerator, *Phys. Rev. Lett.* 105(21): 215007.

Buck, A., Nicolai, M., Schmid, K., Sears, C. M. S., Sävert, A., Mikhailova, J. M., Krausz, F., Kaluza, M. C. & Veisz, L. (2011). Real-time observation of laser-driven electron acceleration, *Nature Phys.* 7: 543–548.

Clayton, C. E., Ralph, J. E., Albert, F., Fonseca, R. A., Glenzer, S. H., Joshi, C., Lu, W., Marsh, K. A., Martins, S. F., Mori, W. B., Pak, A., Tsung, F. S., Pollock, B. B., Ross, J. S., Silva, L. O. & Froula, D. H. (2010). Self-guided laser wakefield acceleration beyond 1 GeV using ionization-induced injection, *Phys. Rev. Lett.* 105(10): 105003.

DesRosiers, C., Moskvin, V., Bielajew, A. F. & Papiez, L. (2000). 150-250 meV electron beams in radiation therapy, *Phys. Med. Biol.* 45(7): 1781–1805.

Dong, P., Reed, S. A., Yi, S. A., Kalmykov, S., Shvets, G., Downer, M. C., Matlis, N. H., Leemans W. P., McGuffey, C., Bulanov, S. S., Chvykov, V., Kalintchenko, G., Krushelnick, K., Maksimchuk, A., Matsuoka, T., Thomas, A. G. R. & Yanovsky, V. (2010a). Formation of optical bullets in laser-driven plasma bubble accelerators, *Phys. Rev. Lett.* 104(13): 134801.

Dong, P., Reed, S. A., Yi, S. A., Kalmykov, S., Li, Z. Y., Shvets, G., Matlis, N. H., McGuffey, C., Bulanov, S. S., Chvykov, V., Kalintchenko, G., Krushelnick, K., Maksimchuk, A., Matsuoka, T., Thomas, A. G. R., Yanovsky, V. & Downer, M. C. (2010b). Holographic visualization of laser wakefields, *New J. Phys.* 12(4): 045016.

Decker, C. D., Mori, W. B., Tzeng, K.-C. & Katsouleas, T. (1996). The evolution of ultra-intense, short-pulse lasers in underdense plasmas, *Phys. Plasmas* 3(5): 2047–2056.

Esarey, E., Schroeder, C. B. & Leemans, W. P. (2009). Physics of laser-driven plasma-based electron accelerators, *Rev. Mod. Phys.* 81(3): 1229–1285.

Fang, F., Clayton, C. E., Marsh, K. A., Pak, A. E., Ralph, J. E., Lopes, N. C. & Joshi, C. (2009). Pump depletion limited evolution of the relativistic plasma wave-front in a forced laser-wakefield accelerator, *Plasma Phys. Control. Fusion* 51: 024003.

Faure, J., Glinec, Y., Santos, J. J., Ewald, F., Rousseau, J.-P., Kiselev, S., Pukhov, A., Hosokai, T. & Malka. V. (2005). Observation of laser-pulse shortening in nonlinear plasma waves, *Phys. Rev. Lett.* 95(20): 205003.

Faure, J., Glinec, Y., Gallot, G. & Malka, V. (2006). Ultrashort laser pulses and ultrashort electron bunches generated in relativistic laser-plasma interaction, *Phys. Plasmas* 13(5): 056706.

Froula, D. H., Clayton, C. E., Döppner, T., Marsh, K. A., Barty, C. P. J., Divol, L., Fonseca, R. A., Glenzer, S. H., Joshi, C., Lu, W., Martins, S. F., Michel, P., Mori, W. B., Palastro, J. P., Pollock, B. B., Pak, A., Ralph, J. E., Ross, J. S., Siders, C. W., Silva, L. O. & Wang T. (2009). Measurements of the critical power for self-injection of electrons in a laser wakefield accelerator, *Phys. Rev. Lett.* 103(21): 215006.

Fuchs, M., Weingartner, R., Popp, A., Major, Zs., Becker, S., Osterhoff, J., Cortrie, I., Zeitler, B., Hörlein, R., Tsakiris, G. D., Schramm, U., Rowlands-Rees, T. P., Hooker, S. M., Habs, D., Krausz, F., Karsch, S. & Grüner, F. (2009). Laser-driven soft-X-ray undulator source, *Nature Phys.* 5: 826–829.

Gaul, E. W., Martinez, M., Blakeney, J., Jochmann, A., Ringuette, M., Hammond, D., Borger, T., Escamilla, R., Douglas, S., Henderson, W., Dyer, G., Erlandson, A., Cross, R., Caird, J., Ebbers, C. & Ditmire., T. (2010). Demonstration of a 1.1 petawatt laser based on a hybrid optical parametric chirped pulse amplification/mixed Nd:glass amplifier *Appl. Opt.* 49(9): 1676–1681.

Glinec, Y., Faure, J., Le Dain, L., Darbon, S., Hosokai, T., Santos, J. J., Lefebvre, E., Rousseau, J. P., Burgy, F., Mercier, B. & Malka, V. (2005). High-resolution γ-ray radiography produced by a laser-plasma driven electron source, *Phys. Rev. Lett.* 94(2): 025003.

Glinec, Y., Faure, J., Malka, V., Fuchs, T., Szymanowski, H. & Oelfke, U. (2006). Radiotherapy with laser-plasma accelerators: Monte Carlo simulation of dose deposited by an experimental quasimonoenergetic electron beam, *Med. Phys.* 33(1): 155–162.

Gorbunov, L. M., Kalmykov, S. Yu. & Mora, P. (2005). Laser wakefield acceleration by petawatt ultrashort laser pulses, *Phys. Plasmas* 12(3): 033101.

Gordienko, S. & Pukhov. A. (2005). Scalings for ultrarelativistic laser plasmas and quasimonoenergetic electrons, *Phys. Plasmas* 12(4): 043109.

Gordon, D. F., Ting, A., Helle, M. H., Kaganovich, D. & Hafizi, B. (2010). Electro-optic shocks from blowout laser wakefields, *New J. Phys.* 12(4): 045026.

Grüner, F., Becker, S., Schramm, U., Eichner, T., Fuchs, M., Weingartner, R., Habs, D., Meyer-ter-Vehn, J., Geissler, M., Ferrario, M., Serafini, L., van der Geer, B., Backe, H., Lauth, W. & Reiche, S. (2007). Design considerations for table-top, laser-based VUV and X-ray free electron lasers, *Appl. Phys. B: Lasers Opt.* 86: 431–435.

Hafizi, B., Ting, A., Sprangle, P. & Hubbard, R. F. (2000). Relativistic focusing and ponderomotive channeling of intense laser beams, *Phys. Rev. E* 62(3): 4120–4125.

Hafz, N. A. M., Jeong, T. M., Choi, I., Lee, S. K., Pae, K. H., Kulagin, V. V., Sung, J. H., Yu, T. J., Hong, K.-H., Hosokai, T., Cary, J. R., Ko, D.-K. & Lee, J. (2008). Stable generation of GeV-class electron beams from self-guided laser-plasma channels, *Nat. Photon.* 2: 571–577.

Hafz, N. A. M., Lee, S. K., Jeong, T. M. & Lee, J. (2011). Evolution of self-injected quasi-monoenergetic electron beams in a plasma bubble, *Nucl. Instrum. Methods in Phys. Res. A* 637: S51–S53.

Hartemann, F. V., Gibson, D. J., Brown, W. J., Rousse, A., Ta Phuoc, K., Malka, V., Faure, J. & Pukhov, A. (2007). Compton scattering x-ray sources driven by laser wakefield acceleration, *Phys. Rev. ST Accel. Beams* 10(1): 011301.

Hein, J., Kaluza, M. C., Bödefeld, R., Siebold, M., Podleska, S. & Sauerbrey, R. (2006). POLARIS: an all diode-pumped ultrahigh peak power laser for high repetition rates, *Lect. Notes Phys.* 694: 47–66.

Helle, M. H., Kaganovich, D., Gordon, D. F. & Ting, A. (2010). Measurement of electro-optic shock and electron acceleration in a strongly cavitated laser wakefield accelerator, *Phys. Rev. Lett.* 105(10): 105001.

Hemker, R. G., Hafz, N. M. & Uesaka, M. (2002). Computer simulations of a single-laser double-gas-jet wakefield accelerator concept, *Phys. Rev. ST Accel. Beams* 5(4): 041301.

Herrmann, D., Veisz, L., Tautz, R., Tavella, F., Schmid, K., Pervak, V. & and Krausz, F. (2009). Generation of sub-three-cycle, 16 TW light pulses by using noncollinear optical parametric chirped-pulse amplification, *Opt. Lett.* 34(16): 2459-2461.

Hidding, B., Königstein, T., Willi, O., Rosenzweig, J. B., Nakajima, K. & Pretzler, G. (2011). Laser-plasma-accelerators – a novel, versatile tool for space radiation studies, *Nucl. Instrum. Methods in Phys. Res. A* 636: 31–40.

Kaganovich, D., Gordon, D. F. & Ting, A. (2008). Observation of large-angle quasi-monoenergetic electrons from a laser wakefield, *Phys. Rev. Lett.* 100(21): 215002.

Kainz, K. K., Hogstrom, K. R., Antolak, J. A., Almond, P. R., Bloch, C. D., Chiu, C., Fomytskyi, M., Raischel, F., Downer, M. & Tajima, T. (2004). Dose properties of a laser accelerated electron beam and prospects for clinical application, *Med. Phys.* 31(7): 2053–2067.

Kalmykov, S., Yi, S. A., Khudik, V. & Shvets, G. (2009). Electron self-injection and trapping into an evolving plasma bubble, *Phys. Rev. Lett.* 103(13): 135004.

Kalmykov, S. Y., Yi, S. A., Beck, A., Lifschitz, A. F., Davoine, X., Lefebvre, E., Pukhov, A., Khudik, V., Shvets, G., Reed, S. A., Dong, P., Wang, X., Du, D., Bedacht, S.,

Zgadzaj, R., Henderson, W., Bernstein, A., Dyer, G., Martinez, M., Gaul, E., Ditmire, T. & Downer, M. C. (2010a). Numerical modelling of a 10-cm-long multi-GeV laser wakefield accelerator driven by a self-guided petawatt pulse, *New J. Phys.* 12(4): 045019.

Kalmykov, S. Y., Beck, A., Yi, S. A., Khudik, V., Shadwick, B. A., Lefebvre, E. & Downer, M. C. (2010b). Electron self-injection into an evolving plasma bubble: the way to a dark current free GeV-scale laser accelerator, *AIP Conf. Proc.* 1299: 174–179.

Kalmykov, S. Y., Yi, S. A., Beck, A., Lifschitz, A. F., Davoine, X., Lefebvre, E., Khudik, V., Shvets, G. & Downer, M. C. (2011a). Dark-current-free petawatt laser-driven wakefield accelerator based on electron self-injection into an expanding plasma bubble, *Plasma Phys. Control. Fusion* 53: 014006.

Kalmykov, S. Y., Beck, A., Yi, S. A., Khudik, V. N., Downer, M. C., Lefebvre, E., Shadwick, B. A. & Umstadter, D. P. (2011b). Electron self-injection into an evolving plasma bubble: quasi-monoenergetic laser-plasma acceleration in the blowout regime, *Phys. Plasmas* 18(5): 056704.

Kneip, S., Nagel, S. R., Martins, S. F., Mangles, S. P. D., Bellei, C., Chekhlov, O., Clarke, R. J., Delerue, N., Divall, E. J., Doucas, G., Ertel, K., Fiúza, F., Fonseca, R., Foster, P., Hawkes, S. J., Hooker, C. J., Krushelnick, K., Mori, W. B., Palmer, C. A. J., Ta Phuoc, K., Rajeev, P. P., Schreiber, J., Streeter, M. J. V., Urner, D., Vieira, J., Silva, L. O. & Najmudin Z. (2009). Near-GeV acceleration of electrons by a nonlinear plasma wave driven by a self-guided laser pulse, *Phys. Rev. Lett.* 103(3): 035002.

Kneip, S., McGuffey, C., Martins, J. L., Martins, S. F., Bellei, C., Chvykov, V., Dollar, F., Fonseca, R., Huntington, C., Kalintchenko, G., Maksimchuk, A., Mangles, S. P. D., Matsuoka, T., Nagel, S. R., Palmer, C. A. J., Schreiber, J., Ta Phuoc, K., Thomas, A. G. R., Yanovsky, V., Silva, L. O., Krushelnick K. & Najmudin Z. (2010). Bright spatially coherent synchrotron X-rays from a table-top source, *Nature Phys.* 6: 980–983.

Korzhimanov, A. V., Gonoskov, A. A., Khazanov, E. A., & Sergeev A. M. (2011). Horizons of petawatt laser technology, *Physics – Uspekhi* 54(1): 9–28.

Kostyukov, I., Nerush, E., Pukhov, A. & Seredov., V. (2009). Electron self-injection in multidimensional relativistic-plasma wake fields, *Phys. Rev. Lett.* 103(17): 175003.

Leemans, W. P., Rodgers, D., Catravas, P. E., Geddes, C. G. R., Fubiani, G., Esarey, E., Shadwick, B. A., Donahue, R. & Smith, A. (2001). Gamma-neutron activation experiments using laser wakefield accelerators, *Phys. Plasmas* 8(5): 2510–2516.

Leemans, W. P., Nagler, B., Gonsalves, A. J., Tóth, Cs., Nakamura, K., Geddes, C. G. R., Esarey, E., Schroeder, C. B. & Hooker S. M. (2006). GeV electron beams from a centimetre-scale accelerator, *Nature Phys.* 2: 696–699.

Li, Z., Zgadzaj, R., Wang, X., Reed, S., Dong, P. & Downer, M. C. (2010). Frequency-domain streak camera for ultrafast imaging of evolving light-velocity objects, *Opt. Lett.* 35(24): 4087–4089.

Li, Z., Zgadzaj, R., Wang, X., Dong, P. & Downer, M. C. (2011). Frequency-domain tomography of evolving light-velocity objects, in *Quantum Electronics and Laser Science Conference*, Baltimore, Maryland, OSA Technical Digest (CD) (Optical Society of America, 2011), paper QTuB2; *www.opticsinfobase.org/abstract.cfm?URI=QELS-2011-QTuB2*

Lifschitz, A. F., Davoine, X., Lefebvre, E., Faure, J., Rechatin, C. & Malka, V (2009). Particle-in-Cell modelling of laser-plasma interaction using Fourier decomposition, *J. Comput. Phys.* 228: 1803–1814.

Liu, J. S., Xia, C. Q., Wang, W. T., Lu, H. Y., Wang, Ch., Deng, A. H., Li, W. T., Zhang, H., Liang, X. Y., Leng, Y. X., Lu, X. M., Wang, C., Wang, J. Z., Nakajima, K., Li, R. X & Xu, Z.

Z. (2011). All-optical cascaded laser wakefield accelerator using ionization-induced injection, *Phys. Rev. Lett.* 107(3): 035001.

Lu, W., Huang, C., Zhou, M., Tzoufras, M., Tsung F. S., Mori, W. B. & Katsouleas, T. (2006). A nonlinear theory for multidimensional relativistic plasma wave wakefields, *Phys. Plasmas* 13(5): 056709.

Lu, W., Tzoufras, M., Joshi, C., Tsung, F. S., Mori, W. B., Vieira, J., Fonseca, R. A. & Silva, L. O. (2007). Generating multi-GeV electron bunches using single stage laser wakefield acceleration in a 3D nonlinear regime, *Phys. Rev. ST Accel. Beams* 10(6): 061301.

Lundh, O., Lim, J., Rechatin, C., Ammoura, L., Ben-Ismail, A., Davoine, X., Gallot, G., Goddet, J.-P., Lefebvre, E., Malka, V. & Faure, J. (2011). Few femtosecond, few kiloampere electron bunch produced by a laser-plasma accelerator, *Nature Phys.* 7: 219–222.

Maksimchuk, A., Reed, S., Naumova, N., Chvykov, V., Hou, B., Kalintchenko, G., Matsuoka, T., Nees, J., Rousseau, P., Mourou, G., Yanovsky, V. (2007). Energy scaling of quasi-monoenergetic electron beams from laser wakefields driven by 40-TW ultra-short pulses, *Appl. Phys B: Lasers Opt.* 89: 201–207.

Malka, V., Faure, J., Marquès, J. R., Amiranoff, F., Rousseau, J. P., Ranc, S., Chambaret, J. P., Najmudin, Z., Walton, B., Mora, P. & Solodov, A. (2001). Characterization of electron beams produced by ultrashort (30 fs) laser pulses, *Phys. Plasmas* 8(6): 2605–2608.

Malka, V., Faure, J., Rechatin, C., Ben-Ismail, A., Lim, J. K., Davoine, X. & Lefebvre, E. (2009). Laser-driven accelerators by colliding pulses injection: A review of simulation and experimental results, *Phys. Plasmas* 16(5): 056703.

Malka, V., Faure, J. & Gauduel, Y. A. (2010). Ultra-short electron beams based spatio-temporal radiation biology and radiotherapy, *Mutat. Res.: Rev. Mutat. Res.* 704(1-3): 145–151.

Mangles, S. P. D., Thomas, A. G. R., Lundh, O., Lindau, F., Kaluza, M. C., Persson, A., Wahlström, C.-G., Krushelnick, K. & Najmudin, Z. (2007). On the stability of laser wakefield electron accelerators in the monoenergetic regime, *Phys. Plasmas* 14(5): 056702.

Martins, S. F., Fonseca, R. A., Lu, W., Mori, W. B. & Silva L. O. (2010). Exploring laser-wakefield-accelerator regimes for near-term lasers using particle-in-cell simulation in Lorentz-boosted frames, *Nature Phys.* 6: 311–316.

Mora, P. & Antonsen Jr., T. M. (1996). Electron cavitation and acceleration in the wake of an ultraintense, self-focused laser pulse, *Phys. Rev. E* 53(3): R2068–R2071.

Mora, P. & Antonsen Jr., T. M. (1997). Kinetic modeling of intense, short laser pulses propagating in tenuous plasmas, *Phys. Plasmas* 4(1): 217–229.

Morshed, S., Antonsen, Jr., T. M. & Palastro, J. P. (2010). Efficient simulation of electron trapping in laser and plasma wakefield acceleration, *Phys. Plasmas* 17(6): 063106.

Mourou, G. A., Tajima, T. & Bulanov, S. V. (2006). Optics in the relativistic regime, *Rev. Mod. Phys.* 78(2): 309–371.

Oguchi, A., Zhidkov, A., Takano, K., Hotta, E., Nemoto, K. & Nakajima, K. (2008). Multiple self-injection in the acceleration of monoenergetic electrons by a laser wake field, *Phys. Plasmas* 15(4): 043102.

Osterhoff, J., Popp, A., Major, Zs., Marx, B., Rowlands-Rees, T. P., Fuchs, M., Geissler, M., Hörlein, R., Hidding, B., Becker, S., Peralta, E. A., Schramm, U., Grüner, F., Habs, D., Krausz, F., Hooker, S. M. & Karsch, S. (2008). Generation of stable, low-divergence electron beams by laser-wakefield acceleration in a steady-state-flow gas cell, *Phys. Rev. Lett.* 101(8): 085002.

Pai, C.-H., Chang, Y.-Y., Ha, L.-C., Xie, Z.-H., Lin, M.-W., Lin, J.-M., Chen, Y.-M., Tsaur, G., Chu, H.-H., Chen, S.-H., Lin, J.-Y., Wang, J. & Chen, S.-Y. (2010). Generation of intense ultrashort midinfrared pulses by laser-plasma interaction in the bubble regime, *Phys. Rev. A* 82(6): 063804.

Pukhov, A. & Meyer-ter-Vehn, J. (2002). Laser wake field acceleration: the highly non-linear broken-wave regime, *Appl. Phys. B: Lasers Opt.* 74: 355–361.

Pukhov, A., an der Brügge, D. & Kostyukov, I. (2010). Relativistic laser plasmas for electron acceleration and short wavelength radiation generation, *Plasma Phys. Control. Fusion* 52: 124039.

Quesnel, B. & Mora, P. (1998). Theory and simulation of the interaction of ultraintense laser pulses with electrons in vacuum, *Phys. Rev. E* 58(3): 3719–3732.

Ralph, J. E., Marsh, K. A., Pak, A. E., Lu, W., Clayton, C. E., Fang, F., Mori, W. B. & and Joshi, C. (2009). Self-guiding of ultrashort, relativistically intense laser pulses through underdense plasmas in the blowout regime, *Phys. Rev. Lett.* 102(17): 175003.

Ramanathan, V., Banerjee, S., Powers, N., Cunningham, N., Chandler-Smith, N. A., Zhao, K., Brown, K., Umstadter, D., Clarke, S., Pozzi, S., Beene, J., Vane, C. R. & Schultz, D. (2010). Submillimeter-resolution radiography of shielded structures with laser-accelerated electron beams, *Phys. Rev. ST Accel. Beams* 13(10): 104701.

Rechatin, C., Faure, J., Davoine, X., Lundh, O., Lim, J., Ben-Ismail, A., Burgy, F., Tafzi, A., Lifschitz, A. F., Lefebvre, E. & Malka, V. (2010). Characterization of the beam loading effects in a laser plasma accelerator, *New J. Phys.* 12(4): 045023.

Reed, S. A., Chvykov, V., Kalintchenko, G., Matsuoka, T., Yanovsky, V., Vane, C. R., Beene, J. R., Stracener, D., Schultz, D. R. & Maksimchuk, A. (2007). Efficient initiation of photonuclear reactions using quasimonoenergetic electron beams from laser wakefield acceleration, *J. Appl. Phys.* 102(7): 073103.

Rosenzweig, J. B., Breizman, B., Katsouleas, T. & Su, J. J. (1991). Acceleration and focusing of electrons in two-dimensional nonlinear plasma wake fields, *Phys. Rev. A* 44(10): R6189–R6192.

Ross, I. N., Collier, J. L., Matousek, P., Danson, C. N., Neely, D., Allott, R. M., Pepler, D. A., Hernandez-Gomez, C. & Osvay, K. (2000). Generation of terawatt pulses by use of optical parametric chirped pulse amplification, *Appl. Opt.* 39(15): 2422–2427.

Rousse, A., Ta Phuoc, K., Shah, R., Fitour, R. & Albert, F. (2007). Scaling of betatron X-ray radiation, *Eur. Phys. J. D* 45: 391–398.

Schlenvoigt, H.-P., Haupt, K., Debus A., Budde, F., Jäckel, O., Pfotenhauer, S., Schwoerer, H., Rohwer, E., Gallacher, J. G., Brunetti, E., Shanks, R. P., Wiggins, S. M. & Jaroszynski, D. A. (2008). A compact synchrotron radiation source driven by a laser-plasma wakefield accelerator, *Nature Phys.* 4: 130–133.

Schroeder, C. B., Esarey, E., Geddes, C. G. R., Benedetti, C. & Leemans, W. P. (2010). Physics considerations for laser-plasma linear colliders, *Phys. Rev. ST Accel. Beams* 1310: 101301.

Sun, G.-Z., Ott, E., Lee, Y. C. & Guzdar, P. (1987). Self-focusing of short intense pulses in plasmas, *Phys. Fluids* 30(2): 526–532.

Semushin, S. & Malka, V. (2001). High density gas jet nozzle design for laser target production, *Rev. Sci. Instrum.* 72(7): 2961–2965.

Spence, D. J. & Hooker, S. M. (2001). Investigation of a hydrogen plasma waveguide, *Phys. Rev. E* 63(1): 015401(R).

Spence, D. E., Kean, P. N. & Sibbett, W. (1991) 60-fsec pulse generation from a self-mode-locked Ti:sapphire laser, *Opt. Lett.* 16(1): 42–44.

Strickland, D. & Mourou, G. (1985). Compression of amplified chirped optical pulses, *Opt. Commun.* 56(3): 219–221.

Suk, H., Barov, N., Rosenzweig, J. B. & Esarey, E. (2001). Plasma electron trapping and acceleration in a plasma wake field using a density transition, *Phys. Rev. Lett.* 86(6): 1011–1014.

Sung, J. H., Lee, S. K., Yu, T. J., Jeong, T. M. & Lee, J. (2010). 0.1 Hz 1.0 PW Ti:sapphire laser, *Opt. Lett.* 35(18): 3021–3023.

Tajima, T. & Dawson, J. M. (1979). Laser electron accelerator, *Phys. Rev. Lett.* 43(4): 267–270.

Thomas, A. G. R., Mangles, S. P. D., Najmudin, Z., Kaluza, M. C., Murphy, C. D. & Krushelnick, K. (2007) Measurements of wave-breaking radiation from a laser-wakefield accelerator, *Phys. Rev. Lett.* 98(5): 054802.

Thomas, A. G. R., Najmudin, Z., Mangles, S. P. D., Murphy, C. D., Dangor, A. E., Kamperidis, C., Lancaster, K. L., Mori, W. B., Norreys, P. A., Rozmus, W. & Krushelnick, K. (2007). The effect of laser focusing conditions on propagation and monoenergetic electron production in laser wakefield accelerators, *Phys. Rev. Lett.* 98(9): 095004.

Thomas, A. G. R., Mangles, S. P. D., Murphy, C. D., Dangor, A. E., Foster, P. S., Gallacher, J. G., Jaroszynski, D. A., Kamperidis, C., Krushelnick, K., Lancaster, K. L., Norreys, P. A., Viskup, R. & Najmudin, Z. (2009). Ultrashort pulse filamentation and monoenergetic electron beam production in LWFAs, *Plasma Phys. Control. Fusion* 51: 024010.

Tsung, F. S., Lu, W., Tzoufras, M., Mori, W. B., Joshi, C., Vieira, J. M., Silva, L. O. & Fonseca, R. A. (2006). Simulation of monoenergetic electron generation via laser wakefield accelerators for 5-25 TW lasers, *Phys. Plasmas* 13(5): 056708.

Tzoufras, M., Lu, W., Tsung, F. S., Huang, C., Mori, W. B., Katsouleas, T., Vieira, J., Fonseca, R. A. & Silva, L. O. (2009). Beam loading by electrons in nonlinear plasma wake, *Phys. Plasmas* 16(5): 056705.

Vieira, J., Fiúza, F., Silva, L. O., Tzoufras, M. & Mori W. B. (2010). Onset of self-steepening of intense laser pulses in plasmas, *New J. Phys.* 12(4): 045025.

Weingartner, R., Fuchs, M., Popp, A., Raith, S., Becker, S., Chou, S., Heigoldt, M., Khrennikov, K., Wenz, J., Seggebrock, T., Zeitler, B., Major, Zs., Osterhoff, J., Krausz, F., Karsch, S. & Grüner, F. (2011). Imaging laser-wakefield-accelerated electrons using miniature magnetic quadrupole lenses, *Phys. Rev. ST Accel. Beams*, 14(5): 052801.

Wang, X., Zgadzaj, R., Henderson, W., Yi, A. S., Kalmykov, S., Khudik, V., d'Avignon, E., Dong, P., Fazel, N., Korzekwa, R., Chang, Y.-Y., Tsai, Hai-En, Dyer, G., Gaul, E., Martinez, M., Borger, T., Aymond, F., Hammond, D., Escamilla, R., Marijanovic, S., Shvets, G., Ditmire, T. & Downer, M. C. (2011). Self-injected petawatt laser-driven plasma electron acceleration in 10^{17} cm^{-3} plasma, in *Quantum Electronics and Laser Science Conference*, Baltimore, Maryland, OSA Technical Digest (CD) (Optical Society of America, 2011), paper QMJ3; *http://www.opticsinfobase.org/abstract.cfm?URI=QELS-2011-QMJ3*

Wu, H.-C., Xie, B.-S., Liu, M.-P., Hong, X.-R., Zhang, S. & Yu, M. Y. (2009). Electron trajectories and betatron oscillation in the wake bubble in laser-plasma interaction, *Phys. Plasmas* 16(7): 073108.

Xu, H., Yu, W., Lu, P., Senecha, V. K., He, F., Shen, B., Qian, L., Li, R., & Xu, Z. (2005). Electron self-injection and acceleration driven by a tightly focused intense laser beam in an underdense plasma, *Phys. Plasmas* 12(1): 013105.

5

Coherent Laser Manipulation of Ultracold Molecules

Elena Kuznetsova[1,2], Robin Côté[1] and S. F. Yelin[1,2]
[1]*Department of Physics, University of Connecticut, Storrs, Connecticut*
[2]*ITAMP, Harvard-Smithsonian Center for Astrophysics, Cambridge, Massachusetts*
United States

1. Introduction

The realization of rovibrationally stable dense samples of ultracold diatomic molecules remains one of the main stepping stones to achieve the next slate of major goals in the field of atomic and molecular physics. Though obtaining diatomic alkali molecules was seen as a logical next step following the optical cooling of atoms, many of the possible applications currently under investigation extend beyond atomic and molecular physics. For example, spectroscopy of ultracold molecules can help in testing extensions of the Standard Model via the search for a permanent electric dipole moment of the electron (1; 2), or the energy difference between enantiomers of chiral molecules (3). Various molecular transitions can be utilized to track the time dependence of fundamental constants, including the fine structure constant and the proton to electron mass ratio (4). They also open the way for cold and ultracold chemistry, where the interacting species and products are in a coherent quantum superposition state (5) and reactions can happen via quantum tunneling. Dipolar ultracold quantum gases promise to show a plethora of new phenomena due to anisotropic long-range dipole-dipole interactions (6). Dipolar molecules in optical lattices can be employed as a quantum simulator of condensed matter systems, and they are predicted to demonstrate new quantum phases such as a dipolar crystal, supersolid, checkerboard and collapse phases (7; 8). Ultracold polar molecules also represent an attractive platform for quantum computation (9). They offer a variety of long-lived states for qubit encoding, including rotational, spin and hyperfine (if electronic and nuclear spins are non-zero), Λ and Ω-doublet states (10) and scalability to a large number of qubits. Polar molecules can be easily controlled by DC electric and magnetic fields, as well as by microwave and optical fields, allowing the design of various traps (11; 12). The main appeal of polar molecules for quantum information processing, however, comes from their permanent electric dipole moment, permitting them to interact via a long-range dipole-dipole interaction. The dipole-dipole interaction offers a tool to construct two-qubit gates, required for universal quantum computation (9; 13).
Ultracold molecules in their ground vibrational state $v = 0$ (and even in specific rotational, hyperfine or Zeeman states) are required for many of these applications since they have a large permanent electric dipole moment and are stable with respect to collisions and spontaneous emission. Currently translationally ultracold (100 nK - 1 mK) molecules are produced by magneto- (14) and photo-association (15) techniques. In a typical photoassociation scheme,

a pair of colliding atoms is photoassociated into a bound electronically excited molecular state that spontaneously decays, forming molecules in the electronic ground state. In magnetoassociation, a magnetic field is adiabatically swept across a Feshbach resonance, converting two atoms from a scattering state into a bound molecular state. These techniques have most successfully been applied to form alkali dimers from ultracold alkali metal atoms (14–16). In both techniques the molecules are translationally cold, but vibrationally hot, since they are formed in high vibrational states near a dissociation limit of the electronic ground state. Therefore, once created, molecules have to be rapidly transfered to the ground rovibrational state.

Basically all successfull methods for cooling vibrational (and rotational) degrees of freedom require refined laser pulse techniques from simple STIRAP pulses to optical control. In the recent past, several methods have been proposed to reach this goal: they can be divided into non-coherent and coherent techniques. Non-coherent methods include the pump-dump technique and radiative vibrational cascade. In the pump-dump technique a pump pulse transfers population from an initial state to some intermediate vibrational level of an electronically excited state, followed by a dump pulse which brings the population from the intermediate to the ground vibrational level (17). Ultracold molecules in the ground vibrational state have been produced using this technique in a photoassociation experiment (16) with a 6% efficiency. The pump-dump technique requires pulses shorter than the excited state lifetime (large intensity is therefore needed to achieve reasonable transfer efficiency). The transfer efficiency is low, unless pulses of specific area (e.g. π pulses) are used. To increase the transfer efficiency and avoid losses due to spontaneous emission from the excited electronic state, a sequence of alternating short pump and dump pulses can be applied, each transferring a small fraction of the population to the target state (18). Pulses in this case can be weak, since each has to transfer only a small fraction of the population. Additional spectral shaping of the pulses can provide population transfer to a desirable target state, where population is coherently accumulated. In this case molecules formed in the ground vibrational state have to be removed from the interaction volume to avoid excitation by subsequent pulses, leading to increase in the duration of the transfer process. In the second non-coherent method molecules in the ground electronic state are allowed to radiatively decay from the initial high vibrational level reaching the ground $v = 0$ state after several decay steps (19). In this case many intermediate vibrational states are populated which would result in loss of molecules from a trap due to vibrationally inelastic collisions with background atoms and formed molecules. Another incoherent technique, named molecular optical pumping (20), allows to transfer molecules from high vibrational states to the ground state using a shaped laser pulse. Molecules are excited to vibrational levels of higher-energy electronic states and spontaneously decay back to the ground one. The excited state vibrational levels are chosen to have a good Franck-Condon factors with the $v = 0$ vibrational level in the ground electronic state. The laser pulse is spectrally shaped so that all frequencies allowing molecules to be excited back from the $v = 0$ state are removed, and a significant fraction of molecules accumulates in the ground vibrational state after a few excitation- spontaneous emission cycles.

The major coherent methods are adiabatic passage and coherent control techniques. The latter one utilizes spectrally shaped broadband optical pulses to transfer the molecules from an initial to the ground vibrational state with high efficiency (21). Stimulated Raman Adiabatic Passage (STIRAP) (22) has recently attracted significant interest as an efficient way to produce

deeply bound molecules, starting from Feshbach molecules (23; 24). It allows to realize high transfer efficiency and preserve the high phase-space density of an initial atomic gas. In STIRAP, the laser pulses, coupling an initial and a final state to an intermediate excited state, are applied in a counter-intuitive sequence where a pump pulse is preceeded by a Stokes pulse. During the transfer, the system stays in a "dark" state, *i.e.*, a coherent superposition of initial and final states, preventing any losses that would otherwise occur from the excited state. By adiabatically changing amplitudes of the laser pulses, the "dark" state evolves from the initial to the final state, resulting in nearly 100% transfer efficiency. A STIRAP transfer from a Feshbach to a next lower vibrational state of a ground electronic potential has been demonstrated in $^{87}Rb_2$ molecules held in an optical lattice to avoid inelastic collisions (25). Recently, heteronuclear $^{40}K^{87}Rb$ molecules have been transferred from a Feshbach to a deeply bound ($>$ 10 GHz binding energy) vibrational state using STIRAP (26). Prospects of STIRAP-based photoassociation of a thermal ensemble of ^{85}Rb atoms have recently been analyzed theoretically in (27). Finally, in breakthrough experiments at JILA and Innsbruck ultracold weakly bound KRb (23) and Cs_2 (24) Feshbach molecules have been successfully brought to their ground rovibrational state via two- and multi-photon STIRAP, respectively.
In principle, STIRAP allows lossless transfer of the population from an initial to the target state with 100 % efficiency. The main difficulty with a two-pulse STIRAP in molecules is to find an intermediate vibrational state of the excited electronic potential which has a good Franck-Condon overlap with both highly delocalized initial high vibrational state and a tightly localized ground vibrational state (28). It is particularly difficult for homonuclear atoms having the ground electronic potential scaling as $1/R^6$ and the first excited potentials as $1/R^3$ with interatomic distance R, resulting in a non-favorable potential curves' overlap. It is less an issue for heteronuclear molecules having both potentials falling off with distance according to the $1/R^6$ law, making the overlap better and enabling a one-step STIRAP (23). It was therefore proposed in (29) to transfer population in several steps down the ladder of vibrational states using a sequence of stimulated optical Raman transitions. In this case the initial and final vibrational levels of each step do not differ significantly, and it is easier to find a suitable intermediate vibrational level in the excited electronic state. In this step-wise approach, however, the population is transfered through a number of vibrational levels in the ground electronic state subject to vibrational relaxation due to inelastic collisions with a background atom or another molecule. The released kinetic energy greatly exceeds the trap depth resulting in loss of both molecules and atoms from the trap. The step-wise transfer therefore has to be faster than the vibrational relaxation time. In the next section we describe a multistate chainwise STIRAP in which molecules are brought to the ground rovibrational state through a series of intermediate states in one run, which allows to minimize collisional losses in intermediate states. In Section III we discuss direct STIRAP conversion of ultracold atoms from a scattering continuum into deeply bound molecules in the presence of a Feshbach resonance. Direct conversion without first forming Feshbach molecules allows to reduce collisional losses during formation of molecules.

2. Theory of multistate chainwise adiabatic passage in the presence of decay

In this section, we present a multistate chainwise STIRAP technique allowing an efficient transfer of a molecule from a high-lying state to the ground vibrational state which minimizes the population loss due to inelastic collisions during the transfer process (30).

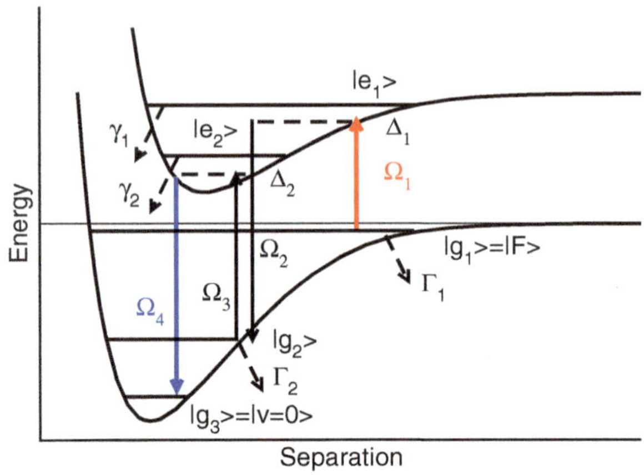

Fig. 1. Schematic showing the multistate chainwise STIRAP transfer of population from the Feshbach $|g_1\rangle$ to the ground $|g_3\rangle$ vibrational state.

We analyze a simple five-level model molecular system with states chainwise coupled by optical fields as illustrated in Fig. 1. The states $|g_1\rangle$, $|g_2\rangle$ and $|g_3\rangle$ are vibrational levels of a ground electronic molecular state, while $|e_1\rangle$ and $|e_2\rangle$ are vibrational states of an excited electronic molecular state. Molecules are formed in a high energy state $|g_1\rangle$, which in the following we assume to be a molecular Feshbach state. The state $|g_3\rangle$ is the deepest bound vibrational state $v = 0$, and $|g_2\rangle$ is an intermediate vibrational state. The goal is to efficiently transfer the population from the state $|g_1\rangle$ to state $|g_3\rangle$. At least two vibrational levels $|e_1\rangle$ and $|e_2\rangle$ in an excited electronic state are required, one which has a good Franck-Condon overlap with $|g_3\rangle$, and the other having a good overlap with the initial Feshbach molecular state $|g_1\rangle$. In the states $|e_1\rangle$ and $|e_2\rangle$, molecules decay due to spontaneous emission and collisions, and in the states $|g_1\rangle$ (for bosonic molecules) and $|g_2\rangle$ they experience fast inelastic collisions with background atoms leading to loss of molecules from a trap. It means that populating the states $|e_1\rangle$, $|e_2\rangle$ and $|g_2\rangle$ has to be avoided when a background atomic gas is present, or the transfer process has to be faster than the collisional relaxation time.

First, we analyze the system neglecting all decays. The wave function of the system is $|\Psi\rangle = \sum_i C_i \exp(-i\phi_i(t))\,|i\rangle$, where $i = g_1, e_1, g_2, e_2, g_3$; $\phi_{g_1} = 0$, $\phi_{e_1} = \nu_1 t$, $\phi_{g_2} = (\nu_2 - \nu_1)t$, $\phi_{e_2} = (\nu_3 + \nu_2 - \nu_1)t$, $\phi_{g_3} = (\nu_4 - \nu_3 + \nu_2 - \nu_1)t$; ν_i is the frequency of the ith optical field. The evolution is then governed by the Schrödinger equation

$$i\hbar \frac{\partial\,|\Psi\rangle}{\partial t} = H(t)\,|\Psi\rangle\,. \tag{1}$$

The time-dependent Hamiltonian of the system in the rotating wave approximation is given by

$$H(t) = \begin{pmatrix} 0 & -\Omega_4(t) & 0 & 0 & 0 \\ -\Omega_4(t) & \Delta_2 & -\Omega_3(t) & 0 & 0 \\ 0 & -\Omega_3(t) & 0 & -\Omega_2(t) & 0 \\ 0 & 0 & -\Omega_2(t) & \Delta_1 & -\Omega_1(t) \\ 0 & 0 & 0 & -\Omega_1(t) & 0 \end{pmatrix}, \tag{2}$$

where $\Omega_1(t) = \mu_1\mathcal{E}_1(t)/2\hbar$, $\Omega_2(t) = \mu_2\mathcal{E}_2(t)/2\hbar$, $\Omega_3(t) = \mu_3\mathcal{E}_3(t)/2\hbar$ and $\Omega_4(t) = \mu_4\mathcal{E}_4(t)/2\hbar$ are the Rabi frequencies of optical fields; $\mathcal{E}_i$ is the amplitude of ith optical field, μ_i is the dipole matrix element along the respective transition, $\Delta_1 = \omega_1 - \nu_1$ and $\Delta_2 = \omega_4 - \nu_4$ are one-photon detunings of the fields, and the ω_i are the molecular frequencies along transition i.

We assumed in Eq. (2) that pairs of fields coupling two neighboring ground state vibrational levels are in a two-photon (Raman) resonance; in this case, the system has a dark state, given by the expression

$$\left|\Phi^0\right\rangle = \frac{\Omega_2\Omega_4\left|g_1\right\rangle - \Omega_4\Omega_1\left|g_2\right\rangle + \Omega_1\Omega_3\left|g_3\right\rangle}{\sqrt{\Omega_4^2\Omega_1^2 + \Omega_1^2\Omega_3^2 + \Omega_2^2\Omega_4^2}}. \tag{3}$$

In c-STIRAP (as in classical STIRAP) the optical fields are applied in a counterintuitive way, *i.e.* at $t = -\infty$ only a combination of the Ω_4, Ω_3, Ω_2 fields, and at $t = +\infty$ only of Ω_3, Ω_2 and Ω_1 is present. As a result the dark state is initially associated with the $|g_1\rangle$ and finally with the $|g_3\rangle$ states. Adiabatically changing the Rabi frequencies of the optical fields so that the system stays in the dark state during its evolution, one can transfer the system from the initial high-lying $|g_1\rangle$ to the ground vibrational $|g_3\rangle$ state with unit efficiency, defined as the population of the $|g_3\rangle$ state at $t = +\infty$. The dark state does not have contributions from the $|e_1\rangle$ and $|e_2\rangle$ excited states, which means that they are not populated during the transfer process. As a result, the decay from these states does not affect the transfer efficiency. Decay from the $|g_1\rangle$, $|g_2\rangle$, $|g_3\rangle$ states will, however, degrade the coherent superposition (3) and result in population loss from the dark state and reduction of the transfer efficiency. In the next two subsections we consider two c-STIRAP schemes which can be used for efficient population transfer to the ground vibrational state with minimal population loss due to intermediate states decay.

2.1 c-STIRAP with two pulses

Two regimes can provide efficient population transfer to the ground state using multiple intermediate states. In the first one, called c-STIRAP, as also proposed in (31), the Stokes pulses Ω_2 and Ω_4 are applied simultaneously followed with a delay by pump pulses Ω_1 and Ω_3, applied at the same time as well. It means that, ideally, only two pulses can be used so that $\Omega_2(t) = \Omega_4(t) = \Omega_s(t)$ are Stokes pulses and $\Omega_1(t) = \Omega_3(t) = \Omega_p(t)$ are pump pulses. To simplify the analysis, we set the one-photon detunings to zero $\Delta_1 = \Delta_2 = 0$, and define an effective Rabi frequency $\Omega(t) = \sqrt{\Omega_p(t)^2 + \Omega_s(t)^2}$ and a rotation angle $\theta(t)$ by $\tan\theta = \Omega_p/\Omega_s$. In this case the Hamiltonian (2) has a zero eigenvalue describing the dark state (3). Four other eigenvalues correspond to bright states and are given in Appendix A along with a rotation matrix W converting adiabatic eigenstates into the bare ones. To study the effect of the decay from the intermediate state $|g_2\rangle$ and the initial state $|g_1\rangle$ on the dark state evolution, we turn to a density matrix description (32). The density matrix equation taking into account decay in the bare state basis is

$$i\hbar\frac{d\rho}{dt} = [H,\rho] - \mathcal{L}\rho, \tag{4}$$

where the Liouville operator $\mathcal{L}$ consists of the usual decays (see exact form in the Appendix A), where only population decays ($\propto T_1^{-1}$) into other vibrational states or the continuum are considered. Initially, all population is assumed to be in state $|g_1\rangle$.

It is convenient to use the adiabatic basis states to study the effect of decay. In the adiabatic basis the density matrix equation (4) transforms into

$$i\hbar\frac{d\rho^a}{dt} = [H^a, \rho^a] - i\hbar\left[W^T\dot{W}, \rho^a\right] - \mathcal{L}^a\rho^a \tag{5}$$

where the density matrix and the Liouville operator in this basis are given by $\rho^a = W^T\rho W$ and $\mathcal{L}^a\rho^a = W^T\mathcal{L}\rho W$, and the Hamiltonian H^a is diagonal; W is the rotation matrix, given by (A1) in the Appendix A. Initial conditions for Eq. (5) read as $\rho^a_{00} = 1$, $\rho^a_{nm} = 0$ for $n, m \neq 0$, where ρ^a_{00} denotes the dark state population. The second term on the right hand side of Eq. (5) is responsible for non-adiabatic couplings. This term results in excitation of nondiagonal density matrix elements due to coupling with the dark state. Since the nondiagonal elements ρ^a_{0i}, $i = 0, ..4$, are excited first, the second term in the r.h.s. of Eq. (5), which is $\propto \dot{\theta}\rho^a_{00}$, has to be much smaller than the first term $\sim |\varepsilon_0 - \varepsilon_i|\rho^a_{0i}$ to keep these non-diagonal elements negligible. Thus, in order to maintain adiabaticity for the system to stay in the dark state during the transfer process, we must have

$$\dot{\theta} \ll |\varepsilon_0 - \varepsilon_{1,2}|, \; |\varepsilon_0 - \varepsilon_{3,4}|. \tag{6}$$

This gives a standard STIRAP adiabaticity requirement $\Omega T_{\text{tr}} \gg 1$, where T_{tr} is the c-STIRAP transfer time.

From Eq. (5), the density matrix equation for the dark state population in terms of density matrix elements in the bare state basis can be written as

$$\begin{aligned} \dot{\rho}^a_{00}/\xi &= -\Gamma_1\cos^4\theta\rho_{g_1g_1} - \Gamma_2\sin^2\theta\cos^2\theta\rho_{g_2g_2} + \\ &+ \frac{1}{2}\Big((\Gamma_2+\Gamma_1)\sin\theta\cos^3\theta\rho_{g_1g_2} + \Gamma_2\cos\theta\sin^3\theta\rho_{g_2g_3} - \Gamma_1\sin^2\theta\cos^2\theta\rho_{g_1g_3} + c.c.\Big) \end{aligned} \tag{7}$$

with $\xi = (1 - \frac{1}{4}\sin^2 2\theta)^{-1}$. Provided that the transfer is adiabatic, so that the system stays in the dark state during the evolution: $\rho^a_{00} \approx 1$, the density matrix elements in the bare state basis appearing in (7) are expressed via the dark state population in the following way:

$$\begin{aligned} \rho_{g_1g_1} &\approx \rho^a_{00}\,\xi\,\cos^4\theta, \\ \rho_{g_2g_2} &\approx \rho^a_{00}\,\xi\,\sin^2\theta\cos^2\theta, \\ \text{Re}\left(\rho_{g_1g_2}\right) = \text{Re}\left(\rho_{g_2g_1}\right) &\approx -\rho^a_{00}\,\xi\,\sin\theta\cos^3\theta, \\ \text{Re}\left(\rho_{g_2g_3}\right) = \text{Re}\left(\rho_{g_3g_2}\right) &\approx -\rho^a_{00}\,\xi\,\cos\theta\sin^3\theta, \\ \text{Re}\left(\rho_{g_1g_3}\right) = \text{Re}\left(\rho_{g_3g_1}\right) &\approx \rho^a_{00}\,\xi\,\sin^2\theta\cos^2\theta. \end{aligned}$$

Inserting these back into Eq. (7), one obtains the equation describing the dark state population decay

$$\dot{\rho}^a_{00} \approx -\left(\Gamma_1\cos^4\theta + \Gamma_2\sin^2\theta\cos^2\theta\right)\xi\rho^a_{00}. \tag{8}$$

As expected, the dark state is not affected by the decay from the excited states $|e_{1,2}\rangle$, but only by the decay from the states $|g_{1,2}\rangle$, which destroys the coherent superposition the dark state is based on.
Taking Gaussian laser pulses with the pump and Stokes pulse Rabi frequencies $\Omega_p = \Omega_0 \exp\left(-(t-\tau/2)/T^2\right)$ and $\Omega_S = \Omega_0 \exp\left(-(t+\tau/2)/T^2\right)$ with $\tan\theta = \exp\left(2t\tau/T^2\right)$, one can see that at the moment $t = 0$ of maximal interaction the rotation angle is $\theta = \pi/4$. The density matrix equation (8) then takes a form

$$\dot{\rho}^a_{00} = -\rho^a_{00}(\Gamma_1 + \Gamma_2)/3. \tag{9}$$

Equation (9) shows that in this regime the decay of both the initial and intermediate vibrational states is not suppressed, *i.e.* directly influences the dark state evolution. This scheme therefore can be applied only if $\Gamma_{1,2}T_{\mathrm{tr}} \ll 1$. It implies that c-STIRAP pulses have to be shorter than the collisional relaxation time. This requirement combined with the adiabaticity condition might result in large Rabi frequencies needed for high transfer effciency. However, as will be shown in the next section, using reasonable values of Rabi frequencies (~ 10 MHz) and pulse durations, rather high transfer efficiencies of the order of 90% and 85% for fermionic and bosonic alkali dimers, respectively, can be realized in this scheme in the presence of collisions. High transfer efficiency with weaker and longer pulses is possible in this scheme if the remaining atoms are removed after the molecules are formed. It can be done by applying a "blast" laser pulse resonant with an atomic, but not a molecular electronic transition. In this case $\Gamma_{1,2}$ are determined by vibrationally inelastic molecule-molecule collisions, which result in slower decay compared to atom-molecule collisions due to a typically smaller density of molecules. Although, it is worth noting that in this case a collision results in two molecules lost from a trap. Another situation when this regime gives high transfer efficiency using weaker pulses can be used is when molecules are formed in an optical lattice with initially two atoms per site. In this case $\Gamma_{1,2}$ will be determined by off-resonant Raman scattering of lattice and STIRAP laser fields (25), which can be sufficiently suppressed.

2.2 Chainwise "straddling" STIRAP

As we discussed in the previous subsection, if the remaining atoms are not removed from the trap, the molecules in the $|g_{1,2}\rangle$ states are subject to vibrationally inelastic atom-molecule collisions. To maximaize the number of molecules transferred to the ground $|g_3\rangle$ state the population of the decaying intermediate state $|g_2\rangle$ has to be minimized. This can be achieved using an extension of the STIRAP technique to multiple chainwise-coupled states, called a "straddling" STIRAP (33). Namely, if $\Omega_2, \Omega_3 \gg \Omega_1, \Omega_4$ and Ω_2, Ω_3 temporally overlap both the Ω_1 and Ω_4 pulses, populations in all intermediate states are minimized. To simplify the analysis, we assume $\Omega_2 = \Omega_3 = \Omega_0$; Ω_0 is independent of time (in practice the corresponding pulses just have to be much longer than $\Omega_1(t)$, $\Omega_4(t)$ and overlap both of them), and $\Omega_0 \gg |\Omega_1|, |\Omega_4|$. As in the previous subsection, we set one-photon detunings to zero $\Delta_1 = \Delta_2 = 0$, and define the effective Rabi frequency $\Omega(t) = \sqrt{\Omega_1(t)^2 + \Omega_4(t)^2}$ and a rotation angle by $\tan\theta(t) = \Omega_1(t)/\Omega_4(t)$. The eigenvalues of the system (2) are $\varepsilon_0 = 0$, corresponding to the dark state, and $\varepsilon_{1,2} = \pm\Omega/\sqrt{2}$, and $\varepsilon_{3,4} = \pm\sqrt{2}\Omega_0$ to bright states. Adiabatic eigenstates $|\Phi\rangle = \{|\Phi^n\rangle\}$, $n = 0, \ldots 4$ and the bare states $|\Psi\rangle = \left\{\left|\Psi^l\right\rangle\right\}$, $l = g_1, e_1, g_2, e_2, g_3$ are transformed via a rotation matrix (A2), given in the Appendix A.

The adiabaticity condition (6) in this case becomes $\dot{\theta} \ll \Omega,\ \Omega_0$. If the condition $\dot{\theta} \ll \Omega$ is satisfied, the dark state will not couple to the $|\Phi^{1,2}\rangle$ states. Coupling to the $|\Phi^{3,4}\rangle$ states will be suppressed even more strongly, since $\Omega \ll \Omega_0$.
The density matrix equation for the dark state population in terms of density matrix elements in the bare state basis in this case is given by

$$\begin{aligned}\dot{\rho}^a_{00} &= \Gamma_1 \cos^2\theta \rho_{g_1g_1} - \Gamma_2 \left(\frac{\Omega \sin 2\theta}{2\Omega_0}\right)^2 \rho_{g_2g_2} + \\ &+ \left(\Gamma_2 \frac{\Omega}{4\Omega_0} \sin 2\theta \sin\theta \rho_{g_2g_3} + \frac{\Gamma_1}{2} \sin\theta\cos\theta\rho_{g_1g_3} + \right. \\ &\left. + (\Gamma_1+\Gamma_2)\frac{\Omega}{2\Omega_0} \sin 2\theta \cos\theta \rho_{g_1g_2} + c.c.\right) . \end{aligned} \tag{10}$$

For adiabatic evolution $\rho^a_{00} \approx 1$, and the density matrix elements in the bare state basis are expressed via the dark state population as:

$$\begin{aligned} \rho_{g_2g_2} &\approx \sin^2 2\theta \frac{\Omega^2}{4\Omega_0^2}\rho^a_{00}, \\ \mathrm{Re}\left(\rho_{g_1g_2}\right) = \mathrm{Re}\left(\rho_{g_2g_1}\right) &\approx -\frac{\Omega}{2\Omega_0} \sin 2\theta \cos\theta \rho^a_{00}, \\ \mathrm{Re}\left(\rho_{g_2g_3}\right) = \mathrm{Re}\left(\rho_{g_3g_1}\right) &\approx -\frac{\Omega}{2\Omega_0} \sin 2\theta \sin\theta \rho^a_{00}, \\ \mathrm{Re}\left(\rho_{g_1g_3}\right) = \mathrm{Re}\left(\rho_{g_3g_1}\right) &\approx \sin 2\theta \rho^a_{00}/2. \end{aligned}$$

The decay of the dark state due to the population loss from the $|g_1\rangle$ and $|g_2\rangle$ states is then described by the equation (keeping only terms up to the Ω^2/Ω_0^2 order)

$$\dot{\rho}^a_{00} \approx -\left((\Gamma_2 + \Gamma_1 \cos^2\theta)\left(\frac{\Omega}{2\Omega_0}\sin 2\theta\right)^2 + \Gamma_1\cos^2\theta\right)\rho^a_{00} . \tag{11}$$

Equation (11) shows that intermediate state decay can be neglected during the transfer time T_{tr} if $(\Gamma_1+\Gamma_2)T_{tr}\,(\sin 2\theta\Omega/2\Omega_0)^2 \ll 1$. From this expression one can see that the intermediate state decay rate is reduced by a factor $(\Omega/\Omega_0)^2 \ll 1$ in this regime. It also follows from Eq. (11) that decay from the initial state $|g_1\rangle$ is not suppressed, so that the transfer process has to be faster than this decay.
Both schemes can be readily extended to a general N-state chainwise-linked system with odd number of states having $(N+1)/2$ ground and $(N-1)/2$ excited levels (33; 34). In the two-pulse STIRAP scheme, considered in the previous subsection, all Stokes pulses are applied simultaneously followed with a delay by pump pulses applied at the same time as well. In the second regime counterintuitively ordered pump and Stokes pulses drive the first and the last transitions in the chain, while intermediate states are coupled by strong CW (or pulsed with durations longer than that of the pump and Stokes pulses) fields. In the

second regime analyzed in this subsection the population of all intermediate states is strongly suppressed, and the transfer efficiency close to unity can be realized.
The major prerequisite for high transfer efficiency in STIRAP is the two-photon resonance between fields coupling vibrational levels in the ground electronic state via Raman transitions. It requires all fields to be phase coherent. In a general case the frequency difference between any fields in the chain can be in the THz range. To maintain phase coherence at these large frequency differences the fields can then be phase locked to an optical frequency comb (35).

2.3 Collisional relaxation rates for fermionic and bosonic alkali dimers

Magneto- and photo-association techniques produce molecules mostly from ultracold Bose, two-spin component Fermi and mixture alkali metal atomic gases. Weakly bound Feshbach molecules rapidly decay due to vibrationally inelastic atom-molecule collisions, which were found to be the major molecule lifetime limiting factor in atomic traps. Depending on the quantum statistics due to the nuclear spin of the constituent atoms, the alkali dimers show different behavior with respect to inelastic atom-molecule and molecule-molecule collisions. Fermionic alkali dimers in the Feshbach state are very stable with respect to collisions. They are particularly stable close to the resonance, where the scattering length is large. The stability of the fermionic molecules has been explained based on the Pauli exclusion principle combined with significantly different length scales associated with the initial and final vibrational states (36). Lifetimes of the Feshbach molecular states of the order of 1 s have been observed experimentally for $^6Li_2$ fermionic molecules (37; 38), giving $\Gamma_1 \sim 1$ s^{-1}, and of the order of 100 ms for $^{40}K_2$ molecules (39), giving $\Gamma_1 \sim 10$ s^{-1}. More deeply bound Feshbach molecules have larger decay rates, with the corresponding collision coefficient $k_{\rm inel} \sim 10^{-11}$ cm^{3}s^{-1}. With typical densities of atoms in traps $n_{\rm at} \sim 10^{11} - 10^{14}$ cm^{-3}, it gives $\Gamma_1 = k_{\rm inel} n_{\rm at} \sim (1 - 10^3)$ s^{-1} for these molecules. To calculate the intermediate state decay rate Γ_2 we use results of a theoretical analysis of collisional stability of low-lying vibrational states of fermionic and bosonic Li_2 molecules (40). In low vibrational states, as was shown, fermionic molecules experience fast vibrational quenching due to collisions with surrounding atoms, leading to loss of both molecules and atoms from a trap. The inelastic atom-molecule collision coefficient for fermionic molecules in these low vibrational states is of the order of $k_{\rm inel} \sim 3 \cdot 10^{-10}$ cm^{3}s^{-1} (calculated in Ref.(40) for fermionic $^6Li_2$ in $v = 1$). The vibrational relaxation rate Γ_2 can then be estimated using $n_{\rm at} \sim 10^{11} - 10^{14}$ cm^{-3}, giving $\Gamma_2 = k_{\rm inel} n_{\rm at} \sim 3 \cdot (10^1 - 10^4)$ s^{-1}.
In contrast to fermionic alkali dimers, bosonic dimers experience fast vibrational quenching due to inelastic atom-molecule collisions, even in their Feshbach state. As was observed experimentally for ^{23}Na (41) and ^{133}Cs (42), the inelastic collision coefficient for bosonic molecules due to atom-molecule collisions is of the order of $k_{\rm inel} \sim 5 \cdot 10^{-11}$ cm^{3}s^{-1} for the Feshbach state. An inelastic atom-molecule collision coefficient $k_{\rm inel} \sim 10^{-10}$ cm^{3}s^{-1} and an elastic collision coefficient of the same order have been theoretically predicted for $^{87}Rb_2$ molecules for magetic fields below the Feshbach 1007.4 G resonance (43). Fast vibrationally inelastic atom-molecule collisions thus limit the lifetime of the molecules in the trap to 100 μs - 1 ms. They also limit the atom-to-molecule conversion efficiency during the magnetic field ramp across the resonance. The lifetime of the bosonic molecules in the trap can be significantly extended if at the end of the magnetic field ramp a "blast" laser pulse is applied, selectively removing atoms from the trap (41). In this case, the main loss mechanism is vibrationally inelastic molecule-molecule collisions. The corresponding collision coefficient

Two-pulse c-STIRAP		"Straddling" STIRAP	
$1 \ll$	$\Omega T_{\rm tr}$	$1 \ll$	$\Omega T_{\rm tr} \ll \frac{\Omega_0^2}{\Omega\Gamma_2}$
$\Omega \sim$	$E_{\rm Fesh}$	$\Omega \sim$	$E_{\rm Fesh}$
$1 \gg$	$\Gamma_{1,2} T_{\rm tr}$	$1 \gg$	$\Gamma_1 T_{\rm tr}$

Table 1

in the Feshbach state was measured for a bosonic $^{23}Na_2$ molecule as $k_{\rm inel} \sim 5.1 \cdot 10^{-11}$ cm^{3}s^{-1} (41). In this experiment, the initial atomic density and the atom-to-molecule conversion efficiency were $n_{\rm at} \sim 1.7 \cdot 10^{14}$ cm^{-3} and 4%, respectively, giving the molecular density of $n_{\rm mol} \sim 6 \cdot 10^{12}$ cm^{-3} and therefore the decay rate $\Gamma_1 = k_{\rm inel} n_{\rm mol} \sim 300$ s^{-1}. The vibrational relaxation rate Γ_2 of intermediate vibrational states for bosonic molecules can be estimated from the inelastic atom-molecule collision coefficient in low vibrational states $k_{\rm inel} \sim 6 \cdot 10^{-10}$ cm^{3}s^{-1} (calculated in Ref.(40) for bosonic ^{7}Li in $v = 1$). For a typical density of atoms in a trap, $n_{\rm at} \sim 10^{11} - 10^{14}$ cm^{-3} and the resulting relaxation rate is $\Gamma_2 \sim 6 \cdot (10^1 - 10^4)$ s^{-1}.
The heteronuclear molecules are formed from a mixture of Bose and Fermi atomic gases, and their collisional properties are expected to differ from pure fermionic and bosonic molecules discussed above. Stability of the KRb molecules with respect to collisions with initial fermionic and bosonic atoms has been recently studied in (44). Vibrationally inelastic relaxation was found to be dominated by atom-molecule collisions, and the corresponding collision coefficients to strongly depend on the quantum statistics of the atoms. Close to the heteronuclear Feshbach resonance the collision coefficient for collisions with indistinguishable fermions (^{40}K in the same hyperfine state) was found to be $k_{\rm inel} < 10^{-11}$ cm^{3}s^{-1}; for collisions with indistinguishable bosons (^{87}Rb in the same hyperfine state) $k_{\rm inel} \sim (2-3) \cdot 10^{-10}$ cm^{3}s^{-1} close to the resonance. Finally, for collisions with distinduishable atoms (^{40}K in a different hyperfine state) the collision coeffcient $k_{inel} \sim (3-5) \cdot 10^{-11}$ cm^{3}s^{-1} was measured. These results are therefore consistent with coefficients of vibrationally inelastic collisions with pure fermionic and bosonic atoms, considered in previous paragraphs.
Let us now estimate the parameters of the optical pulses providing maximal transfer efficiency. As follows from Eqs. (3) and Eq. (11), in both STIRAP schemes the decay of the Feshbach state strongly affects the transfer efficiency, and the condition $\Gamma_1 T_{\rm tr} \ll 1$ has to be satisfied to minimize molecular loss. At the same time the adiabaticity condition requires $\Omega T_{\rm tr} \gg 1$. Weakly bound Feshbach states are very close to a dissociation threshold, a typical binding energy $E_{\rm Fesh}$ being tens kHz -tens MHz (~ 1 μK - 1 mK). At these small binding energies the first pulse in the chain (Ω_p or Ω_1 in the two STIRAP schemes), coupling the Feshbach state $|g_1\rangle$ to the first excited state $|e_1\rangle$ can lead to back-action, *i.e.* back transfer of the molecule to the scattering contunuum, thus molecular dissociation, via a stimulated Raman process. This effect is minimized if the binding energy of a molecule $E_{\rm Fesh}$ is much larger than the effective Rabi frequency corresponding to the coupling between the $|e_1\rangle$ state and the scattering continuum. Typically, the dipole moment of the bound-bound transition greatly exceeds the dipole moment of the bound-continuum transition. Therefore, the Rabi frequency of the $\mathcal{E}_1$ field between the bound-bound transition $|g\rangle_1 - |e\rangle_1$ will be much larger than the effective Rabi frequency for the same field corresponding to coupling between the $|e_1\rangle$ and the scattering continuum. It means that choosing $\Omega \sim E_{\rm Fesh}$ one can make the back transfer process negligible. Finally, we have the following requirements for the Rabi frequencies and durations of the STIRAP pulses (amplitudes and durations are assumed the same for the Stokes and the pump pulse to maximize the transfer efficiency).

The set of requirements in the table 1 allows one to obtain a general range of Rabi frequencies and pulse durations, providing optimal population transfer. Since the goal is the production of dense molecular gases with a large number of molecules, we consider a high initial atomic density $n_{at} \sim 10^{14}$ cm^{-3} in a trap in our estimates. For bosonic molecules the inelastic atom-molecule collision coefficient differs at resonance and for deeper bound Feshbach molecules by about a factor of two, giving $\Gamma_1 \sim 5 \cdot 10^3 - 10^4$ s^{-1} at this density. One can see that the c-STIRAP transfer time has be $T_{tr} \ll 10^{-4}$ s. Adiabaticity typically requires $\Omega T_{tr} \sim 10^2$, giving a lower limit on the Rabi frequencies of the c-STIRAP pulses $\Omega \gg 10^6$ s^{-1}. As our analysis shows transfer efficiencies $> 90\%$ can be achieved with pulse durations of several μs and Rabi frequencies of $5 - 10$ MHz. It means that to minimize dissociation of molecules due to the back transfer deeper bound Feshbach molecules with $E_{Fesh} > 1$ MHz are prefered. For deeper bound fermionic molecules the decay rate $\Gamma_1 \sim 10^3$ s^{-1} at this high atomic density, resulting in the same pulse durations and Rabi frequencies, maximizing the transfer efficiency, as for bosonic molecules. At resonance, fermionic molecules have significantly smaller decay rates $\Gamma_1 \sim 1 - 10$ s^{-1}. In this case high transfer efficiencies $> 90\%$ can be realized with longer and weaker pulses of hundred μs duration and Rabi frequencies ~ 1 MHz.

Next we illustrate the technique with numerical simulations of a transfer process based on the system (4) for model seven-state bosonic and fermionic molecular systems. A seven-state system shown in Fig. 2a is easier to realize experimentally, *i.e.* to find transitions with good Franck-Condon factors, than the five-level scheme, analyzed in the previous subsection. For example, a seven-state chainwise path from the Feshbach to the ground vibrational state was found in $^{87}Rb_2$. The Feshbach state is experimentally formed after a magnetic field crosses the Feshbach resonance at 1007.4 G. Far from the resonance at 973 G the Feshbach state binding energy is 24 MHz (25). In the first step it can be coupled to an electronically excited pure long range molecular state $\left|0_g^-, v, J = 0\right\rangle$, located close to the $5S_{1/2} + 5P_{3/2}$ dissociation asymptote (45). For example, $v = 31$ vibrational level can be chosen located 6.87 cm^{-1} (≈ 206 GHz) below the asymptote. The corresponding Rabi frequency scales with the field intensity as $\Omega_1 = 2\pi \times 2.9\sqrt{I(W/cm^2)}$ s^{-1}, giving the transition dipole moment $\mu_1 \sim 0.3$ D. The second transition of the STIRAP scheme in (25) was to the second-to-last bound vibrational state, located 637 MHz below the ground electronic state dissociation asymptote. The corresponding Rabi frequency scales as $\Omega_2 = 2\pi \times 6\sqrt{I(W/cm^2)}$ s^{-1}, giving the transition dipole moment $\mu_2 \sim 0.6$ D. The authors mention that the Franck-Condon factors from the excited $\left|0_g^-, v = 31, J = 0\right\rangle$ state to the ground state vibrational levels down to the $X\,^1\Sigma_g^+(v = 116)$ are similar to the second-to-last vibrational state. This includes the $X\,^1\Sigma_g^+(v = 119)$ from where the ground vibrational state can be reached in five steps (29):

$$
\begin{aligned}
X\,^1\Sigma_g^+(v = 119, J = 0) &\rightarrow A^1\Sigma_u^+(v' = 185, J = 1),\\
A\,^1\Sigma_u^+(v' = 185, J = 1) &\rightarrow X^1\Sigma_g^+(v = 52, J = 0),\\
X\,^1\Sigma_g^+(v = 52, J = 0) &\rightarrow A^1\Sigma_u^+(v' = 24, J = 1),\\
A\,^1\Sigma_u^+(v' = 24, J = 1) &\rightarrow X^1\Sigma_u^+(v = 0, J = 0).
\end{aligned}
$$

The results of the numerical solution of the density matrix equation (4) for a fermionic molecular system are given in Fig. 2. The left column presents the maximal transfer efficiency case for the the two-pulse c-STIRAP scheme, and the right column for the "straddling" STIRAP scheme. In the two-pulse c-STIRAP scheme, the states $|g_1\rangle - |e_1\rangle$; $|g_2\rangle - |e_2\rangle$; $|g_3\rangle - |e_3\rangle$ are coupled by the pump field $\Omega_p = \Omega_p^{\max}(1 + \tanh{(t - \tau/2)/T})/2$, while the states $|e_1\rangle - |g_2\rangle$; $|e_2\rangle - |g_3\rangle$; $|e_3\rangle - |g_4\rangle$ are coupled by the Stokes field $\Omega_S = \Omega_S^{\max}(1 - \tanh{(t + \tau/2)/T})/2$. In the "straddling" STIRAP scheme the states $|e_1\rangle - |g_2\rangle$; $|g_2\rangle - |e_2\rangle$, $|e_2\rangle - |g_3\rangle$; and $|g_3\rangle - |e_3\rangle$ are coupled by CW laser fields with a Rabi frequency Ω_0, the first transition $|g_1\rangle - |e_1\rangle$ and the last transition $|e_3\rangle - |g_4\rangle$ in the chain are coupled by the fields $\Omega_1 = \Omega_1^{\max}(1 + \tanh{(t - \tau/2)/T})/2$ and $\Omega_4 = \Omega_4^{\max}(1 - \tanh{(t + \tau/2)/T})/2$, respectively. The transfer efficiency does not strongly depend on the form of optical pulses in this case, the same transfer efficiency was obtained using Gaussian pulses. In the numerical analysis a deeper bound Feshbach state with $E_{\mathrm{Fesh}} \sim$ tens MHz was assumed, which was the case in the $^{87}Rb_2$ STIRAP experiment (25). The Rabi frequencies of STIRAP fields were chosen to satisfy the second condition of the 1, and the pulse duration T and delay τ were varied to find the maximal transfer efficiency. To estimate the decay rate of intermediate vibrational states, the atomic density $n_{\mathrm{at}} \sim 10^{14}$ cm^{-3} was used along with the inelastic collision coefficient $k_{\mathrm{inel}} \sim 3 \cdot 10^{-10}$ cm^{3}s^{-1}, giving $\Gamma_{2,3} = 3 \cdot 10^4$ s^{-1}. Lifetimes of vibrational states of an excited electronic state of the order of 30 ns ($\gamma_{1,2,3} = 3 \cdot 10^7$ s^{-1}) were also assumed.

The numerical analysis demonstrates that $> 90\%$ of population in the case of two-pulse c-STIRAP (see Fig. 2d) and $> 96\%$ in the case of "straddling" STIRAP (see Fig. 2e) can be transfered from the Feshbach to the ground vibrational state for the chosen Rabi frequencies using the two-pulse and "straddling" STIRAP schemes, respectively, even in the presence of fast collisional decay of the Feshbach state. This transfer efficiency is realized using STIRAP pulses much shorter than the Feshbach state lifetime. Thus the influence of the decay of this state is significantly reduced.

Results of a similar analysis for a model seven-level bosonic molecular system are shown in Fig. 3. Transfer efficiency of the order of 85% and 92% can be realized with the two-pulse and "straddling" STIRAP schemes, respectively. In this case the form of the STIRAP pulses plays an important role due to the fast decay of the Feshbach state. Using Gaussian pulses instead of tanh pulses results in significantly smaller transfer efficiency since by the time the Stokes pulse arrives the Feshbach state experiences noticable decay. With tanh pulses it is, however, possible to make the delay time between the moment of molecule formation and the start of the transfer process reasonably small to minimize the Feshbach state decay.

We can now estimate intensities of CW and pulsed fields corresponding to Rabi frequencies used in our calculations. Typical dipole moments of electric dipole-allowed transitions between molecular electronic states are of the order of 1 D (Debye) and larger. Assuming that the chosen transitions have reasonably large Franck-Condon factors, we use an estimate of transition dipole moments between vibrational levels in the ground and excited electronic state $D_{v,v'} \sim 1$ D. Taking the peak Rabi frequency of the pump and Stokes fields $\Omega_{1,4}^{\max}{}^{max} = 3 \cdot 10^7$ s^{-1}, the corresponding intensity is $I_{1,4}^{\mathrm{peak}} = c\mathcal{E}_{1,4}^2/4\pi = c(\Omega_{1,4}^{\max}\hbar/D_{v,v'})^2/4\pi \sim 0.2$ W/cm^2; for CW fields with a Rabi frequency $\Omega_0 = 6 \cdot 10^7$ s^{-1} the corresponding intensity is $I_{2,3} \sim 0.9$ W/cm^2.

To conclude this section, we analyzed a method to coherently transfer ultracold molecules formed in high-lying vibrational states to the ground vibrational state, based on the multilevel chainwise STIRAP technique. Molecules are transfered from a high vibrational state into a ground rovibrational state $v = 0, J = 0$ using Raman transitions via several intermediate vibrational states in the ground electronic state. The former one has lower transfer efficiency

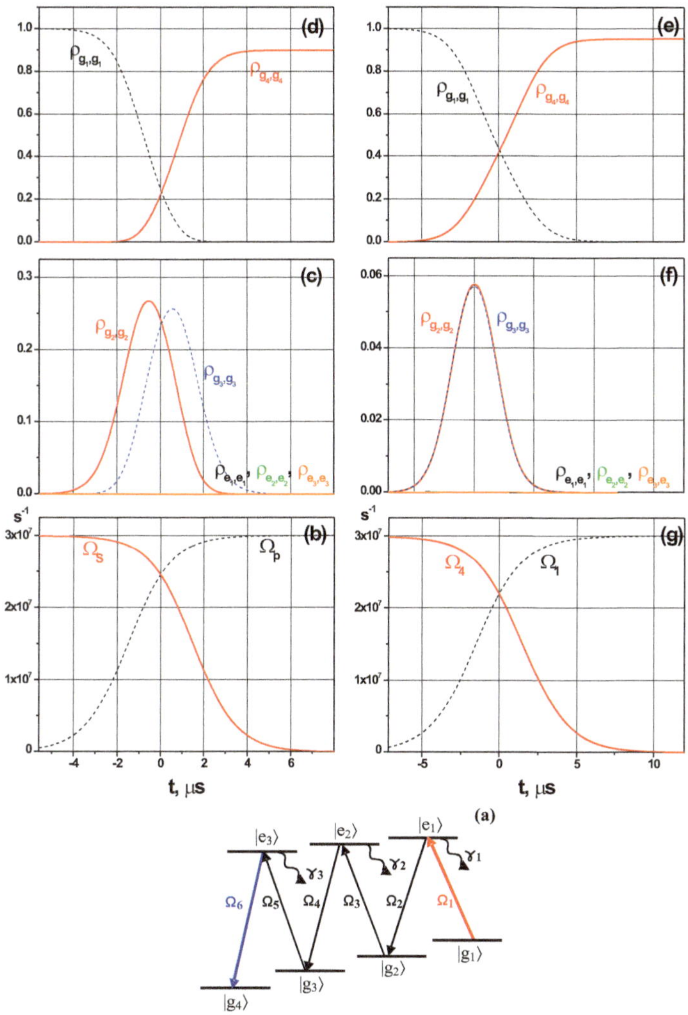

Fig. 2. Results of numerical solution of Eq. (4) for a seven-state model fermionic molecular system, shown in (a). Left colunm (figures (b),(c),(d)) and right column (figures ((e),(f),(g))) present results for a two-pulse and "straddling" STIRAP schemes, respectively. Parameters used: $\Gamma_1 = 10^3\ s^{-1}, \Gamma_2 = \Gamma_3 = 3 \cdot 10^4\ s^{-1}, \gamma_1 = \gamma_2 = \gamma_3 = 3 \cdot 10^7\ s^{-1}$; for a two-pulse STIRAP $\Omega_p^{max} = \Omega_S^{max} = 3 \cdot 10^7\ s^{-1}$, $T = 2\ \mu s$, $\tau = -3\ \mu s$; for "straddling" STIRAP $\Omega_1^{max} = \Omega_4^{max} = 6 \cdot 10^6\ s^{-1}, \Omega_0 = 6 \cdot 10^7\ s^{-1}, T = 3\ \mu s, \tau = -3\ \mu s$.

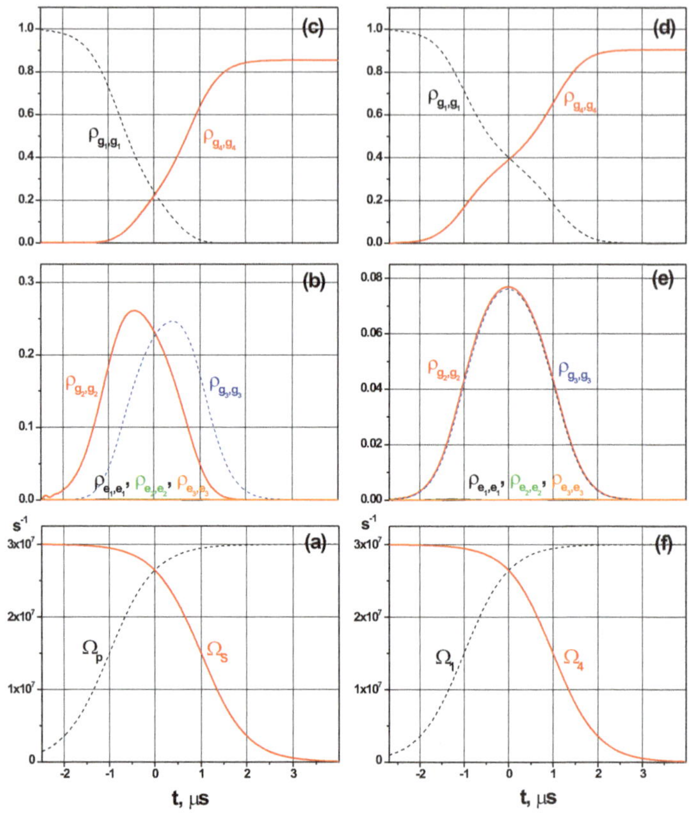

Fig. 3. Results of numerical solution of Eq. (4) for a model seven-state bosonic molecular system. Parameters used: $\Gamma_1 = 10^4$ s^{-1}, $\Gamma_2 = \Gamma_3 = 6 \cdot 10^4$ s^{-1}, $\gamma_1 = \gamma_2 = \gamma_3 = 3 \cdot 10^7$ s^{-1}; for a two-pulse STIRAP $\Omega_p^{max} = \Omega_S^{max} = 3 \cdot 10^7$ s^{-1}; for "straddling" STIRAP $\Omega_1^{max} = \Omega_4^{max} = 3 \cdot 10^7$ s^{-1}, $\Omega_0 = 6 \cdot 10^7$ s^{-1}, $T = 1$ μs, $\tau = -2$ μs for both schemes.

for both fermionic and bosonic molecules compared to the "straddling" scheme, but is experimentally simpler, since it can be realized with two pulses, while the "straddling" scheme requires at least three optical fields (two pulsed and one CW). Numerical analysis of the transfer process for a typical bosonic and fermionic molecular system in a trap with an atomic density $n_{at} \sim 10^{14}$ cm^{-3} shows that transfer efficiencies $\sim$ 92% and $\sim$ 96% respectively are possible even in the presence of fast collisional relaxation of the Feshbach molecular state.
Multistate chainwise STIRAP, as described in subsection 2.2, has been recently used in the Innsbruck experiment to transfer Cs_2 molecules from a Feshbach to the ground rovibrational state. The transfer efficiency of 55% has been achieved, limited by insufficient laser field intensities resulting in imperfect adiabaticity of the transfer and finite laser linewidth.
The multistate chainwise STIRAP technique allows one to use various transitions, coupled by *e.g.* rf fields and DC interactions. It can therefore be used in combination with the recently demonstrated resonant association method, where a pair of atoms is converted into

a molecule using a magnetic field modulated at a frequency close to a binding frequency of a Feshbach molecule (46). Another possibility, which we discuss in the next section, is to use the magnetic field dependent DC interchannel coupling between an entrance and a closed channel states as a first transition in the STIRAP chain (47) followed by optical transitions to the ground vibrational state. The first transition in these cases will directly couple the continuum states of colliding atoms and either a Feshbach molecular state or a bound state in the closed channel. As a result the overall atom-to-molecule conversion efficiency is expected to be higher compared to the two-step conversion sequence, when first a Feshbach molecular state is created, from where a molecule is transferred to the ground vibrational state. The chainwise STIRAP can be applied to resonant photoassociation as well, then the first transition in the STIRAP chain will couple the continuum states to a high energy vibrational state in the ground electronic state (48).

3. Efficient formation of ground state ultracold molecules via STIRAP from the continuum at a Feshbach resonance

In this section we describe photoassociative Stimulated Raman Adiabatic Passage (STIRAP) near a Feshbach resonance in a thermal atomic gas (49). We show that it is possible to use *low intensity* laser pulses to directly excite pairs of atoms in the continuum near a Feshbach resonance and to transfer nearly the entire atomic population to the lowest rovibrational level of the molecular ground state. This differs from the STIRAP techniques used in creation of ground state KRb (23) and Cs_2 (24) molecules in that the formation process starts directly from the atomic scattering continuum, avoiding formation of Feshbach molecules.

Feshbach molecules, used in the STIRAP transfer of KRb and Cs_2 to the ground rovibrational state (23; 24) are usually short-lived because of inelastic collisions with background atoms or other molecules. This is especially true for those produced from bosonic or mixed bosonic/fermionic atoms, for which collisions are not suppressed by the Fermi statistics at ultralow temperatures. In a dense atomic gas of density $n_{at} \sim 10^{13} - 10^{14}$ cm^{-3}, the collisional decay rate can be up to $\sim 10^4$ s^{-1} (with inelastic rate coefficient $K_{\text{inel}} \sim 0.5 - 1.0 \times 10^{-10}$ cm^{3}s^{-1} for Feshbach molecules (50; 51)). The STIRAP transfer must therefore be fast enough to avoid loosing molecules by inelastic decay. To alleviate this problem, we propose to start the STIRAP process directly from the scattering continuum without first forming Feshbach molecules. Using this approach with many STIRAP pulses and a fast repetition rate would also allow the conversion of nearly an entire atomic ensemble into ground state molecules, not only those atoms that were first transferred to a Feshbach molecular state.

Efficient adiabatic passage from the continuum requires laser pulses shorter than the coherence time of the continuum (27; 52; 53). The adiabaticity condition of STIRAP, $\Omega T \gg 1$, where T is the transfer time, therefore implies a large effective Rabi frequency Ω for the pulses. In addition, dipole matrix elements between the continuum and the bound state are usually small, and so the pump pulse that couples the continuum and the excited state would require a very high intensity, which proves impractical. Thus the previous STIRAP experiments (23), being restricted by the very short coherence time of the continuum, used a Feshbach molecular state as an initial state.

The small continuum-bound dipole matrix elements can be dramatically increased by photoassociating atoms in the vicinity of a Feshbach resonance. It has been shown, both theoretically and experimentally, that the photoassociation rate increases in the presence

of a Feshbach resonance by several orders of magnitude (54–57). This can be explained by considering that delocalized scattering states acquire some bound-state character due to admixture of a bound level associated with a closed channel, resulting in a large increase of the Franck-Condon factor between the initial scattering state and the final excited state. The recently proposed Feshbach Optimized Photoassociation (FOPA) technique (57) relies on this enhancement to directly reach deeply bound ground state vibrational levels from the scattering continuum. Consequently, photoassociation in the vicinity of a Feshbach resonance is expected to increase molecular formation rate up to 10^6 molecules/s (57).
Here we show that the approach used in FOPA can be combined with STIRAP for reducing the required pulse intensity. We predict highly efficient transfer of an entire atomic ensemble into the lowest rovibrational level in the molecular ground state. We note that a proposal, where the admixing of a short-range potential to a longer-range excited electronic potential, was recently suggested for improving a two-color pump-dump photoassociation scheme (58). The scattering continuum states have good overlap with the long-range potential, while the admixed short-range potential provides a good overlap with tightly bound vibrational levels of the ground electronic state, greatly improving the efficiency of photoassociation.
This section is organized as follows. In the next subsection, we derive a theoretical model of a combined atomic and molecular system. Fano theory is used to describe the interaction of a bound molecular state with the scattering continuum, represented as closed and open channel, respectively. The resulting continuum states are coupled by two laser fields to the vibrational target state in the ground state via the intermediate excited molecular electronic vibrational state. Next, we describe STIRAP-assisted conversion of a pair of colliding atoms into a deeply bound molecule. In subsection B, we present the results of numerical solutions of the model described in subsection A and in the Appendix B, using typical parameters of alkali dimers. We find optimal Rabi frequencies and profiles of STIRAP pulses. In subection C we average the pair-of-atoms STIRAP transfer efficiency over a thermal atomic ensemble, calculate a fraction of atoms that can be converted into molecules by one STIRAP sequence, and the number of pulses and overall time required to convert an entire atomic ensemble into molecules.

3.1 Model

We consider a three level system plus a continuum as shown in Figure 4, representing scattering states of two colliding atoms and bound states of a molecule. The ground level labeled $|1\rangle$ is the final product state to which a maximun of population must be transfered. Typically, this level will be the lowest vibrational level ($v'' = 0, J'' = 0$) of a ground molecular potential. This ground level is coupled to an excited bound level $|2\rangle$ of an excited molecular potential via a "Stokes" pulse depicted by the blue down-arrow in Figure 4. This level $|2\rangle$ is itself coupled via a pump pulse (red up-arrow) to an initial continuum of unbound scattering states $|\Psi_\epsilon\rangle$ of energies ϵ (shaded area in Figure 4). If we denote $C_1(t)$, $C_2(t)$ and $C(\epsilon, t)$ the time dependent amplitudes associated to the final, intermediate, and initial states $|1\rangle$, $|2\rangle$, and $|\Psi_\epsilon\rangle$, respectively, then the total wave function $|\Phi\rangle$ of the system is given by:

$$|\Phi\rangle = C_1(t)\,|1\rangle + C_2(t)\,|2\rangle + \int d\epsilon\; C(\epsilon, t)\,|\Psi_\epsilon\rangle\,. \tag{12}$$

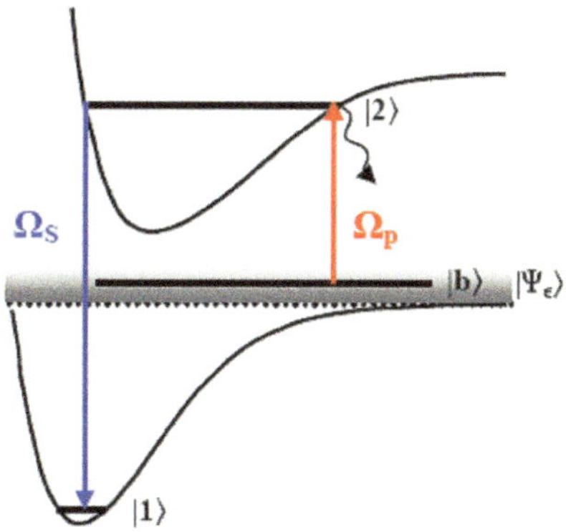

Fig. 4. Schematics: population from the initial state $|\Psi_\epsilon\rangle$ is transferred to a final target state $|1\rangle$ via an intermediate state $|2\rangle$. Both $|\Psi_\epsilon\rangle$ and $|1\rangle$ are coupled to $|2\rangle$ by a pump and a Stokes pulse, respectively labeled Ω_P and Ω_S. A bound level $|b\rangle$ corresponding to a closed channel can be imbedded in the continuum.

We assume that the levels associated with states $|1\rangle$ and $|2\rangle$ are well-isolated, and that there are no off-resonant laser couplings to other levels: this ensures the sufficient accuracy of the three-state model (see e.g. (23; 24; 59)).

No restriction applies to the definition of the continuum state $|\Psi_\epsilon\rangle$ as it can be associated to either a single-channel or a multi-channel scattering state. In this work, we consider the multi-channel case in which a bound level $|b\rangle$ associated to a closed channel is embedded in the continuum of scattering states $|\epsilon'\rangle$ of an open channel. When the energy of $|\epsilon'\rangle$ coincides with that of $|b\rangle$, a Feshbach resonance (14) occurs. These are common in binary collisions of alkali atoms due to hyperfine mixing and the tuning of the Zeeman interaction by an external magnetic field, hence the possibility to control interatomic interactions with a magnetic field. Following the Fano theory presented in Ref.(60), the scattering state $|\Psi_\epsilon\rangle$ can be expressed as:

$$|\Psi_\epsilon\rangle = a(\epsilon)\,|b\rangle + \int d\epsilon'\; b(\epsilon,\epsilon')\,|\epsilon'\rangle\;, \tag{13}$$

with

$$a(\epsilon) = \sqrt{\frac{2}{\pi\Gamma(\epsilon)}}\sin\Delta\,, \tag{14}$$

and

$$b(\epsilon,\epsilon') = \frac{1}{\pi}\sqrt{\frac{\Gamma(\epsilon')}{\Gamma(\epsilon)}}\frac{\sin\Delta}{\epsilon-\epsilon'} - \cos\Delta\,\delta(\epsilon-\epsilon')\,. \tag{15}$$

Here, $\Delta = -\arctan(\frac{\Gamma}{2(\epsilon-\epsilon_F)})$ is the phase shift due to the interaction between $|b\rangle$ and the scattering state $|\epsilon\rangle$ of the open channel. We assume $\Delta \in [-\pi/2, \pi/2]$. The width of the Feshbach resonance, $\Gamma = 2\pi|V(\epsilon)|^2$, is weakly dependent on the energy, while $V(\epsilon)$ is the interaction strength between the open and closed channels. The position of the resonance, $\epsilon_F = E_b + P\int \frac{|V(\epsilon')|^2 d\epsilon'}{\epsilon-\epsilon'}$, includes an interaction induced shift from the energy of the bound state E_b.

If we label E_i the energy of the state $|i\rangle$, the total Hamiltonian H is given by:

$$H = \sum_{i=1,2} E_i |i\rangle\langle i| + \int d\epsilon\ \epsilon |\Psi_\epsilon\rangle\langle\Psi_\epsilon| + V_{\text{light}}\ . \tag{16}$$

The light-matter interaction Hamiltonian V_{light} takes the form:

$$\begin{aligned} V_{\text{light}} &= -\left[\vec{\mu}_{21}|2\rangle\langle 1| + \text{H.c.}\right]\cdot\left(\vec{\mathcal{E}}_p + \vec{\mathcal{E}}_S + \text{c.c.}\right) \\ &\quad - \int d\epsilon\ \left[\vec{\mu}_{2\Psi_\epsilon}|2\rangle\langle\Psi_\epsilon| + \text{H.c.}\right]\cdot\left(\vec{\mathcal{E}}_p + \vec{\mathcal{E}}_S + \text{c.c.}\right)\ , \end{aligned} \tag{17}$$

where $\vec{\mathcal{E}}_{p,S} = \hat{e}_{p,S}\mathcal{E}_{p,S}\exp(-i\omega_{p,S}t)$ are the pump and Stokes laser fields of polarization $\hat{e}_{p,S}$, respectively, while $\vec{\mu}_{21}$ and $\vec{\mu}_{2\Psi_\epsilon}$ are the dipole transition moments between the states $|2\rangle$ and $|1\rangle$, and $|2\rangle$ and $|\Psi_\epsilon\rangle$, respectively. In this form, the Hamiltonian already takes into account the mixing between the bound state of the closed channel and the scattering states of the open channel. The Schrödinger equation describing the STIRAP conversion of two atoms into a molecule gives:

$$i\hbar\frac{\partial C_1}{\partial t} = E_1\, C_1 - \vec{\mu}_{21}^{*}\cdot\vec{\mathcal{E}}_S^{*}\, C_2, \tag{18}$$

$$i\hbar\frac{\partial C_2}{\partial t} = E_2\, C_2 - \vec{\mu}_{21}\cdot\vec{\mathcal{E}}_S\, C_1 \tag{19}$$

$$-\int_{\epsilon_{th}}^{\infty} d\epsilon\ \vec{\mu}_{2\Psi_\epsilon}\cdot\vec{\mathcal{E}}_p\, C(\epsilon,t),$$

$$i\hbar\frac{\partial C(\epsilon,t)}{\partial t} = \epsilon\, C(\epsilon,t) - \vec{\mu}_{2\Psi_\epsilon}^{*}\cdot\vec{\mathcal{E}}_p^{*}\, C_2. \tag{20}$$

For simplicity, we set the origin of the energy to be the position of the ground state $|1\rangle$, and use the rotating wave approximation with $C_1 = c_1$, $C_2 = c_2 e^{-i\omega_S t}$, and $C(\epsilon,t) = c(\epsilon,t)e^{-i(\omega_S-\omega_P)t}$. Eqs.(18)-(20) then become:

$$i\frac{\partial c_1}{\partial t} = -\Omega_S c_2, \tag{21}$$

$$i\frac{\partial c_2}{\partial t} = \delta c_2 - \Omega_S c_1 - \int_{\epsilon_{\text{th}}}^{\infty} d\epsilon\ \Omega_\epsilon c(\epsilon,t), \tag{22}$$

$$i\frac{\partial c(\epsilon,t)}{\partial t} = \Delta_\epsilon c(\epsilon,t) - \Omega_\epsilon^{*} c_2, \tag{23}$$

where $\delta = E_2/\hbar - \omega_S$, $\Delta_\epsilon = \epsilon/\hbar - (\omega_S - \omega_p)$, and ϵ_{th} is the dissociation energy of the ground electronic potential with respect to the state $|1\rangle$. The Rabi frequencies of the fields are $\Omega_S = \vec{\mu}_{21}\cdot\hat{e}_S\mathcal{E}_S/\hbar$ (assumed real), $\Omega_\epsilon = \vec{\mu}_{2\Psi_\epsilon}\cdot\hat{e}_p\mathcal{E}_p/\hbar$.

The previous system of three equations can be reduced into a two-equation system by eliminating the continuum amplitude $c(\epsilon,t)$ in Eq.(23). Introducing a solution in the form of $c(\epsilon,t) = s(\epsilon,t)\exp\left(-i\Delta_\epsilon t\right)$ into Eq.(23), we get

$$s = i\int_0^t dt'\ \Omega_\epsilon^{*}(t')c_2(t')e^{i\Delta_\epsilon t'} + s(\epsilon, t=0), \tag{24}$$

where $t = 0$ is some moment before the collision of the two atoms. The resulting continuum amplitude is

$$c(\epsilon,t) = i\int_0^t dt'\ \Omega_\epsilon^*(t')c_2(t')e^{i\Delta_\epsilon(t'-t)} + s(\epsilon,t=0)e^{-i\Delta_\epsilon t}. \tag{25}$$

Inserting this result into Eq. (22), we obtain a final system of equations for the amplitudes of the bound states:

$$i\frac{\partial c_1}{\partial t} = -\Omega_S c_2\,, \tag{26}$$

$$\begin{aligned} i\frac{\partial c_2}{\partial t} &= (\delta - i\gamma)c_2 - \Omega_S c_1 - \int_{\epsilon_{\text{th}}}^{\infty} d\epsilon\ \Omega_\epsilon(t)s(\epsilon,t=0)e^{-i\Delta_\epsilon t} \\ &\quad + i\int_{\epsilon_{\text{th}}}^{\infty} d\epsilon\ \Omega_\epsilon(t)\int_0^t dt'\ \Omega_\epsilon(t')^* c_2(t')e^{i\Delta_\epsilon(t'-t)} \\ &\equiv (\delta - i\gamma)c_2 - \Omega_S c_1 - S + T\,, \end{aligned} \tag{27}$$

where we introduced a spontaneous decay term γc_2 in Eq.(27).
The third term of Eq.(27), labelled S, corresponds to the source function, wheareas the last term, labelled T, corresponds to the "back-stimulation" term (or back-conversion) which accounts for the transfer of the bound molecules back into the continuum. The initial amplitude of the continuum wave function $s(\epsilon, t = 0)$ appearing in the source term has been discussed in various contributions (27; 52; 53). A Gaussian wavepacket provides the most classical description of a two-atom collision characterized by a minimal uncertainty relation between the energy bandwidth δ_ϵ of the wavepacket and the duration of the collision:

$$s(\epsilon,t=0) = \frac{1}{(\pi\delta_\epsilon^2)^{1/4}}e^{-\frac{(\epsilon-\epsilon_0)^2}{2\delta_\epsilon^2}+\frac{i}{\hbar}(\epsilon-\epsilon_0)t_0}, \tag{28}$$

where t_0 is the moment of the collision and ϵ_0 is the central energy of the wavepacket.
Futhermore, the Rabi frequency of the field coupling continuum states $|\Psi_\epsilon\rangle$ to the state $|2\rangle$ is given by (60)

$$\Omega_\epsilon = \frac{\vec{\mu}_{2\epsilon}\cdot\hat{e}_p\mathcal{E}_p}{\hbar}\frac{q\Gamma/2+\epsilon-\epsilon_F}{\sqrt{(\Gamma/2)^2+(\epsilon-\epsilon_F)^2}}\text{sgn}(\epsilon-\epsilon_F), \tag{29}$$

where $\vec{\mu}_{2\epsilon}$ is the dipole matrix element between an unperturbed scattering state $|\epsilon\rangle$ and the state $|2\rangle$, and q is the Fano parameter, expressed as:

$$q = \frac{(\vec{\mu}_{2b}\cdot\hat{e}_p) + P\int\frac{V(\epsilon')(\vec{\mu}_{2\epsilon'}\cdot\hat{e}_p)d\epsilon'}{\epsilon-\epsilon'}}{\pi V^*(\epsilon)(\vec{\mu}_{2\epsilon}\cdot\hat{e}_p)}, \tag{30}$$

where $\hat{e}_p$ is the polarization vector of the pump field, and $\vec{\mu}_{2b}$ is the dipole matrix element between bound states $|2\rangle$ and $|b\rangle$. The q factor is essentially the ratio of the dipole matrix elements from the state $|2\rangle$ to the bound state $|b\rangle$ (modified by the continuum) and to an unperturbed continuum state $|\epsilon\rangle$. This factor can be made much larger than unity, and as will be shown below, the total dipole matrix element from the continuum can be enhanced by this factor in the presence of the resonance. The magnitude of q can be controlled by the choice of the vibrational state $|2\rangle$. Selecting a tightly bound excited vibrational state will increase the

bound-bound and decrease the continuum-bound dipole matrix elements, resulting in larger q, whereas choosing a highly excited state close to a dissociation threshold will decrease q.
Using the expressions given in Eqs.(28), (29), and (30) for the initial amplitude of the continuum wave function, the Rabi frequency between the continuum state $|\Psi_\epsilon\rangle$ and the excited bound state $|2\rangle$, and the Fano parameter, respectively, we obtain the following complete expression for the source term:

$$S = S_0 \int_{\epsilon_{\rm th}}^{\infty} d\epsilon \, g(q,\epsilon) \, \mathrm{sgn}(\epsilon - \epsilon_F) e^{-\frac{(\epsilon-\epsilon_0)^2}{2\delta_\epsilon^2} + \frac{i(\epsilon-\epsilon_0)t_0}{\hbar}} e^{-i\Delta_\epsilon t}, \tag{31}$$

with $S_0 = \vec{\mu}_{2\epsilon} \cdot \hat{e}_p \mathcal{E}_p / \hbar (\pi \delta_\epsilon)^{1/4}$, and where the function $g(q,\epsilon)$ is defined as

$$g(q,\epsilon) \equiv \frac{q + \frac{2}{\Gamma}(\epsilon - \epsilon_F)}{\sqrt{1 + \frac{4}{\Gamma^2}(\epsilon - \epsilon_F)^2}}. \tag{32}$$

We assume that the unperturbed continuum is structureless and that the corresponding Rabi frequency $\vec{\mu}_{2\epsilon} \cdot \hat{e}_p \mathcal{E}_p / \hbar$ depends only weakly on the energy. We also extend $\epsilon_{\rm th}$ to $-\infty$ to have the initial continuum wavefunction normalized to unity[1]: $\int_{-\infty}^{\infty} d\epsilon |C(\epsilon)|^2 = 1$.
We can as well obtain a complete expression for the back-stimulation term T. We have:

$$T = \left| \frac{\vec{\mu}_{2\epsilon} \hat{e}_p}{\hbar} \right|^2 \mathcal{E}_p(t) \int_{\epsilon_{th}}^{\infty} d\epsilon \, g^2(q,\epsilon) \int_0^t dt' \, c_2(t') \mathcal{E}_p(t') e^{i\Delta_\epsilon (t'-t)}. \tag{33}$$

Extending the lower integration limit[2] allows for an analytical solution for the integrals over energy and time, leading to the following expression for the back-stimulation term:

$$\begin{aligned} T = \left| \frac{\vec{\mu}_{2\epsilon} \hat{e}_p}{\hbar} \right|^2 & \left[\pi \hbar \mathcal{E}_p^2(t) c_2(t) + \frac{\pi \Gamma}{2} (q - i)^2 \mathcal{E}_p(t) \right. \\ & \left. \times \int_0^t dt' \, c_2(t') \mathcal{E}_p(t') e^{[\Gamma/2\hbar + i(\epsilon_F/\hbar - \omega_S + \omega_p)](t'-t)} \right]. \end{aligned} \tag{34}$$

3.2 Results of STIRAP transfer for a pair of atoms

In this subsection, we consider two different cases: first, when $\Gamma \gg \delta_\epsilon$, *i.e.*, when the width Γ of the Feshbach resonance is much larger than the thermal energy spread δ_ϵ of the colliding atoms, and second when $\Gamma \ll \delta_\epsilon$. By considering these two limiting cases of broad and narrow

---

[1] Extension of $\epsilon_{\rm th}$ to $-\infty$ in the source term (31) can be justified by the sharp reduction of the Gaussian term $\exp(-(\epsilon_{\rm th} - \epsilon_0)^2/2\delta_\epsilon^2)$ for $\epsilon_0 - \epsilon_{th} > \delta_\epsilon$. For ϵ_0 close to $\epsilon_{\rm th}$, this approximation is less accurate, but as will be shown in Section IV, these low energies give negligible contribution to transfer efficiency averaged over an atomic ensemble due to their small weight in the Maxwell-Boltzmann distribution.

[2] The extension of the lower integration limit to $-\infty$ in the back-stimulation term (33) can be explained by the following argument. For an optimal transfer the duration of laser pulses has to be of the order of the coherence time of the populated continuum, given by $1/\delta_\epsilon$. The period of the exponent $\exp(i\Delta_\epsilon(t'-t))$ on the other hand is given by $2\pi/\Delta_\epsilon$. The integral over time therefore quickly goes to zero if this period is smaller then the duration of pulses. As a result, the time integral is non-zero only for energies $|\Delta_\epsilon| = |\epsilon/\hbar - (\omega_S - \omega_p)| < \delta_\epsilon$. In the case of the pump laser resonant with the center of the thermal distribution $|\epsilon_{\rm th} - (\omega_S - \omega_p)| \sim \delta_\epsilon$, and the extension of the energy integration to $-\infty$ is well-justified.

Reso-nance	δ_ϵ μK	Γ μK	Ω_S^0 10^8 s^{-1}	I_S W/cm^2	I_p W/cm^2	T_S μs	T_p μs	τ_S μs	τ_p μs
None	10	—	0.72	62	4×10^5	1.5	3	0.75	1.0
Broad	10	1000	0.74	65	4000	1.4	3.4	0.65	1.0
Narrow	100	1	2.24	600	400	0.157	0.3	0.1	0.207

Table 2. Parameters of the Stokes and pump photoassociating pulses providing optimal population transfer for a pair of atoms shown in Fig.5. We use $q = 10$, $\gamma = 10^8$ s^{-1}, and $\mu_{2b} = \mu_{21} = 0.1$ D. Rabi frequencies are modeled by Gaussians $\Omega_{S,p} = \Omega_{S,p}^0 \exp\left(-(t - t_0 \pm \tau_{S,p})\right)/T_{S,p}^2$, where $\pm$ refers to the Stokes and pump pulse, respectively.

resonances, more practical expressions for both the source term S and the back stimulation term T can be found. The derivation of the final system of equations used in numerical solutions is given in Appendix B. Here, we describe the solutions of these systems for both broad and narrow resonances.

We note that during the transfer an initial incoherent mixture of atomic scattering states is converted into a pure internal state, which seems to decrease the entropy of the system. However, the entropy is transferred to the center-of-mass motion of the created molecules, which can lead to a slight translational heating of the sample.

Using Eqs.(A3)-(A4) and (A7)-(A8) with the parameters listed in Table 2 for a broad ($\Gamma = 1$ mK) and a narrow ($\Gamma = 1$ μK) Feshbach resonance, we obtain the results for the STIRAP transfer of an atom pair, depicted in Fig. 5. Here the left column corresponds to the broad resonance, and the right column to the narrow resonance. The top row shows the variation of the Rabi frequencies over the time period required for the population transfer along with population in the intermediate state $|2\rangle$ (middle row) and final state $|1\rangle$ (bottom row).

For the broad case, we considered a Feshbach resonance with a width $\Gamma = 1$ mK (typical for broad resonances), and a thermal atomic ensemble with an energy bandwidth $\delta_\epsilon = 10$ μK. We see that the transfer efficiency can reach $\sim 97\%$ of the continuum state into the target state $|1\rangle$ (see Fig. 5 c). The parameters of the Gaussian laser pulses used (optimized Rabi frequencies, durations and delays of laser pulses) are given in Table 2: the peak intensities of the Stokes and pump fields were calculated from Rabi frequencies as $I_S = c\mathcal{E}_S^2/8\pi = c(\Omega_S^0\hbar)^2/8\pi\mu_{21}^2$ and $I_p = c\mathcal{E}_p^2/8\pi = c(\Omega_p^0)^2\delta_\epsilon/32\pi^{3/2}\mu_{2\epsilon}^2$, where we use Eq.(30) to estimate the continuum-bound dipole matrix element $\mu_{2\epsilon} \approx \mu_{2b}/q\pi V(\epsilon) = \sqrt{2}\mu_{2b}/q\sqrt{\pi\Gamma}$, resulting in $I_p = q^2 c(\Omega_p^0)^2\delta_\epsilon\Gamma/64\sqrt{\pi}\mu_{2b}^2$.

When comparing the results for a broad resonance to that of the unperturbed continuum (*i.e.*, far from the resonance), we find that the source term S is enhanced by the factor $g(q, \epsilon_0)$ (see Eq. (38) in Appendix B):

$$g(q, \epsilon_0) = \frac{q + \frac{2}{\Gamma}(\epsilon_0 - \epsilon_F)}{\sqrt{1 + \frac{4}{\Gamma^2}(\epsilon_0 - \epsilon_F)^2}}. \tag{35}$$

This factor has a maximum at $2(\epsilon_0 - \epsilon_F)/\Gamma = 1/q$, with the corresponding maximum value $\sqrt{1+q^2} \approx q$ for $q \gg 1$: hence, the source amplitude is enhanced q times. In this limit, all populated continuum states experience the same transition dipole matrix element enhancement factor to the state $|2\rangle$, so that the system essentially reduces to the case of a flat

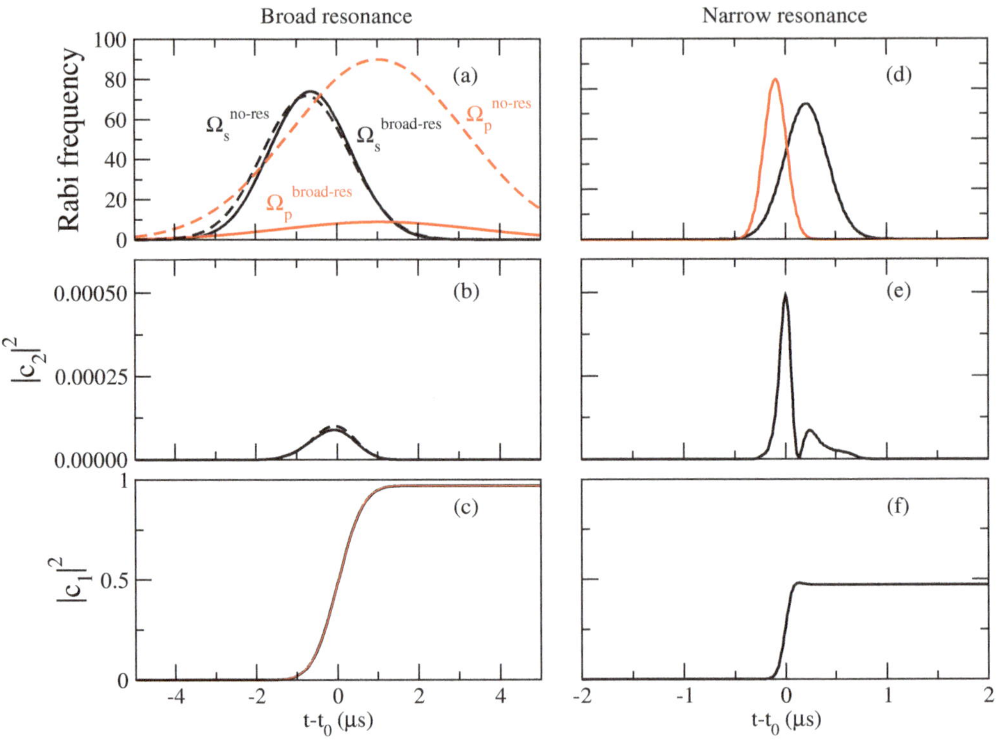

Fig. 5. Time-dependence of the Stokes and pump pulses (top row) and population in state $|2\rangle$ (middle row) and target state $|1\rangle$ (bottom row) for the STIRAP transfer of a pair of atoms within the center of the thermal distribution. The left column is for a broad Feshbach resonance, while the right column is for a narrow resonance (see Table 2 for values of parameters used). The dashed blue lines in the left column are the results obtained without resonance, when the parameters are adjusted to obtain the same overall transfer efficiency as for the broad resonance. The Stokes Rabi frequency is in units of 10^6 s^{-1}, while the pump Rabi frequency is in dimensionless units $(16\pi/\delta_\epsilon)^{1/4}\,\vec{\mu}_{2\epsilon}\hat{e}_p\mathcal{E}_p$ in the broad resonance limit and $(2\pi/\Gamma)^{1/2}\,\vec{\mu}_{2\epsilon}\hat{e}_p\mathcal{E}_p$ in the narrow resonance limit. Note that the scale for the Rabi frequencies in the narrow resonance case is 40 times the scale for the broad resonance, and the magnitude of the pump Rabi frequency is enlarged 10 times for better visibility.

continuum with an uniformly enhanced transition dipole matrix element. One thus expects that in this limit, the adiabatic passage should be efficient, requiring less pump laser intensity when compared to the unperturbed (*i.e.* without resonance) scattering continuum. This is clearly demonstrated in Fig. 5 (left column, dashed lines): to reach the same $\sim$ 97% transfer efficiency achieved with the broad resonance, a very large pump laser intensity ($\sim$ 100 times larger) is required if there is no resonance in the continuum (Fig. 5 a), while the Stokes laser intensity is basically the same. Considering the intensity used in this particular example, this would lead to intensities in the range of 5×10^5 W/cm^2, making STIRAP from the continuum technically impossible to achieve without a resonance. This is consistent with the analysis of photoassociative adiabatic passage from an unstructured continuum (27), and the above

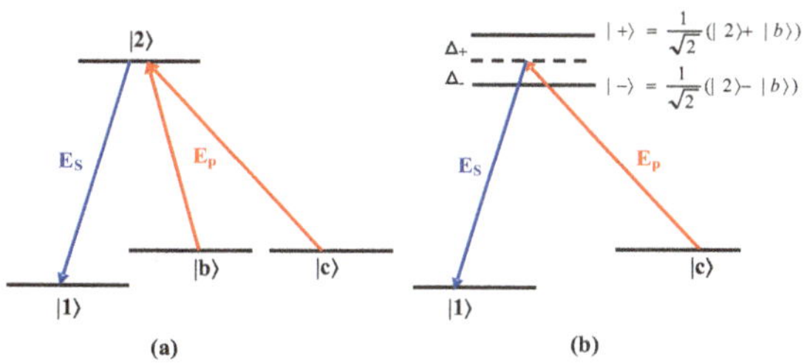

Fig. 6. Illustration of the reduction of STIRAP transfer efficiency due to destructive quantum interference for a narrow resonance: (a) a simplified level scheme where the scattering continuum is modeled by a single state $|c\rangle$ and the interaction between the continuum and the Feshbach state $|b\rangle$ is neglected; (b) an equivalent scheme, where the strong coupling between the Feshbach state $|b\rangle$ and the excited state $|c\rangle$ by the pump field forms "dressed" states $|\pm\rangle$. We have the Rabi frequency $\Omega_{2c} = \Omega_{+c}^2/\Delta_+ + \Omega_{-c}^2/\Delta_- = 0$, since $\Omega_{+c} = \Omega_{-c}$ and $\Delta_+ = -\Delta_-$.

prediction that in the presence of a wide resonance the required pump laser intensity is reduced by a factor of $\sim 1/q^2$.

Results of adiabatic passage for a pair of atoms in a narrow resonance limit are shown in Fig. 5 (right column). We considered typical values for a narrow resonance width $\Gamma = 1\ \mu$K and the ensemble energy bandwidth $\delta_\epsilon = 100\ \mu$K. Again, we give the parameters providing the optimal transfer in Table 2. In this limit, the transfer efficiency is lower: in the specific case analyzed here, it does not exceed 47%. The reason for this lower efficiency compared to a wide resonance is the destructive quantum interference which leads to electromagnetically induced transparency (61) in the transition from the continuum to the excited state. It can be explained using the following argument (see Fig. 6). The limit of a narrow Feshbach resonance corresponds to a weak coupling between the bound Feshbach state and the scattering continuum, and thus can be neglected in this simplified explanation. The system then can be viewed as consisting of bound and continuum states $|b\rangle$ and $|c\rangle$ having the same energy, which are coupled by the pump field to a molecular state $|2\rangle$, itself coupled to the state $|1\rangle$ by the Stokes field. Assuming that initially all the population is in the state $|c\rangle$, due to the small interaction strength between $|b\rangle$ and $|c\rangle$, we can eliminate the state $|b\rangle$, taking into account its coupling to $|2\rangle$ by the pump laser as the formation of "dressed" states $|\pm\rangle = (|2\rangle \pm |b\rangle)/\sqrt{2}$. If the dipole matrix element of the $|b\rangle \to |2\rangle$ transition is much larger than that of the $|c\rangle \to |2\rangle$ transition, the detuning of the "dressed" states satisfies $|\Delta_\pm| = \Omega_p^{2b} \gg \Omega_p^{2c}$, Ω_S. As a result, the one-photon coupling of $|c\rangle$ to the excited state, as well as two-photon coupling to $|1\rangle$ vanishes, preventing the adiabatic transfer. This mechanism is similar to the Fano interference effect, the difference is that the continuum is initially populated. One can therefore view it as an inverse Fano effect. The effective dipole matrix element of the $|c\rangle \to |2\rangle$ transition is $\mu_{2c} \sim \mu_{2b}/q\sqrt{\xi}$. In the case we analyzed, $q = 10$, $\xi = \Gamma/\sqrt{2}\delta_\epsilon = 0.01$, and $\mu_{2c} \approx \mu_{2b}$, which gives $\sim 50\%$ transfer efficiency.

The transfer efficiency increases if the Feshbach state is far detuned from the populated continuum. Our calculations show that for a Feshbach state detuning $\epsilon_F/\hbar - (\omega_S - \omega_p) \gg |\Omega_{2b}|^2/\gamma$, the transfer efficiency reaches 70% using the laser pulse parameters in Table 2. We note that the smaller intensity of the pump pulse used for the narrow resonance, as compared to the broad resonance, is due to the fact that we used the same $q = 10$ and assumed $\mu_{2b} = 0.1$

Reso-nance	δ_ϵ μK	Γ μK	Ω_S^0 10^8 s^{-1}	I_S W/cm^2	I_p W/cm^2	T_S μs	T_p μs	τ_S μs	τ_p μs
None	10	—	0.50	30	1.7×10^5	1.5	3.3	0.75	1.3
Broad	10	1000	0.60	40	2500	1.3	3.2	0.7	1.25
Narrow	100	1	2.24	600	400	0.157	0.3	0.1	0.207

Table 3. Parameters of the Stokes and pump photoassociating pulses providing optimal population transfer shown in Fig.7 for averaging over a Maxwell-Boltzmann distribution of energies. We use $q = 10$, $\gamma = 10^8$ s^{-1}, and $\mu_{2b} = \mu_{21} = 0.1$ D (1 D=10^{-18} esu cm = 0.3934 ea_0).

D for both resonances. From the definition of q, it means that the continuum-bound dipole matrix element $\mu_{2\varepsilon}$ is higher in the narrow than in the broad resonance we considered. This explains the smaller resulting pump pulse intensity. The overall conclusion for a narrow resonance is that, as opposed to a broad resonance, the presence of the Feshbach resonance prevents one from realizing high transfer efficiencies. It should be noted, however, that the destructive quantum interference effect is based on negligible interaction between the Feshbach and continuum states during the transfer time, since $T < \delta_\varepsilon^{-1} \ll \Gamma^{-1}$. This argument shows that already for $\Gamma \geq \delta_\epsilon$, there is enough interaction to neutralize the effect of destructive interference. Therefore, we expect that the broad resonance limit can be extended down to $\Gamma \sim \delta_\epsilon$, making it applicable to a wide variety of atomic species.

3.3 Conversion of atomic ensembles into ground state molecules

The results of Fig. 5 were obtained for a pair of atoms having a specific mean collision energy $\epsilon_0 = \hbar(\omega_S - \omega_p)$. Such a situation could be realized in very tight traps, *e.g.*, in tight optical lattices. For a system with a wider energy distribution, one would like to find an ensemble averaged transfer efficiency, and thus one needs to calculate the transfer probability $P(\epsilon_0) = |c_1|^2$ for all central wavepacket energies ϵ_0 within the thermal spread of energies, and perform the averaging as

$$P_{\text{avg}} = \frac{2}{\sqrt{\pi}(k_B T)^{3/2}} \int_0^\infty e^{-\epsilon_0/k_B T} \sqrt{\epsilon_0} P(\epsilon_0) d\epsilon_0, \tag{36}$$

where we assume a Maxwell-Boltzmann energy distribution, the pump laser resonant with the center of the distribution at $\langle\epsilon\rangle = 3/2 k_B T$, and set the bandwidth of the distribution at $\delta_\epsilon = \sqrt{\langle(\Delta\epsilon)^2\rangle} = \sqrt{3/2} k_B T$. The results are shown in Fig.7: while the maximum transfer efficiency in the broad resonance case is $\sim 70\%$, it can be achieved with lower laser intensities than in the case of a pair of atoms of Fig.5.

Given the adiabatic photoassociation probability $P(\epsilon)$ for two colliding atoms with relative energy ϵ, we can calculate the number of atoms photoassociated during the time overlap τ of the Stokes and pump pulses. During this time, the atom with the energy $\epsilon = \mu v^2/2$, where μ is the reduced mass, will collide with atoms in the volume $\pi b^2 v \tau$, where πb^2 is the collision cross-section. The impact parameter for the collision corresponding to a partial wave with angular momentum ℓ is $b = (\ell + 1/2)\hbar/p = (\ell + 1/2)\hbar/\sqrt{2\mu\epsilon}$. The number of collisions that atoms with a relative energy in the interval $(\epsilon, \epsilon + d\epsilon)$ will experience during the transfer time is therefore $N(\epsilon)d\epsilon = \pi b^2 v \tau \rho(\epsilon) d\epsilon$, where $\rho(\epsilon) = 2\rho \exp(-\epsilon/k_B T)\sqrt{\epsilon}/\sqrt{\pi}(k_B T)^{3/2}$ is the spectral density of the atoms (ρ is the density of the sample). Finally, $\ell = 0$ for ultracold s-wave collisions, and the fraction of atoms in the energy interval $(\epsilon, \epsilon + d\epsilon)$ photoassociated

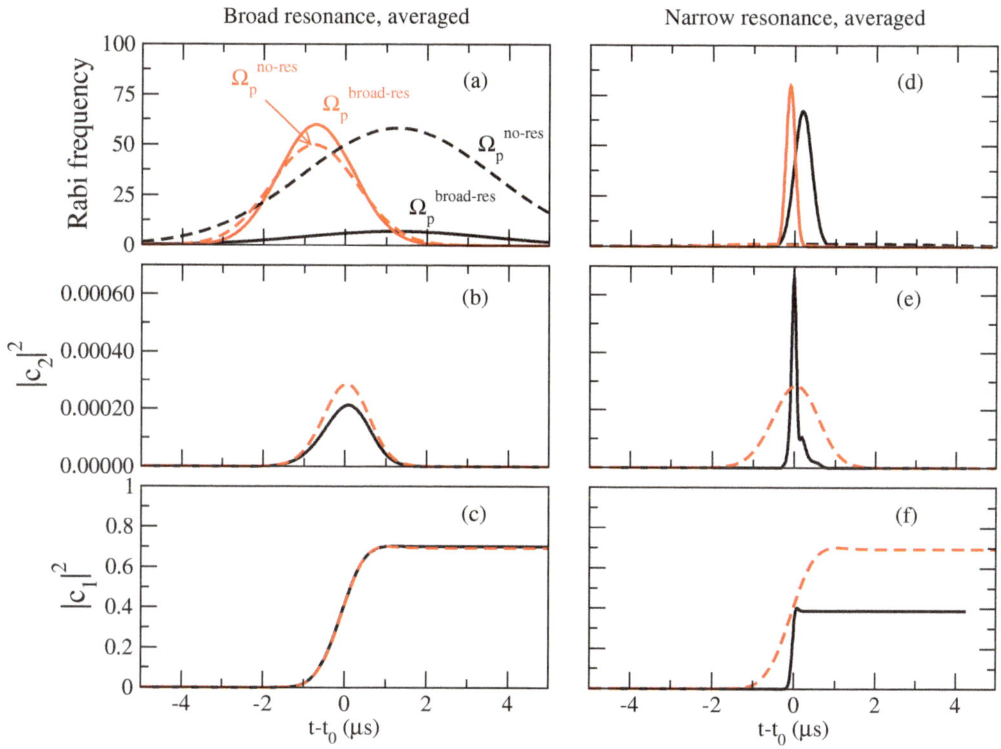

Fig. 7. Same as Fig. 5, but for the energy averaged transfer. The parameters are listed in Table 3.

by the two pulses is $f(\epsilon) = P(\epsilon)N(\epsilon)$, or

$$f(\epsilon) = \frac{\sqrt{2\pi}\hbar^2}{4(\mu k_B T)^{3/2}} \tau\rho P(\epsilon) \exp\left(-\epsilon/k_B T\right). \tag{37}$$

The total fraction of atoms photoassociated by a pair of pulses is $f = \int_0^\infty d\epsilon \; f(\epsilon) \approx P_{\text{avg}}\rho\sqrt{2\pi}\tau\hbar^2/4\mu^{3/2}\sqrt{k_B T}$, where we assumed that $P(\epsilon)$ does not significantly vary within the ensemble, and approximated it by the averaged value P_{avg}. Considering as an example ^{6}Li atoms at $T = 100\ \mu$K with an atomic density $\rho = 10^{12}$ cm^{-3}, an overlap time $\tau \sim 1\ \mu$s, and assuming $P_{\text{avg}} = 0.7$, the fraction of atoms photoassociated by the Stokes and pump pulses is $f \sim 2.5 \times 10^{-4}$: for heavier atoms $f \sim 10^{-6} - 10^{-5}$. It will therefore require $\sim 10^4 - 10^6$ pairs of pulses to convert an entire atomic ensemble into deeply bound molecular levels.

As was shown in (62), in the limit of a narrow resonance longer pulses with durations $T_S, T_p \sim 1/\Gamma$ can be used. The reason is that population gets "trapped" in the bound state $|b\rangle$ for a time $\sim 1/\Gamma$ (it can be seen from the expression for the narrow resonance source function A.6), and as a result coherent transfer is still possible. The fraction of atoms associated per pulse pair in this case is comparable to the case of a wide resonance, since the smaller transfer efficiency $P(\epsilon)$ is compensated by a larger pulse overlap τ. The long pulse duration results in

its narrow bandwidth $\sim \Gamma$, much smaller than the thermal ensemble energy δ_ϵ. Conversion efficiency per pulse pair in this case might be increased by simultaneously chirping the Feshbach resonance energy and the pump pulse frequency, i.e. by tuning ϵ_F and ω_P in time keeping the two-photon resonance condition $\epsilon_F = \omega_S - \omega_p$ satisfied.
While only a small fraction of atoms can be transferred to $|1\rangle$ by a pair of STIRAP pulses, a train of pulse pairs can be applied to photoassociate the entire atomic ensemble. To prevent excitation of molecules in $|1\rangle$ back to the continuum by subsequent pulses, they have to be removed before the next pair of pulses is applied. This could be realized by applying, after each pair of Stokes and pump pulses, a relatively long pulse resonant to a transition from $|1\rangle$ to some other vibrational level in the excited electronic potential which decays spontaneously to a deep vibrational state in the ground electronic potential. This long pulse would optically pump molecules out of the state $|1\rangle$ to deeper vibrational states in the ground electronic potential. It therefore has to be longer than the spontaneous decay time of the excited state. The excited state has to be chosen carefully so that it does not decay back into the scattering continuum. This would empty the $|1\rangle$ state and deposit molecules into ground potential vibrational states according to Franck-Condon factors before the next pair of pulses arrives. Finally, after all atoms have been converted into molecules the recently demonstrated optical pumping for molecules method (20) can be applied, which would transfer molecules from all populated vibrational states into the ground level $v = 0$.
The optimal strategy is to actually choose an excited state that decays mostly to the $v = 0$ level. This would allow one to avoid storing molecules in unstable vibrational states and using the optical pumping method. If such a state cannot be directly reached from $|1\rangle$, a four-photon STIRAP transfer can be applied (30), which provides efficient transfer to deeply bound molecular states. It allows one to choose the final state $|1\rangle$, from which the excited state decaying predominantly to $v = 0$ can be easily reached. In this case rotational selectivity can also be preserved, since only $v = 0, J = 0$ and $v = 0, J = 2$ states will be populated.
The total time required to photoassociate the whole atomic ensemble and transfer it to the $v = 0$ level can be estimated as follows. As the numerical results show, the adiabatic passage requires ~ 5 μs, the follow-up pulse emptying state $|1\rangle$ can have a ~ 100 ns duration, if the excited state lifetime is tens of ns, resulting in the whole sequence ~ 6 μs. Then the train of $10^4 - 10^6$ pulse pairs will take $\sim 0.1 - 10$ s. The final step, optical pumping to the $v = 0$ level, requires $\sim$ hundred μs, so the overall formation time is $\sim 0.1 - 10$ s. Given an illuminated volume $\sim 10^{-2} - 10^{-3}$ mm^3 and an atomic density $\rho \sim 10^{12}$ cm^{-3}, the resulting production rate is expected to be $10^5 - 10^8$ molecules/s. This compares well with recent experiments on STIRAP production of ground state KRb molecules starting from the Feshbach bound state, where $\sim 3 \times 10^4$ ground state molecules are produced during the entire cycle, including creation of Feshbach molecules, taking $\sim 10 - 30$ s (23).
Alternatively, back-stimulation of formed molecules into the continuum by subsequent STIRAP pulses can be avoided by placing them in a moving optical lattice, holding molecules but not atoms (27). Another way to avoid back-stimulation, applicable to polar molecules, is to overlap the atomic trap with a gradient of a DC electric field. It will leave dipoleless atoms unaffected, while shifting molecules out of STIRAP laser beams.
To summarize, combining both photoassociation and coherent optical transfer to rovibrational levels of the ground electronic molecular potential can allow one to convert an entire atomic ensemble into deeply bound molecular states, and to produce an ultracold molecular gas with high phase-space density. Photoassociative adiabatic passage in a thermal ultracold

atomic gas can be greatly facilitated by a Feshbach resonance. The presence of a bound state imbedded in and resonant with scattering continuum states strongly enhances the continuum-bound transition dipole matrix element to an excited electronic state, thus requiring less laser intensity for efficient transfer. In the limit of a wide resonance, when compared to the thermal spread of collision energies, the dipole matrix element is enhanced by the Fano parameter q. By choosing a tightly bound excited vibrational state, q can be made much larger than unity, resulting in the intensity of the pump pulse required for efficient adiabatic passage to be $\sim 1/q^2$ times smaller than in the absence of the resonance. Numeical modeling of the adiabatic passage using typical parameters of alkali dimers shows that intensities of the pump pulse, coupling the continuum to an excited state, of kW/cm^2 are sufficient for optimal transfer, which is $\sim$ 100 times smaller than without resonance. Optimal pulse durations are several μs, resulting in energies per pulse $\sim$ 10 μJ for a focus area of 1 mm^2.

If the Feshbach resonance is narrow compared to the thermal energy spread of colliding atoms, adiabatic passage is hindered by destructive quantum interference. The reason is that electromagnetically induced transparency significantly reduces the transition dipole matrix element from the scattering continuum to an excited state in the presence of the bound Feshbach state. In the narrow resonance limit, photoassociative adiabatic passage is therefore more efficient if the resonance is far-detuned.

Due to low atomic collision rates at ultracold temperatures, only a small fraction of atoms can be converted into molecules by a pair of photoassociative pulses. To convert an entire atomic ensemble, a train of pulse pairs can be applied. We estimate that $10^4 - 10^6$ pulse pairs will associate an atomic gas of alkali dimers with a density 10^{12} cm^{-3} in an illuminated volume of $10^{-2} - 10^{-3}$ mm^3 in $0.1 - 10$ s, resulting in extremely high production rates of $10^5 - 10^8$ molecules/s. High transfer efficiencies combined with low intensities of adiabatic photoassociative pulses also make the broad resonance limit attractive for quantum computation. For example, a scheme proposed in (63) can be realized, where qubit states are encoded into a scattering and a bound molecular states of polar molecules. To perform one and two-qubit operations, this scheme requires a high degree of control over the system, which our model readily offers.

Finally, marrying FOPA and STIRAP is a very promising avenue to produce large amounts of molecules, for a variety of molecular species. In fact, although we described here examples based on magnetically induced Feshbach resonances, such resonances are extremely common, and can be induced by several interactions, such as external electric fields or optical fields. Even in the absence of hyperfine interactions, other interactions can provide the necessary coupling, such as in the case of the magnetic dipole-dipole interaction in ^{52}Cr (64; 65).

4. Conclusions

Precise control over internal and external degrees of freedom of molecules will open the way for new fundamental studies and applications in physics and chemistry. As has been clearly seen with atoms in the recent decades, well-controlled laser fields offer an exquisite control tool over atomic internal and external states, including laser cooling and trapping, coherent manipulation of atomic quantum states and in particular various techniques used for quantum information applications, atomic spectroscopy. Recent years have witnessed mastering of single atom manipulation in individual traps, including optical dipole traps and atom chips, and optical lattices, with most manipulation techniques relying on laser fields. There is a great incentive in the atomic and molecular optics community to extend the precise control techniques developed for atoms to molecules. We have outlined in this chapter some experimentally relatively simple laser pulse techniques that can accomplish this task.

A prerequisite for many of the new studies is a high phase space density molecular sample in a stable internal state, specifically in the ground rovibrational state and preferably in the lowest hyperfine sublevel. We have in particular discussed two examples of coherent laser control of molecular states, multistate chainwise STIRAP and photoassociatice adiabatic passage near Feshbach resonance, which provide efficient transfer of molecules to the ground rovibrational state. In chainwise STIRAP the transfer is based on a generalized dark state, which is a superposition of all ground vibrational levels involved in the process. Selecting a special time order of the laser pulses coupling vibrational states and optimizing durations and intensities transfer efficiencies $> 90\%$ are predicted even in the presence of fast collisional decay of intermediate vibrational states. This technique has recently been applied to transfer Cs_2 Feshbach molecules to the ground rovibrational state with 55% efficiency, limited by technical issues. Additionally, we outlined how the step from the atomic scattering continuum to the ground rovibrational molecular state can be done in one coordinated step. In the presence of a Feshbach resonance delocalized scattering states acquire some bound-state character due to admixture of a bound level associated with a closed channel. It strongly enhances the Franck-Condon factor between the initial scattering state and a bound intermediate excited molecular state, a technique named Feshbach Optimized Photoassociation. We analyzed the transfer efficiency and intensities of the laser pulses required for optimal transfer both with and without the resonance and found that $> 70\%$ efficiencies are possible with relatively low intensity pulses of several W/cm^2 in the presence of the resonance.

5. Acknowledgments

We gratefully acknowledge finantial support from NSF and AFOSR under the MURI award FA9550-09-1-0588.

6. Appendix

A. Rotation and dephasing matrices

The Hamiltonian (2) in the case of the two-pulse STIRAP scheme, discussed in Section 2.1 has a zero eigenvalue $\varepsilon_0 = 0$, describing the dark state, and four eigenvalues, $\varepsilon_{1,2} = \pm\Omega\sqrt{1-\sin 2\theta/2}$ and $\varepsilon_{3,4} = \pm\Omega\sqrt{1+\sin 2\theta/2}$, corresponding to bright states. Adiabatic eigenstates $|\Phi\rangle = \{|\Phi^n\rangle\}$, $n = 0,...4$ and the bare states $|\Psi\rangle = \left\{\left|\Psi^l\right\rangle\right\}$, $l = g_1, e_1, g_2, e_2, g_3$ are transformed as $\left|\Psi^l\right\rangle = \sum_n W_{ln}|\Phi^n\rangle$, $|\Phi^n\rangle = \sum_l W_{ln}\left|\Psi^l\right\rangle$ via an orthogonal ($W^{-1} = W^T$) rotation matrix, given by the expression

$$W = \frac{1}{2}\begin{pmatrix} 2c^+c^- & s^- & s^- & s^+ & s^+ \\ 0 & 1 & -1 & -1 & 1 \\ 2s^-c^+ & -(s^-+c^-) & -(s^-+c^-) & (s^++c^+) & (s^++c^+) \\ 0 & -1 & 1 & -1 & 1 \\ -2s^+s^- & c^- & c^- & c^+ & (s^+-c^+) \end{pmatrix},$$

where $c^{\pm} = \cos\theta/\sqrt{1 \pm \sin 2\theta/2}$, $s^{\pm} = \pm\sin\theta/\sqrt{1 \pm \sin 2\theta/2}$.
In the "straddling" STIRAP scheme the rotation matrix reads as:

$$W = \frac{1}{2}\begin{pmatrix} -2\cos\theta & 0 & \sin 2\theta\frac{\Omega}{\Omega_0} & 0 & -2\sin\theta \\ -\sqrt{2}\sin\theta & 2 & -\cos 2\theta\frac{\Omega}{\sqrt{2}\Omega_0} & -1 & \sqrt{2}\cos\theta \\ -\sqrt{2}\sin\theta & -1 & -\cos 2\theta\frac{\Omega}{\sqrt{2}\Omega_0} & 1 & \sqrt{2}\cos\theta \\ \sin\theta\frac{\Omega}{\Omega_0} & -1 & \sqrt{2} & -1 & \cos\theta\frac{\Omega}{\Omega_0} \\ \sin\theta\frac{\Omega}{\Omega_0} & 1 & \sqrt{2} & 1 & \cos\theta\frac{\Omega}{\Omega_0} \end{pmatrix},$$

where terms of the order of $O(\Omega^2/\Omega_0^2)$ and higher are neglected.
The Liouville operator in the bare state basis has a form

$$\mathcal{L}\rho = \frac{1}{2}\begin{pmatrix} 2\Gamma_1\rho_{g_1g_1} & (\gamma_1+\Gamma_1)\rho_{g_1e_1} & (\Gamma_2+\Gamma_1)\rho_{g_1g_2} & (\gamma_2+\Gamma_1)\rho_{g_1e_2} & \Gamma_1\rho_{g_1g_3} \\ (\gamma_1+\Gamma_1)\rho_{e_1g_1} & 2\gamma_1\rho_{e_1e_1} & (\gamma_1+\Gamma_2)\rho_{e_1g_2} & (\gamma_1+\gamma_2)\rho_{e_1e_2} & \gamma_1\rho_{e_1g_3} \\ (\Gamma_2+\Gamma_1)\rho_{g_2g_1} & (\gamma_1+\Gamma_2)\rho_{g_2e_1} & 2\Gamma_2\rho_{g_2g_2} & (\gamma_2+\Gamma_2)\rho_{g_2e_2} & \Gamma_2\rho_{g_2g_3} \\ (\gamma_2+\Gamma_1)\rho_{e_2g_1} & (\gamma_1+\gamma_2)\rho_{e_2e_1} & (\gamma_2+\Gamma_2)\rho_{e_2g_2} & 2\gamma_2\rho_{e_2e_2} & \gamma_2\rho_{e_2g_3} \\ \Gamma_1\rho_{g_3g_1} & \gamma_1\rho_{g_3e_1} & \Gamma_2\rho_{g_3g_2} & \gamma_2\rho_{g_3e_2} & 0 \end{pmatrix}.$$

We include the decay from the Feshbach and the intermediate state using a rate Γ_1 and Γ_2, respectively, and from the excited states $|e_{1,2}\rangle$, given by $\gamma_{1,2}$, and assume that all decay is due to population loss out of the system, e.g. to other vibrational levels or continuum. We also assume that molecules in the ground vibrational state $|g_1\rangle$ do not decay.

B. Adiabatic passage in the limits of broad and narrow Feshbach resonances

In this appendix, we discuss Eqs.(26) and (27) for various relative widths of the Feshbach resonance Γ with respect to the thermal energy spread δ_ϵ of the colliding atoms. We first describe the case of a broad resonance, *i.e.* when the width of the Feshbach resonance greatly exceeds the thermal energy spread ($\Gamma \gg \delta_\epsilon$), and second consider the opposite situation of a narrow resonance ($\Gamma \ll \delta_\epsilon$). Finally, we briefly present the case where there is no resonance.

B.1 Limit of a broad Feshbach resonance

The typical thermal energy spread for colliding atoms in photoassociation experiments with non-degenerate gases is $\delta_\epsilon \sim 10-100$ μK. The broad resonance case occurs for resonances having a width of several Gauss (~ 1 mK), for which we have $\Gamma/\delta_\epsilon \sim 10-100$. A wide variety of systems exhibit broad resonances. For instance, they can be found in collision of ^{6}Li atoms at 834 G for the $|f=1/2, m_f=1/2\rangle \otimes |f=1/2, m_f=-1/2\rangle$ entrance channel ($\Gamma = 302$ G= 40 mK) and in ^{7}Li at 736 G for the $|f=1, m_f=1\rangle \otimes |f=1, m_f=1\rangle$ entrance channel ($\Gamma = 145$ G = 19 mK). We note here that these values of Γ are slightly different than the "magnetic" width ΔB usually given and based on the modelling of the scattering length.
The source function can be readily calculated from Eq.(31) by noticing that the Rabi frequency term can be set at $\epsilon = \epsilon_0$ corresponding to the maximum of the Gaussian function in the integrand. Using the function $g(q,\epsilon)$ defined in Eq.(32), the result takes the form

$$\begin{aligned} S_w &= S_0\sqrt{2\pi}\delta_\epsilon g(q,\epsilon_0)\text{sgn}(\epsilon_0-\epsilon_F)e^{-(t-t_0)^2\delta_\epsilon^2/2\hbar^2 - i(\epsilon_0/\hbar-(\omega_S-\omega_p))t} \\ &= S_{\text{no-res}}g(q,\epsilon_0)\text{sgn}(\epsilon_0-\epsilon_F), \end{aligned} \tag{38}$$

where $S_{\text{no-res}}$ is the source function without a resonance given below in Eq.(47). Strictly speaking, this expression is valid for $|\epsilon_F - \epsilon_0| \geq \delta_\epsilon$, but since $\Gamma \gg \delta_\epsilon$ Eq.(38) is a good approximation for a wide range of detunings $\epsilon_F - \epsilon_0$.
The back-stimulation term (34) can be further simplified in the limit of a broad resonance. In this case, both $c_2(t)$ and $\mathcal{E}_p(t)$ change on a time scale $\sim 1/\delta_\epsilon$, *i.e.* slowly compared to the decay time $\sim \hbar/\Gamma$ of the exponent. Therefore, we can rewrite (34) as:

$$\left|\frac{\vec{\mu}_{2\epsilon}\hat{e}_p}{\hbar}\right|^2 \pi\hbar \left[1 + \frac{(q-i)^2}{1+2i(\epsilon_F - \hbar(\omega_S - \omega_p))/\Gamma}\right] c_2(t)\mathcal{E}_p^2(t). \tag{39}$$

The system (26)-(27) in the case of a broad resonance becomes:

$$i\frac{\partial c_1}{\partial t} = -\Omega_S c_2, \tag{40}$$

$$\begin{aligned} i\frac{\partial c_2}{\partial t} &= -\Omega_S c_1 - S_w + (\delta - i\gamma)c_2 \\ &\quad -i\pi\hbar|\Omega_{\text{no-res}}(t)|^2 \left[1 + \frac{(q-i)^2}{1+2i(\epsilon_F - \hbar(\omega_S - \omega_p))/\Gamma}\right] c_2, \end{aligned} \tag{41}$$

where $\Omega_{\text{no-res}} = \vec{\mu}_{2\epsilon}\hat{e}_p\mathcal{E}_p/\hbar$ is the continuum-bound Rabi frequency in the absence of resonance. We also added a spontaneous decay term γc_2, assuming that the excited molecules dissociate into high energy continuum states and the resulting atoms leave a trap. From Eq.(38), one can see that in a broad resonance case, the source amplitude is enhanced by the factor $g(q,\epsilon_0) = (q + 2(\epsilon_0 - \epsilon_F)/\Gamma)/\sqrt{1+4(\epsilon_0-\epsilon_F)^2/\Gamma^2}$ when compared to the unperturbed continuum case. This factor has a maximum at $2(\epsilon_0 - \epsilon_F)/\Gamma = 1/q$, with the corresponding maximum value $g_{\max} \sim q$ for $q \gg 1$.

B.2 Limit of a narrow Feshbach resonance

This situation occurs when the width of the resonance is of the order of a few micro-Gauss or less. Examples of narrow resonances include ^{6}Li^{23}Na at 746 G for the $|f_1 = 1/2, m_{f1} = 1/2\rangle|f_2 = 1, m_{f2} = 1\rangle$ channel (Γ = 7.8 mG = 1 μK) (66), or ^{6}Li^{87}Rb at 882 G for the $|f_1 = 1/2, m_{f1} = 1/2\rangle|f_2 = 1, m_{f2} = 1\rangle$ channel (*p*-wave, Γ = 10 mG = 1.3 μK).
We note that the source term expressed in Eq.(31) can be rewritten in a time representation:

$$\begin{aligned} S &= S_0\sqrt{2\pi}\delta_\epsilon e^{-i(\epsilon_0/\hbar - (\omega_S - \omega_p))t} \\ &\times \left[e^{-(\tau-\tau_0)^2} + \xi e^{2iD - D^2}\int_{-\infty}^{\infty} e^{-(\tau' - iD)^2}(I_1(\xi|\tau - \tau_0 - \tau'|)\right. \\ &-L_{-1}(\xi|\tau - \tau_0 - \tau'|) - iq(I_0(\xi|\tau - \tau_0 - \tau'|) \\ &\left.-L_0(\xi|\tau - \tau_0 - \tau'|))\text{sgn}(\tau - \tau_0 - \tau')d\tau'\right], \end{aligned} \tag{42}$$

where we introduced the dimensionless variables $\tau = t\delta_\epsilon/\sqrt{2}\hbar$, $D = (\epsilon_F - \epsilon_0)/\sqrt{2}\delta_\epsilon$, $\xi = \Gamma/\sqrt{2}\delta_\epsilon$; $I_{0,1}$ and $L_{0,-1}$ are modified Bessel and Struve functions. One can see from this expression that the source function is a sum of the pure source function of the unperturbed continuum, given by the first term in square brackets, and of the admixed bound state, given by the integral. The coefficient $\xi = \Gamma/\sqrt{2}\delta_\epsilon$, which is the ratio of the Feshbach resonance

width to the width of the thermal energy spread, gives the ratio of contributions from the bound state and the unperturbed continuum, respectively.
It is then easier to notice that in the limit of a narrow resonance, the Gaussian function in the integrand of Eq.(42) is much narrower than the Bessel and Struve functions, which change on the time scale $\sim 1/\xi$. Therefore the source term can be aproximated as:

$$\begin{aligned} S_n = & S_0\sqrt{2\pi}\delta_\epsilon e^{-i(\epsilon_0/\hbar-(\omega_S-\omega_p))t}[e^{-(\tau-\tau_0)^2} \\ & +\xi\sqrt{\pi}e^{2iD-D^2}(I_1(\xi|\tau-\tau_0|)-L_{-1}(\xi|\tau-\tau_0|) \\ & -iq(I_0(\xi|\tau-\tau_0|)-L_0(\xi|\tau-\tau_0|))\mathrm{sgn}(\tau-\tau_0))]. \end{aligned} \quad (43)$$

Since $\xi \ll 1$, the real part of the source function is given by the first term in the square brackets, which is a pure continuum source function, while the imaginary part is due to the admixed bound state and its magnitude depends on the product ξq. Using asymptotic expansions of modified Bessel and Struve functions $I_0(x) - L_0(x) \to -2/\pi x$, $I_1(x) - L_{-1}(x) \to -2/\pi x^2$, it is seen from Eq.(43) that the contribution to the source function from the bound state decays on the time scale $|\tau - \tau_0| \sim 1/\xi$, while the contribution from the unperturbed continuum decays on the time scale $|\tau - \tau_0| \sim 1 \ll 1/\xi$.
In the limit of a narrow resonance the system (26)-(27) becomes:

$$i\frac{\partial c_1}{\partial t} = -\Omega_S c_2, \quad (44)$$

$$\begin{aligned} i\frac{\partial c_2}{\partial t} = & -\Omega_S c_1 - S_n + (\delta - i\gamma)c_2 \\ & -i\left|\frac{\vec{\mu}_{2\epsilon}\hat{e}_p}{\hbar}\right|^2\left[\pi\hbar\mathcal{E}_p^2 c_2 + \frac{\pi\Gamma}{2}(q-i)^2\mathcal{E}_p(t)\right. \\ & \left.\times\int_0^t dt'\, c_2(t')\mathcal{E}_p(t')e^{\Gamma(t'-t)/2\hbar+i(\epsilon_F/\hbar-(\omega_S-\omega_p))(t'-t)}\right]. \end{aligned} \quad (45)$$

B.3 Continuum without resonance

Finally, let us consider the case of a continuum without resonance. In this case the continuum-bound Rabi frequency Eq.(29) is:

$$\Omega_\epsilon = \Omega_{\text{no-res}} = \vec{\mu}_{2\epsilon}\cdot\hat{e}_p\,\mathcal{E}_p/\hbar, \quad (46)$$

and the source function is

$$S_{\text{no-res}} = S_0\sqrt{2\pi}\delta_\epsilon e^{-(t-t_0)^2\delta_\epsilon^2/2\hbar^2-i(\epsilon_0/\hbar-(\omega_S-\omega_p))t}. \quad (47)$$

The back-stimulation term (34) reduces to

$$|\vec{\mu}_{2\epsilon}\cdot\hat{e}_p/\hbar|^2\,\pi\hbar\mathcal{E}_p^2 c_2 = \pi\hbar\,|\Omega_{\text{no-res}}(t)|^2\,c_2, \quad (48)$$

and the system (26)-(27) takes the simple form:

$$i\frac{\partial c_1}{\partial t} = -\Omega_S c_2, \tag{49}$$

$$i\frac{\partial c_2}{\partial t} = -\Omega_S c_1 + (\delta - i\gamma)c_2 - i\pi\hbar|\Omega_{\text{no-res}}(t)|^2 c_2 - S_{\text{no-res}}. \tag{50}$$

C. References

[1] D. DeMille, F. Bay, S. Bickman, D. Kawall, D. Krause Jr., S. E. Maxwell, L. R. Hunter, Phys. Rev. A 61, 052507 (2000).
[2] J. J. Hudson, B. E. Sauer, M. R. Tarbutt, E. A. Hinds, Phys. Rev. Lett. 89, 023003 (2002).
[3] D. W. Rein, J. Mol. Evol. 4, 15 (1974); V.S.Letohov, Phys. Lett. A 53, 275 (1975).
[4] V. V. Flambaum, M. G. Kozlov, Phys. Rev. Lett. 99, 150801 (2007).
[5] R. V. Krems, Int. Rev. Phys. Chem. 24, 99 (2005).
[6] M. A. Baranov, Phys. Rep. 464, 71 (2008).
[7] K. Goral, L. Santos, M. Lewenstein, Phys. Rev. Lett. 88, 170406 (2002).
[8] C. Trefzger, C. Menotti, B. Capogrosso-Sansone, M. Lewenstein, arxiv:1103.3145 (2011).
[9] D. DeMille, Phys. Rev. Lett. 88 067901 (2002).
[10] L. D. Carr, D. DeMille, R. V. Krems, J. Ye, New J. Phys. **11**, 055049 (2009).
[11] M. Lysebo, L. Veseth, Phys. Rev. A 83, 033407 (2011).
[12] A. Andre et al., Nat. Phys. **2**, 636 (2006).
[13] S. F. Yelin, K. Kirby, R. Côté, Phys. Rev. A **74**, 050301 (2006).
[14] T. Köhler, K. Góral, P. S. Julienne, Rev. Mod. Phys. 78, 1311 (2006).
[15] K. M. Jones, E. Tiesinga, P. D. Lett, P. S. Julienne, Rev. Mod. Phys. 78, 483 (2006).
[16] J. M. Sage, S. Sainis, T. Bergeman, D. DeMille, Phys. Rev. Lett. 94, 203001 (2005).
[17] A. Vardi, D. Abrashkevich, E. Frishman, M. Shapiro, J. Chem. Phys. 107, 6166 (1997).
[18] A. Pe'er, E. A. Shapiro, M. C. Stowe, M. Shapiro, J. Ye, Phys. Rev. Lett. 98, 113004 (2007); E. A. Shapiro, M. Shapiro, A. Pe'er, J. Ye, Phys. Rev. A 78, 029903 (2008).
[19] E. Juarros, P. Pellegrini, K. Kirby, R. Côté, Phys. Rev. A 73, 041403(R) (2006).
[20] M. Viteau, A. Chotia, M. Allegrini, N. Bouloufa, O. Dulieu, D. Comparat, P. Pillet, Science 321, 232 (2008).
[21] C. P. Koch, J. P. Palao, R. Kosloff, F. Masnou-Seeuws, Phys. Rev. A 70, 013402 (2004).
[22] V. N. Vitanov, M. Fleischhauer, B. W. Shore, K. Bergmann, Adv. At. Mol. Opt. Phys. 46, 55 (2001).
[23] K.-K. Ni, S. Ospelkaus, M. H. G. de Miranda, A. PeŠer, B. Neyenhuis, J. J. Zirbel, S. Kotochigova, P. S. Julienne, D. S. Jin, J. Ye, Science 322, 231 (2008).
[24] J. G. Danzl, M. J. Mark, E. Haller, M. Gustavsson, R. Hart, J. Aldegunde, J. M. Hutson, H. C. Nagerl, Nat. Phys. 6, 265 (2010).
[25] K. Winkler, F. Lang, G. Thalhammer, P. v. d. Straten, R. Grimm, J. H. Denschlag, Phys. Rev. Lett. 98, 043201 (2007).
[26] S. Ospelkaus, A. Pe'er, K. - K. Ni, J. J. Zirbel, B. Neyenhuis, S. Kotochigova, P. S. Julienne, J. Ye, D. S. Jin, arXiv: atom-ph: 0802.1093.
[27] E. A. Shapiro, M. Shapiro, A. Pe'er, J. Ye, Phys. Rev. A 75, 013405 (2007).
[28] W. C. Stwalley, Eur. Phys. J. D 31, 221 (2004).
[29] D. Jaksch, V. Venturi, J. I. Cirac, C. J. Williams, P. Zoller, Phys. Rev. Lett. 89, 040402 (2002).

[30] E. Kuznetsova, P. Pellegrini, R. Côté, M. D. Lukin, S. F. Yelin, Phys. Rev. A 78, 021402 (2008).
[31] B. W. Shore, K. Bergemann, J. Oreg, S. Rosenwaks, Phys. Rev. A 44, 7442 (1991).
[32] P. A. Ivanov, N. V. Vitanov, K. Bergmann, Phys. Rev. A 70, 063409 (2004).
[33] V. S. Malinovsky, D. J. Tannor, Phys. Rev. A 56, 4929 (1997).
[34] N. V. Vitanov, B. W. Shore, K. Bergemann, Eur. Phys. J. D 4, 15 (1998).
[35] S. T. Cundiff, J. Ye, Rev. Mod. Phys. 75, 325 (2003).
[36] D. S. Petrov, C. Salomon, G. V. Shlyapnikov, Phys. Rev. Lett. 93, 090404 (2004); D. S. Petrov, C. Salomon, G. V. Shlyapnikov, Phys. Rev. A 71, 012708 (2005).
[37] J. Cubizolles, T. Bourdel, S. J. J. M. F. Kokkelmans, G. V. Shlyapnikov, C. Salomon, Phys. Rev. Lett. 91, 240401 (2003).
[38] S. Jochim, M. Bartenstein, A. Altmeyer, G. Hendl, C. Chin, J. Hecher Denschlag, R. Grimm, Phys. Rev. Lett. 91, 240402 (2003).
[39] C. A. Regal, M. Greiner, D. S. Jin, Phys. Rev. Lett. 92, 083201 (2004).
[40] J. M. Hutson, P. Soldan, Int. Rev. Phys. Chem. 26, 1 (2007).
[41] T. Mukaiyama, J. R. Abo-Shaeer, K. Hu, J. K. Chin, W. Ketterly, Phys. Rev. Lett. 92, 180402 (2004).
[42] C. Chin, T. Kraemer, M. Mark, J. Herbig, P. Waldburger, H. - C. Hagerl, R. Grimm, Phys. Rev. Lett. 94, 123201 (2005).
[43] G. Smirne, R. M. Godun, D. Cassettari, V. Boyer, C. J. Foot, T. Volz, N. Syassen, S. Durr, G. Rempe, M. D. Lee, K. Goral, T. Köhler, Phys. Rev. A 75, 020702(R) (2007).
[44] J. J. Zirbel, K. - K. Ni, S. Ospelkaus, J. P. D'Incao, C. E. Wieman, J. Ye, D. S. Jin, arXiv: cond-mat: 0710.2479.
[45] A. Fioretti, C. Amiot, C. M. Dion, O. Dulieu, M. Mazzoni, G. Smirne, C. Gabbanini, Eur. Phys. J. D 15, 189 (2001).
[46] S. T. Thompson, E. Hodby, C. E. Williams, Phys. Rev. Lett. 95, 190404 (2005); T. M. Hanna, T. Köhler, K. Burnett, Phys. Rev. A 75, 013606 (2007).
[47] H. Y. Ling, H. Pu, B. Seaman, Phys. Rev. Lett. 93, 250403 (2004).
[48] E. Juarros, P. Pellegrini, K. Kirby, R. Cote, Phys. Rev. A 73, 041403(R) (2006).
[49] E. Kuznetsova, M. Gacesa, P. Pellegrini, S. F. Yelin, R. Côté, New J. Phys. 11, 055028 (2009).
[50] T. Mukaiyama, J. R. Abo-Shaeer, K. Xu, J. K. Chin, and W. Ketterle, Phys. Rev. Lett. 92, 180402 (2004).
[51] C. Chin, T. Kraemer, M. Mark, J. Herbig, P. Waldburger, H.-C. Nägerl, and R. Grimm, Phys. Rev. Lett. 94, 123201 (2005).
[52] A. Vardi, D. Abrashkevich, E. Frishman, M. Shapiro, J. Chem. Phys. 107, 6166 (1997).
[53] A. Vardi, M. Shapiro, K. Bergmann, Optics Express 4, 91 (1999).
[54] Ph. Courteille, R. S. Freeland, D. J. Heinzen, F. A. van Abeelen, B. J. Verhaar, Phys. Rev. Lett. 81, 69 (1998).
[55] F. A. van Abeelen, D. J. Heinzen, B. J. Verhaar, Phys. Rev. A 57, R4102 (1998).
[56] M. Junker, D. Dries, C. Welford, J. Hitchcock, Y. P. Chen, R. G. Hulet, Phys. Rev. Lett. 101, 060406 (2008).
[57] P. Pellegrini, M. Gacesa, R. Côté, Phys. Rev. Lett. 101, 053201 (2008).
[58] C. P. Koch and R. Moszyński, Phys. Rev. A 78, 043417 (2008).
[59] F. Lang, K. Winkler, C. Strauss, et al., Phys. Rev. Lett. 101, 133005 (2008);
[60] U. Fano, Phys. Rev. 124, 1866 (1961).

[61] M. Fleischhauer, A. Imamoglu, J.P. Marangos, Rev. Mod. Phys. 77, 633 (2005).
[62] A. C. Han, E. A. Shapiro, M. Shapiro, arxiv:1104.1480 (2011).
[63] C. Lee, E.A. Ostrovskaya, Phys. Rev. A 72, 062321 (2005).
[64] A. Griesmaier, J. Werner, S. Hensier, J. Stuhler, T. Pfau, Phys. Rev. Lett. 94, 160401 (2005).
[65] Z. Pavlovic, R. V. Krems, R. Côté, H. R. Sadeghpour, Phys. Rev. A 71, 061402 (2005).
[66] M. Gacesa, P. Pellegrini, R. Côté, Phys. Rev. A 78, 010701(R) (2008).

6

Time-Resolved Laser Spectroscopy of Semiconductors - Physical Processes and Methods of Analysis

T. Brudevoll[1], A. K. Storebo[1], O. Skaaring[2], C. N. Kirkemo[3], O. C. Norum[4], O. Olsen[4] and M. Breivik[5]
[1]FFI (Norwegian Defence Research Establishment), Kjeller
[2]Kongsberg Defence and Areospace, Kjeller
[3]Department of Physics, University of Oslo, Oslo
[4]Department of Physics, Norwegian University of Science and Technology, Trondheim
[5]Department of Electronics and Telecommunications
Norwegian University of Science and Technology, Trondheim
Norway

1. Introduction

Time-resolved laser spectroscopy has become an important method for extracting optical and transport parameters of semiconductors and semiconductor nanostructures. In many of the spectroscopic techniques using lasers the material is brought into a state far away from thermal equilibrium. Interpreting the non-equilibrium state of the material often constitutes a considerable challenge. Many different types of quasiparticles are involved, free electrons and different flavors of holes as well as excitons in many variants, all coupled to lattice modes or hybrid lattice/charge-carrier modes. Spin and magnetic systems are included among the cases considered important. A simplification that can be made under low-level laser irradiation using long pulses is to assume that carrier and lattice temperatures are equal with no excess sample heating. Once this 'isothermal' assumption is broken, sophisticated analysis tools must be used, even for seemingly simple problems and regardless of laser pulse length.

There is now a vast literature dealing with the results from time-resolved laser spectroscopy experiments. In particular, the book by Shah (Shah, 1999) covers the fundamentals of the subject. However, this field is in very rapid progress, and therefore the text will mainly emphasize later works. We describe new developments and survey the current state of the art regarding methods used for analyzing experiments, assuming some basic à priori knowledge of key experimental techniques. We shall do this survey by taking particular examples from the literature which in our opinion defines some of the main trends. By extracting the essence of selected papers and adding some of our own recent results, we hope to provide a useful guide for those interested in the subject of laser-matter interactions. Obviously, this type of interaction lies at the heart of many future technologies. Because of its central position, laser spectroscopy has become a mixture of old and new theoretical tools, many of which takes years to learn and do well. Accordingly, any presentation of these tools runs a great risk of being bogged down with complicated mathematical relations and a time-consuming, heavy burden of pages. To avoid this, we restrict ourselves wherever possible to key ideas and relations actually needed for the experiments that we discuss, and present them in the context of their immediate use. Naturally, this comes at the expense of some loss of theoretical completeness, but in this way it quickly becomes clear what kind of theory is needed and how to continue with a narrow, well-targeted search for further background material.

2. Main spectroscopic techniques

Carrier dynamics in semiconductors can be explored by means of nonlinear laser spectroscopic techniques such as optical pump and probe spectroscopy, four-wave mixing, and transient gratings. In addition, time-resolved photoluminescence has been used to study hot-carrier relaxation and exciton formation. The main method used in time-resolved laser spectroscopy is the pump-probe technique. A strong pump pulse brings the material into a non-equilibrium state. Energetic excess charge carriers are generated by the pump pulse, and a weak probe pulse at specific delay times after the pump pulse instantly probes the state of the material. The delay of the probe beam provides the time resolution of the experiment.

Many different realizations exist. Development of new mid-infrared ultrafast laser sources makes photons with lower energies available, being well suited to probe intraband transitions. The response of free carriers and intra-exciton transitions can be studied with sub-picosecond temporal resolution. Mid-infrared probing of electron-hole (*e-h*) droplet formation in the direct gap semiconductor CuCl is a particularly striking example (Nagai et al. 2001). It is possible to extract information by utilizing either the reflected or the transmitted probe signal. On longer timescales the pump beam can give rise to luminescence from recombining excitons or electron-hole pairs. This luminescence can be analyzed, and e.g. the time evolution of the carrier temperature can be extracted. By using circularly polarized light instead of linearly polarized light, it is possible to induce imbalances in the spin (spin polarization) of the photoexcited carriers, producing, say, more spin up than spin down electrons. An accurate analysis of spin systems is possible by time-resolved two-photon photoemission, Faraday effect, and differential transmission spectroscopy (DTS). Also the prospect of controlling nuclear spins using optics is currently investigated (Makhonin et al., 2010). A particularly interesting case to study as semiconductor device dimensions shrink would be carrier dynamics and scattering occurring at material boundaries. Recent progress in experimental techniques suggests that five-wave mixing setups will provide access to such processes, (Voelkmann, 2004, Meier 2005).

Another important recent development is the introduction of combined ultrafast laser and terahertz THz methods, such as Terahertz time-domain spectroscopy (THz-TDS). Here an ultrafast laser is the pump, and the THz beam is the probe. To an even larger degree than mid-infrared lasers, a THz beam can probe the intraband properties of a semiconductor because the photon energy is much too low to generate band to band transitions. Unlike optical methods, THz-TDS enables the determination of the complex dielectric function (or optical conductivity) from which one can quantitatively evaluate the carrier density and the scattering time of the photoexcited carriers without resorting to the Kramers-Kronig relation. THz radiation can be generated in many ways, see (Blanchard et al., 2011) for an overview.

Furthermore, in semiconductors with medium or wide band gaps the excitonic effects are often significant, and here THz probing becomes very useful. A laser pump beam can create excitons, either directly by exciting an electron from the valence band into the 1*s* hydrogenic state of the exciton, or indirectly by exciting the electron from the valence band high into the conduction band. After a while, as the excess energy of the electron is dissipated by phonon scattering, it condenses into a hydrogenic-like state with a valence band hole, having a total energy less than the semiconductor band gap. The THz field can now directly probe the intraexciton transitions.

Finally, knowledge of the cooling dynamics in photoexcited *e-h* plasmas may also be of some interest in the study of low-temperature quantum-degenerate effects (Butov et al., 2002).

3. Laser pulselength regimes

3.1 General

In laser experiments involving band-to-band transitions, two types of carriers are generated, electrons and holes. In semiconductors electrons usually have much smaller effective masses than holes, and therefore electrons initially receive most of the laser energy. A simplified band structure for a direct gap semiconductor is shown in Fig. 1.

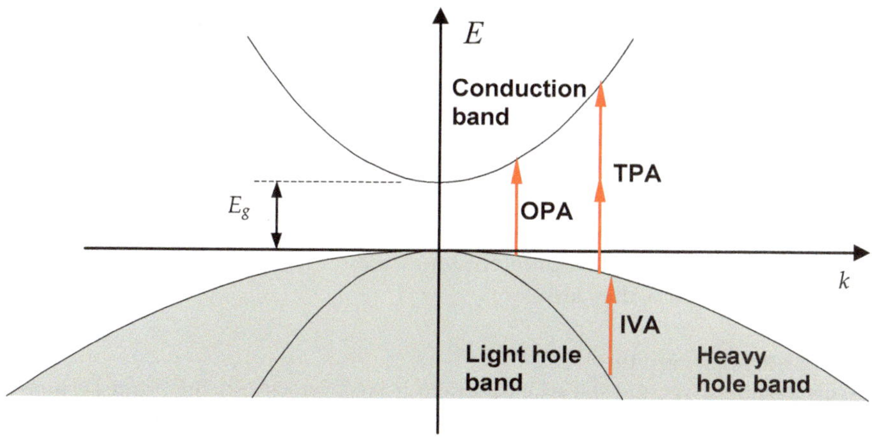

Fig. 1. Illustration of the band structure and important absorption processes in a direct gap semiconductor: OPA - one-photon absorption, TPA - two-photon absorption, IVA - inter-valence band absorption. In addition there will be absorption of radiation by free carriers, and there will be radiative or non-radiative recombination of electrons and holes. The shaded area indicates that the valence band is essentially fully occupied at thermal equilibrium, while the conduction band is empty. Below the heavy and light hole bands there is a split-off band, which in cases of strong spin-orbit splitting is separated from the former bands by a considerable energy gap (not shown).

Short pulses and short time delays between pump and probe sometimes help us to avoid the estimation of recombination effects and provide a very direct insight into carrier relaxation and hot phonon generation phenomena by isolating these processes from subsequent effects that occur on longer timescales. Energy transfer due to long, high fluence laser pulses on the other hand, often increases the overall lattice temperature, but makes possible the investigation of many additional physical effects. In particular, if the photon energy of the incident radiation just exceeds the band gap, absorption can be highly non-linear and depend strongly on wavelength, radiation intensity, carrier temperatures, and lattice temperatures (Storebo et al., 2010).

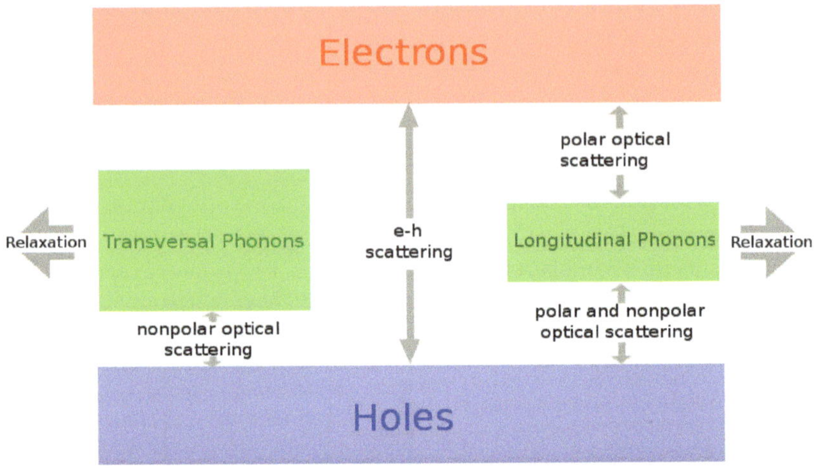

Fig. 2. Laser energy pathways between the different carrier ensembles. Electrons and holes initially pick up the laser energy. The ensembles are directly connected through electron-hole scattering, as well as indirectly through the longitudinal phonon distribution. The excess energy of the system can exit through both longitudinal and transversal phonons.

Laser energy pathways between the different carrier ensembles are illustrated in Fig. 2. We shall now make a tour through the different timescales of the laser experiments and discuss the physics-related issues as they arise.

3.2 Coherent ultrafast regime

In an ideally coherent regime the excitations of a laser irradiated quantum system would maintain a well-defined phase relation with the laser field. This coherence is rapidly destroyed by various scattering or collision processes, and therefore these influences can never be completely disregarded, but should rather be incorporated into the theory. A collision has a finite duration defined by the oscillation period of the energy quantum exchanged during the collision. In GaAs, the oscillation period of optical phonons is ~ 115 fs and the oscillation period of the electron plasma (plasmons are longitudinal collective carrier oscillations) is ~ 150 fs at a carrier density of $5 \cdot 10^{17} \mathrm{cm}^{-3}$. The interval between collisions can often be 100-200 fs. The dynamics of the coherent regime can be explored using differential transmission spectroscopy (DTS) (Shah, 1999) with laser pulses much shorter than the characteristic times given above, that is, on the order of 10 fs or shorter. Bulk materials, semiconductor microcavities, and nanostructures are investigated. Schemes for avoiding de-coherence is also an important aspect of future quantum information systems.

3.2.1 Semiconductor Bloch Equations (SBE)

Much of the early analysis on this subject matter has been carried out using the Semiconductor Bloch Equations (SBE), which we shall examine next. We note that SBE are most frequently used in situations where the optical properties are prominent, typically for excitations near the band gap, and for describing the dynamics of excitons and semiconductor lasers. If we treat the charge carriers by quantum mechanics and the radiation field classically, we arrive at the following standard equations for the polarization density, electron density, and hole density (Haug & Koch, 2009)

$$\begin{aligned}\frac{\partial \mathrm{P}_{k}}{\partial \mathrm{t}} &= -\mathrm{i}\left(\mathrm{e}_{\mathrm{e},k}+\mathrm{e}_{\mathrm{h},k}\right)\mathrm{P}_{k}-\mathrm{i}\left(\mathrm{f}_{\mathrm{e},k}+\mathrm{f}_{\mathrm{h},k}-1\right)\omega_{\mathrm{R},k}+\left(\frac{\partial \mathrm{P}_{k}}{\partial \mathrm{t}}\right)_{\mathrm{coll}} \\ \frac{\partial \mathrm{f}_{\mathrm{e},k}}{\partial \mathrm{t}} &= -2\,\mathrm{Im}\left(\omega_{\mathrm{R},k}\mathrm{P}_{k}^{*}\right)+\left(\frac{\partial \mathrm{f}_{\mathrm{e},k}}{\partial \mathrm{t}}\right)_{\mathrm{coll}} \\ \frac{\partial \mathrm{f}_{\mathrm{h},k}}{\partial \mathrm{t}} &= -2\,\mathrm{Im}\left(\omega_{\mathrm{R},k}\mathrm{P}_{k}^{*}\right)+\left(\frac{\partial \mathrm{f}_{\mathrm{h},k}}{\partial \mathrm{t}}\right)_{\mathrm{coll}}\end{aligned} \tag{1}$$

We now proceed to explain the meaning of the various terms occurring in the above equations. In separating out the time variation of the polarization variables P_{k} and the carrier occupation variables f_{ik} (where $i = e,h$) due to collisions as we have done here it has been assumed that the non-collision terms correspond to the time-dependent Hartree-Fock approximation and that the collision terms are corrections to Hartree-Fock. This is reflected in the expressions for the single-particle energies $\mathrm{e}_{\mathrm{e,k}}$ and $\mathrm{e}_{\mathrm{h,k}}$ defined as

$$\hbar \mathrm{e}_{\mathrm{i},k} = \hbar\varepsilon_{\mathrm{i},k} + \mathrm{i}\,\Sigma_{\mathrm{exch,i}}(\boldsymbol{k}) = \hbar\varepsilon_{\mathrm{i},k} - \sum_{q}\mathrm{V}_{|k-q|}\mathrm{f}_{\mathrm{i},q} \tag{2}$$

In band structure terminology the single particle energies in the Hartree-Fock approximation includes a part of the contribution from carrier-carrier interactions called exchange (Ashcroft & Mermin, 1976, Ch. 17). The strength of the carrier-carrier interactions are given by the parameter V_{q}. The exchange energy Σ_{exch} renormalizes the single particle energies $\hbar\varepsilon_{\mathrm{i,k}}$. If further interaction terms are included, this is called correlation (or collision) terms.

When an electron is excited from the valence band a hole is created. This electron and the corresponding hole create a microscopic polarization in the form of a dipole to which the incoming radiation field $\boldsymbol{E}$ couples. Propagation effects of the radiation field must explicitly be taken into account by simultaneously solving the wave equation, e.g. by a numerical scheme such as the Finite Difference Time-Domain method. Electrons and holes can either form excitons or free electron-hole pairs. The factor $-(\mathrm{f}_{\mathrm{e,k}}+\mathrm{f}_{\mathrm{h,k}}-1)=\mathrm{f}_{\mathrm{v,k}}-\mathrm{f}_{\mathrm{c,k}}$ represents the population inversion at $\boldsymbol{k}$, that is, the difference between electron occupation of the valence band state $\boldsymbol{k}$ and electron occupation of the conduction band state $\boldsymbol{k}$. This factor influences the absorption spectrum of the material by state filling effects. The strength of the interband dipole coupling to the radiation field is given by the generalized Rabi frequency,

$$\omega_{\mathrm{R},k} = \frac{1}{\hbar}\left[\boldsymbol{d}_{\mathrm{c,v}}\cdot\boldsymbol{E} + \sum_{q\neq k}\mathrm{V}_{|k-q|}\mathrm{P}_{q}\right] \tag{3}$$

where $d_{c,v}$ is the transition dipole moment between initial and final states, obtained by a summation over the positions of the charge carriers i,

$$\boldsymbol{d} = e\left(\sum_i x_i, \sum_i y_i, \sum_i z_i\right) \tag{4}$$

and calculating the matrix element of the vector $\boldsymbol{d}$.

Remember that the Rabi frequency is a semiclassical concept representing the frequency of population oscillation between the levels of a resonantly illuminated two-level system. When the system is illuminated it will cyclically absorb photons and reemit them by stimulated emission. The Rabi frequency is a measure of the coupling strength between the electronic system and the radiation field. We see from the form of the equation for $\omega_{R,k}$ that the electron-hole system reacts to the combination of the applied field $\boldsymbol{E}(t)$ and the polarization field of the generated electron-hole pairs. Absorption of radiation and generation of electrons and holes are described by the factor $-2\,\mathrm{Im}\left[\omega_{R,k} P_k^*\right]$.

At first glance, the equations for electron occupation and hole occupation appear very similar. If scattering is ignored, the rate of change in electron and hole populations are identical, and the two equations do not seem to mix different momentum states $\boldsymbol{k}$. But a closer look reveals that Coulomb terms in the generalized Rabi frequency and in the exchange energy in fact lead to such a coupling.

The original two-level SBE assume a scalar field and were not designed to account for the degrees of freedom connected with carrier spin or with the transversality of the exciting radiation. Spin is related to the concept of exchange, which we will encounter several times in the following sections. The charge carriers are fermions, and as such they must have antisymmetric wavefunctions under exchange of identical particles. Antisymmetry can either come from the spin or the spatial part of the wavefunction, and is responsible for how particles arrange themselves. This can be viewed as a kind of particle interaction, and including exchange is therefore vital in many cases. Exchange effects are a manifestation of the Pauli exclusion principle.

Spin and the transversality of the radiation field are somehow connected. Optical excitation with circular polarization generates spin-polarized carrier populations (optical spin alignment) and can be used to investigate spin dynamics. Research on spin-dependent phenomena in semiconductors is progressing in many different directions. Much of the emphasis within spintronic device research is related to long-lived electron-spin polarization and long-range transport properties. But outside the mainstream one still finds many interesting effects to be studied. For example, in materials lacking inversion symmetry, the resulting small splitting of degenerate single-particle electronic states is connected to the circular photogalvanic effect, CPGE (Ganichev, 2000, 2001). Spin relaxation times can be measured by noticing that the nonlinear absorption is different for linear and circular polarized excitation. (Ganichev, 2002).

In spin systems, not only the electrons are important. The hole spin may also be used to carry information. Holes play an important role for magnetic correlations in diluted magnetic semiconductors, as we shall see in a later section. From a fundamental point of view, holes in III–V semiconductors are a model system for spin relaxation with pronounced spin-orbit (SO) coupling. The SO coupling leads to a strong momentum dependent mixing of spin and orbital-momentum eigenstates, so that scattering processes can change spin and orbital angular momentum. Although both electrons and holes are affected by this relaxation mechanism, it causes a much faster spin relaxation for the *p*-like holes than the *s*-like electrons, because the former experience the SO coupling directly, while the mainly *s*-like conduction band only achieves an orbital momentum by a weak coupling to remote bands with nonzero orbital momentum.

The mechanisms of spin relaxation for bulk and for semiconductor quantum wells are quite similar. The dominant ones for most semiconductor structures under investigation are the D'yakonov-Perel' (DP) mechanism, related to the spin-splitting of the single-particle states in systems lacking inversion symmetry, and the Bir-Aronov-Pikus (BAP) mechanism, caused by spin flip electron-hole exchange scattering processes. In addition, we have the phonon- or impurity-mediated spin flip Elliot-Yafet mechanism.

A detailed understanding of the processes contributing to spin relaxation on ultrashort timescales is impossible using simplified relaxation-time approximations. And we have seen that we must consider holes as well as electrons. Before we mention alternative approaches to include spin and the transversality of the radiation, it is useful first to take a look at the difficulties involved. It is important to remember that the field of spin dynamics is an aspect of carrier dynamics. Historically, numerical transport theory schemes used for the calculation of carrier dynamics did not resolve the spin of the charge carriers. But it is perfectly possible to incorporate spin into these schemes to analyze ultrafast spin-dependent optical experiments. Therefore we shall first take a short step back and consider spin from a traditional transport theory viewpoint.

In the transport theory approach, band structures are first calculated and used as input for the calculation of the momentum and time-dependent carrier distribution functions including the relevant interaction mechanisms.

By combining a realistic band structure with simulations of the evolving carrier distribution function, we can achieve a quantitative physical picture of the hole-spin dynamics, which in turn can be confirmed by optical experiments. A typical model system for such experiments is GaAs, having rather strong spin-orbit coupling. Krauss et al. (Krauss, 2008) obtained theoretical spin-relaxation times in quantitative agreement with measured hole-spin relaxation times, but it also turned out that a careful analysis of the experimental setup was necessary to achieve this agreement. They showed that different optical techniques for the measurement of hole-spin dynamics gave different results, in contrast to the case of the much simpler electron-spin dynamics. In particular, the relation between spin and band eigenstates (quasiparticle states) is not as simple as for electrons. Therefore different relaxation times for the quasiparticle polarization and the spin polarization have to be defined, and further related to the polarization extracted from differential transmission measurements. In electron dynamics, all these quantities coincide and can reveal the degree of spin polarization.

Extension of the SBE formalism in order to treat the dynamics of holes including the spin and polarization degrees of freedom can be found in e.g. (Rössler, 2003). Here a six band model is used for the heavy hole, light hole, and spin-orbit split off band, each of these three bands containing two spin degrees of freedom (Fig. 1). Undoubtedly, treating hole dynamics with the SBE is complicated. We only mention here a long standing issue regarding the polarization-dependent four wave mixing (FWM) quantum beats, which have been observed after simultaneous excitations of two optical transitions, i.e. heavy hole and light hole. The signal

magnitude and its beat phase depend on the polarization of the pump and probe with respect to each other, and this can in principle be analyzed with Bloch equations (SBE) in a six-band model. However, such a theory would predict identical FWM intensities for the two polarization configurations, with pump and probe having either parallel or perpendicular polarizations. Much work has been devoted to explain this phenomenon theoretically. One of the successful, but disputed (Wang al. 2006) candidates for explanation is the bi-exciton theory, which includes exciton-exciton interactions not included in standard SBE.
In the literature, we now find specific SBE formulations for quantum wells, wires, and dots (Kira et al., 1999, Haug & Koch, 2009, Feldtmann, 2009). We especially mention formulations for carbon nanotubes (Hirtschulz et al. 2008).
In carbon nanotubes (CNTs), many-body effects such as electron-electron and electron-phonon interactions play an important role, and excitons couple to vibronic modes. Much of the work on CNTs so far has dealt with the steady-state response of these excitations to cw fields. Frequency domain descriptions using Hubbard models (Ma & Yuan 1998, Lopez-Sancho et al. 2001) or *ab initio* density functional theory and the Bethe-Salpeter equation (Chang et al. 2004) have been developed. The Bethe-Salpeter equation is well suited for describing excitonic effects, but such calculations are linked to *ab initio* electronic structure methods and this combination can be extremely computer intensive. See e.g. (Christensen et al., 2011) for an update on the merits of such approaches. In contrast, Hirtschulz et al. describe a compact dynamical method for CNT properties which seems very useful for the study of ultrashort laser-induced nonlinear optical response. Many-body density matrix theory was combined with tight-binding band structure calculations of CNTs. It was shown that the derived carbon nanotube Bloch equations contained excitons as elementary optical excitations, both for low-level excitation as well as for optical nonlinearities and gain induced at higher excitations. They illustrated the temporal response to ultrashort pulses by calculating the time dependence of the electron density and the optical Stark effect.

3.2.2 Simplification by the use of excitonic bases

Several approaches have been developed to study the nonlinear optical response of semiconductors. Non-equilibrium Green's function and density matrix techniques are among the most common (Haug & Jauho, 1996, Haug & Koch, 2009, Mahan, 2007). Semiconductor Bloch equations (SBEs) are based on the density matrix, as are the dynamics controlled truncation (DCT) equations, which are an alternative to the Hartree-Fock treatment of the SBE mentioned above. Standard SBE have some drawbacks. Coulomb interaction between free electrons and holes require the Hartree-Fock procedure, which readily give the excitons but cannot describe exciton-exciton interactions. In addition, the solution of coupled SBE equations is a rather demanding numerical task.
In recent years it has become clear that using a different basis can greatly simplify this task for cases where excitonic effects dominate. Instead of using free electrons and holes as the starting point of the theory, one can use correlated pairs, i.e. excitons as basic operators. In ordinary SBE, the Coulomb interactions between carriers are cut by the HF method we saw above to get a closed, but numerically heavy set of equations. This truncation scheme leads to a spurious divergence in the carrier-carrier scattering rate, which must be fixed by a manual procedure. Higher-order Coulomb correlations cannot be incorporated unless other computationally intensive methods are chosen, such as the Bethe-Salpeter Equation or non-equilibrium Greens functions.

Excitons being composite bosons, do not have strict commutation rules. It becomes impossible to treat interactions between excitons through an effective exciton Hamiltonian $H_{ex} + U_{exex}$, where U_{exex} is the interaction potential between excitons regarded as elementary bosons. As a consequence, standard many-body techniques will not work, because they all depend on a perturbative expansion in the interaction potential U_{exex}.

By using exciton based SBE (Haug & Koch, 2009, Feldtmann, 2009) or other exciton based approaches (Combescot et al., 2003, Wang, 2009, Wang & Dignam, 2009), to study specific problems, we can achieve substantial simplifications. The method of Combescot et al. leads to an analytic expression for the polarization induced by the laser pulse, with the two sources for the polarization, Pauli blocking and Coulomb interaction becoming tracable troughout the calculation.

An excitonic basis requires a certain mapping (or pairing) of electrons to holes. This mapping of the electron-hole basis onto the excitonic basis is not one-to-one. For example, if we have two electrons and two holes in specific single-particle states, there are two different ways in which these electrons could be paired with the holes.

If we introduce a certain pairing scheme, this ambiguity may be resolved. Two such schemes will be described here. The first is to restrict an electron-hole pair to be in a given internal exciton state. In most cases it will be most profitable to choose the 1*s* state. We then allow the center-of-mass momentum $\mathbf{K}_{CM}$ of the pair to take any value, and call this choice the state-restricted exciton basis. The second approach is to pair electrons and holes that have opposite momenta, meaning that the resulting exciton will have $\mathbf{K}_{CM} = \mathbf{0}$ (remember that photons carry a very small momentum and will therefore create $\mathbf{K}_{CM} \sim \mathbf{0}$ excitons directly). Now these momentum-restricted excitons can have any internal state 1*s*,2*s*,2*p*, including continuum exciton states.

In order to find the best scheme for a given problem, we must consider their strengths and weaknesses. For example, a 1s state-restricted approach requires resonant optical excitation of 1*s* excitons and cannot account for THz-induced transitions between different internal excitonic states e.g., 1*s* to 2*p*. But the freedom in momentum allows us to study four-wave mixing and examine the time evolution of exciton momenta, including scattering of excitons to optically inactive excitonic states with nonzero $\mathbf{K}_{CM}$. The state-restricted scheme can also to a certain degree account for the formation of biexcitons.

The fixed-momentum approach cannot describe four-wave mixing and the formation of biexcitons. Interactions between excitons are limited by the restriction in momentum, and include only the exchange effect, which sometimes can be dominant. However, because it includes all internal states, it allows for excitation to states other than the 1s state, and can account for scattering of the optically active excitonic states into optically inactive excitonic states. Furthermore, Terahertz induced transitions between excitonic internal states can be modeled, as well as state-filling effects. In the coherent regime, most of the excitons still have $\mathbf{K}_{CM} = \mathbf{0}$ momentum. Including only $\mathbf{K}_{CM} = \mathbf{0}$ excitons is therefore not a serious limitation on short timescales.

A central problem with excitonic bases is how to deal with state-filling effects associated with the composite nature of excitons. State-filling results in non-bosonic carrier density-dependent commutation relation for the excitonic operators. At low carrier densities, state-filling is negligible and the excitons will obey boson commutation rules. At moderate excitation densities this is a reasonable approximation. Another alternative is to "bosonize" the excitons, resulting in creation and annihilation operators following strict boson commutation rules. Bosonization schemes are still debated, with difficulties defining an effective Hamiltonian (Combescot et al., 2003).

On the other hand, it is sometimes possible to just accept and use the non-bosonic exciton creation and annihilation operators, and to explicitly take their exact density-dependent commutation relations into account, as explained in the paper by Wang & Dignam (Wang & Dignam, 2009). This can be done by first defining a set of quasi-boson states. These are electron-hole pair states consisting of the natural electron and hole single-particle eigenstates of the semiconductor structure, paired so that the total momentum $\mathbf{K}_{CM} = \mathbf{0}$. Based on quasi-boson commutation rules and a quasi-boson Hamiltonian. the dynamical equations for the quasi-boson operators are obtained. The true excitons are a superposition of quasi-boson states, and this superposition then defines the actual transformation between quasi-bosons and true excitons. Thus, dynamic equations for the true excitons finally appear by a transformation from the corresponding quasi-boson dynamic equations.

Wang & Dignam used a momentum-restricted $\mathbf{K}_{CM} = \mathbf{0}$ excitonic basis including state-filling and exchange to study the excitonic influence on the ultrafast nonlinear absorption in quantum wells. The optical pulse was resonant on the 1*s* exciton. They found that, the momentum-restriction notwithstanding, resonant excitation of the 1*s* exciton state at moderate carrier densities $\sim 1.3 \cdot 10^{10}\,\mathrm{cm}^{-2}$ could be described by further restricting the basis to 1*s* exciton states only. However, for higher carrier densities $\sim 5.0 \cdot 10^{10}\,\mathrm{cm}^{-2}$, the 1*s* peak was reduced and blueshifted and the band edge was redshifted. These effects on the absorption can only be described correctly if the coupling of the 1*s* states to both optically active and inactive higher exciton states is taken into account. Again, characteristic for the use of excitonic bases, it was possible to obtain analytical expressions for the density-dependent blueshift and bleaching of the 1*s* excitonic resonance for the case of moderate carrier densities.

To summarize, the approximations we have seen here are made possible by a close scrutiny of the excitation conditions, finding that only a small portion of the exciton states are relevant for a given problem.

3.2.3 Systems with electrical contacts

Another issue when probing optoelectronic structures is the need to account for injection from electrical contacts, i.e. when the semiconductor has open spatial boundaries to exchange charge carriers with an environment (e.g. for investigation of free carrier external field induced drift velocity optical nonlinearities). Contact or injection model problems are natural ingredients in methods arising from transport theory, e.g. non-equilibrium Greens functions, Wigner formalism, and Monte Carlo methods. As we have already mentioned, such methods are also used to analyze short pulse laser experiments. SBE, on the other hand, is a theory that has appeared directly from the analysis of optical experiments, and carrier injection/open boundaries have obviously not been the primary concern.

In nanoscale devices both coherent - i.e. scattering-free - and incoherent - i.e., phase-breaking -processes must be treated simultaneously. In spite of the quantum mechanical behavior of carrier dynamics in the device active region, the overall behavior is often the result of an interaction between phase coherence and energy relaxation/dephasing, the latter being primarily due to the presence of spatial boundaries or contacts/charge reservoirs. A microscopic theory for the description of quantum-transport phenomena in systems with open boundaries was proposed by (Proietti Zaccaria & Rossi, 2003). They showed that the application of the conventional Wigner-function quantum transport formalism to this problem leads to unphysical results, such as injection of coherent electronic states from the contacts. A generalization of the Wigner-function formulation was made, able to describe both the active region and the incoherent nature of carrier injection more akin to the phenomenological injection models already used for open quantum devices.

A generalization to systems with open boundaries of the SBE has also been proposed (Rossi, 1998). However, the emphasis was mainly on phase coherence and energy relaxation within the active region, and to a lesser extent on the carrier-injection from the electrical contacts into the device.

3.2.4 Semiconductor Luminescence Equations (SLE)

Recombination of electron-hole pairs and luminescence are fundamental processes in semiconductors. The luminescence in atomic systems is modified when atoms are optically coupled or when they are located within a cavity. With an array of quantum wells a similar case can be established for semiconductors. Radiation from individual QWs is partially reflected, transmitted, or absorbed by the neighbors. By placing the quantum wells in a semiconductor microcavity, further modifications occur. For low photon energy excitations the semiconductor material shows excitonic resonances below the fundamental absorption edge. In a microcavity designed so that the exciton resonance is strongly coupled to the single longitudinal cavity mode one obtains a double peaked mode coupling spectrum which is revealed in transmission, reflection, and photoluminescence. Like in atomic systems, strong coupling effects and suppressed or enhanced spontaneous emission are a result of the high-quality optical resonance.

Pump-probe experiments can usually be explained in sufficient detail by a classical description of the electromagnetic field. A quantum treatment only leads to small corrections as long as the classical fields exceed the vacuum fluctuations. Therefore a full quantum theory of radiation interacting with semiconductor materials is necessary only in situations where these fluctuations really make a difference. Experimentally this is the case for the analysis of luminescence, i.e., for radiation spontaneously emitted when excited electrons return from the conduction to the valence band (radiative electron-hole recombination). Spontaneous emission is a pure quantum effect and no classical radiation field needs to be present. Analysis of statistical properties and correlations of the electromagnetic field requires a full quantum theory as well. A study of luminescence properties together with the presence of classical fields should therefore both be challenging and give valuable insight into the coupling of matter and radiation.

Within a classical description of radiation, a major problem arises from the consistent inclusion of carrier-carrier Coulomb interactions. Excitonic effects manifest themselves through inclusion of the Coulomb interaction between carriers. In this review, a theory for the semiconductor luminescence of electron-hole pairs is presented which includes many-body effects. We shall treat the interacting carrier-photon system in the electron-hole picture. The operator equations are presented for the case of a single quantum well (QW), but the theory is part of a more general approach for investigating quantum properties of radiation in semiconductors. The theory is valid both in the linear and nonlinear regimes and can be used not only to analyze stationary emission under quasi-steady state conditions but also for the temporal emission in transient non-equilibrium situations. Mixing of incoherent and coherent fields can also be investigated. In the presentation by Kira et al. (Kira et al., 99), further results explaining the emission of radiation from single-quantum wells, quantum-well arrays, and quantum wells inside high-finesse optical cavities are summarized.

Here we shall mainly concentrate on incoherent luminescence resulting from a population of excited electrons and holes. As a consequence of the attractive interband interaction, strong correlations between conduction-band electrons and valence-band holes are to be expected. We have already mentioned earlier that for low excitation densities and low carrier temperatures these correlations lead to excitonic (bound electron-hole pair) resonances below the fundamental absorption edge, and that these resonances often dominate the absorption and emission spectra. The semiconductor luminescence equations (SLE) provide a description of incoherent excitonic photoluminescence. They are based on a generalization of the Hartree-Fock decoupling scheme and lay the foundations for relations between photoluminescence and absorption, the build-up of radiation from excitons, and nonlinear radiation effects. The equations bear some resemblance to the semiclassical Semiconductor Bloch equations describing the coherent excitation dynamics.

When coherent sources resonantly excite the QW (optical pumping), we have to extend the analysis to include also a classical driving field. In this case the semiconductor luminescence equations are coupled to the semiconductor Bloch equations leading to a mixing of coherent excitation and luminescence. So, in essence, coherent field effects are described classically, and incoherent field effects are treated quantum mechanically. A typical situation where the SLE can be used is after an ultrashort laser pulse has excited electrons high up in the conduction band, and we want to study the luminescence in a narrow spectral region near the exciton peaks which appears as the hot electrons have cooled down. The SLE equations for a quantum well in the absence of a coherent driving field, following the notation of Kira et al. (Kira et al., 1999) are,

$$\begin{aligned} i\hbar\frac{\partial}{\partial t}\left\langle b^{+}_{q_z,\boldsymbol{q}_{\parallel}} b_{q'_z,\boldsymbol{q}_{\parallel}}\right\rangle &= \hbar\left(\omega_{\boldsymbol{q}'}-\omega_{\boldsymbol{q}}\right)\left\langle b^{+}_{q_z,\boldsymbol{q}_{\parallel}} b_{q'_z,\boldsymbol{q}_{\parallel}}\right\rangle + i\xi_{\boldsymbol{q}} u_{QW,\boldsymbol{q}}\left\langle b_{\boldsymbol{q}'} P^{+}_{QW}\left(\boldsymbol{q}_{\parallel}\right)\right\rangle \\ &+ i\xi_{\boldsymbol{q}'} u^{*}_{QW,\boldsymbol{q}}\left\langle b^{+}_{\boldsymbol{q}} P_{QW}\left(\boldsymbol{q}'_{\parallel}\right)\right\rangle \end{aligned} \tag{5}$$

$$\begin{aligned} i\hbar\frac{\partial}{\partial t}\left\langle b^{+}_{q_z,\boldsymbol{q}_{\parallel}} P_{\boldsymbol{k}_{\parallel}}\left(\boldsymbol{q}_{\parallel}\right)\right\rangle &= \left[\varepsilon^{c}_{\boldsymbol{k}_{\parallel}+\boldsymbol{q}_{\parallel}} - \varepsilon^{v}_{\boldsymbol{k}_{\parallel}} - \hbar\omega_{\boldsymbol{q}} - \Sigma\left(\boldsymbol{k}_{\parallel},\boldsymbol{q}_{\parallel}\right)\right]\hbar\left(\omega_{\boldsymbol{q}'}-\omega_{\boldsymbol{q}}\right)\left\langle b^{+}_{\boldsymbol{q}} P_{\boldsymbol{k}_{\parallel}}\left(\boldsymbol{q}_{\parallel}\right)\right\rangle \\ &- \left(1 - f^{e}_{\boldsymbol{k}_{\parallel}+\boldsymbol{q}\parallel} - f^{h}_{\boldsymbol{k}_{\parallel}}\right)\Omega_{ST}\left(\boldsymbol{k}_{\parallel},\boldsymbol{q}\right) + f^{e}_{\boldsymbol{k}_{\parallel}+\boldsymbol{q}_{\parallel}} f^{h}_{\boldsymbol{k}_{\parallel}}\Omega_{SE}\left(\boldsymbol{q}\right) \end{aligned} \tag{6}$$

$$i\hbar\frac{\partial}{\partial t} f^{e}_{\boldsymbol{k}_{\parallel}} = 2i\sum_{q_z,\boldsymbol{q}_{\parallel}} \mathrm{Im}\left[-i d^{*}_{cv}\left(\boldsymbol{q}_{\parallel}\right)\xi_{\boldsymbol{q}} u^{*}_{QW,\boldsymbol{q}}\left\langle b^{+}_{q_z,\boldsymbol{q}_{\parallel}} P_{\boldsymbol{k}_{\parallel}-\boldsymbol{q}_{\parallel}}\left(\boldsymbol{q}_{\parallel}\right)\right\rangle\right] \tag{7}$$

$$i\hbar\frac{\partial}{\partial t} f^{h}_{\boldsymbol{k}_{\parallel}} = 2i\sum_{q_z,\boldsymbol{q}_{\parallel}} \mathrm{Im}\left[-i d^{*}_{cv}\left(\boldsymbol{q}_{\parallel}\right)\xi_{\boldsymbol{q}} u^{*}_{QW,\boldsymbol{q}}\left\langle b^{+}_{q_z,\boldsymbol{q}_{\parallel}} P_{\boldsymbol{k}_{\parallel}}\left(\boldsymbol{q}_{\parallel}\right)\right\rangle\right] \tag{8}$$

where the renormalized free-particle energy is given by the expression

$$\Sigma\left(\boldsymbol{k}_{\|},\boldsymbol{q}_{\|}\right)=\sum_{\boldsymbol{k}_{\|}'}\mathrm{V}_{\boldsymbol{k}_{\|}-\boldsymbol{k}_{\|}'}\left(\mathrm{f}^{\mathrm{e}}_{\boldsymbol{k}_{\|}'+\boldsymbol{q}_{\|}}+\mathrm{f}^{\mathrm{h}}_{\boldsymbol{k}_{\|}'}\right) \tag{9}$$

with

$$\mathrm{V}_{\boldsymbol{k}_{\|}}=\frac{\mathrm{e}^2}{2\varepsilon_0\left|\boldsymbol{k}_{\|}\right|}\int \mathrm{g(z')g(z)e}^{-\boldsymbol{k}_{\|}|\mathrm{z-z'}|}\mathrm{dzdz'} \tag{10}$$

and where $\mathrm{g(z)}$ is the quantum-well confinement function (product of the envelope functions of the electrons in the conduction and valence bands), the (b^+/b) are photon creation/annihilation operators, $\xi_q=\sqrt{\frac{\hbar\omega_q}{2\varepsilon_0}}$, $\mathrm{u}_{\mathrm{QW},q}=\int \mathrm{g(z)u}_q\mathrm{(z)dz}$ with $\mathrm{u}_q\mathrm{(z)}$ and ω_q denoting the stationary radiation eigenmode amplitude and frequency, respectively. A QW polarization can then be defined by

$$\mathrm{P_{QW}}\left(\boldsymbol{q}_{\|}\right)=\sum_{\boldsymbol{k}_{\|}}\left[\boldsymbol{d}^*_{\mathrm{cv}}\mathrm{P}_{\boldsymbol{k}_{\|}}\left(\boldsymbol{q}_{\|}\right)+\boldsymbol{d}_{\mathrm{cv}}\mathrm{P}^+_{\boldsymbol{k}_{\|}}\left(-\boldsymbol{q}_{\|}\right)\right] \tag{11}$$

Here the electron picture is used in both the conduction and valence bands for the definition of the dimensionless operators $\mathrm{P}^+_{\boldsymbol{k}_{\|}}\left(\boldsymbol{q}_{\|}\right)$ and $\mathrm{P}_{\boldsymbol{k}_{\|}}\left(\boldsymbol{q}_{\|}\right)$. The conduction band electron creation/annihilation (c^+/c) and valence band electron creation/annihilation (v^+/v) operators define

$$\mathrm{P}^+_{\boldsymbol{k}_{\|}}\left(\boldsymbol{q}_{\|}\right)=c^+_{\boldsymbol{k}_{\|}+\boldsymbol{q}_{\|}}v_{\boldsymbol{k}_{\|}},\mathrm{P}_{\boldsymbol{k}_{\|}}\left(\boldsymbol{q}_{\|}\right)=v^+_{\boldsymbol{k}_{\|}}c_{\boldsymbol{k}_{\|}+\boldsymbol{q}_{\|}} \tag{12}$$

$\mathrm{P}^+_{\boldsymbol{k}_{\|}}\left(\boldsymbol{q}_{\|}\right)$ is a dimensionless microscopic polarization operator which creates an *electron* in the conduction band and destroys an *electron* in the valence band, i.e., it creates an electron-hole pair, whereas $\mathrm{P}_{\boldsymbol{k}_{\|}}\left(\boldsymbol{q}_{\|}\right)$ destroys (recombines) the same electron-hole pair. If this recombination occurs radiatively, the emitted photon will have momentum $\boldsymbol{q}_{\|}$. In the electron-hole picture, remembering that a hole at $-\boldsymbol{k}_{\|}$ actually represents a missing valence electron at $\boldsymbol{k}_{\|}$, the center of mass of this electron-hole pair moves with the momentum $\boldsymbol{q}_{\|}$. The actual dipole density can now be written as a sum over the electron-hole pair momenta $\boldsymbol{q}_{\|}$

$$\mathrm{P}(\boldsymbol{r})=\frac{1}{\Lambda}\sum_{\boldsymbol{q}_{\|}}\mathrm{P}_{QW}\left(\boldsymbol{q}_{\|}\right)\mathrm{e}^{-\mathrm{i}\boldsymbol{q}_{\|}\cdot\boldsymbol{r}_{\|}}\mathrm{g(z)} \tag{13}$$

Where, as mentioned above, $\mathrm{g(z)}$ is the quantum-well confinement function (product of the envelope functions of the electron in the conduction and valence bands), and Λ is the quantum well normalization area. This represents a closed set of equations with the renormalized stimulated contribution

$$\Omega_{\mathrm{ST}}\left(\boldsymbol{k}_{\|},\boldsymbol{q}\right)=\mathrm{d}_{\mathrm{cv}}^{*}\left(\boldsymbol{q}_{\|}\right)\left(\sum_{q'z}\left[\mathrm{i}\xi_{\boldsymbol{q}'_{z},\boldsymbol{q}_{\|}}\mathrm{u}_{\mathrm{QW},q'_{z},\boldsymbol{q}_{\|}}\left\langle \mathrm{b}_{\boldsymbol{q}}^{+}\mathrm{b}_{q_{z},\boldsymbol{q}_{\|}}\right\rangle-\frac{\mathrm{g}_{\mathrm{QW}}}{\varepsilon_{0}\mathrm{n}^{2}\Lambda}\sum_{\boldsymbol{k}'_{\|}}\mathrm{d}_{\mathrm{cv}}^{*}\left(\boldsymbol{q}_{\|}\right)\left\langle \mathrm{b}_{\boldsymbol{q}}^{+}\mathrm{P}_{\boldsymbol{k}'_{\|}}\left(\boldsymbol{q}_{\|}\right)\right\rangle\right]\right) + \sum_{\boldsymbol{k}'_{\|}}\mathrm{V}_{\boldsymbol{k}'_{\|}-\boldsymbol{k}_{\|}}\left\langle \mathrm{b}_{\boldsymbol{q}}^{+}\mathrm{P}_{\boldsymbol{k}'_{\|}}\left(\boldsymbol{q}_{\|}\right)\right\rangle \tag{14}$$

where $\mathrm{g}_{\mathrm{QW}}=\int \mathrm{g(z)g(z)dz}$ and n is the quantum well index of refraction. In $\Omega_{\mathrm{ST}}\left(\boldsymbol{k}_{\|},\boldsymbol{q}\right)$ the first term is a source term due to field correlations and the second and third terms describe its renormalization due to the dipole self-energy and the Coulomb interaction, respectively. In the SLE equation describing the radiation-matter correlation $\left\langle \mathrm{b}_{\boldsymbol{q}}^{+}\mathrm{P}_{\mathrm{k}_{\|}}\right\rangle$ the term proportional to $1-\mathrm{f}^{\mathrm{e}}_{\boldsymbol{k}_{\|}+\boldsymbol{q}_{\|}}-\mathrm{f}^{\mathrm{h}}_{\boldsymbol{k}_{\|}}$ represents either absorption or stimulated emission depending on the carrier occupation of the various quantum well electronic states. The strength of the spontaneous emission,

$$\Omega_{\mathrm{SE}}\left(\boldsymbol{q}\right)=\mathrm{i}\xi_{\boldsymbol{q}}\mathrm{u}_{\mathrm{QW},\boldsymbol{q}}\mathrm{d}_{\mathrm{cv}}\left(\boldsymbol{q}_{\|}\right) \tag{15}$$

is determined by the dipole matrix element d_{cv} and the effective mode strength $\mathrm{u}_{\mathrm{QW}_{\mathrm{q}}}$. The term $\left\langle \mathrm{b}_{\boldsymbol{q}}^{+}\mathrm{P}_{\mathrm{k}_{\|}}(\mathrm{q}_{\|})\right\rangle$ represents an electron-hole pair with center of mass momentum $\hbar\mathrm{q}_{\|}$ recombining by emitting a photon with the same in-plane momentum $\mathrm{q}_{\|}$. When excited carriers are present in the QW, this correlation builds up even if the field-particle and the field-field correlations are initially absent, because the source term $\mathrm{f}^{\mathrm{e}}_{\boldsymbol{k}_{\|}+\boldsymbol{q}_{\|}}\mathrm{f}^{\mathrm{h}}_{\boldsymbol{k}_{\|}}\Omega_{\mathrm{SE}}\left(\boldsymbol{q}\right)$ in the equation is nonzero. Thus, $\left\langle \mathrm{b}_{\mathrm{q}_{z},\boldsymbol{q}_{\|}}^{+}\mathrm{P}_{\boldsymbol{k}_{\|}}\left(\boldsymbol{q}_{\|}\right)\right\rangle$ describes spontaneous photon emission and electron-hole recombination.

Considering the product $\mathrm{f}^{\mathrm{e}}_{\boldsymbol{k}_{\|}+\boldsymbol{q}_{\|}}\cdot\mathrm{f}^{\mathrm{h}}_{\boldsymbol{k}_{\|}}$, spontaneous recombination occurs only if an electron at $\mathrm{k}_{\|}+\mathrm{q}_{\|}$ and a hole at $\mathrm{k}_{\|}$ are simultaneously present. As field correlations begin to build up, the stimulated contribution $\Omega_{\mathrm{ST}}\left(\boldsymbol{k}_{\|},\boldsymbol{q}\right)$ becomes able to influence the photoluminescence spectrum. What we observe as photoluminescence is therefore a result of an interaction of field-field and field-particle correlations affected by both spontaneous and stimulated emission.

Luminescence is connected to the carrier and photon-number operators of the system. The total change in photon and carrier numbers follow from the SLE as

$$\frac{\partial}{\partial \mathrm{t}}\sum_{\boldsymbol{q}}\left\langle \mathrm{b}_{\boldsymbol{q}}^{+}\mathrm{b}_{\boldsymbol{q}}\right\rangle=-\sum_{\mathrm{k}_{\|}}\mathrm{f}^{\mathrm{e,h}}_{\boldsymbol{k}_{\|}} \tag{16}$$

meaning that any time an electron and a hole recombines radiatively, a photon is emitted.

Regarding spontaneous emission as the source of the photoluminescence, a reasonable question at this stage would be whether or not radiation field modes of different in-plane momentum $\mathrm{q}_{\|}$ can be coupled. It turns out that within the Hartree-Fock approximation, spontaneous emission from an incoherent carrier distribution (we call this incoherent

spontaneous emission) is only allowed when the recombining electron-hole pair and the emitted photon have the same in-plane momentum. Therefore photoluminescence from different $q_{||}$ can couple only via the carrier-occupation dynamics.
If, in addition to the spontaneous component there is a non-zero coherent (optical pumping) component, quantum correlations of the type

$$\Delta\left\langle b^{+}_{q_z,\boldsymbol{q}_{||}} P_{\boldsymbol{k}_{||}}\left(\boldsymbol{q}_{||}\right)\right\rangle = \left\langle b^{+}_{q_z,\boldsymbol{q}_{||}} P_{\boldsymbol{k}_{||}}\left(\boldsymbol{q}_{||}\right)\right\rangle - \left\langle b^{+}_{q_z,\boldsymbol{q}_{||}}\right\rangle\left\langle P_{\boldsymbol{k}_{||}}\left(\boldsymbol{q}_{||}\right)\right\rangle \quad (17)$$

will contain coherent contributions from both the semiconductor Bloch equations as well as from the incoherent processes described by the semiconductor luminescence equations. When the coherent component is absent initially, it can be shown that the expectation values of the field and polarization operators in the second term on the RH side of the above equation are zero, and in this case the SLE given above directly gives the expectation value of the correlation.
Stimulated emission also conserves the in-plane momentum, meaning that modes with different in-plane momentum do not couple radiatively. We just saw that this also holds for spontaneous emission. Incoherent photoluminescence is thereby greatly simplified, since field-field and field-matter correlations only contain quantities with the same $q_{||}$. Thus, in summary, photoluminescence contributions with different $q_{||}$ can only affect each other via the carrier-occupation.
Under stationary optical pumping the analysis could become even simpler since electron-hole pairs lost due to recombination are replaced by the pump, keeping carrier occupations constant. But coherent pumping introduces additional correlation terms leading to a coupling of modes with different $q_{||}$, which is discussed in Kira et al. (Kira et al., 1999).

3.2.5 Plasmons and coupled plasmon LO phonon effects

Charge-density waves (plasmons) in the carrier ensemble and lattice vibrations (phonons) are basic collective excitations in solids. In polar semiconductors, longitudinal optical (LO) phonons carry long range electric fields which couple to the electric fields of the longitudinal charge-density waves. Plasmons and LO phonons then form two new hybrid modes at different frequencies. This phenomenon was first observed by incoherent Raman scattering of continuous laser radiation from GaAs (Mooradian & Wright, 1966). The LO phonon-plasmon coupled hybrid excitations exhibit fundamentally different properties compared to the bare resonances. The mixed resonances still convey the carrier-carrier as well as the carrier-lattice interaction. Therefore, they are of fundamental importance for transport and relaxation dynamics in polar materials.
Free carriers in semiconductors can be introduced either by doping or by optical excitation. Under optical excitation, both electrons and holes appear simultaneously. Because of the large mass of heavy holes compared to conduction band electrons, holes can often be viewed as stationary while electrons are subjected to the plasmon oscillation.
The longitudinal mode frequencies of the coupled plasmon-phonon system are zeros of the total dielectric response function in the weak damping, long wavelength limit, i.e., where (Ridley, 1999)

$$\varepsilon_T = \varepsilon_L + \varepsilon_e - \varepsilon_0 = 0 \quad (18)$$

The free space permittivity ε_0 is subtracted once to avoid double counting, as it is included in both the lattice permittivity ε_L and the electronic permittivity ε_e. These can be given as simple Drude models,

$$\varepsilon_{\mathrm{L}} = \varepsilon_{\infty} \frac{\omega^2 - \omega_{\mathrm{LO}}^2}{\omega^2 - \omega_{\mathrm{TO}}^2} \tag{19}$$

$$\varepsilon_{\mathrm{e}} = \varepsilon_0 - \frac{\omega_{\mathrm{p}}^2}{\omega^2} \tag{20}$$

Here ω_{LO} and ω_{TO} are the longitudinal optical and transverse optical bare phonon frequencies, with the bare plasmon frequency for a carrier density n defined as $\omega_p^2 = ne^2/m^*\varepsilon_\infty$. The trick of defining an optical effective mass m^* accounts for a non-parabolic conduction band; m^* is essentially a fitting parameter which changes as the doping density increases. The result is that the longitudinal phonon mode and pure plasmon mode are replaced by the two coupled modes designated by ω_+ and ω_-. Their frequencies are given by

$$\omega^4 - \omega^2\left(\omega_{\mathrm{LO}}^2 + \omega_{\mathrm{p}}^2\right) + \omega_{\mathrm{TO}}^2\omega_{\mathrm{p}}^2 = 0 \tag{21}$$

In Fig. 3 we have calculated the eigenfrequencies for the two hybrid modes ω_+ and ω_- in $Cd_{0.28}\,Hg_{0.72}Te$ as a function of carrier density at long wavelengths (**q ~ 0**) of the modes. We notice that at low carrier densities the upper mode ω_+ resembles a bare LO phonon mode with a fixed resonance frequency, whereas the lower mode ω_- resembles a bare plasmon with a linearly increasing resonance frequency as a function of carrier density. At low carrier densities the oscillation period of the plasmon is very long, but at higher carrier densities its oscillation period starts to approach that of the LO phonon-like upper mode. At this level of carrier density the effect of mode coupling becomes strong, and must be accounted for when dealing with transport problems or relaxation dynamics. This coupling region is in fact the most difficult domain to model. If the carrier density is increased even more, the upper mode changes character and becomes plasmon-like while the lower mode becomes LO phonon like. Again, we have a plasmon-like and a LO phonon-like mode, but now it is the upper hybrid mode that is plasmon-like and the lower hybrid mode that is LO phonon-like. A small artefact with the lower hybrid mode clearly seen in the figure is that it oscillates at the *tranverse* optical phonon frequency at high carrier densities, rather that at the LO phonon frequency. LO phonons generally have a larger restoring force than do TO phonons, and they therefore oscillate at a slightly higher frequency. This difference is due to the long range electric field that LO phonons carry. At high carrier densities this field is (over)screened by the now rapidly oscillating plasmon charge carriers, and thus the oscillation will occur at (or below) the bare TO frequency.

The physics described above is based on an assumption of quasi steady-state or quasi equilibrium. We know that oscillating charge carriers generally emit radiation. If the material can be brought far away from thermal equilibrium, radiation from the hybrid modes should become clearly discernible. In addition, the formation time from bare resonances to the fully developed hybrid modes could be revealed if the number of charge carriers is suddenly increased from a very low value to a value where coupling is expected to occur.

We shall first deal with the issue of radiation from the hybrid modes, that is, free-space radiation originating from longitudinal modes of a polar semiconductor as a function of coupling. It is perhaps to be expected that plasmons are the strongest radiators, and that the radiation spectrum is only significantly disturbed when entering the coupling region. If the semiconductor is excited with a short laser pulse, we readily obtain an inhomogeneous, hot electron-hole plasma. A colder, well thermalized plasmon-phonon hybrid would clearly be a more ideal object of study. Therefore the sample should have a background donor doping (giving an n-type semiconductor) in which the colder plasmon-phonon hybrid can be established. Steady-state non-equilibrium of this colder hybrid is ensured if it is interacting with a hot, laser-induced electron-hole plasma. The density of the hot plasma can be several orders of magnitude higher than the background doping density. Interactions between the cold hybrid mode and the laser-induced hot plasma are rather subtle. Samples for this kind of experiments have a surface space-charge region. An ultrashort laser pulse creates electron-hole pairs, suddenly screening out the surface field. Higher doping density results in a shorter space-charge layer, resulting in faster screening of the surface field. Coulomb coupling of the ultrafast surface field transient to both the electronic and the lattice subsystems then initiate the collective oscillations which form the hybrid modes. Both

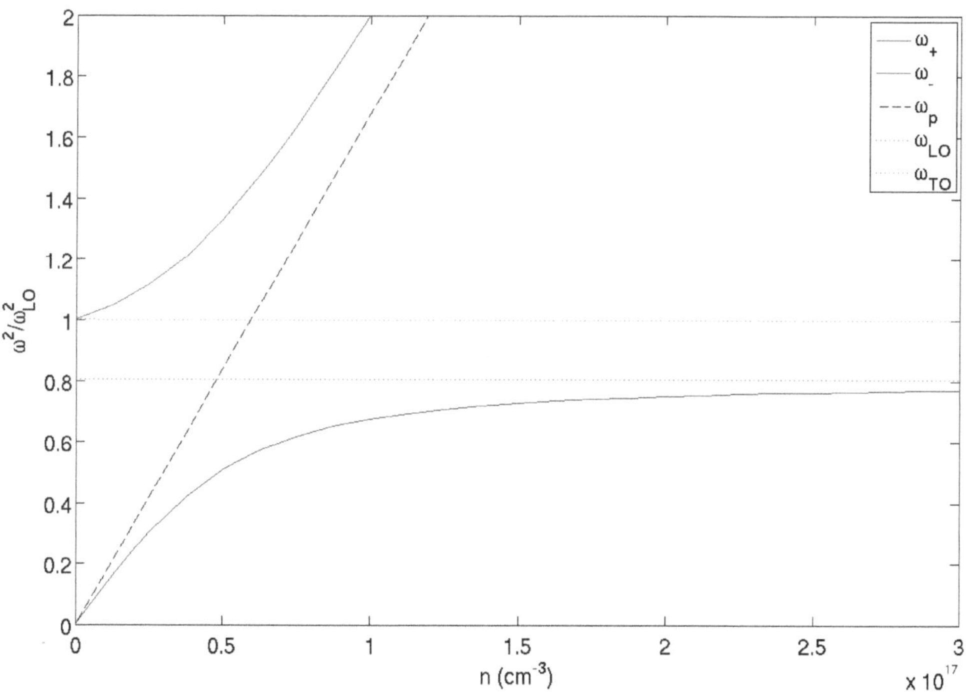

Fig. 3. A plot of the frequencies associated with the hybrid modes. Calculations done for $Cd_{0.28}Hg_{0.72}Te$ at 300 K. The band gap is 0.271 eV.

components of the cold hybrid mode have intrinsic damping rates, for LO phonons a decay time of 1-2 ps describes the disintegration of the LO mode into acoustic phonon modes, and for electrons sub-ps momentum relaxation times connected to various carrier scattering processes define the damping rate of the plasmon. Presence of a hot, laser induced plasma of high density in the same sample will change the cold plasma carrier scattering rates (especially the rates concerning donor electron scattering with laser induced hot holes) and enhance the damping of the cold hybrid mode. The altered damping rates can be extracted from the radiation emitted by the hybrid mode, and compared with rates extracted from electric mobility measurements of the doped sample.

Time-resolved measurements of hybrid modes in polar semiconductors can be made by reflective electro-optic sampling (REOS) (Dekorsy et al., 2000). Charge carriers are generated by an ultrafast laser pulse, and the hybrid modes weakly modulate the surface field as observed by reflection. Among the drawbacks of REOS is a lack of sensitivity to oscillations beyond the surface field layer and less distinction between photocarrier and background doping carrier densities.

Two other detection techniques are optically gated antennas and free-space electro-optic sampling, which give the phase of the THz pulse relative to the excitation pulse. Phase data can in principle reveal details of the starting mechanism of the hybrid oscillations, but they are technically difficult to obtain due to dispersion and absorption effects in the semiconductor (Leitenstorfer et al,. 1999, Kono et al., 2000).

Ignoring this phase information, Hasselbeck et al. (Hasselbeck et al., 2002) used a broadband interferometric method with two 30 fs pump pulses directed at nearby regions in InAs to extract transient THz radiation signals interfering in the far field. The angle of incidence for the Ti:sapphire laser beams was 45° and the pump pulses exceeded the band gap by 1.2 eV. Excited carrier density in the pump absorption region was estimated to be $>10^{18}\,\mathrm{cm}^{-3}$, with an absorption depth of the pump > 200 nm, and the surface accumulation field depth was ~ 5 nm. This ensures that the pump penetrated well into the neutral bulk region where the hybrid oscillations are created. Response times of the surface field screening were 52 - 76 fs, decreasing with increased doping.

By adjusting the time-delay between the pump pulses an interferogram is created, which could be Fourier transformed and analyzed in order to extract vital data regarding the hybrid modes.

In (Kersting et al, 1997) a simple model is presented which can be used to calculate the power spectrum of the emitted THz radiation of the experiment, assuming homogeneous broadening (Hasselbeck et al., 2002);

$$
\begin{aligned}
&\mathrm{P}(\omega) \sim \omega^4 \left|\mathrm{E}(\omega)\right|^2 \cdot \left| \frac{\omega_\mathrm{p}^2\left(\varepsilon_\infty \mathrm{D}_1(\omega) - 4\pi\cdot\Omega\right) + \Omega\left[\mathrm{D}_2(\omega) - \omega_\mathrm{p}^2\right]}{\varepsilon_\infty \mathrm{D}_1(\omega)\mathrm{D}_2(\omega) - 4\pi\cdot\Omega\omega_\mathrm{p}^2} \right|, \\
&\mathrm{D}_1(\omega) = -\omega^2 + \mathrm{i}\gamma_\mathrm{ph}\omega + \omega_\mathrm{LO}^2 \\
&\mathrm{D}_2(\omega) = -\omega^2 + \mathrm{i}\gamma_\mathrm{el}\omega + \omega_\mathrm{p}^2
\end{aligned}
\tag{22}
$$

with $\Omega = \omega_{TO}^2(\varepsilon_s - \varepsilon_\infty)/4\pi$ (ε_s is the static dielectric constant) and γ_{ph}, γ_{el} as LO phonon decay and electron momentum relaxation rates, respectively. At the edge of the space-charge layer the surface electric field transient is modeled as $E(t) = E_0[\tanh(t/\tau)+1]$, giving $|E(\omega)|^2 \sim \mathrm{csch}^2(\pi\omega\tau/2)$, where τ is an analytic approximation to a calculated response time of the

surface field screening. The center frequencies of the spectral peaks are quite insensitive to a change in optical excitation power, since they are determined by the cold plasma density.
The laser-generated, hot charge carriers do not emit coherently due to their inhomogeneous, i.e., quasi-exponential distribution in the bulk region adjacent to the surface. Distinguishable contributions from both ω_+ and ω_- hybrid modes can be directly observed in the radiated signals. By varying the donor doping, the plasma frequency changes, and we can study the different regions of hybrid mode coupling similar to those depicted in Fig. 3.
Having discussed the emission of radiation from non-equilibrium hybrid modes, we shall now turn our attention to the formation of hybrid modes after a sudden increase in carrier density. Femtosecond laser pulses create a dense, excited carrier plasma in an intrinsic semiconductor within a time shorter than a typical oscillation cycle of the bare LO phonon and plasmon resonances. It is to be expected that the hybrid mode resonances will not be established instantaneously in such a strongly non-equilibrium situation. Transition from an LO phonon resonance to a fully coupled phonon-plasmon hybrid will be governed by many-body quantum correlations. Screened Coulomb and phonon interactions must therefore be considered self-consistently within the same approach.
We shall briefly outline the key ingredients of a quantum kinetic theory in the framework of Keldysh non-equilibrium Greens functions (Haug & Jauho, 1996). The analysis can be viewed as a generalization of the steady-state theory presented above, where we found the longitudinal hybrid eigenmode resonances as the roots of the dielectric function. Manybody system response can quite generally be obtained via the dielectric function, usually expressed via its inverse form ε^{-1} (Ridley, 1999, Mahan, 2007). For a NIR pump-THz probe experiment we are interested in the inverse dielectric function $\varepsilon_q^{-1}(t_d,\omega)$ where t_d is the delay between pump and probe, and $\boldsymbol{q}$, ω are the wavevector and frequency of the hybrid phonon-plasmon resonance, respectively. $\mathrm{Re}(1/\varepsilon_q)$ essentially describes the screening effects of the manybody system, in particular the resonant overscreening of the long range electrostatic restoring force of the LO phonon, which was seen earlier to reduce the ω_- oscillation frequency to lie below the TO phonon frequency at high carrier densities (Fig. 3). The imaginary part of the dielectric function reflects the energy loss of a carrier interacting with the many-particle system. Sustained longitudinal electromagnetic eigenmodes are identified from the peaks of $\mathrm{Im}(1/\varepsilon_q)$.
The two-time particle interaction is calculated self-consistently with the semiconductor Bloch equations, and the Dyson equation determines the effective screened interaction potential,

$$\mathrm{W}_q(1,2) = \mathrm{W}_q^0(1,2) + \mathrm{W}_q^0(1,3)\mathrm{L}_q(3,4)\mathrm{W}_q(4,2) \tag{23}$$

where the polarization function $L_q(t;t')$ is based on the random-phase approximation (RPA). The effective screened interaction potential $\mathrm{Wq}^0(1,2)$ describes the combined Coulomb interaction and the interaction due to LO phonon scattering. Exponential damping is introduced into the phonon propagator, and the corresponding two-time Greens functions are modeled according to the generalized Kadanoff-Baym ansatz. Then an incomplete Fourier transform of the resulting effective interaction potential $\mathrm{Wq}^r(t_1,t_2)$ is taken, considering the relative time $t_1 - t_2$, and the inverse dielectric function is finally obtained by setting $t_D = t_1$ (Huber et al., 2005b, Vu & Haug, 2000),

$$W_q^r(t_1,\omega) = \int_{-\infty}^{t_1} dt_2 W_q^r(t_1,t_2) e^{i\omega(t_1-t_2)},$$
$$\varepsilon_q^{-1}(t_D,\omega) = W_q^r(t_D,\omega) / V_q \tag{24}$$

with V_q denoting the strength of the bare Coulomb interactions.
Huber et al. (Huber et al., 2005b) investigated the formation of hybrid modes in intrinsic InP using a 12-fs Ti:sapphire laser pump with a single-cycle THz probe at 28 THz, delayed with the time t_D. Real-time evolution of the probe THz electric field was directly measured by means of the ultrabroadband electro-optic sampling technique with sampling delay time t_s. The measured data allowed access to both real and imaginary parts of $\varepsilon_q(t_D,\omega)$ in a 7 - 60 THz window.
Delaying the probe with respect to the pump resulted in characteristic trailing oscillations long after the original probe signal had died out. The trailing oscillations represent a change in the probe signal which was monitored by varying the sampling delay time t_s. Increasing the delay between pump and probe resulted in more trailing oscillation cycles. These oscillation cycles are manifestations of a quantum beating between the newly developing upper and lower hybrid modes ω_+ and ω_- in Fig. 3.
The hybrid resonances can be followed by a Fourier transformation with respect to t_s of the change in the real-time probe signal for various pump-probe delays t_D, thereby extracting the inverse dielectric function $\varepsilon_q^{-1}(t_d,\omega)$ in the long wavelength limit $\boldsymbol{q} = \mathbf{0}$. For an unexcited semiconductor only the bare LO phonon resonance is found, which vanishes after photoexcitation and then two new hybrid modes arise after a while. By varying the carrier densities the position of the peaks in of $\mathrm{Im}(1/\varepsilon_{q=0})$ can be followed. As a reference, TO phonons in InP oscillate at $\omega_{TO}/2\pi = 9.7$ THz, and LO phonons at $\omega_{LO} = 10.3$ THz. In the time-domain, it was demonstrated both experimentally and theoretically that the time for establishment of fully developed hybrid modes depends on the carrier density as $\tau_H = 1.6\omega_+ / 2\pi$. This means that the buildup takes shorter time in denser systems, c.f. Fig 3. Typical values of τ_H for InP would be ~ 140 fs at a carrier density of $10^{18}\,\mathrm{cm}^{-3}$.

3.2.6 Magnetic systems

Diluted ferromagnetic semiconductors are currently explored for use in information processing and storage devices. By introducing Mn atoms in GaAs or other III-V semiconductors a new, ferromagnetic material is obtained in which fast spin manipulation seems possible. Other candidate material classes investigated in the literature are chrome spinels, manganese oxides, transition metals, rare earths, pyrochlore, EuO, and EuS.
Spin manipulation is a basic ingredient in spintronics, spin-photonics and quantum computation applications. Carrier density induced ferromagnetic effects can either be controlled by light or current via electrical contacts and gates. Magnetic field pulses and spin-currents can control the spin on the picosecond timescale, but for femtosecond manipulation, ultrashort laser pulses will have to be introduced. For the investigation of candidate materials a particular experimental method is used, ultrafast pump-probe magneto-optical spectroscopy. Here the pump optical pulse creates carrier populations

whose subsequent interactions starts the magnetization dynamics which can be followed as a function of time via Faraday or Kerr rotation.
Natural timescales for phonon mediated spin relaxation and magnetization precession lie in the picosecond range. This was considered to be the ultimate limit for the attainable speed of magnetization switching prior to investigations with fs pulsed lasers. Then ultrafast demagnetization with the aid of fs lasers was demonstrated and this opened up new possibilities for high-speed magnetic devices. We shall first consider the metals because a large body of results has been obtained here and the corresponding theory has matured considerably in the last few years.
The ultrafast demagnetization process seen in transition metals such as Ni, Co, Fe (delocalized magnetism of *d*-electrons) and the rare-earth metal Gd (magnetism of *f* -electrons), is related to three characteristic timescales: 1) a femtosecond demagnetisation with timescale τ_E , 2) a picosecond magnetization recovery with timescale τ_M, and 3) a hundred picosecond-nanosecond magnetization precession.
Finding a satisfactory theory from first principles of the ultrafast physics going on under 1) is not straightforward. Instead, it seems at the moment more profitable to start with simpler models in order to describe the influences of the different subsystems (charge carriers, spin, photons, and phonons). For example, it is known that spin-orbit interactions are important, but modeling of all three temporal regimes on the same footing is currently not possible using quantum theory.
Early attempts assumed a three-temperature (3T) phenomenological model with rate equations for electron, phonon, and spin temperatures (energies). Using a spin temperature is quite unsatisfactory, considering that the spin system is not in equilibrium on the femtosecond timescale. If the spin is instead coupled to a two-temperature (2T) model for phonon and electron temperatures, we can obtain some improvement. Such a model deals with an energy flow concept, interpreting the ultrafast demagnetization as a non-coherent, "thermal" process. The energy goes from photon to electron and then to the spin system, without the need for stating any underlying quantum mechanism behind the spin flip.
Among the candidates for underlying quantum mechanisms are Elliott-Yafet (EY) electron-phonon scattering, EY electron scattering with impurities or other electrons and the electron-electron inelastic exchange scattering. The EY model for phonon scattering seems to be a strong candidate in many of the metals.
Further progress in the field of phenomenological models (Atxitia & Chubykalo-Fesenko, 2010) has resulted in three central branches: 1) Langevin dynamics based on the Landau-Lifshitz-Gilbert (LLG) equation and classical Heisenberg Hamiltonian for localized atomic spin moments, 2) the Landau-Lifshitz-Bloch (LLB) micromagnetics model, and 3) the Koopmans's magnetization dynamics model (M3TM).
The LLB model includes the dynamics governed by both the atomistic LLG model and the M3TM model, with both classical and quantum versions. The LLB equation for a quantum spin S is based on the density matrix approach. For simple quantum spin systems modeled a s a two level system with spin-up and spin-down bands (i.e. $S = \pm 1/2$) it has been shown to be equivalent to the M3TM model (Atxitia & Chubykalo-Fesenko, 2010). The LLB equation for classical spins is equivalent to an ensemble of exchange-coupled atomistic spins modeled by LLG equations.
Quantum mechanisms responsible for the ultrafast demagnetization in the LLB model are described by the coupling parameter λ, which defines the rate of the spin flip. Relations

between such parameters in the phenomenological models are a key to understand whether or not a given mechanism can act on more than one of the three characteristic timescales of the demagnetization process mentioned above.

The LLB equation can handle large spatial scales, which is a prerequisite for modeling multiscale magnetization dynamics. Coupling the spin to the electron temperature from the 2T model, we automatically include a purely carrier-based spin flip process. If both electron and phonon temperatures are coupled to the spin dynamics we automatically include the carrier-phonon scattering induced Elliott-Yafet mechanism. Coupled to the 2T model, the LLB equation has recently been shown to describe correctly all three stages of the ultrafast demagnetization processes: the sub-picosecond demagnetization, the picosecond magnetization recovery and the nanosecond magnetization precession (Atxitia & Chubykalo-Fesenko, 2010).

As regards semiconductors, similar magnetization effects have been demonstrated (Kapetanakis et al., 2009). Using density matrix equations of motion a nonequilibrium theory of ultrafast magnetization reorientation in ferromagnetic (Ga,Mn)As was presented. Interactions and coherent nonlinear optical effects were treated in a similar way as the Semiconductor Bloch equations. Both resonant and non-resonant photoexcitation were included, taking into account the relevant bands. A femtosecond collective spin tilt was induced by nonlinear, near-ultraviolet (3 eV), coherent photoexcitation with linearly polarized light. The magnetization was initiated by the interaction of non-thermal itinerant carriers (holes) with local Mn spins. The dynamics generated a subsequent uniform magnetic precession. This was interpreted as a possibility of non-thermal magnetization control by tuning the laser frequency and polarization direction.

3.3 Incoherent ultrafast regime

We shall now consider ultrafast phenomena after coherence has been completely destroyed by scattering. This is the particular branch of scattering dominated carrier relaxation dynamics which is representative for traditional microelectronics devices.

When electrons are photoexcited high into the conduction band, scattering processes are initiated and the initial configuration of excited charge carriers in the bands starts to change. First the charge carriers thermalize among themselves within 1 ps due to carrier-carrier scattering. Thermalization means that the charge carriers obtain a common temperature which is different from the lattice temperature. Then the cooling of the carriers towards the lattice temperature starts, and this can take tens to hundreds of picoseconds depending on the amount of photoexcited carriers and how high above the band gap the electrons were excited. Screening of the carrier-phonon scattering mechanisms at high photoexcitation densities can reduce the energy transfer rate to the lattice. Considerable feedback from the lattice to the carriers are related to bottleneck effects when the most active phonon types receive too much energy in a short time. These hot phonons obtain much higher temperatures than the less active phonons, which tend to have temperatures very close to the initial lattice temperature. Usually the energy of the pump pulse is very small in these kinds of experiments, so the hot phonons are essentially due to poor energy distribution between different phonon modes. Consequences of having hot phonons are a substantially prolonged carrier cooling time. After the cooling has resulted in equilibration of carrier and lattice temperatures, a final 'condensation' process where electrons and holes recombine completes the photoexcitation cycle. Before

we embark on a study of the carrier cooling and hot phonon effects, we shall investigate another type of condensation. As the photoexcited carriers start to cool, the free carrier electron-hole pairs begin to condensate into excitons. Both excitions and free electron-hole pairs have rather long lifetimes, often in the μs range. Recombination can happen within a free electron-hole pair or go via exciton recombination.
The Saha equation (Kaindl et al., 2009, Suzuki & Shimano 2011), gives the equilibrium ratio between the density of excitons N_{ex} and free carriers (i.e. when these two populations are reciprocally thermalized) in the Boltzmann limit. For bulk materials it reads,

$$\frac{N_{eh}^2}{N_{ex}} = \frac{1}{4}\left(\frac{2\mu k_B T_c}{\hbar \pi}\right)^{\frac{3}{2}} e^{\left(-E_0/k_B T_c\right)} \tag{25}$$

where μ and E_0 are the exciton reduced mass and the binding energy, respectively. For a given total e-h pair density $N = N_{eh} + N_{ex}$, it yields the relationship between carrier temperature T_c and free-carrier density N_{eh}.
Free electron hole pair populations obtain thermal equilibrium with excitons within 2 ps. If we consider long timescales of some ps to hundreds of ps after the pump pulse with no external disturbance, this could be considered an instant equilibrium. The Saha equation implies that hot carriers have a low exciton density while cold carriers contain a high exciton density. However, the above equation only represents an exclusive equilibrium between free electron-hole pairs and excitons. As we saw above the carriers will still not be in equilibrium with the phonon system within a 2 ps timescale. What can happen under the carrier cooling phase is a thermal re-ionization of excitons by the phonons to create free-electron-hole pairs. The result will be a slow formation time of excitons on the order of several hundreds of ps when lattice and excitons eventually reach an equilibrium. We can consider the temporal region of carrier cooling to be free from recombination lifetime effects. After the cooling phase a final equilibrium of all subsystems is reached by recombination of excitons and free carriers.
This effect has been investigated in Si by Suzuki et al. (Suzuki & Shimano 2011) using optical pump and THz probe measurements. They evaluated the time-dependent fraction of free carriers and excitons. The waveforms of the THz pulse transmitted after the sample with and without the optical pump were recorded by electro-optic sampling. The complex transmittance change was then obtained by Fourier transformation, thereby yielding a photoinduced change in the complex dielectric function. To estimate the free e-h pair density at each delay time, the data were fitted with the free-carrier based Drude model in the low photon energy region (2.0–7.8 meV), where the Drude component dominates the spectrum. The Drude component decreases with increasing delay time, and an exciton component (Kaindl et al., 2009) emerged as a peak at ~ 10-12 meV, an energy that corresponds to the 1s-2p transition of excitons in Si. Thus the formation dynamics of excitons was revealed through the observation of the 1s-2p transition at ~ 3 THz. Contrary to a free-carrier behavior, the exciton density, as indicated by the 1s-2p absorption at 12 meV, gradually increased after the photoexcitation and reached a constant value after 400 ps for an initial lattice temperature of 30 K. Long after this period, recombination eventually will influence the populations.
As we have just seen, the interaction between the hot free electron-hole plasma and the phonons constitutes a very important basis for the formation of a stable exciton

population. In (Suzuki & Shimano 2011), cooling of the hot carriers was studied using a set of cooling rate expressions. A very flexible, but more computationally heavy alternative for studying this particular cooling dynamics would be to use a particle-based Monte Carlo (MC) simulation. The flexibility lies in the fact that very few assumptions have to be made about the system at hand, so that not only special cases can be considered. In the following we present some of our recent MC results in order to demonstrate the physics behind the dissipation of short laser pulses. In this case we shall investigate a low-gap semiconductor, $Hg_{0.72}Cd_{0.28}Te$ (MCT). As shown in Fig. 2, laser energy can be exchanged between carriers and phonons in several ways before finally escaping through the phonon distributions. Plots of the rate of energy exchange for $Hg_{0.72}Cd_{0.28}Te$ due to the interactions are shown in Fig. 4, for coupled MC simulations with hot phonons and exact screening.

Longitudinal optical phonons constitute the most important route for energy relaxation when carrier densities are low. Almost all the energy relaxed out of the system goes by this route, through polar optical interaction with high energy electrons. Relaxation through transversal phonons occurs via the non-polar deformation potential interaction and is usually less important.

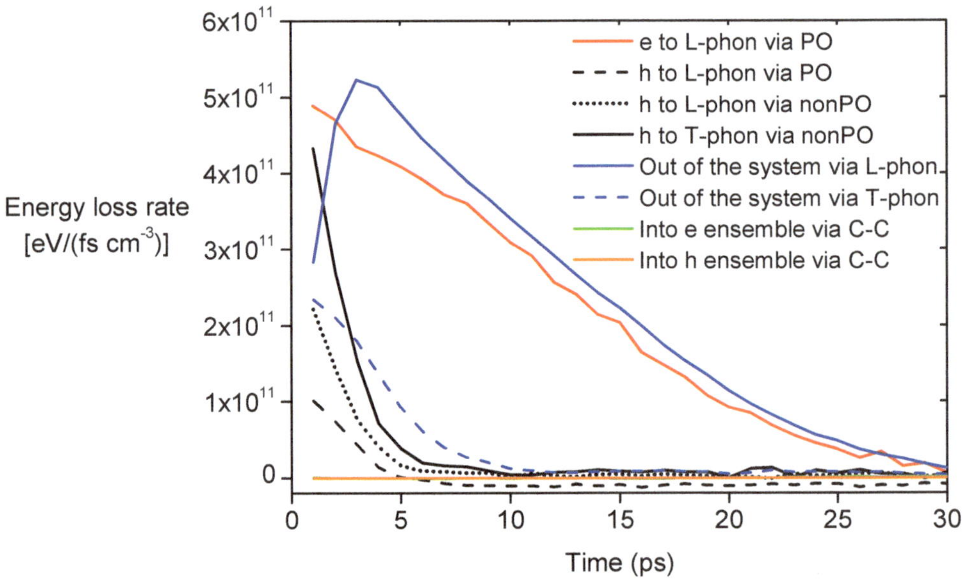

Fig. 4. Energy transfer rates of the different mechanisms at a carrier density of $n = 10^{16}\,\text{cm}^3$. Hot phonons are simulated and the screening length is calculated from the simulated distributions.

We also see from the figures that the green graph representing the hole energy loss rate to longitudinal optical phonons goes slightly negative after 5 ps. This means that some energy is flowing into the hole ensemble indirectly via the mainly electron-heated hot longitudinal optical phonons. This can help energy relaxation and even out the temperature difference between electrons and holes. For the particular case investigated here however, the flow is not large. Electrons in MCT will mainly interact with small wavevector longitudinal optical phonons, since electrons occupy a very narrow band in k-space with small electron wave vectors. Thus, only longitudinal optical phonons with small wavevectors will become really "hot", as illustrated in Fig. 5.

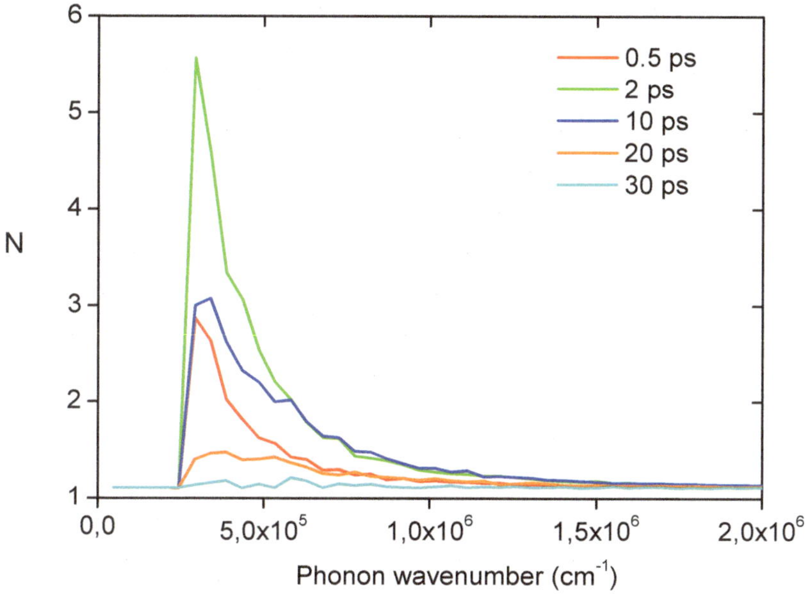

Fig. 5. Longitudinal phonon occupation at select times in simulations of coupled electron and hole ensembles at $n = 10^{16}\,\mathrm{cm}^3$, with hot phonons accounted for.

Longitudinal optical phonon modes with larger wavevectors have a phonon occupancy corresponding more closely to the lattice temperature.

Transversal phonons do not become hot either, as we can see from Fig. 6. Holes receive less of the laser energy and occupy a much broader range of wave vectors and will therefore interact with a much broader range of phonon wave vectors. Because the phonons that are hot have small wave vectors these phonons will not interact as efficiently with the broad-wavevector hole ensemble as one could expect.

At high laser fluences carrier densities will increase. Then the interaction with polar optical phonons tends to be weakened by carrier induced screening, and transversal phonons becomes a more important relaxation route. Another complication at high carrier densities and laser fluences is the occurrence of plasmons or longitudinal hybrid mode oscillations. An individual carrier can also be scattered by plasmons in elemental semiconductors such as Ge, Si etc., but not by hybrid modes.

In polar materials however, such as the compound semiconductors GaAs and MCT the plasmon mode and the longitudinal optical mode combine via their respective long range electrical fields, as we saw earlier. The combined modes replace the plasmon and polar optical mode as scattering agents of the charge carriers, and can be simulated by the same method as we have outlined here.

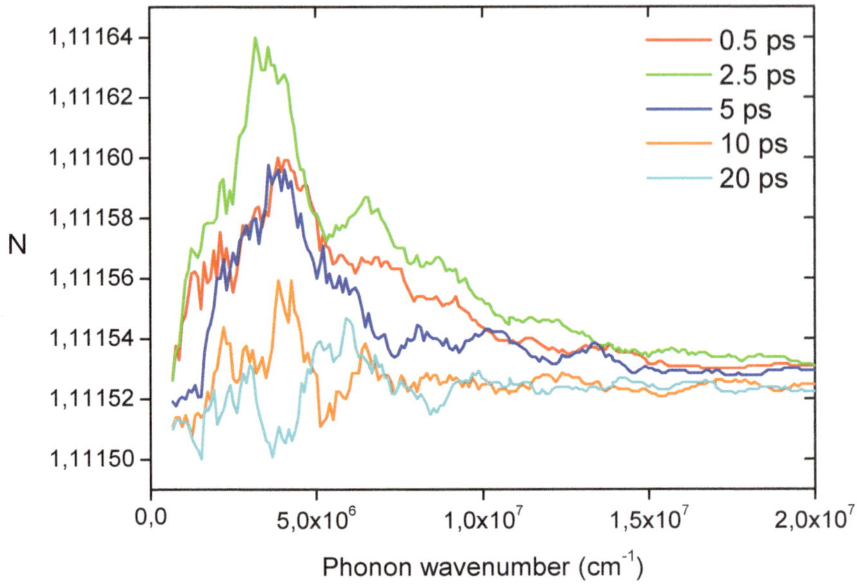

Fig. 6. Transversal phonon occupation at select times in simulations of coupled electron and hole ensembles at $n = 10^{16}\,\mathrm{cm}^3$, with hot phonons accounted for.

3.4 Long pulse regime

We shall devote some space (albeit small) also to this case, because of all the non-linear optical effects that appear. Using infrared (IR) wavelengths, intraband free carrier (FCA) and inter-valence band (IVA) absorption between e.g. the light and heavy hole bands can be studied, see Fig. 1. Although the total number of carriers generated in the conduction band can be comparable to that of a short pulse experiment, the majority of the carriers are well thermalized at temperatures which lie close to the overall lattice temperature. Nevertheless, differences between carrier and lattice temperature smaller than 100 K can have a major impact on processes occurring near the band gap, according to (Storebo et al., 2010). For example, it was found that a slightly elevated carrier temperature can strongly reduce the Pauli-induced saturation or "bleaching" of one-photon absorption across the band gap, or affect other long-timescale parameters, like carrier lifetimes and impact ionization rates. The difficulties within the modeling in this regime lies not only in the solution of many coupled differential equations, but also to a large degree on severe inaccuracies in recombination rates and TPA, FCA, IVA absorption coefficients which has now lasted for decades. The discrepancies have persisted because it has been experimentally very difficult to isolate and quantify all the simultaneous absorption and recombination effects. Theoretical estimates could be used to eliminate the uncertainties, but the methods used in the past have proven much too primitive for this task. With the advent of *ab-initio* band structure numerical codes it will become possible to obtain accurate estimates of each process, and examples of progress in this direction will become available in the near future. An important point which we would like to emphasize in this context is that even if the data is obtained numerically, it is crucial for the usefulness of the data that analytical expressions are extracted, so that the results can be incorporated in laser irradation simulations. In addition, the data must also take into account the possibility of different carrier and lattice temperatures, a feature which we have seldom seen considered for cases involving low-intensity laser irradiation.

4. Conclusions

As we have seen, no single post evaluation tool or method will suffice to cover all pulselength ranges, and not all tools are suited for a given experiment. Combining time resolved laser experiments with satisfactory analysis of the results can therefore become a demanding task. When evaluating potential analysis tools, we will take the view that it is vital to have access to flexible methods, even if we see that this often comes at the expense of heavy numerics. It is true that the analysis of e.g. a semiconductor device by simple analytical models often benefits the physical understanding. But in the field of laser spectroscopy many processes contribute in unpredictable ways, and to have some of these ways sporadically blocked by a priori assumptions or limitations within the method of analysis can become a frustrating experience. Some of the methods reviewed in this chapter introduce substantial simplifications, and should be used, but not without focusing on their limitations and assuring the access to more flexible tools. As regards the future prospects of laser spectroscopy, it is clear that the techniques developed in this field will play a crucial role in achieving quantum coherent control of information bearing variables such as spin as well as continuing to be a central method for device and material characterization.

5. References

Ashcroft, N.W. & Mermin, N.D. (1976). *Solid State Physics*, Holt-Saunders Japan LTD, Tokyo

Atxitia U. & Chubykalo-Fesenko O. (2010). Ultra-fast magnetization rates within the Landau-Lifshitz-Bloch model. *ArXiv:1054v1 [cond-mat.mtr-sci]* (23 Nov 2010)

Axt, V.M. & Stahl, A. (1994). A Dynamics-Controlled Truncation Scheme for the Hierarchy of Density Matrices in Semiconductor Optics. *Z. Phys. B Condens. Matter*, Vol.93, pp. 195

Butov, L.V.; Lai, C.W.; Ivanov, A. L., Gossard, A.C. & Chemla, D.S. (2002). Experimental approaches for exciton Bose-Einstein condensation. *Nature*, Vol.417, pp. 47

Blanchard, F.; Sharma, G.; Razzari, L.; Ropagnol, X.; Bandulet, H.C.; Vidal, F.; Morandotti, R.; Kieffer, J.C.; Ozaki, T.; Tiedje, H.; Haugen, H.; Reid, M. & Hegmann, F. (2011). Generation of Intense Terahertz Radiation via Optical Methods. *IEEE Journal of Selected Topics in Quantum Electronics*, Vol.17, No.1, January/February 2011

Chang, E; Bussi, G.; Ruini, A & Molinari, E. (2004). Excitons in Carbon Nanotubes: An *Ab Initio* Symmetry-Based Approach. *Phys. Rev. Lett.*, Vol.92, 196401

Christensen, N.E.; Svane, A.; Cardona, M.; Chantis, A.; Laskowski, N.R.; van Schilfgaarde, M. & Kotani, T. (2011). Calculations of quasi-particle spectra of semiconductors under pressure. *Phys. Status Solidi B*, 1-6

Combescot, M.; Betbeder-Matibet, O. & Leuenberger, M.N. (2009). Analytical approach to semiconductor Bloch equations. *Europhys. Lett.*, Vol.88, 57007

Dekorsy, T.; Cho, G.C. & Kurz, H. (2000), In *Light Scattering in Solids VIII*, M. Cardona & G. Guntherodt, (Eds.), Springer

Feldtmann, T. (2009). Influence of Phonons on Semiconductor Quantum Emission, Dissertation available at http://deposit.ddb.de/cgi-bin/dokserv?idn=999828991&dok_var=d1&dok_ext=pdf&filename=999828991.pdf

Ganichev, S.D.; Ketterl H; Prettl, W; Ivchenko, E.L & Vorbjev, L.E. (2000). Circular Photogalvanic Effect induced by Monopolar Spin Orientation in p-GaAs/AlGaAs multiple Quantum Wells. *Appl. Phys. Lett.* Vol. 77, pp. 3146

Ganichev,S.D.; Ivchenko, E.L; Danilov, S.N; Eroms, J.; Wegscheider, W,; Weiss, D. & Prettl, W. (2001). Conversion of Spin into Directed Electric Current in Quantum Wells. *Phys. Rev. Lett.*,Vol.86, pp. 4358

Ganichev, S.D.; Danilov, S.N; Bel'kov, V.V; Ivchenko, E.L; Bichler, M.; Wegscheider, W.; Weiss, D. & Prettl, W. (2002). Spin-Sensitive Bleaching and Monopolar Spin Orientation in Quantum Wells. *Phys. Rev. Lett.*, Vol.88, 057401

Griffin, C; Snoke, D.W. & Stringari, S. (1995). *Bose-Einstein Condensation*, Cambridge University Press, Cambridge, UK

Hanamura, E. & Haug, H. (1977). Experimental approaches for exciton Bose-Einstein condensation. *Phys. Rep., Phys. Lett.*, Vol.33, pp. 209

Hasselbeck, M. P.; Stalnaker, D.L.; Schlie, A.; Rotter, T. J.; Stintz, A. & Sheik-Bahae, M. (2002). Emission of terahertz radiation from coupled plasmon-phonon modes in InAs. *Phys. Rev. B* Vol.65, 233203

Haug, H. & Koch, S.W. (2009). *Quantum Theory of the Optical and Electronic Properties of Semiconductors*, 5th ed., World Scientific, Singapore

Haug, H & Jauho, A.P. (1996). *Quantum Kinetics in Transport and Optics of Semiconductors*, Springer, Berlin.

Hirtschulz, M.; Milde, F.; Malić, E.; Butscher, S.; Thomsen, C.; Reich, S.; & Knorr, A. (2008). Carbon nanotube Bloch equations: A many-body approach to nonlinear and ultrafast optical properties. *Phys. Rev. B*, Vol.77, 035403

Huber, R.; Kaindl, R.A.; Schmid, B.A & Chemla, D.S. (2005a). Broadband terahertz study of excitonic resonances in the high-density regime in GaAs/$Al_xGa_{1-x}As$ quantum wells. *Phys. Rev. B*, Vol.72, 161314(R)

Huber, R.; Kubler, C.; Tubel, S.; Leitenstorfer, A.; Vu, Q. T.; Haug, H.; Kohler F. & Amann, M.C. (2005b). Femtosecond Formation of Coupled Phonon-Plasmon Modes in InP: Ultrabroadband THz Experiment and Quantum Kinetic Theory, *Phys. Rev. Lett.*, Vol.94, 027401

Kaindl, R. A.; Hagele, D.; Carnahan, M. A. & Chemla, D.S. (2009). Transient terahertz spectroscopy of excitons and unbound carriers in quasi-two-dimensional electron-hole gases. *Phys. Rev. B*, Vol. 79, 045320

Kapetanakis, M. D.; Perakis, I. E.; Wickey, K. J.; Piermarocchi, C. & Wang, J. (2009). Femtosecond Coherent Control of Spins in (Ga,Mn)As Ferromagnetic Semiconductors using Light. *Phys. Rev. Lett.*, Vol.103, 047404

Kersting, R.; Unterrainer, K.; Strasser, G.; Kauffmann, H. & Gornik, E. (1997). Few-Cycle THz Emission from Cold Plasma Oscillations. *Phys. Rev. Lett.*, Vol.79, pp. 3038

Kira, M.; Jahnke, F.; Hoyer, W. & Koch, S.W. (1999). Quantum theory of spontaneous emission and coherent effects in semiconductor microstructures. *Progress in Quantum Electronics*, Vol.23, pp. 189

Kono, S.; Tani, M.; Gu, P. & Sakai, K. (2000). Detection of up to 20 THz with a low-Temperature-grown GaAs Antenna gated with 15 fs Light Pulses. *Appl. Phys. Lett.*, Vol.77, pp. 4104

Krauß, M.; Aeschlimann, M. & Schneider, H.C. (2008). Ultrafast Spin Dynamics Including Spin-Orbit Interaction in Semiconductors, *Phys. Rev. Lett.*, Vol.100, 256601

Leitenstorfer, A.; Hunsche, S.; Shah, J.; Nuss, M.C.; & Knox, W.H. (1999). Detectors and Sources for Ultrabroadband Electro-optic Sampling: Experiment and Theory. *Appl. Phys. Lett.*, Vol.74, pp. 1516

Lindberg, M. & Koch, S.W. (1988). Effective Bloch equations for Semiconductors. *Phys. Rev. B*, Vol.38, No.5, pp. 3342

Lopez-Sancho, M.P.; Muñoz, M.C. & Chico, L. (2001). Coulomb Interactions in Carbon Nanotubes. *Phys. Rev. B*, Vol.63, 165419

Ma, J. & Yuan, R.K. (1998). Electronic and Optical Properties of finite zigzag Carbon Nanotubes with and without Coulomb Interaction. *Phys. Rev. B*, Vol.57, pp. 9343

Mahan, G.D. (2007). *Many Particle Physics* 3rd ed, Springer

Makhonin, M.N.; Chekhovich, E.A.; Senellart, P.; Lemaître, A.; Skolnick, M.S. & Tartakovskii, A.I. (2010). Optically tunable Nuclear Magnetic Resonance in a single Quantum Dot. *Phys. Rev. B*, Vol.82, 161309(R)

Meier, T.; Reichelt, M.; Koch, S.W. & Höfer, U. (2005) Femtosecond time-resolved five-wave mixing at silicon surfaces, *J.Phys. Condens. Matter, Vol.*17, S221

Mooradian, A. & Wright, G.B. (1966). Observation of the Interaction of Plasmons with Longitudinal Optical Phonons in GaAs. *Phys. Rev. Lett.*, Vol.16, pp. 999

Nagai, M.; Shimano, R. & Kuwata-Gonokami, M. (2001). Electron-Hole Droplet Formation in Direct-Gap Semiconductors Observed by Mid-Infrared Pump-Probe Spectroscopy. *Phys. Rev. Lett.*, Vol. 86, pp.5795

Ogawa, T.; Tomio, Y. & Asano, K. (2007). Quantum Condensation in Electron-hole Systems: Excitonic BEC-BCS Crossover and Biexciton Crystallisation. *J. Phys. Condens. Matter*, Vol.19, 295205.

Perebeinos, V. ; Tersoff, J. & Avouris, P. (2004). Scaling of Excitons in Carbon Nanotubes. *Phys. Rev. Lett.*, Vol.92, 257402

Proietti Zaccaria, R. & Rossi. F. (2003). On the problem of generalizing the Semiconductor Bloch equation from a closed to an open System. *Phys. Rev. B, Vol.67*, 113311

Ridley, B.K. (2000). *Quantum Processes in Semiconductors*, Qxford

Rossi, F. & Kuhn, T. (2002). Theory of Ultrafast Phenomena in Photoexcited Semiconductors. *Rev. Mod. Phys.*, Vol.74, pp. 895

Rossi, F.; Di Carlo, A & Lugli, P. (1998). Microscopic Theory of Quantum-Transport Phenomena in Mesoscopic Systems: A Monte Carlo Approach. *Phys. Rev. Lett., Vol.80*, pp. 3348

Rössler, U; Tejedor, C. & Vina, L. (2003). Semiconductor Bloch Equations including Spin and Polarization Degrees of Freedom, in *Physics of Semiconductors 2002. Institute of Physics Conference Serie*s 171, A.R. Long & J.H. Davies, (Eds.), IOP Publishing Bristol, UK

Sandhu, J.S; Heberle, A.P.; Baumberg, J.J. & Cleaver, J.R.A. (2001). Gateable Suppression of Spin Relaxation in Semiconductors. *Phys. Rev. Lett.*, Vol.86, pp. 2150

Shah, J. (1999). *Ultrafast Spectroscopy of Semiconductors and Semiconductor Nanostructures*, Springer

Spataru, J.R.A; Ismail-Beigi, S; Benedict, L.X & Louie, S.G. (2004). Excitonic Effects and Optical Spectra of Single-Walled Carbon Nanotubes. *Phys. Rev. Lett., Vol.92*, 077402

Storebo, A.K; Brudevoll, T. & Stenersen, K. (2010). Numerical Modeling of IR-Laser-Irradiated HgCdTe. *Journal of Electronic Materials*, Vol.39, No.10, pp. 2220

Suzuki, T. & Shimano R. (2011). Cooling Dynamics of photoexcited Carriers in Si studied using optical Pump and terahertz Probe Spectroscopy. *Phys. Rev. B*, Vol.83, 085207

Voelkmann, C.; Reichelt, M.; Meier, T.; Koch, S.W. & Höfer, U. (2004). Five-Wave-Mixing Spectroscopy of Ultrafast Electron Dynamics at a Si(001) Surface, *Phys. Rev. Lett.*, Vol.92, 127405

Vu, Q.T. & Haug, H. (2000). Time-dependent Screening of the Carrier-Phonon and Carrier-Carrier Interactions in nonequilibrium Systems. *Phys Rev. B*, Vol.62, pp. 7179

Wang, D. & Dignam, M.M. (2009). Excitonic Approach to the ultrafast Optical Response of Semiconductor Quantum Wells. *Phys Rev.B*, Vol.79, 165320

Wang, D. (2008). Ph.D. thesis. Queen's University, 2008. Available at URL: http://hdl.handle.net/1974/1593.

Wang, W.; Zhang, J. & Lenstra, D. (2006). Semiconductor Optical Bloch Equations explain polarization dependent FourWave Mixing Quantum Beats in bulk Semiconductors. *Proceedings, Symposium IEEE/LEOS Benelux Chapter*, pp. 249, Eindhoven (2006)

7

Lasers in Atomic Collisions, Cold Plasma and Cold Atom Physics

R. Cabrera-Trujillo, J. Jiménez-Mier and A. M. Juárez
Universidad Nacional Autónoma de México
México

1. Introduction

In the more than 50 years since its conception, the laser and its variety of applications has covered most of the scientific and technological areas of science, from fundamental physics to biophysics and medicine. Probably, the scientific area that has benefited the most is the same area from where the laser originally came: Atomic, Molecular, and Optical (AMO) Physics. The purpose of this chapter is to present, through a selection of examples, how the applications of the laser in atomic collisions, cold plasmas and cold atom physics is still opening new areas of research. The examples chosen are a sample of what is the current research at the Instituto de Ciencias Nucleares (Nuclear Science Institute) and the Instituto de Ciencias Físicas (Physics Science Institute) at the Universidad Nacional Autónoma de México. We hope that, in choosing these examples, the reader will have the opportunity to get a wide view of three main areas of research in applications of lasers to AMO physics.
The chapter is separated in three sections. Section 2 deals with applications of numerical calculations related to laser assisted collisions, in particular the chapter deals with Finite Difference methods to the calculation of charge transfer cross sections. It shows the bits and pieces of how this numerical technique can be implemented. Section 3 shows an example of a calculation for experimental work with cold atoms. It discusses the technique of stimulated Raman adiabatic passage to produce highly excited Rydberg states. Finally, Sec. 4 shows the reader applications of the Optogalvanic Effect, and in particular to the new emergence of applications that are coming about due to the availability of solid state sources in the UV, and mid infrared. The applications of this technique go from simple basic physics research, to National Security, Biophysics and Medical applications.

2. Laser assisted charge transfer collisions

The study of ion-atom and ion-molecule interactions in the presence of an intense laser is a new and exciting field in atomic physics. Laser intensities up to $I = 3.5 \times 10^{16}$ W/cm^2 that correspond to the atomic unit of electric field $E = 5.1 \times 10^9$ V/cm are so intense that compete with the Columbic interaction force in the control of the electron dynamics. As a result, atoms and molecules in the presence of intense laser fields exhibit new properties and behavior which can be studied by means of multiphotonic processes. Those properties generate new behaviors of matter in the presence of intense lasers, with applications that go from the study of ultra-fast phenomena, the development of lasers of high frequency (XUV and X-rays), the investigation of plasma properties and condensed matter under extreme conditions of temperature and pressure, up to the control of atomic and molecular reactions in the presence of intense lasers.

When an asymmetric collision occurs between an ion and an atom in the presence of an intense laser, the collision cross-section for the charge transfer induced by the laser can be several orders of magnitude greater than when it occurs without the presence of the laser (Anis et al. (2006); Cabrera-Trujillo (2009); Copeland & Tang (1976)).

In the first theoretical works on charge transfer induced by laser in a collision, the attention focused in resonant effects where the frequency of the laser was tuned to a transition in the colliding system (Copeland & Tang (1976); Ferrante et al. (1981); Gudzenko & Yakovenko (1972); Pindzola et al. (2003)). These calculations were carried out for very small intensities, or by means of perturbative calculations or for systems that could be considered coupled by two states in the presence of a laser field (Errea et al. (1983)). More recent work Kirchner (2002; 2004; 2005) has focused on laser intensities above 10^{13} W/cm^2 by using the basis generator method (BGM) with a CW laser. The BGM is a nonperturbative approach that has been successful in the study of field-free ion-atom collisions.

The objective of this section is to show the reader how the technique of finite differences can be used in problems of atomic collisions assisted by an intense laser as well as in the calculation of the cross-section for the charge transfer.

2.1 The problem

As an example, let us consider the case of a α particle (He^{2+}) colliding with a hydrogen atom. For this, we will consider favorable conditions to the experiment, that is to say, a laser of Ti:Sapphire of wavelength around 780 nm commonly used in experimental laboratories with an intermediate intensity of 3.5×10^{12} W/cm^2. In atomic units this corresponds to an electric field of $E_0 = 0.01$ a.u. and frequency $\omega = 0.057$ a.u.. These values guarantee that the effects of ionization and excitation due to the laser can be neglected.

A complete calculation based on quantum mechanics for the charge transfer induced by a laser is a very difficult problem, even computationally. For sufficiently high collisions energies, semi-classical models that consider classic trajectories and quantum electrons can be used. In addition, the interaction with the laser can be represented by means of the dipole approach. Thus, the electronic wave function satisfies the Time-Dependent Schröedinger equation (in atomic units)

$$i\frac{\partial}{\partial t}\Psi(\mathbf{r},t) = \left[T + V(t)\right]\Psi(\mathbf{r},t), \tag{1}$$

where

$$T = -\frac{1}{2}\nabla^2, \tag{2}$$

$$V(\mathbf{r},t) = -\frac{Z_T}{|\mathbf{r}|} - \frac{Z_P}{|\mathbf{r}-\mathbf{R}_P(t)|} - \mathbf{E}(t)\cdot\mathbf{d}\,. \tag{3}$$

are the kinetic energy operator and the interaction potential, respectively. In this expression, Z_T and Z_P are the charges of the target and projectile, $\mathbf{R}_P(t)$ is the trajectory of projectile and $\mathbf{d} = -\mathbf{r}$ is the dipole moment of the electron. In this case, we will use the approach of rectilinear trajectories for the projectile along the z-axis, that is to say,

$$\mathbf{R}_P(t) = b\hat{\mathbf{x}} + vt\hat{\mathbf{z}}\,, \tag{4}$$

where $b > 0$ is the impact parameter, v is the speed of the projectile commensurate to the collision energy, and $\hat{\mathbf{x}}$ and $\hat{\mathbf{z}}$ are the unitary vectors throughout the $x-$ and $z-$axes, respectively. For the laser we consider a pulse that reaches its maximum value at $t = 0$ and that agrees with the point of maximum approach between the projectile and the target. Explicitly, we used a Gaussian laser pulse given by:

$$\mathbf{E}(t) = \mathbf{E}_0 e^{-\left(\frac{t}{\tau}\right)^2} \cos(\omega t + \varphi)\,, \tag{5}$$

where $\mathbf{E}_0$ is the amplitude of the electric field, $\tau\sqrt{4\log 2}$ defines the Full Width Half Maximum of the pulse (FWHM), ω is the frequency of the laser, and φ are the phase of the laser at the point of maximum approach in the collision. From this, it is clear that the dynamics of the collision is governed by a differential equation of second order [Eq. (1)], which we required to solve as a function of time.

2.2 Physics on a numerical lattice

The physics of many time-dependent problems can be described by a systems of coupled partial differential equations (generally nonlinear), complemented by the initial and boundary conditions of the system. An example is the Schröedinger equation (in 1-D)

$$\left\{-\frac{\hbar^2}{2\mu}\frac{\partial^2}{\partial x^2} + V(x)\right\}\Psi(x,t) = i\hbar\frac{\partial}{\partial t}\Psi(x,t)\,, \tag{6}$$

or in the time-independent case

$$\left\{-\frac{\hbar^2}{2\mu}\frac{\partial^2}{\partial x^2} + V(x)\right\}\Psi(x) = E\Psi(x)\,. \tag{7}$$

Here μ is the mass of the particle under the effect of the potential $V(x)$. To solve this system of differential equations, we use the finite difference method. The concept of the finite difference approximation to systems that evolve in the time can be illustrated if we consider a problem in 1+1 dimensions (x,t). A numerical lattice is shown in the Fig. 1 which consists of a system of fixed positions x_k, that may or may not be spaced uniformly, and that are defined in the discreet time t^n.

2.2.1 Functions

If the value of a function $f(x,t)$ in the point (x,t) is associated to a point in the lattice, then

$$f(x,t) = f(x_k,t^n) \rightarrow f_k^n\,. \tag{8}$$

Note that we use subscripts to represent spatial dependence in the lattice point and superscript to denote temporal dependence.

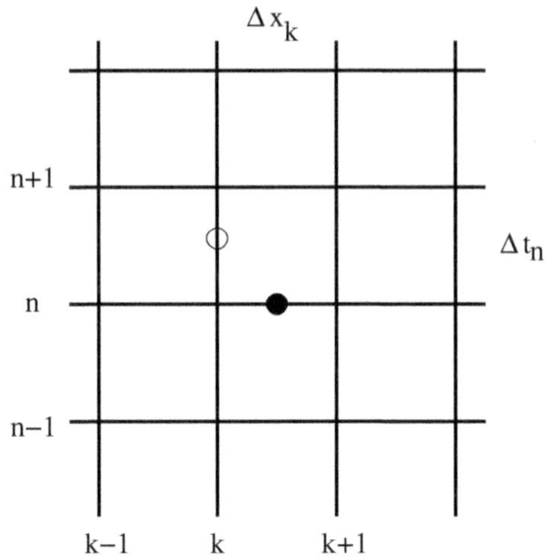

Fig. 1. Uniform numerical lattice in space-time with constant $\Delta x = \Delta x_k^n$ and $\Delta t = \Delta t_k^n$.

2.2.2 Partial derivatives

The finite difference approximation of the partial derivative of $f(x,t)$ can be constructed from the Taylor expansion around x_0. This gives

$$f(x_0 + \Delta x, t) = f(x_0) + \left.\frac{\partial f}{\partial x}\right|_{x_0} \Delta x + \frac{1}{2}\left.\frac{\partial^2 f}{\partial x^2}\right|_{x_0} \Delta x^2 + O(\Delta x^3) \tag{9}$$

and

$$f(x_0 - \Delta x, t) = f(x_0) - \left.\frac{\partial f}{\partial x}\right|_{x_0} \Delta x + \frac{1}{2}\left.\frac{\partial^2 f}{\partial x^2}\right|_{x_0} \Delta x^2 + O(\Delta x^3)\,. \tag{10}$$

From Eq. (9) to first order, we obtain what is known as a forward difference equation

$$\left.\frac{\partial f}{\partial x}\right|_{x_0} = \frac{f(x_0 + \Delta x) - f(x_0)}{\Delta x} + O(\Delta x^2)\,. \tag{11}$$

Similarly, from Eq. (10) we obtain what is know as a backward difference equation

$$\left.\frac{\partial f}{\partial x}\right|_{x_0} = \frac{f(x_0) - f(x_0 - \Delta x)}{\Delta x} + O(\Delta x^2)\,. \tag{12}$$

An approximation to second order for the second derivative is obtained by adding Eqs. (9) and (10):

$$\left.\frac{\partial^2 f}{\partial x^2}\right|_{x_0} = \frac{f(x_0 + \Delta x) - 2f(x_0) + f(x_0 - \Delta x)}{\Delta x^2} + O(\Delta x^3)\,. \tag{13}$$

Let us suppose that $f(x,t)$ correspond to a point in the spatial lattice: $x_0 \to x_k$, $x_0 + \Delta x \to x_{k+1}$, and $f(x,t) \to f_k^n$, then from Eq. (11)

$$\left.\frac{\partial f}{\partial x}\right|_{x_k} = \frac{f_{k+1} - f_k}{x_{k+1} - x_k} + O(\Delta x^2), \tag{14}$$

which represent a forward approximation in space to $(\partial f/\partial x)_{x_k}$ in the k-th lattice point. In the same way, from Eq. (12) the backward approximation is

$$\left.\frac{\partial f}{\partial x}\right|_{x_k} = \frac{f_k - f_{k-1}}{x_k - x_{k-1}} + O(\Delta x^2). \tag{15}$$

And for the second order partial derivative of $f(x,t)$ in the uniform spatial lattice point x_k one obtains:

$$\left.\frac{\partial^2 f}{\partial x^2}\right|_{x_k} = \frac{f_{k+1} - 2f_k + f_{k-1}}{(\Delta x)^2} + O(\Delta x^3). \tag{16}$$

2.2.3 Integrals

If we are required to calculate an integral, for example, when normalizing the solution to the Schrödinger equation, we can use the Riemann definition for an integral, that is

$$\int f(x)dx = \sum_{k=1}^{N-1} f_k(x_{k+1} - x_k), \tag{17}$$

that for a uniform lattice becomes

$$\int f(x)dx = \sum_{k=1}^{N-1} f_k \Delta x. \tag{18}$$

2.3 Time-independent case

In the particular case in which we are required to solve the Time-Independent Schrödinger equation the approximation to the differential equation in finite differences is obtained replacing Eq. (16) into Eq. (7). Therefore, for a uniform lattice with $x_{k+1} - x_k = \Delta x$

$$-\frac{\hbar^2}{2\mu}\left\{\frac{\Psi_{k+1} - 2\Psi_k + \Psi_{k-1}}{\Delta x^2}\right\} + V_k\Psi_k = E\Psi_k \tag{19}$$

where we have omitted the temporal dependency.
Since $k = \{1, ..., N\}$, where N is the number of points in the lattice, then Eq. (19) has a complete solution when we specify the initial and boundary conditions. In our case, $\Psi(x,t) \in \mathcal{L}^2$, that is, it is square integrable such that in the boundary $\{-\infty, \infty\}$ the function is zero. In our case, we do not have a lattice that covers an infinite extension of the space, so we adjust ourselves to a finite lattice where the initial and final points form the boundaries. From here, $\Psi_{k=0} = 0$ and $\Psi_{k=N+1} = 0$ such that

$$-\frac{\hbar^2}{2\mu\Delta x^2}\left\{\Psi_2 - 2\Psi_1\right\} + V_1\Psi_1 = E\Psi_1 \tag{20}$$

and

$$-\frac{\hbar^2}{2\mu\Delta x^2}\Big\{-2\Psi_N+\Psi_{N-1}\Big\}+V_N\Psi_N=E\Psi_N\,. \tag{21}$$

Equation (19) together with the boundary conditions (20) and (21) can be written in matrix form by making a vector array $\vec{\mathbf{\Psi}}=\{\Psi_k\}$. Thus

$$\mathbf{H}\vec{\mathbf{\Psi}}=E\vec{\mathbf{\Psi}} \tag{22}$$

where $\mathbf{H}$ has the matrix elements given by

$$H_{i,i}=\frac{\hbar^2}{\mu\Delta x^2}+V_k\,,$$
$$H_{i,i+1}=-\frac{\hbar^2}{2\mu\Delta x^2}\,, \tag{23}$$
$$H_{i,i-1}=-\frac{\hbar^2}{2\mu\Delta x^2}\,,$$

which is a Hermitian tridiagonal matrix of the form:

$$\mathbf{H}=\begin{pmatrix} H_{1,1} & H_{1,2} & 0 & 0 & \cdots & 0\\ H_{2,1} & H_{2,2} & H_{2,3} & 0 & \cdots & 0\\ 0 & H_{3,2} & H_{3,3} & H_{3,4} & \cdots & 0\\ \vdots & \vdots & \vdots & \ddots & \vdots & \vdots\\ 0 & \cdots & 0 & H_{N-1,N-2} & H_{N-1,N-1} & H_{N-1,N}\\ 0 & \cdots & 0 & 0 & H_{N,N-1} & H_{N,N}\end{pmatrix}. \tag{24}$$

Equation (22) thus represents an eigenvalue problem in matrix algebra with the advantage of being a tridiagonal matrix. The number of eigenvectors and eigenvalues depends on the number of points in the numerical lattice.

2.4 Time-dependent case

For the time-dependent case, we use Eq. (6) and applying the finite difference approximation we get

$$-\frac{\hbar^2}{2\mu}\Big\{\frac{\Psi^n_{k+1}-2\Psi^n_k+\Psi^n_{k-1}}{\Delta x^2}\Big\}+V_k\Psi^n_k=i\hbar\frac{\Psi^{n+1}_k-\Psi^n_k}{t^{n+1}-t^n}\,. \tag{25}$$

Thus, for a constant increment in time $\Delta t=t^{n+1}-t^n$ we have that the wave function at time $t=t^n+\Delta t$ is given by

$$-\frac{\hbar\Delta t}{i2\mu\Delta x^2}\Big\{\Psi^n_{k+1}-2\Psi^n_k+\Psi^n_{k-1}\Big\}-\frac{i\Delta t}{\hbar}V_k\Psi^n_k+\Psi^n_k=\Psi^{n+1}_k\,. \tag{26}$$

This equation is an example of a explicit method, since the solution at a later time is given in terms of the previous time. The last equation can be written as

$$\vec{\Psi}^{n+1} = \left\{1 - \frac{i\Delta t}{\hbar}\mathbf{H}\right\}\vec{\Psi}^{n}. \tag{27}$$

But from Eq. (6) we know that the time-dependent solution for the time-independent Hamiltonian $\hat{\mathbf{H}}$ is

$$\Psi(x,t) = e^{-i\frac{(t-t_0)}{\hbar}\hat{\mathbf{H}}}\Psi(x,t_0), \tag{28}$$

thus we can use the previous equation together with the finite difference method to obtain a improved solution in time for the wave function.

2.4.1 Crank-Nicolson method

By using the split-operator technique, we can write Eq. (28) as

$$\Psi(x,t) = e^{-i\frac{\Delta t}{\hbar}\{\hat{\mathbf{T}}+V(x)\}}\Psi(x,t_0) \tag{29}$$

where $\Delta t = t - t_0$. Separating the operator for the kinetic energy

$$\Psi(x,t) \simeq e^{-i\frac{\Delta t}{2\hbar}\hat{\mathbf{T}}}e^{-i\frac{\Delta t}{2\hbar}\hat{\mathbf{T}}}e^{-i\frac{\Delta t}{\hbar}V(x)}\Psi(x,t_0), \tag{30}$$

therefore

$$e^{i\frac{\Delta t}{2\hbar}\hat{\mathbf{T}}}\Psi(x,t) \simeq e^{-i\frac{\Delta t}{2\hbar}\hat{\mathbf{T}}}e^{-i\frac{\Delta t}{\hbar}V(x)}\Psi(x,t_0). \tag{31}$$

Defining

$$f(x,t,t_0) = e^{-i\frac{\Delta t}{\hbar}V(x)}\Psi(x,t_0), \tag{32}$$

we get, to first order in the exponential,

$$\left\{1 + i\frac{\Delta t}{2\hbar}\hat{\mathbf{T}}\right\}\Psi(x,t) = \left\{1 - i\frac{\Delta t}{2\hbar}\hat{\mathbf{T}}\right\}f(x,t,t_0). \tag{33}$$

Since the kinetic energy operator is $\hat{\mathbf{T}} = -\frac{\hbar^2}{2\mu}\frac{\partial^2}{\partial x^2}$, then applying the finite difference method to the previous equation, we get

$$\left\{\Psi_k^{n+1} - i\frac{\hbar\Delta t}{4\mu\Delta x^2}\left(\Psi_{k+1}^{n+1} - 2\Psi_k^{n+1} + \Psi_{k-1}^{n+1}\right)\right\} = \left\{f_k + i\frac{\hbar\Delta t}{4\mu\Delta x^2}\left(f_{k+1} - 2f_k + f_{k-1}\right)\right\}, \tag{34}$$

or in matrix form

$$\mathbf{A}^{+}\vec{\Psi}^{n+1} = \mathbf{A}^{-}\vec{f}, \tag{35}$$

therefore

$$\vec{\Psi}^{n+1} = (\mathbf{A}^{+})^{-1}\mathbf{A}^{-}\vec{f}. \tag{36}$$

This is the implicit Crank-Nicolson method, where in both sides of the equation the final time appears. The matrices $\mathbf{A}^{+}$ y $\mathbf{A}^{-}$ are tridiagonal matrices with elements given by

$$\begin{aligned} A_{i,i}^{\pm} &= 1 \pm 2\nu, \\ A_{i,i+1}^{\pm} &= \mp\nu, \\ A_{i,i-1}^{\pm} &= \mp\nu, \end{aligned} \tag{37}$$

where $\nu = \frac{i\hbar\Delta t}{4\mu\Delta x^2}$. That is

$$\mathbf{A}^{\pm} = \begin{pmatrix} 1\pm 2\nu & \mp\nu & 0 & 0 & \cdots & 0 \\ \mp\nu & 1\pm 2\nu & \mp\nu & 0 & \cdots & 0 \\ 0 & \mp\nu & 1\pm 2\nu & \mp\nu & \cdots & 0 \\ \vdots & \vdots & \vdots & \ddots & \vdots & \vdots \\ 0 & \cdots & 0 & \mp\nu & 1\pm 2\nu & \mp\nu \\ 0 & \cdots & 0 & 0 & \mp\nu & 1\pm 2\nu \end{pmatrix}. \tag{38}$$

Thus, the temporal evolution of a system can be solved by means of linear algebra approaches, for example, the LU matrix decomposition (Press et al. (1992)).
The extension to a 2- or 3-D case is straightforward since the procedure can be nested in several loops, each one for a dimension.

2.5 Implementation

Although the finite difference method is straightforward, the calculations are computationally intensive and we must take care in obtaining stable and precise results. One of the problems that arises is when the charged projectile moves in the numerical lattice, since when the projectile passes over the $\mathbf{R}_P = \mathbf{r}$, i.e. a point on the lattice, there appears a singularity in the potential and the numerics will behave as if a collision with a wall. Inclusively, if the projectile passes close, there is a periodic oscillating effect when the projectile approaches each point of the lattice as it moves. In order to solve this problem, a damping or soft-core parameter can be introduced in the Coulomb potential. Nevertheless, one finds that this procedure overestimates the charge transfer. A simple solution to this problem consists of placing the trajectory of projectile in the center of the squares defined by the points of the lattice in the plane defined by the collision. One second approach is to fix the position of the He^{2+} ion to the origin to diminish fluctuations due to the large nuclear charge of the projectile. Thus, the hydrogen atom becomes the projectile in this reference system (Galilean transformation). This change only introduces a phase factor in the wave function. In this work we opted for the center of the lattice method.
The numerical lattice used in our implementation of finite differences covers the region $[-6 : 15]_x \times [-6 : 6]_y \times [-25 : 25]_z$ u.a. The size of this lattice was chosen in such a way that it balances the time of calculation and the precision of the physical results. The size of the lattice was chosen of $\Delta x = \Delta y = \Delta z = 0.2$ u.a., which gives a ground energy for the hydrogen atom of $E_H = -0.490$ u.a and of $E_{He^+} = -1.90$ u.a for the ground state of the Helium ion.
Our calculations cover the temporal range from $t_i = -200$ to $t_f = 200$ u.a which gives us sufficient time to separate clearly, the projectile and target at the end of the collision for a collision energy of 1 keV. The size of the temporary step was chosen as 0.06 u.a in such a way that it balances the time that the dynamics take and to preserve the precision of the calculations. It was chosen on the basis of several tests of energy conservation.

2.5.1 Charge transfer

We calculate the charge transfer probability by means of the integration of the electronic charge density over a box Ω_T, that encloses the projectile at the final time of the collision

t_f,

$$P = \int_{\Omega_T} |\Psi(\mathbf{r}, t_f)|^2 d^3\mathbf{r} \equiv \int_{z_T}^{z_{max}} \rho_z(z, t_f) dz \,, \tag{39}$$

where z_T defines the mid point distance between target and projectile and where the charge density $\rho_z(z,t)$ is given by

$$\rho_z(z,t) = \int_{x_{min}}^{x_{max}} \int_{y_{min}}^{y_{max}} |\Psi(x,y,z,t)|^2 \, dx \, dy \,.$$

Therefore it is required that the contributions of the target and the projectile to the wave function are clearly separated in the space, as is shown in the example of the Fig 2. Once the capture probability of electrons is obtained as a function of the impact parameter, we can calculate the cross-section for charge transfer. From conservation of the number of particles in the collision, we have that if I is the intensity of colliding particles per unit area, unit time (that is to say, the particle flux), then $I \, b \, db \, d\phi$ is the number of particles per unit time that interact with the target. This number must be equal to the number of scattered particles in collision per unit of solid angle, $I(d\sigma/d\Omega)d\Omega$, that is to say, the number of scattered particles is proportional to the number of colliding particles per unit of solid angle. Here, the constant of proportionality, $d\sigma/d\Omega$, is the scattering probability and has units of area, and therefore the name of cross-section of the collision. Therefore, since $d\Omega = \sin(\theta) d\theta d\phi$ then

$$\frac{d\sigma}{d\Omega} = \frac{b}{\sin(\theta)} \left| \frac{db}{d\theta} \right|$$

is the differential scattering cross section.

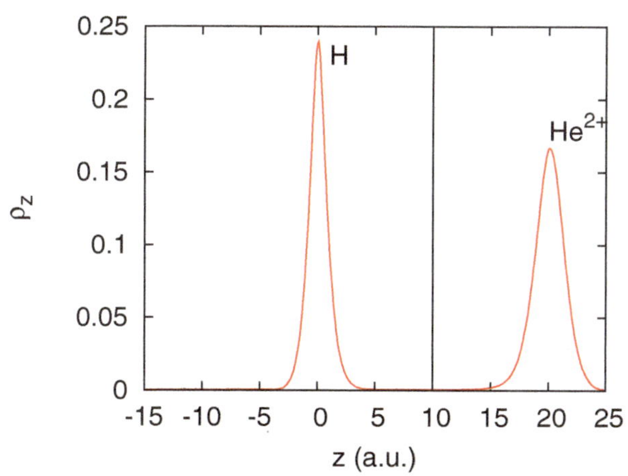

Fig. 2. Example of electronic density ρ_z for He^{++}+H at $b = 0.4$ u.a., $E_0 = 0$ (No field) and a collision energy of 1 keV/amu. The dotted line at $z = 10$ a.u. separates the boundary Ω_T for the projectile (see text).

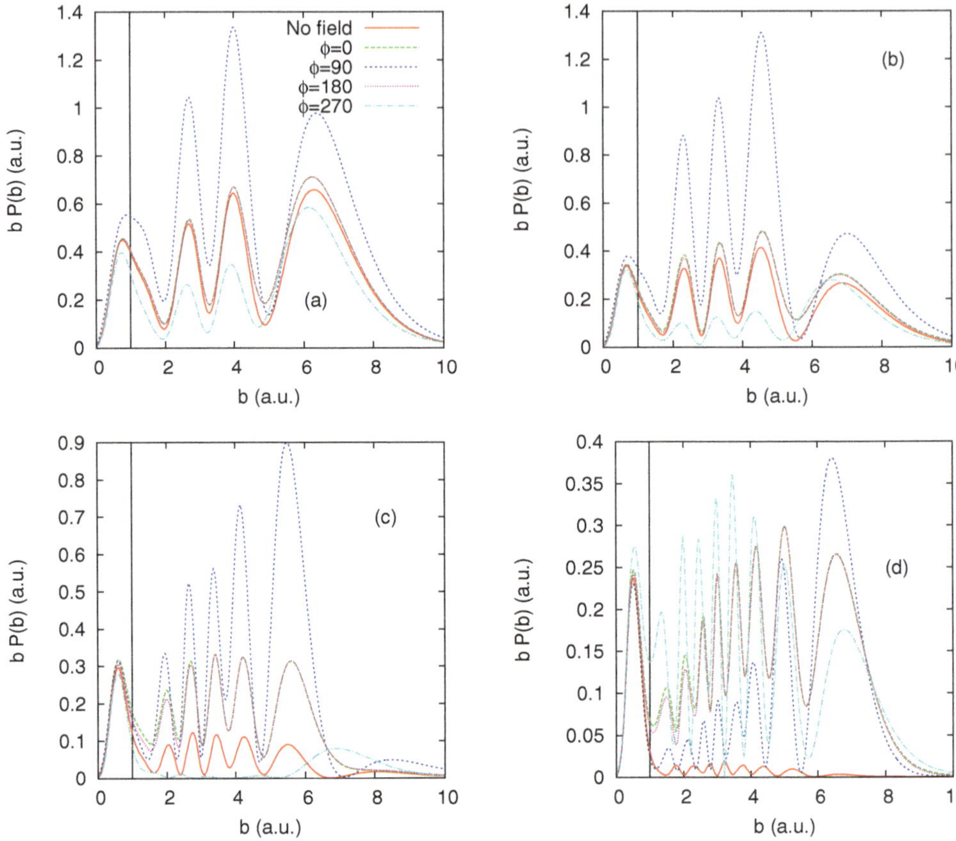

Fig. 3. Impact parameter weighted probability for charge transfer, $bP(b)$, as a function of the impact parameter for laser phase $\phi = 0, 90, 180$, and 270 deg. The results are shown for collision energies of a).- $E = 1.5$; b).- $E = 1.0$; c).- $E = 0.5$; and d).- $E = 0.25$ keV/amu. The results are for a laser field of intensity $I = 3.5 \times 10^{12}$ W/cm^2, wavelength $\lambda = 780$ nm, and FWHM of 6 fs with polarization parallel to the axis of the collision.

For the case of a processes with probability $P(b)$ for each impact parameter or scattering angle, the differential cross section is given by

$$\frac{d\sigma}{d\Omega} = \frac{b\,P(b)}{\sin(\theta)}\left|\frac{db}{d\theta}\right| .$$

Integrating over all the scattering angles we find the total scattering cross section is:

$$\sigma = 2\pi \int_0^\infty b\,P(b)\,db\,. \tag{40}$$

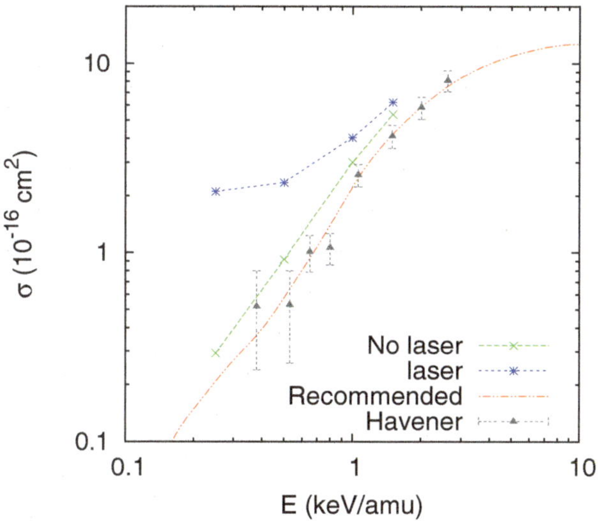

Fig. 4. Electron charge transfer cross section as a function of the projectile energy. The dashed line is the finite difference result without laser. The dotted line is the result with laser field averaged over the phase ϕ. The up-triangle symbols are the experimental data from from Havener (Havener et al. (2005)), as well as the dot-dashed line that corresponds to their recommended laser-free theoretical results, respectively.

2.6 Results

In Fig. 3, we show the results for the weighted electron capture probability as a function of the impact parameter for a laser carrier phase of 0, 90, 180, and 270 deg and electric field parallel to the axis of the collision and for collision energies of $E = 1.5$ keV/amu (Fig. 3a), $E = 1.0$ keV/amu (Fig. 3b), $E = 0.5$ keV/amu (Fig. 3c), and $E = 0.25$ keV/amu (Fig. 3d). In the same figure, we show the results without laser field (solid line). The effect of the laser field occurs for impact parameter larger than $b > 2.0$ a.u. in the so called radial region (Stolterfoht et al. (2007)). For impact parameters $b < 2.0$ the effect is small, in the so called rotational region. This two regions are shown by the vertical line in Fig. 3. In the high energy cases, the largest contribution comes from $\phi = 90$ deg, while the lower contribution is from $\phi = 270$ deg. At high energies, they average mostly to the case without laser field. For low collision energies, however, the electron capture can be up to one order of magnitude larger for the collision energies considered in this work. The total cross-section for the charge transfer induced by the field laser is shown in fig 4 for the case in that the electric field of the laser parallel to the axis of the collision and we compared with the results without field laser as a function of the collision energy. In the same figure, we compare our results to the experimental data without laser field of Havener (Havener et al. (2005)) as well as their recommended theoretical values. First we noticed a good agreement between the experimental results and our cross sections obtained by the finite difference approach for the case of no laser field. In the case of laser field we observe an increase in the cross-section for charge transfer by a factor of up to ~ 10 for low collision energies.

2.7 Conclusions

The presence of an intense laser field in the atom interaction produces a great variety of effects that allow us to the control processes such as generation of harmonics, above threshold ionization, Coulomb explosion, etc. The dynamics of these processes is governed by the time-dependent Schrödinger equation. In this section we have shown how the finite difference method can be used to obtain dynamic numerical solutions and thus to study the interaction process. In the case of atomic collisions in the presence of a laser we have found that for the case of He^{2+} colliding with atomic hydrogen the cross-section of electronic capture intensifies up to a factor of six in average, for a 10 fs Gaussian laser pulse and intensity of $3,5 \times 10^{12}$ W/cm^2 for energies of collision in the region of keV.

3. Laser pulses for the coherent manipulation of quantum states in cold atoms.

A controlled sequence of laser pulses can be used to manipulate quantum states of matter. In particular, the technique of stimulated Raman adiabatic passage (STIRAP) uses a counterintuitive pulse sequence to efficiently transfer population between two long lived states of a system (Carroll & Hioe (1988); Kuklinski et al. (1989); Oreg et al. (1985; 1984)). The basic STIRAP scheme involves three states, the initial state $|0\rangle$, the intermediate state $|1\rangle$ and the final state $|2\rangle$, excited by two laser pulses that partially overlap in time. Usually the intermediate state $|1\rangle$ is short lived, and it may decay into other states of the system. The goal is to transfer the entire population initially in state $|0\rangle$, to the target state $|2\rangle$. This is accomplished by applying a first (Stokes) pulse in near resonance with the energy difference between states $|1\rangle$ and $|2\rangle$ followed by a pump pulse in near resonance with the $|0\rangle$ to $|1\rangle$ transition. If this counterintuitive sequence of pulses is applied maintaining adiabatic evolution conditions, a trapped or dark state is produced and it evolves from the initial state $|0\rangle$ into the final state$|2\rangle$ without ever passing through the intermediate state.

There are several examples of both theoretical and experimental studies of three-level two-pulse STIRAP (Bergmann et al. (1998)). There are also detailed theoretical studies that extend the analysis of STIRAP processes to more than three levels (Malinovsky & Tannor (1997); Oreg et al. (1992); Smith (1992); Vitanov (1998); Vitanov et al. (1998)). Here a clear difference can be established between systems with an odd number of levels, for which a dark state can be immediately identified, and systems with even number of levels, that strictly never have such state (Smith (1992); Vitanov (1998)).

Cold atoms are the perfect laboratory to test coherent state manipulation using light. The cooling and trapping process allows one to produce a cloud of atoms in a specific quantum state (Metcalf & van der Straten (1999)). The translational temperature can range between a few mK to a few nK, and therefore Doppler effects in the interaction between the atoms and electromagnetic radiation are strongly reduced. At these temperatures the atoms move so slowly that in the interaction time they suffer no collisions. The first experiments to study STIRAP with cold atoms used the $5s \rightarrow 5p \rightarrow 5d$ ladder excitation in rubidium (Süptitz et al. (1997)). STIRAP to produce coherent excitation of cold atoms into Rydberg states has also been demonstrated in rubidium magneto-optical traps (Cubel et al. (2005); Deiglmayr et al. (2006)). Cold rubidium atoms have also been used to study in great detail the population dynamics in a STIRAP process (Gearba et al. (2007)). In all these examples the basic three-level and two-pulse STIRAP scheme was used.

Cold Rydberg atoms are at the core of very interesting proposals for quantum information processes with neutral atoms. A recent review was published by Saffman *et al.* (Saffman et al. (2010)). The central idea is to use long range interactions between Rydberg atoms to produce quantum gates. An important requirement posed by these quantum gate protocols is a precise coherent excitation and de-excitation of Rydberg states (Saffman et al. (2010)). The widely used scheme is to excite into a Rydberg state by two-photon excitation starting from the atomic ground state. For example, the excitation $5s \rightarrow 5p_{3/2} \rightarrow n\ell$ in rubidium requires a 780 nm photon for the first step and a wavelength of about 480 nm for the second step, in order to reach the Rydberg state. These were the excitations used to show efficient population transfer into Rydberg states using STIRAP (Cubel et al. (2005); Deiglmayr et al. (2006)). The transition matrix element for excitation into a Rydberg state $n\ell$ is small and it decreases with increasing quantum number according to the power law $n^{-3/2}$ (Saffman et al. (2010)). Using the $5p_{3/2}$ level in rubidium for STIRAP production of Rydberg states therefore requires high laser powers and tight focusing into the cold atom cloud. Also, to extend the STIRAP efficiency to higher n's the laser power has to be scaled by n^3.

In this section we present a theoretical analysis of an alternative, four-level STIRAP scheme to transfer the population of the 5s atomic ground state into long-lived Rydberg states of cold rubidium atoms. It considers as intermediate states the $5p_{3/2}$ and 5d atomic levels and therefore has the advantage of requiring the use of lower power diode lasers to produce the STIRAP pulses (Thoumany et al. (2009)). Furthermore, because the 5d radial wave function extends further out, one expects the transition matrix elements connecting to the np or nf Rydberg states to be larger than the corresponding transition matrix elements between the $5p_{3/2}$ and the ns or nd states. We also show the results of a calculation of these electric dipole matrix elements. The overall study predicts an efficient population transfer and provides a time scale for the STIRAP pulse sequence that ensures permanence within the adiabatic regime.

3.1 Overview of three- and four-level STIRAP.

For completeness a brief review of three-level STIRAP (Bergmann et al. (1998)) is presented here. Consider a system with three energy levels and two radiation fields. A schematic representation of a system in a ladder configuration is shown in Fig. 5(a). It is assumed that the Stokes radiation field is detuned by Δ_S with respect to the frequency of the dipole allowed transition between levels $|1\rangle$ and $|2\rangle$, and that the pump beam is detuned by Δ_P with respect to the $|0\rangle$ to $|1\rangle$ electric dipole transition. This also implies that the direct $|0\rangle \rightarrow |2\rangle$ transition is forbidden. One can write the Hamiltonian matrix H that describes the time evolution of the system:

$$H = \frac{\hbar}{2}\begin{pmatrix} 0 & \Omega_P(t) & 0 \\ \Omega_P(t) & 2\Delta_P & \Omega_S(t) \\ 0 & \Omega_S(t) & 2(\Delta_P + \Delta_S) \end{pmatrix} \tag{41}$$

where $\Omega_S(t)$ and $\Omega_P(t)$ are, respectively, the Rabi frequencies of the Stokes and pump beams. Furthermore, it will also be assumed that the whole system is in resonance, that is, $\Delta_P = -\Delta_S$. The time dependent Schroedinger equation reads

$$i\hbar\frac{\partial \vec{c}}{\partial t} = H\vec{c} \tag{42}$$

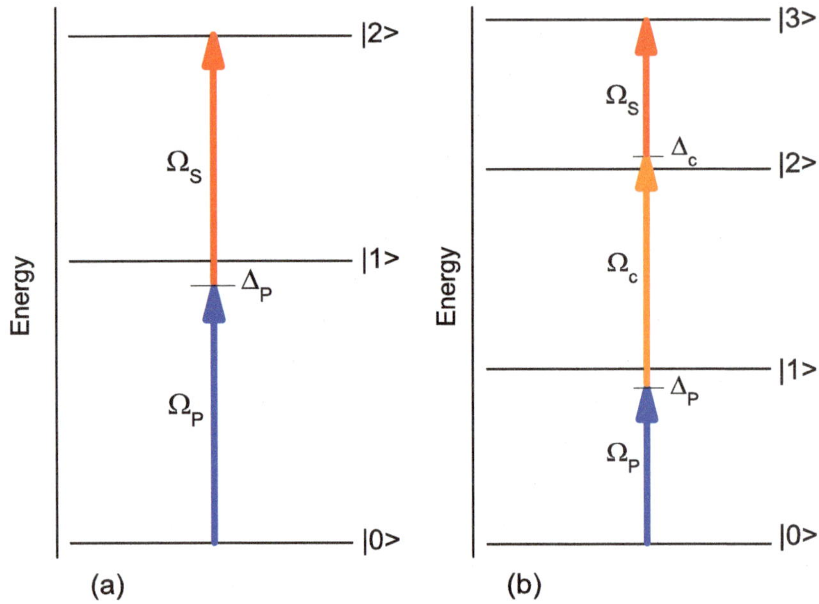

Fig. 5. Schematic representation of a STIRAP process. (a) Three-level STIRAP; (b) four-level STIRAP.

The vector $\vec{c}$ is formed with the expansion coefficients of the state under consideration. We seek a solution to this equation that allows an efficient transfer of population between the initial and final states.

If the time variation of the radiation fields is slow enough the process satisfies the adiabatic condition (Bergmann et al. (1998)) the coefficient vector $\vec{c}$ is an instantaneous eigenvector of the Hamiltonian matrix (41). A direct calculation gives the following eigenvalues for the Hamiltonian:

$$\omega^{+} = \Delta_P + \sqrt{\Delta_P^2 + \Omega_S^2 + \Omega_P^2}$$
$$\omega^{-} = \Delta_P - \sqrt{\Delta_P^2 + \Omega_S^2 + \Omega_P^2}$$
$$\omega^{0} = 0 \tag{43}$$

The eigenstate that corresponds to the last eigenvalue is:

$$\left|v^0\right\rangle = \cos\theta\,|0\rangle - \sin\theta\,|2\rangle \tag{44}$$

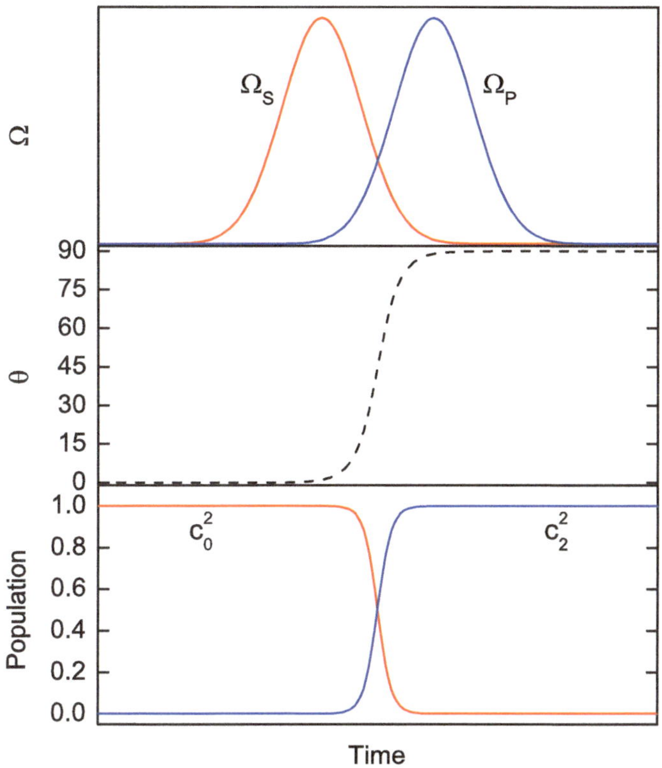

Fig. 6. Pulse sequence (top), mixing angle (middle) and state populations (bottom) in a three-level STIRAP process.

and it is the only one that at all times has no component of the intermediate state $|1\rangle$. The time dependent mixing angle is

$$\tan\theta = \frac{\Omega_P(t)}{\Omega_S(t)} \tag{45}$$

It is now clear that by using light pulses in a sequence in which the Stokes pulse precedes the pump pulse, the mixing coefficient can be varied continuously between 0 and $\pi/2$ and therefore the atomic population can be completely transferred from state $|0\rangle$ to state $|2\rangle$, without ever passing through the intermediate state. This pulse sequence and its effect in the mixing coefficients are graphically represented (Bergmann et al. (1998)) in Fig. 6.

For the system to remain in the dark state throughout the pulse sequence it is necessary that the projection of the rate of change of the expansion coefficient $\vec{c}_0$ onto the other states is small compared to the difference in eigenvalues (Bergmann et al. (1998)):

$$\left|\frac{d\vec{c}_0}{dt}\cdot\vec{c}_\pm\right| << \left|\omega^\pm - \omega^0\right| \tag{46}$$

The extension of STIRAP to systems with more than three levels has also been discussed extensively (Bergmann et al. (1998); Oreg et al. (1992); Smith (1992); Vitanov (1998); Vitanov et al. (1998)). In the case that the intermediate state detunings are zero the behavior of systems with an odd number of states is similar to the one already discussed (Smith (1992); Vitanov (1998)). The Hamiltonian matrix always has a zero eigenvalue and its eigenvector contains only even states in our notation (Smith (1992)). Before all pulses are turned on the corresponding eigenvector is equal to the initial state, and it evolves into the target state as the pulse sequence is applied. The process is robust, in the sense that adiabatic passage is not very sensitive to the detailed forms and delays between pulses, as long as the first and last ones are applied in the counter-intuitive order.

The situation is quite different for an even level system (Oreg et al. (1992); Smith (1992); Vitanov (1998)). There is no adiabatic channel for zero intermediate detunings. In the general case the Hamiltonian matrix has no zero eigenvalue, but it is possible to establish conditions that lead to an eigenstate for adiabatic population transfer (Vitanov (1998)). The conditions require intermediate detunings different from zero and the scheme now depends on the actual pulse sequence because the populations of the intermediate states are different from zero. If the intermediate states can decay by spontaneous emission they may leak population out of the STIRAP process.

Four state STIRAP, shown schematically in Fig. 5(b), is particularly interesting to test these ideas. The Hamiltonian matrix for the system is:

$$H = \frac{\hbar}{2}\begin{pmatrix} 0 & \Omega_P(t) & 0 & 0 \\ \Omega_P(t) & 2\Delta_P & \Omega_c(t) & 0 \\ 0 & \Omega_c(t) & 2\Delta_c & \Omega_S(t) \\ 0 & 0 & \Omega_S(t) & 0 \end{pmatrix} \tag{47}$$

Here Ω_c is the Rabi frequency of a laser that couples the intermediate states $|1\rangle$ and $|2\rangle$ and is detuned from resonance by Δ_{12}. Also, the intermediate detuning $\Delta_c = \Delta_P + \Delta_{12}$ and the assumption is made of overall resonance. This Hamiltonian allows a rather complicated general analytic solution (Oreg et al. (1992)). However, a much simpler condition for the existence of an adiabatic channel can be obtained (Vitanov (1998)). The condition can be extended to general n-level STIRAP processes (Vitanov (1998)). In the case of four levels it is directly expressed in term of the frequency detunings as $\Delta_P\Delta_c > 0$. Therefore, an adiabatic channel exists if there are nonzero intermediate detunings that satisfy this condition (Vitanov (1998)).

Among the schemes proposed for four-level STIRAP there is the production of an intermediate dressed state through the resonant coupling of states $|1\rangle$ and $|2\rangle$ (Smith (1992); Vitanov (1998); Vitanov et al. (1998)). In this case, it is convenient to make $\Delta_{12} = 0$ and the condition for existence of an adiabatic channel, $\Delta_P^2 > 0$ is met as long as the pump beam (and therefore the Stokes beam) is detuned from resonance. It is assumed that the Stokes and pump lasers are turned on in the counter-intuitive sequence while the the coupling laser intensity does not vary in time. Also, to reduce the population of the intermediate states the coupling laser intensity is assumed much larger than both Stokes and pump laser intensities. In order to propose a model for adiabatic population transfer in rubidium it is important to use sensible values for the Hamiltonian parameters. Among them, the Rabi frequencies of the three steps play a significant role. The pump and coupling pulses connect, respectively, the 5s to the $5p_{3/2}$ and the $5p_{3/2}$ to the 5d states. Both transitions are easily saturated with extended cavity diode lasers. We will therefore assume that Ω_P can be controlled to make it comparable to Ω_S, and that Ω_c can be made several times larger than Ω_P. Thus, we must obtain reasonable values for Ω_S.

3.2 Calculation of 5d$\rightarrow$nℓ transition matrix elements.

Rabi frequencies are the product of an electric dipole transition matrix times the electric field amplitude (Metcalf & van der Straten (1999)):

$$\hbar\Omega = \mu_{12}E(t). \tag{48}$$

The electric dipole matrix element μ_{12} is also a product of a geometric factor times the radial matrix element

$$r_{n\ell,n'\ell'} = \int_0^\infty P_{n\ell}(r) r P_{n'\ell'}(r) dr \tag{49}$$

where $P_{n\ell}(r)$ and $P_{n'\ell'}(r)$ are, respectively, the radial wave functions of the initial and final states.

A standard procedure to numerically calculate radial wavefunctions for the valence electron of alkali atoms uses the Numerov algorithm (Gallagher (1994); Zimmerman et al. (1979)). It requires knowledge of the electron potential and the level energy. Starting at large values of r it recursively generates numerical values of the radial wave function for a logarithmic radial mesh. If it is simultaneously applied to initial and final wave functions it allows a direct evaluation of the radial integral. This procedure was used to calculate radial matrix elements for transitions into ns and nd Rydberg states starting from the $5p_{3/2}$ state (Cubel et al. (2005); Deiglmayr et al. (2006); Saffman et al. (2010)). The resulting transition matrix elements were found to be in good agreement with measurements of the Rabi frequencies through the Autler-Townes splitting induced in the lower transition by the upper laser in a two-photon experiment (Cubel et al. (2005); Deiglmayr et al. (2006)).

We followed the same procedure to calculate the Rydberg radial integrals originating in the 5d state. For the Numerov algorithm we used a model potential for alkali atoms (Marinescu et al. (1994)) and the energy levels obtained from a quantum defect analysis (Gallagher (1994); Lorenzen & Niemax (1983)). For comparison we also calculated radial integrals between the $5p_{3/2}$ and the ns, np Rydberg states, with n ranging between 32 and 64. The results are shown in Fig. 7, where we only plot the absolute values of the radial integrals. For each individual series we fitted power expressions of the form $C_\ell \times n^{*-3/2}$, where $n^* = n - \delta_{n\ell}$ is the effective principal quantum number (Gallagher (1994); Lorenzen & Niemax (1983)). The results for the coefficients C_ℓ are shown in Table 1. Our values for the ns and nd series are in good agreement with the previous calculations (Deiglmayr et al. (2006); Saffman et al. (2010)). As expected, our results show that transitions originating from the 5d state have larger radial integrals compared to the transitions originating from the $5p_{3/2}$ state. The C_p coefficient is a factor of 1.9 larger than C_s, and C_f is a factor of 3.1 larger than C_d.

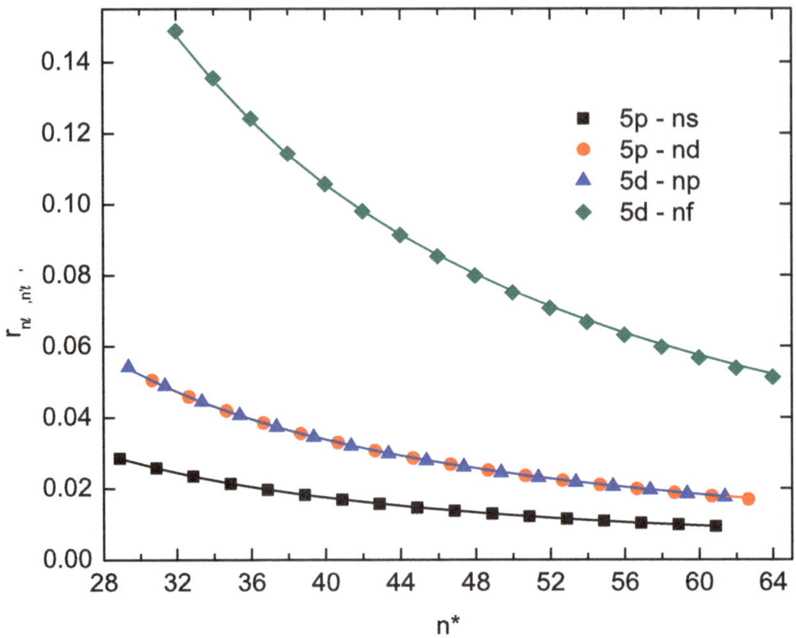

Fig. 7. Results for the absolute values of the $5p_{3/2} \rightarrow n\ell$ and $5d \rightarrow n\ell$ electric dipole radial matrix elements. The symbols are the results of the calculation and the continuous lines are $n^{*-3/2}$ power fits (for details see text).

Rydberg series Coefficient/a_0	$5p_{3/2} \rightarrow$ ns C_s	$5p_{3/2} \rightarrow$ nd C_d	$5d \rightarrow$ np C_p	$5d \rightarrow$ nf C_f
This work	4.414	8.521	8.537	26.692
Deiglmayr et al. (2006)	4.508	8.475		
Saffman et al. (2010)	4.950	8.485		

Table 1. Results for coefficients C_ℓ of Rydberg radial integrals.

3.3 Four-level STIRAP to produce cold Rydberg rubidium atoms.

Three-level STIRAP has already been used to coherently produce Rydberg states in cold rubidium atoms (Cubel et al. (2005); Deiglmayr et al. (2006)). The atomic states are the 5s, $5p_{3/2}$ and ns or np, and the wavelengths of the pump and Stokes lasers are, respectively 780 nm and 480 nm. Because of its small dipole matrix element, the transition between the $5p_{3/2}$ intermediate state and the target Rydberg level usually requires a high intensity Stokes laser tightly focused into the atomic cloud (Cubel et al. (2005); Deiglmayr et al. (2006)). Recently it was shown (Thoumany et al. (2009)) that continuous excitation of Rydberg states in a vapor cell can be achieved using three relatively inexpensive diode lasers, connecting the $5s \rightarrow 5p_{3/2} \rightarrow 5d \rightarrow$ np or nf ladder sequence. Their wavelengths are, respectively, 780 (pump),

776 (coupling) and about 1259 (Stokes) nm.[1] Also, a three-photon process has the added advantage of being able to excite into Rydberg states of higher orbital angular momentum nf. For the four-level STIRAP calculation we consider the time dependent Stokes and pump fields

$$\begin{aligned}\Omega_S(t) &= \Omega_S^0 \exp\left[-(t-\tau_S)^2/T^2\right] \\ \Omega_P(t) &= \Omega_P^0 \exp\left[-(t-\tau_P)^2/T^2\right]\end{aligned} \tag{50}$$

where both pulses have the same width at half maximum $T^* = 2\sqrt{\ln(2)}T$. For convenience the time scale will be expressed in terms of this pulse width. If one uses the times

$$\begin{aligned}\tau_S &= -\frac{\tau}{2} + \frac{T^2}{2\tau}\ln\frac{\Omega_P^0}{\Omega_S^0} \\ \tau_P &= \frac{\tau}{2} + \frac{T^2}{2\tau}\ln\frac{\Omega_P^0}{\Omega_S^0}\end{aligned} \tag{51}$$

then the origin of the time scale occurs when the two pulses cross, and they are are separated a delay τ. Furthermore, for three level STIRAP it was found that the optimum pulse separation is equal to the pulse width ($\tau_S + \tau_P = T^*$) (Gaubatz et al. (1990)). The coupling field Ω_c will be assumed to be on resonance ($\Delta_{12} = 0$), its amplitude to remain constant throughout the STIRAP process, and its value to be larger than both Ω_S^0 and Ω_P^0. This constant coupling field *dresses* states $|1\rangle$ and $|2\rangle$, producing frequency shifts of $\pm\Omega_c/2$ (Vitanov et al. (1998)). The only entry left in the Hamiltonian matrix (Eq. 47) is the frequency shift Δ_P. It will be taken sufficiently close to $\Omega_c/2$ so that the pump step is nearly resonant with one of the dressed states. This will have the effect of moving one of the eigenstates out of the picture and the system to behave as in three-level STIRAP.
The Rabi frequencies chosen for these simulations are shown in Table 2, where we also give the values of the transition matrix elements and the required laser power densities. The Rabi frequency can be expressed in terms of the power density I as (Metcalf & van der Straten (1999))

$$\frac{\Omega}{2\pi} = \mu_{12}\sqrt{\frac{2I}{\epsilon_0 c h^2}} \tag{52}$$

where ϵ_0 is the vacuum permitivity, c is the speed of light and h is Planck's constant. For the upper 5d$\rightarrow$ nf transition matrix element we use the transition matrix element to produce the state with $n = 50$. Two cases are presented. In case A the Rabi frequency of the Stokes pulse is approximately equal to the Rabi frequency for the $5p_{3/2}$ $\rightarrow$45s transition studied by Deiglmayr et al. (Deiglmayr et al. (2006)). The gain in the transition matrix element reduces the required laser power density and puts the experiment within reach, even using a relatively low power diode laser. The Rabi frequency of the pump field was made equal to twice Ω_S^0 and there should be no major requirements in terms of the laser power density. The coupling field Rabi frequency is about three times Ω_P. It can also be obtained from a diode laser, and its power density means that some focusing will be required. In case B the Rabi frequencies of both Stokes and pump pulses are equal, both within reach of our laboratory. It will be shown later that this symmetric configuration simplifies the analysis of the time dependent

[1] The actual wavelength of the Stokes laser depends of the target Rydberg state.

Rabi frequency	$\Omega_S^0/2\pi$	$\Omega_P^0/2\pi$	$\Omega_c/2\pi$
μ_{12}/a_0	.075	5.1	2.1
Case A	15 MHz	30 MHz	100 MHz
I (W/m^2)	3.6×10^5	2.8×10^2	1.9×10^4
Case B	25 MHz	25 MHz	100 MHz
I (W/m^2)	1.0×10^6	1.9×10^2	1.9×10^4

Table 2. Parameters used in the STIRAP simulations.

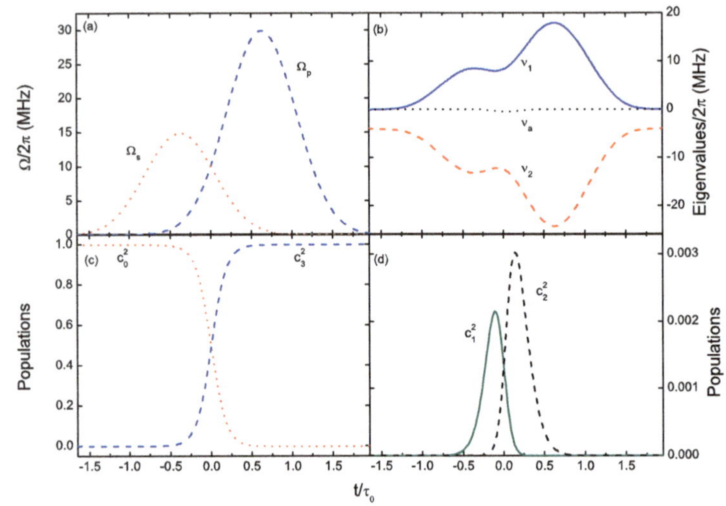

Fig. 8. Results of the simulation of four-level STIRAP for case A. We took equal widths for both Stokes and pump pulses. The pulse delay is equal to the pulse width. (a) pulse sequence; (b) adiabatic eigenvalues ν_0 and its closest eigenvalues ν_1 and ν_2; (c) populations of the adiabatic channels; (c) populations of the intermediate states.

condition for adiabatic following. In both cases the frequency shift $\Delta_P = 48$ MHz, which is slightly smaller than $\Omega_c/2$.

For both cases the eigenvalues and eigenvectors of the Hamiltonian matrix (Eq. 47) were calculated. The adiabatic state corresponds to the eigenvalue that during the STIRAP process is closest to zero. The result of the calculation is shown in Fig. 8 for case A, and in Fig. 9 for case B. The figures include plots of the pulse sequence, with the Stokes pulse preceding the pump pulse, of the three closest eigenvalues and of the populations of the adiabatic state. The eigenvalues show the same qualitative behavior. The adiabatic eigenvalue ω_0 and one of the other eigenvalues (ω_1) tend to zero for large times both before and after the STIRAP process. The other eigenvalue (ω_2) tends to a constant, negative value. Both cases clearly show an adiabatic population transfer between states $|0\rangle$ and $|3\rangle$. But unlike in tree-level STIRAP, there is a small transient population in the intermediate levels $|1\rangle$ and $|2\rangle$. In both cases these transient populations are under 0.004 and last only one pulse width T^*.

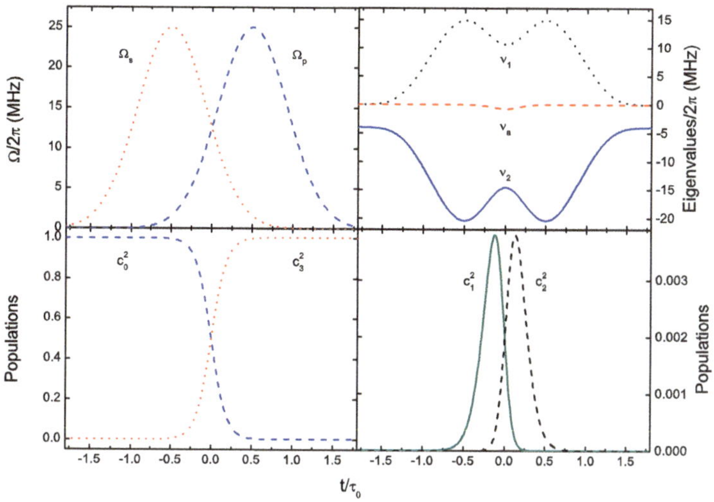

Fig. 9. Results of the simulation of four-level STIRAP for case B. We took equal Rabi frequencies and widths for both Stokes and pump pulses. The pulse delay is equal to the pulse width. (a) pulse sequence; (b) adiabatic eigenvalue ν_0 and its closest eigenvalues ν_1 and ν_2; (c) populations of the adiabatic channels; (c) populations of the intermediate states.

A detailed analysis of the effect of the pulse delay τ in both intermediate state population and the condition for adiabatic following (see below), indicates that the optimum pulse sequence corresponds, as in the three-level case, to $\tau = T^*$. Also, the separation of the adiabatic eigenvalue to the other two eigenvalues, which is one of the determining factors of the adiabatic condition, is largest for $\Delta_P = \Omega_c/2$. For larger values of Δ_P the adiabatic condition severely worsens. However, the numerical analysis is complicated for this value of the shift because the other two eigenvalues start to cross. We found that for values of Δ_P between 30 and 49 MHz we obtained reasonable conditions for adiabatic transfer and at the same time the transient populations of the intermediate states remained small.
Recalling that our unit of time is a pulse width allows us to write the time dependent adiabatic condition as:

$$A_j = \left|\frac{d\vec{c_0}}{dt} \cdot \vec{c_j}\right| \times \left|\omega_0 - \omega_j\right|^{-1} << T \tag{53}$$

where the projection is made onto the non-adiabatic eigenstates. To calculate A_j a numerical time derivative of the adiabatic eigenvector was projected onto the other eigenstates. The results for both case A and case B are shown in Fig. 10. In both cases the curves for A_1 and A_2 have a local maximum close to $t = 0$. However, the projections A_1 have large contributions towards the wings. This occurs because in both $t \to \pm\infty$ limits the adiabatic eigenvalue ω_0 and the nearest eigenvalue ω_1 both tend to zero and the denominator in Eq. (53) rapidly becomes smaller than the numerator.

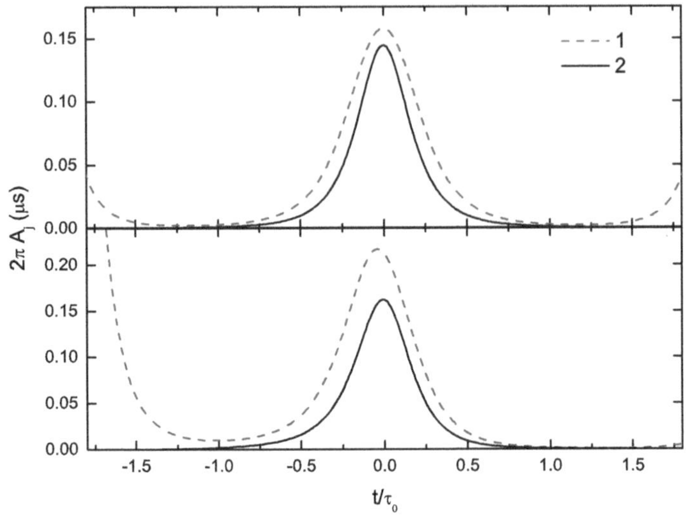

Fig. 10. Time dependent adiabaticity criteria (Eq. 46) for the two closest eigenvalues. Upper panel: case B; lower panel: Case A.

The adiabatic ratio is thus an indication of the time scale of the transfer process. In order to analyze the best conditions for adiabatic transfer one has to consider, on the one hand, the adiabatic condition, an on the other the population of the intermediate, short-lived states. It is convenient to have small values of the adiabatic ratio A_j so that the time scale of the whole process is small, comparable to the shortest lifetime in the system. It is also desirable to have small values for the intermediate state populations. These two quantities depend on the delay between pulses τ. We calculated the adiabatic ratio A_j as function of both time and delay. We took as an indication of the population of the intermediate states the sum of coefficients $c_1^2 + c_2^2$ that also depends on the delay τ. Contour plots for these two quantities are shown in Fig. 11. Both plots were obtained for case B, in which the Rabi frequencies of both Stokes and pump pulses are equal. Similar behavior was found for case A. For low values of the delay the adiabatic ratio near $t = 0$ takes the smaller values and the contribution from the wings is largest. As the delay increases the wing contribution diminishes, but now the values near $t = 0$ start to grow. At the upper end, near $\tau = 1.5$ the adiabatic ratio at $t = 0$ is close to one, and this would impose the need for transfer times that are much larger than the intermediate state lifetime. The behavior of the intermediate state population sum is the opposite. It takes its largest value, close to 0.1 for small delays, and then it decreases. The condition $\tau = T$ is the one that results in small enough values of A_j and at the same time small intermediate state populations. This is precisely the condition that was used to obtain figures 8 and 9. Thus one can use the adiabatic ratios in Fig. 10 to obtain the condition for the pulse width

$$T >> \frac{0.16}{2\pi}\ \mu s \tag{54}$$

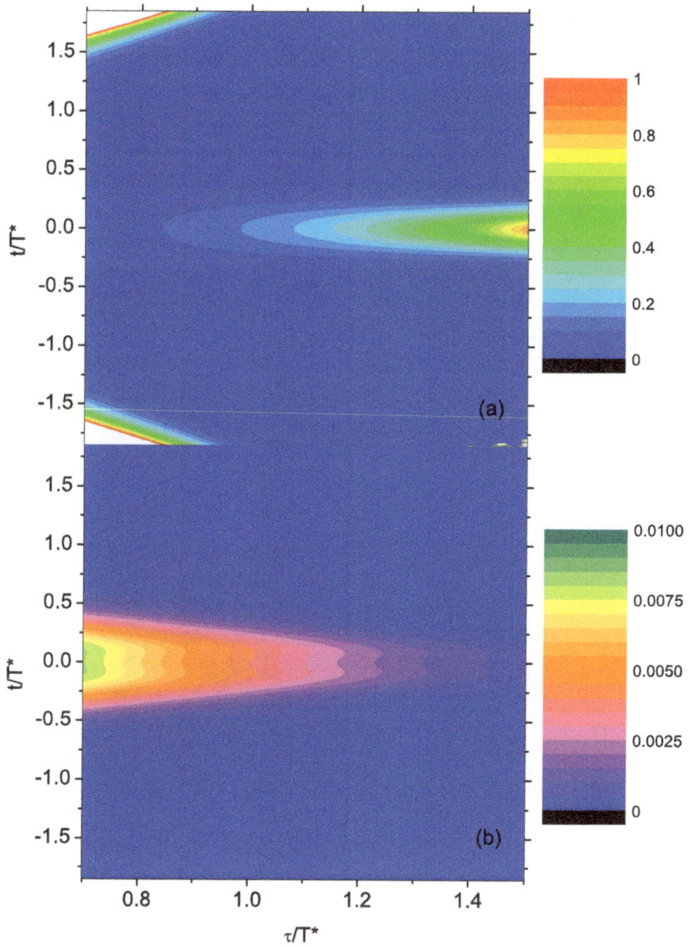

Fig. 11. Adiabatic ratio (top) and populations of the intermediate states as functions of time and delay between pulses. These plots correspond to case B. The abscissae correspond to reduced delay τ/T and the ordinates to the reduced time t/T in the pulse sequence.

This is satisfied with 500 ns pulses which are about a factor of 20 longer than the lifetime of the $5p_{3/2}$ state. At the same time it is possible to obtain them in the laboratory using standard acusto-optic modulators.

3.4 Summary

In summary, in this section we presented a calculation of the 4-level STIRAP process to efficiently transfer population between the 5s ground state in rubidium to nf Rydberg states. The transition matrix elements between the second intermediate state 5f and the Rydberg state result in a gain of a factor of 3 in the Rabi frequency. The whole process should be possible in the laboratory, with commercial diode lasers and in time scales that are at the same time attainable with acusto-optic modulators and not much longer than the shortest lifetime in the four level system.

4. Laser interactions with glow plasmas

Low temperature (2-20 eV in electronic temperature) plasmas provide one of the richest natural sources of radical, excited, neutral, metaestable and clustered atomic and molecular species Marcus &Broekaert (2003). When a laser pulse is made to interact with a plasma of these characteristics, practically any of the processes allowed by molecular or atomic selection rules may take place within the plasma. In this respect, a plasma-laser experiment provides, in an integrated fashion, a very comprehensive set of possibilities to study molecular processes in a single experimental set up. In these experiments one may study and quantify processes such as electron photo-detachment of anions, photo-ionization of metastable states, molecular predissociation and resonant excitation of short lived species. This serendipity provided by plasma laser interaction is hampered, however, by the fact that these processes take place simultaneously within a relatively dense medium. For this reason, it takes some ingenuity to extract and unravel a specific part of information and to make quantitative assessment of a particular process induced by the laser Baribieri et al. (1990). A well established technique that has been used for many years to extract spectroscopic information from plasma-laser interactions is the optogalvanic effect (OGE). In this section a brief on the fundamental characteristics and features of the optogalvanic process will be presented. In addition to this, a discussion on the fundamental processes which give rise to impedance changes within the plasma, as induced by laser light will be commented upon. Based on this basic principles, both the slow and fast optogalvanic effects will be presented and described. Since the slow optogalvanic effect has been studied and used for many years, the focus of the examples on applications in this section will be devoted to the fast optogalvanic effect. To illustrate the power and versatility of the fast optogalvanic effect, an example on the study of atomic photoionization from metastable states will be discussed. The final part of this section will be devoted to outlining the recent trends in optogalvanic studies, especially those related to applied use of novel light sources in the mid infrared spectral ranges. Some selected actual and potential applications in biology as well as the most salient applications in medicine will be presented, as derived from OGE studies.

4.1 The Optogalvanic effect (OGE)

The optogalvanic effect was first observed (Penning, 1926) when, during the observation of the current characteristics of a neon discharge, a change in the impedance was observed when irradiated by an emission of a nearby neon discharge. However, systematic, extensive and practical applications of this interesting effect were only possible when tuneable dye lasers were available in the decade of 1960. In 1970 the technique was mature enough to allow studies of electron excitation from metastable states with high resolution (Green et al. (1980)) and the spectroscopic studies of refractory elements in hollow cathode discharges. Incidentally, the study of refractory elements had been hampered in the past due to the need of complex ovens of high temperatures. Plasmas in hollow cathode discharges provide a good source of refractory materials in the gas phase as these species are produced with

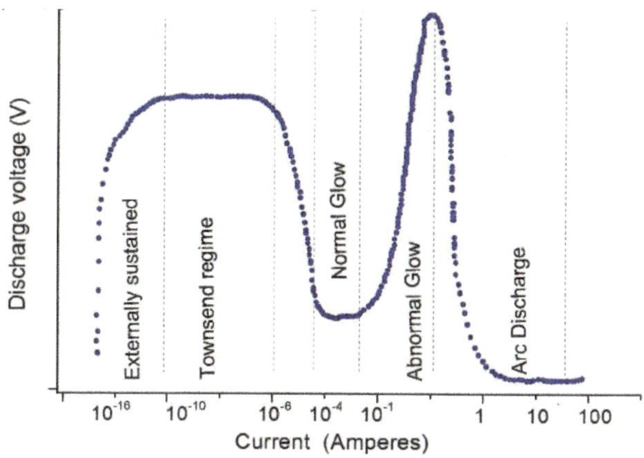

Fig. 12. Voltage current curve illustrating the different discharge regimes that may occur in a gaseous discharge, as a function of the discharget's current.

ion sputtering processes, within the discharge. The technique has grown stronger with the coming decades and is now widely used in a wide range of applications ranging from trace gas detection, studies of autoionizing states, isotopic ratio analysis, and laser stabilization applications, to name just a few.

The optogalvanic effect consists, briefly, in a change of the conductivity properties of a plasma, as induced by a pulse of electromagnetic radiation. This change in conductivity may be incremental or decremental, and presents a strong dependence with wavelength. In addition to this, the OGE can be fast, i.e. the conductance changes are as fast as the exciting laser pulse width, or slow, where the response can be 2 or 3 orders of magnitude slower than the exciting laser pulse. Before presenting a brief description of this process, it is convenient to revise briefly the basics of steady low-pressure gaseous discharges. Gas discharges can be separated in different regimes according to the voltage-current characteristic curve presented in Fig. 12. Although the optogalvanic effect has been observed both in the Townsend region (Kravis & Haydon (1981)) and in the arc region the most interesting and relevant studies have been carried out in the normal glow discharge. This is in part due to the fact that the geometry of a discharge in the glow regime is well established and can be understood based on simple atomic collision models.

Figure 13 shows the typical layout of a glow discharge between parallel plates. This schematic representation corresponds to a direct-current discharge in a tube filled with a rare gas, in a pressure range between 0.1 to 1 Torr. The physical explanation of these differentiated regions can be consulted in depth elsewhere (Engel (1965)). For the slow optogalvanic effect, the most relevant region corresponds to the positive column. The slow response of the conductivity changes as compared to the pulse width occur if the laser pulse is focused in this region. In this region, the axial component of the electric field is nearly constant and several orders of magnitude smaller as compared to that of the dark space. The fast optogalvanic effect takes place if the laser is focused in the small region depicted in Fig. 13 and labeled as the Aston space. This region is characterized by the fact that the largest electric field gradient takes place in this region. In this region, the presence of a bimodal electron energy distribution of electrons slowed by inelastic collisions with the gas, and fast electrons whose energy is too

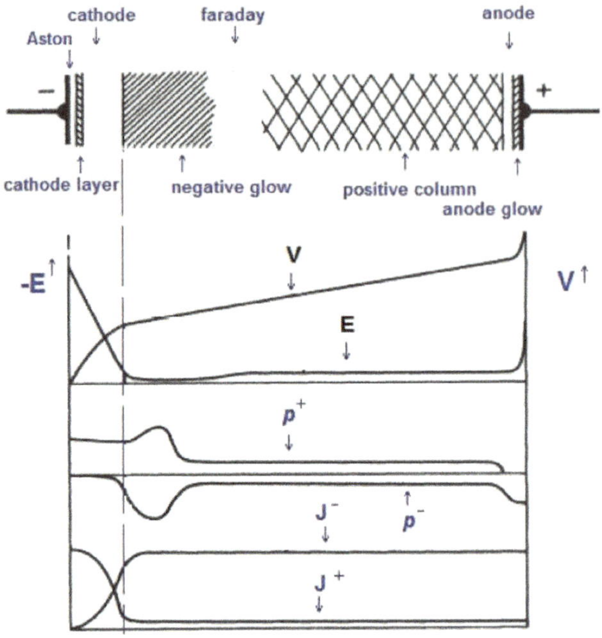

Fig. 13. Spatial distribution of a discharge between two parallel plates in the glow regime. The upper figure shows the electric field (E) distribution, space charge ρ^+, ρ^- and current densities J^+ and J^-.

high to contribute to the electron impact cross section. The above mentioned characteristic is the responsible for the differences in time response of the slow and fast optogalvanic effects. The fast optogalvanic effect has its origin in photoionization events. Due to the bimodal structure of the electron energy distribution, this region is rich in excited states. A laser pulse, even in the visible energy range, can induce a transition from one of these excited states into the continuum. Due to the high electric field present in the Crookes space, any photoelectron produced induces the production of a cascade of electrons in a nanosecond time scale. The slow optogalvanic process, on the other hand, is due to the resonant absorption of light, and its subsequent conversion of electromagnetic energy into kinetic energy. This conversion is mediated by super-elastic collisions which transfer the internal energy of the photon, molecule or radical, into kinetic energy. The scale of time for the propagation of the slow OGE is of the order of micro or milliseconds, depending on the size and geometry of the vessel containing the discharge. Experimentally, the simplest set up required to carry out optogalvanic measurements is depicted in Fig. 13.

In its simplest case, the experimental arrangement to measure optogalvanic transitions in plasmas is depicted in Fig. 14. A CW laser, tunable in wavelength is modulated in intensity with a mechanical chopper. The modulated light source is made then to interact with a discharge in the normal glow regime, either in the positive column or close to the Aston dark space. The variations in the plasma current is recorded with a phase sensitive lock in amplifier. These variations are recorded as a function of wavelength and the resulting spectrum can then

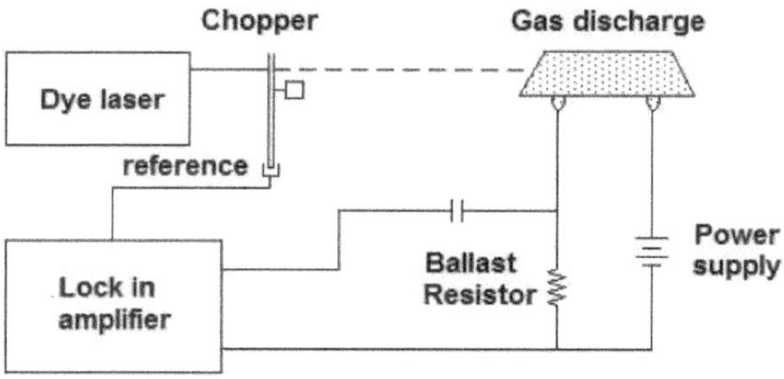

Fig. 14. Schematic of the most basic experimental arrangement required to carry out optogalvanic measurements. Most advanced schemes include a pulsed laser source and a boxcar integrator.

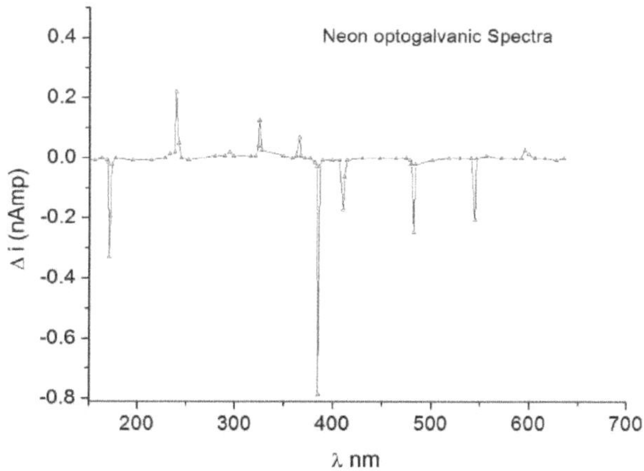

Fig. 15. Experimental optogalvanic spectra of neon illustrating the decrease and increase of conductivity (hence of current) as a function of wavelength.

be analyzed. A spectra of the optogalvanic effect in neon, as obtained with such a simple experimental arrangement is shown in Fig. 15.

In figure 15 one may notice a characteristic feature of the optogalvanic effect: the current can vary positively or negatively. The increase is possible because the resonant interaction of light with the plasma may increment both the density of ionized atoms or molecules and the increase in conductivity due to inelastic transfer of light to atomic or molecular kinetic energy. The decrease is usually produced by the quenching on the metastable population due to optical pumping, followed by optical emission in a cascade of dipole allowed transitions down to the ground state. This quenching effectively decreases the density of ionized targets and, hence the resonant transfer of electromagnetic energy into heat or ion electron pairs. This latter effect is, in turn reflected on the decrease of current as a function of wavelength of the laser light.

4.2 Photoionization studies with the Fast OGE

The use of Resonance Ionization Spectroscopy (RIS) has been very useful in the study of several interesting physical processes. With this technique it has been possible to study highly excited Rydberg states (Babin & Gagné (1992) and references therein), carry out isotopic effect studies and isotope separation (Letokhov & Chebotayev (1977)) and reach the single atom detection limit (Hurst et al. (1979)). However, the usefulness of this powerful technique is hampered by the need of a relatively complex experimental arrangement. The use of RIS usually requires the availability of an atomic or molecular beam two or more laser systems to reach intermediate resonance states prior to ionization and fast charged particle detectors to obtain the photoion or photoelectron yield as a function of wavelength. The fast optogalvanic effect (FOGE) may offer an interesting alternative to these techniques, with the advantage over the previously mentioned RIS techniques of its inherent experimental simplicity. The FOGE method was initiated with the observation of (Broglia et al. (1987)), of a rapid optogalvanic signal 10^{-10} s in the dark space of a hollow cathode discharge (HCD). This fast pulse arises solely from photoionization events which take place in the Aston dark space. In this scheme, the discharge is used as a photoelectron detector as well as a neutral atomic vapor generator via cathodic sputtering and is thus particularly appropriate to the study of refractory elements. The FOGE scheme has been used to study high resolution photoionization transitions of uranium (Gagné et al. (1983)) and to carry out fundamental studies of the dark space properties of a hollow cathode (Babin & Gagné (1992)) as well as the photoionization spectroscopy of the copper atom (Widiger et al. (1994)). The value of this method is its capability of measuring the photoionization cross section as well as the volume density of the initial state of the sample under study. This can be obtained directly from the time dependent quantification of the evolution of the fast optogalvanic signal generated in the discharge Aston dark space (Stockhausen et al. (1996)).

4.2.1 Experimental arrangement of the FOGE

Figure 16 shows a schematic representation of the simplest FOGE set up that can be used to carry out photoionization measurements. In the example shown the laser pulse is produced by a pulsed dye laser pumped with a nitrogen laser, with mean pulse energy and duration of 200 μJ and a pulse width of 5 ns respectively. The beam radius at the waist, positioned in the dark space and grazing the cathode, is 0.17 mm and is focused by a set of cylindrical lenses not shown in the schematic representation. The laser pulse width and energy delivers a fluence of 5 x 10^{25} photons/cm^2/s. The FWHM of the laser is 0.6 cm^{-1}, but this width is not significant as most of the line broadening in the photoionization experiments arises from Doppler broadening. The discharge is maintained with a current-stabilized voltage source, to which we added a 550 ballast resistor mounted in series with the lamp. This resistance value was chosen in order to limit the discharge's current and prevent it from becoming an arc. A 0.1 μF capacitor is connected in parallel with the discharge in order to maintain the voltage drop during the laser pulse. A current divider directs the photoionization current towards a boxcar, which is protected from the high voltage DC of the lamp and high frequency instabilities by a decoupling capacitor and a pair of fast diodes mounted in parallel.

In order to maximize the atomic vapor production, one should keep the gas pressure as low as possible whatever the carrier gas used. Operating the discharge at a high current may cause premature erosion of the cathode and the formation of secondary zones of variable conductivity, which would ruin the stability of the discharge. Recall that the hollow

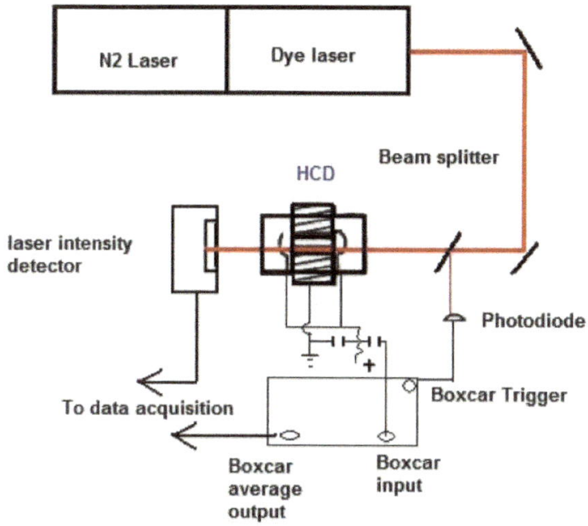

Fig. 16. Experimental arrangement for photoionization measurements employing the Fast Optogalvanic Effect (FOGE). More complex arrangements may involve the use of another laser for ground state pumping used to simplify spectroscopic analysis. HCD stands for Hollow Cathode Discharge.

cathode geometry and the operating conditions must be so that the negative glows overlap in the middle of the discharge. At the same time, the fast electrons accelerated on one side of the hollow cathode and going across the plasma must not strongly perturb the facing dark space. Figure 16 shows the simplest experimental arrangement that can be used for FOGE measurements. One of the most useful uses of the FOGE effect is on the study of photoionization of refractory materials. The usual acronym used for this use of the FOGE technique is PIOG, standing for Photoionization Optogalvanic effect. Fig. 17 shows a part of the PIOG spectrum (Levesque et al. (1997)) recorded using a 150 mTorr Xe discharge. On the same plot a slow optogalvanic effect spectrum is recorded for calibration purposes. The wavelength scale is not shown in the figure but can easily be inferred by using the simultaneously acquired slow optogalvanic spectrum, along with the Fabry- Perot fringes. Instead, the scanning time is recorded in the x axis in order to illustrate that, due to the nonlinear nature of the scanning of the dye lasers, it is convenient to record a Fabry-Perot interferogram along with the PIOG and TOG. Using such a reference, the nonlinear wavelength scanning of the dye laser is evident here. Those scanning speed variations restrict the wavelength calibration reproducibility of the PIOG spectra to 0.3 cm^{-1}. However, when care is taken to minimize these non-linearities with a closed loop positioning control, the PIOG technique can easily yield, on a specific spectrum, consistent from line to line within 0.1 cm^{-1}.

One also observes that the PIOG and TOG spectra are really different. Even though it frequently happens that strong photoionization lines regroup around an intense TOG line, as shown on figure 17 (lines marked A and B), other intense lines are not related to TOG lines, or, if so, to very weak ones, as exemplified by the lines marked C and D. In summary, the

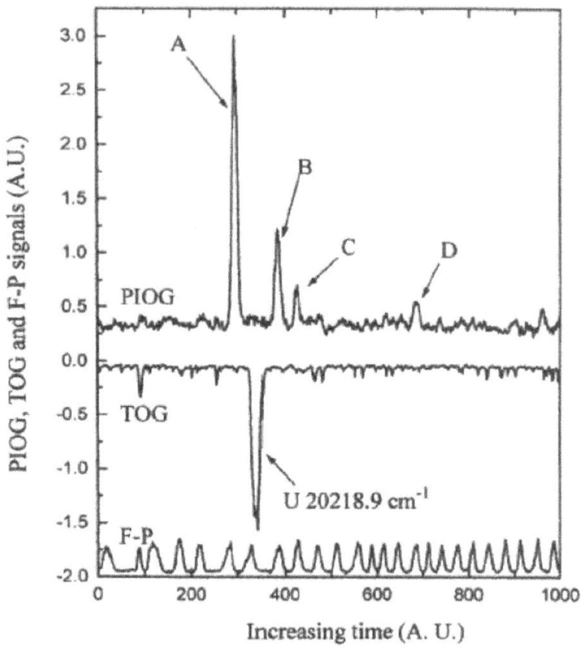

Fig. 17. PIOG Spectra of a Xenon discharge, along with a thermal OGE spectra and Fabry-Perot interferogram. The nonlinear scanning properties of dye lasers are made evident. (with permission from IOP publishing, Levesque et al. (1997))

FOGE technique provides a useful tool to assess the photoionization yield of gas phase and refractory elements, and, combined with other techniques, may yield very useful results.

4.3 New developments and future trends of OGE

The optogalvanic effect and techniques is a relatively old and well established technique. However, with the development of new, simple and affordable light sources of solid state and in wavelengths previously inaccessible, the technique is still producing a wealth of new applications. In this section a brief outline of these new applications and new trends will be presented. One of the most salient features is its great sensitivity. The efficiency of detection of the optogalvanic technique can be very competitive, reaching the part per billion detection level. This technique can then be favorably compared to other trace gas detection alternatives such as Cavity Ring Down, optoacustic spectroscopy or Resonance Ionization spectroscopy. Trace gas sensing, in turn, is of great interest in numerous areas. These include atmospheric chemistry, volcanic activity, agriculture, industrial process and workplace surveillance, gas certification, medical diagnostics, etc. The competence of the OG effect to provide a tool for trace detection allows foreseeing that its use will continue to be intensive in the future.

4.3.1 Laser sources for use with OGE

In the past, the only choice of lasers to implement the OGE and FOGE studies were pulsed N2 or Nd:YAG lasers. With the recent development of external cavity diode lasers, or ECDL,

Laser	wavelength range	Power range
External Cavity Diode Laser	800-1600 nm	1-200 mW
CO laser	2.7-4, 5.1-6, μm	mW-Watt
CO2 laser	9-11 μm	Watt
Lead salt diode	4-30 μm	0.1 mW
Quantum Cascade Laser	4-24 μm	mW
Optical parametric Oscillator	3-16 μm	<1W
Difference frequency generator	3-16 μm	mW

Table 3. Typical ranges and parameters of laser sources available for optogalvanic and spectroscopic studies.

the capabilities and potential of the optogalvanic studies is more appealing today. For mid infrared application there are alternatives provided by quantum cascade lasers, as well as nonlinear optical devices, based on four wave mixing, but operating with very low powers (Markus (2003)). This very recent developments on mid infrared off the shelf optics, based on photonic band gap hollow fibers allow to forecast a very intensive use of the technique in trace detection in the mid infrared. Table 3 briefly summarizes the current situation related to the current accessible wavelengths, tuning characteristics, typical power range, and operation parameters of laser sources available for optogalvanic and spectroscopic studies in general.

4.3.2 Application of OGE spectroscopy in biology and medicine

Trace gas detection is currently a very active area of applied research. With the high sensitivity provided by the optogalvanic spectroscopy, there is a great potential for this technique in the trace gas detection area. In this final part of the section devoted to the OGE some illustrative applications in the area of biology and medicine will be presented. Of particular interest for these applications is the Infrared spectral range. The applications range from the monitoring industrial processes, specially of flames and catalytic reduction of pollutants, to air quality monitoring in great cities, the quantification of CO_2 sinks and sources, to national security to the monitoring of biological processes in living organisms (including the human body). In all of these applications, the use of trace detection spectrometers coupled with laser-based spectroscopic gas are indispensable (Sigrist (2003)). The most intense absorption bands of organic molecules lie in the wavelength region between 2.5 and 20 μm, where these molecules have their fundamental rovibrational transitions. The areas of interest where the optogalvanic spectroscopy could find a future niche will involve the detection of organic molecules which absorb in this range of wavelengths. To be more specific, the detections such as ethylene, methane, CO_2 and CH_4 could open up a wealth of possibilities for the technique (Cristescu et al. (2008)).

In the case of ethylene, it is a well know plant hormone that is conspicuous in many biological processes of the plant and fruits, such as death cell signaling, nitrogen fixation processes, circadian clock system of several plants, molecular alarm systems of fungi infection of fruits and tomatoes and even in the metabolic changes of human physiology induced by the light from the sun (Cristescu et al. (2008)). Acetone is a biomarker the increase of which in the human breath signals the onset of diabetes. A sensitive detector based on optogalvanic spectroscopy to quantify the presence of this molecule would have potentially a great use in hospitals and small clinics (Aman & Smith (2005)). Nitric oxide detection in the part per billion sensitivity could be of great clinical use in the early detection of lung pathologies and organ rejection monitoring, hepatopulmonary diagnostics, malaria and even gastritis, through the detection of the helicobacter pillory bacteria through its metabolic signature in the human breath. The scope and importance of these applications and several others make the area of research of new OGE detectors and techniques a very promising area of research in the future.

5. Acknowledgments

We knowledge support from grants DGAPA-UNAM IN-116309 (J.J.M.), DGAPA-UNAM IN-101611-3 (R.C.T.), and DGAPA-PAPIIT IN113910 (AMJ).

6. References

Aman, A. & Smith, D. (2005). *Breath Analysis for clinical diagnosis and therapeutic monitoring*, World Scientific Publishing Co. Pte Ltd., Singapore.

Anis, F., Roudnev, V., Cabrera-Trujillo, R. & Esry, B. D. (2006). Laser assisted charge transfer in $He^{2+} \rightarrow H$ collisions, *Phys. Rev. A* 73: 043414.

Babin, F. & Gagné, J. (1992). Hollow Cathode Discharge (HCD) dark space diagnostics with laser photoionization and galvanic detection, *Appl. Phys. B* 54: 35.

Baribieri, B., Beverini, N., & Sasso, A. (1990). Optogalvanic spectroscopy, *Rev. Mod. Phys.* 62: 603.

Bergmann, K., Theuer, H. & Shore, B. W. (1998). Coherent population transfer among quantum states of atoms and molecules, *Rev. Mod. Phys.* 70: 1003 and references therein.

Broglia, M., Catoni, F., Montone, A. & Zampetti, P. (1987). Galvanic detection of laser photoionization in hollow-cathode discharges: Experimental and theoretical study, *Phys. Rev. A* 36: 705.

Cabrera-Trujillo, R. (2009). Laser and isotope effects in charge transfer processes in atomic collisions, *Rad. Effects Defects Solids* 164: 402.

Carroll, C. E. & Hioe, F. T. (1988). Three-state systems driven by resonant optical pulses of different shapes, *J. Opt. Soc. Am. B* 5: 1335.

Copeland, D. A. & Tang, C. L. (1976). Photon-assisted nonresonant charge exchange: A simple molecular model, *J. Chem. Phys.* 65: 3161.

Cristescu, S., Persijn, S., Lintel, H. & Harrend, F. (2008). Laser-based systems for trace gas detection in life sciences, *Appl. Phys. B* 92: 343.

Cubel, T., Teo, B., Malinovsky, V., Guest, J., Reinhard, A., Knuffman, B., Berman, P. & Raithel, G. (2005). Coherent population transfer of ground-state atoms into Rydberg states, *Phys. Rev. A* 72: 023405.

Deiglmayr, J., Reetz-Lamour, M., Amthor, T., Westermann, S., de Oliveira, A. & Weidemüller, M. (2006). Coherent excitation of Rydberg atoms in an ultracold gas, *Opt. Commun.* 264: 293.

Engel, A. v. (1965). *Ionized Gases*, second edition, Claredon, Oxford, England.

Errea, L. F., Méndez, L. & Riera, A. (1983). On laser-induced charge transfer in the HeH^{2+} quasimolecule, *J. Chem. Phys.* 79: 4221.

Ferrante, G., Cascio, L. L. & Spagnolo, B. (1981). Laser-assisted symmetric charge transfer in atomic collisions, *J. Phys. B* 14: 3961.

Gagné, J., Demers, Y., Pianarosa, P. & Dréze, C. (1983). Effet optogalvanique dans une décharge á cathode creuse: Mécanisme et dosage isotopique de l'uranium, *Journal de physique (Paris)* C7-44: 355.

Gallagher, T. (1994). *Rydberg Atoms*, Cambridge University Press, Cambridge, UK.

Gaubatz, U., Rudecki, P., Becker, M., Schiemann, S. & Bergmann, K. (1990). Population transfer between molecular vibrational levels by stimulated Raman scattering with partially overlapping laser fields. a new concept and experimental results, *Chem. Phys. Lett.* 92: 5363.

Gearba, M., Camp, H., Trachy, M., Veshapidze, G., Shah, M., Jang, H. & DePaola, B. (2007). Measurement of population dynamics in stimulated Raman adiabatic passage, *Phys. Rev. A* 76: 013406.

Green, R., Havrilla, G. & Trask, T. (1980). Laser-enhanced ionization spectrometry: Characterization of electrical interferences, *Appl. Spectrosc.* 34: 561.

Gudzenko, L. I. & Yakovenko, S. I. (1972). Radiative collisions, *Sov. Phys. JETP* 35: 877.

Havener, C. C., Rejoub, R., Krstić, P. S. & Smith, A. C. H. (2005). Charge transfer in low-energy collisions of He^{2+} with atomic hydrogen, *Phys. Rev. A* 71: 042707.

Hurst, G., Payne, S., Kramer, S. & Young, J. (1979). Resonance ionization spectroscopy and one-atom detection, *Rev. Mod. Phys.* 51: 767.

Kirchner, T. (2002). Manipulating ion-atom collisions with coherent electromagnetic radiation, *Phys. Rev. Lett.* 89: 093203.

Kirchner, T. (2004). Laser-field-induced modifications of electron-transfer processes in ion-atom collisions, *Phys. Rev. A* 69: 063412.

Kirchner, T. (2005). Pauli blocking and laser manipulation of electron dynamics in atomic collisions, *Nucl. Instr. Meth. B* 233: 151.

Kravis, S. & Haydon, S. (1981). Laser-induced optogalvanic effects under prebreakdown conditions in neon, *J. Phys. D: Appl. Phys.* 14: 151.

Kuklinski, J., Gaubatz, U., Hioe, F. & Bergmann, K. (1989). Adiabatic population transfer in a three-level system driven by delayed laser pulses, *Phys. Rev. A* 40: 6741.

Letokhov, V. & Chebotayev, V. (1977). *Nonlinear Spectroscopy*, Springer Verlag, Berlin.

Levesque, S., Gagne, J. M. & Babin, F. (1997). Optogalvanic photoionization spectroscopy, *J. Phys B: At. Mol. Opt Phys* 30: 1331.

Lorenzen, C.-J. & Niemax, K. (1983). Quantum defects of the $n^2p_{1/2,3/2}$ levels in ^{39}K and ^{85}Rb, *Phys. Scr.* 27: 300.

Malinovsky, V. & Tannor, D. (1997). Simple and robust extension of the stimulated Raman adiabatic passage technique to n-level systems, *Phys. Rev. A* 56: 4929.

Marinescu, M., H.R., S. & Dalgarno, A. (1994). Dispersion coefficients for alkali-metal dimers, *Phys. Rev. A* 49: 982.

Marcus R.K. and Broekaert J.A.C., (2003). *Glow Discharge Plasmas in Analytical Spectroscopy*, John Wiley &Sons, Ltd. ISBN: 0-471-60699-5, London, England

Markus, W. (2003). Trace gas monitoring by laser photoacoustic spectroscopy and related techniques, *Rev. Sci. Instrum.* 74: 486.

Metcalf, H. & van der Straten, P. (1999). *Laser Cooling and Trapping*, Springer, New York.

Oreg, J., Bergmann, K., Shore, B. & Rosenwaks, S. (1992). Population transfer with delayed pulses in four-state systems, *Phys. Rev. A* 45: 4888.

Oreg, J., Hazak, G. & Eberly, J. (1985). Multilevel inversion schemes in and beyond the adiabatic limit, *Phys. Rev. A* 32: 2776.

Oreg, J., Hioe, F. & Eberly, J. (1984). Adiabatic following in multilevel systems, *Phys. Rev. A* 29: 690.

Pindzola, M. S., Minami, T. & Schultz, D. R. (2003). Laser-modified charge-transfer processes in proton collisions with lithium atoms, *Phys. Rev. A* 68: 013404.

Press, W. H., Teukolsky, S. A., T.Vetterling, W. & Flannery, B. P. (1992). *Numerical recipes*, 2nd edition, Cambridge University Press, New York, USA.

Saffman, M., Walker, T. & Molmer, K. (2010). Quantum information with Rydberg atoms, *Rev. Mod. Phys.* 82: 2313.

Sigrist, M. (2003). Trace gas monitoring by laser photoacoustic spectroscopy and related techniques, *Rev. Sci. Instrum.* 74: 486.

Smith, A. V. (1992). Numerical studies of adiabatic population inversion in multilevel systems, *J. Opt. Soc. Am. B* 9: 1543.

Stockhausen, G., Mende, W. & Kocj, M. (1996). Laser-induced photoionization in the dark space of a hollow cathode, *J. Phys B: At. Mol. Opt Phys* 29: 665.

Stolterfoht, N., Cabrera-Trujillo, R., Öhrn, Y., Deumens, E., Hoekstra, R. & Sabin, J. R. (2007). Strong isotope effect on the charge transfer in slow collisions of He^{2+} with atomic hydrogen, deuterium and tritium, *Phys. Rev. Letters* 99: 103201.

Süptitz, W., Duncan, B. & Gould, P. L. (1997). Efficient 5d excitation of trapped Rb atoms using pulses of diode-laser light in the counterintuitive order, *J. Opt. Soc. Am. B* 14: 1001.

Thoumany, P., Germann, T., Hänsch, T., Stania, G., Urbonas, L. & T., B. (2009). Spectroscopy of rubidium Rydberg states with three diode lasers, *J. Mod. Opt.* 56: 2055.

Vitanov, N. (1998). Adiabatic population transfer by delayed laser pulses in multistate systems, *Phys. Rev. A* 58: 2295.

Vitanov, N., Shore, B. & Bergmann, K. (1998). Adiabatic population transfer in multistate chains via dressed intermediate states, *Eur. Phys. J. D* 4: 15.

Widiger, G., Willke, B. & Kock, M. (1994). Laser photoionization of copper atoms in the dark space of a hollow cathode discharge, *J. Phys B: At. Mol. Opt Phys* 27: 899.

Zimmerman, M., Littman, M., Kash, M. & Klepner, D. (1979). Stark structure of the Rydberg states of alkali-metal atoms, *Phys. Rev. A* 20: 2251.

8

Ultrafast Time-Resolved Spectroscopy

László Nánai[1], Szabolcs Beke[1] and Koji Sugioka[2]
[1]*University of Szeged, Department of General and Environmental Physics*
Laser Technology Laboratory
[2]*RIKEN - The Institute of Physical and Chemical Research, Wako, Saitama,*
[1]*Hungary*
[2]*Japan*

1. Introduction

In this chapter we will focus on the subject of spectroscopical investigation of matter (gas, liquid, solid, low dimension materials, etc) using ultrashort (ps, fs) laser pulses in different wavelength region (visible, infrared, UV, X-ray). We will point out the rigorous difference in aims of classical spectroscopy using highly narrowed electromagnetic waves with high temporal and spatial coherence compared to the wide narrowed spectral structures of ultrashort light pulses.

We will discuss briefly the main characteristics of light-matter interaction on the base of the linear and nonlinear optics assumption. Some outlook of facilities of classical spectroscopy will be made with attempt of time resolved techniques e.g., time of flight and pump and probe techniques. We will also demonstrate some interesting results of ultrafast time resolved spectroscopy in material science and solid state physics.

2. Elements of linear and nonlinear optics of light-matter interaction

The study of phenomena induced by laser beam (continuous (CW) and pulsed) on solid surfaces is a widely investigated subject of modern solid state physics and chemistry. From the advent of the first lasers (late 60s) a huge number of scientific papers had been devoted to the investigations of different kinds of laser - matter processes such as laser induced damage, plasma formation, phase transition in different circumstances, micro- and macroprocessing, laser-induced chemical reaction at solid - gas and solid - liquid surfaces etc. The laser became a very useful tool for initiating unique chemical reactions to produce advanced materials (ultra hard ones for example). Number of technological applications, for example laser induced deposition of metals on porous materials and semiconductor surfaces are already in service of high-tech industry (micro- and nanoelectronics). Production with tailored properties of different nanostructures and components serves in chemical industry as catalysators, etc. [1]

Some preliminaries of the light-matter interaction are treated in a quasi-classical approximation:

Semi-classical approximation means that the electromagnetic wave is described classically on the base of Maxwell's equations while the material is described on quantum mechanical level (taking into account the electronic, vibration and rotation energy levels of the atoms).

3. Linear optics approximation

From Maxwell's equations we get:

$$\overline{D} = \varepsilon_0 \overline{E} + \overline{P}(\overline{E}) \quad \text{and} \quad \overline{B} = \mu_0 \overline{H} + \mu_0 \overline{M}(\overline{H}) \tag{1}$$

where $\overline{D}$ - electric flux density, $\overline{E}$ - electric field vector, $\overline{B}$ - magnetic flux density, $\overline{H}$ - magnetic field vector, ε_0 - dielectric constant of vacuum = $8.85 \cdot 10^{-12}$ $C^2N^{-1}m^{-2}$ [F/m], μ_0 - permeability of free space = $4\pi \cdot 10^{-7}$ H/m (we will care materials for which $\overline{M} = 0 \to \overline{B} = \mu_0 \overline{H}$).
The polarization

$$\overline{P}(\overline{E}) = \varepsilon_0 \chi \overline{E} \tag{2}$$

where χ - electric susceptibility.
The wave equation for electric field:

$$\nabla^2 \overline{E} = \mu_0 \varepsilon_0 \frac{\partial^2 \overline{E}}{\partial t^2} + \mu_0 \varepsilon_0 \chi \frac{\partial^2 \overline{E}}{\partial t^2} = \mu_0 \varepsilon_0 \varepsilon_r \frac{\partial^2 \overline{E}}{\partial t^2} \tag{3}$$

where $\varepsilon_r = (1 + \chi)$ is the relative dielectric constant.
It is well known from optics that the frequency ν of electromagnetic wave and the material refractive index n are related to c_0 (speed of light in vacuum)

$$\nu^2 = \frac{1}{\mu_0 \varepsilon_0} \frac{1}{\varepsilon_r} = \frac{c_0^2}{\varepsilon_r} \quad \text{and} \quad n = \frac{c}{\nu} \tag{4}$$

where $c_0 = 3 \cdot 10^8$ m/s.
The solution of the wave equation can be expressed:

$$E(z,t) = \mathrm{Re}\{\overline{E}(z,\omega)\exp(-i\overline{k}\overline{r} + i\omega t)\} \tag{5}$$

where z is the direction of propagation, $\omega = 2\pi\nu$ the circular frequency and k is the wave vector ($k = 2\pi / \lambda$, $\lambda = c / \nu$ the wavelength).
For the materials we have the

- group velocity

$$\upsilon_g = \frac{d\omega}{dk} \tag{6}$$

- phase velocity

$$\upsilon_{ph} = \frac{\omega}{k} = \frac{c}{n} \tag{7}$$

and the Beer-Lambert law

$$I(z) = I_0 e^{-\alpha z} \tag{8}$$

where I_0 is the intensity of the light at the incidence plane, $I(z)$ the intensity in the depth level z, and α is the absorption coefficient.
The real (ε') and imaginary (ε'') parts of dielectric constants at given frequency are related to optical material parameters n and k through Kramers-Kronig relations.

$$n = \sqrt{\frac{\sqrt{(\varepsilon')^2 + (\varepsilon'')^2} + \varepsilon'}{2}}, \quad k = \sqrt{\frac{\sqrt{(\varepsilon')^2 + (\varepsilon'')^2} - \varepsilon'}{2}}. \tag{9}$$

The above-mentioned equations describe the light - matter interaction events for relatively low beam intensities where the material constants do not depend on the intensity, but only on the frequency.

4. Nonlinear optics approximation

For the laser beam intensities comparable to the electric field strength inside the atoms we have to take into account the higher-order (nonlinear) terms in the field to get the polarization $\overline{P}$. (For H we got $E \sim 10^9 V / cm$ and $I \sim 10^{16} W / cm^2$)

$$\overline{P}(t) = \varepsilon_0 \left(\chi^{(1)} \overline{E} + \chi^{(2)} \overline{E}^2 + \chi^{(3)} \overline{E}^3 + \ldots \right) \tag{10}$$

where $\chi^{(n)}$ is the nth order of the susceptibilities of the medium.
As a result we can get different phenomena e.g., 2nd and higher order harmonic generation, sum- and difference frequency generation, parametric generation, etc.
We note that in all processes the conservation laws for photon energy and momentum (phase matching) should be fulfilled as

$$\omega = \omega_1 + \omega_2 + \omega_3 + \ldots \quad and \quad \overline{k} = \overline{k}_1 + \overline{k}_2 + \overline{k}_3 + \ldots \tag{11}$$

Non-linear optical effects usually called as N wave mixing processes, where N is the number of photons participating in reactions. The more photons the weaker the probability of the effect. At very high laser intensities e.g., at very short pulses very high order effects might have been realized. The materials usually have different refractive indices for different frequencies and polarizations; therefore the phase matching for them will be satisfied only for special directions e.g., second harmonics birefringence.
A number of special methods have been developed (due to nonlinear crystals to produce effects such as second harmonic generation (SHG), optical parametric generation (OPG), optical parametric oscillation (OPO), optical parametric amplification (OPA), quick switching features as electrooptical Kerr and Pockel cells [2-4].

5. The spectroscopy

5.1 General remarks

Spectroscopy in general is the study of the interaction between light and matter such as determination of quantum energy levels in substrates (gas, solid, liquid, etc.). In this "traditional" simple form, one measures the spectroscopic components of the reflected beam

after transmitted through the medium or emitted from it due to the external excitation of the energetic levels. In a classical his performance one uses - normally - spectrally narrow light beam. This beam may be tuned across discrete energetic levels of the studied target. Different regions of the electromagnetic spectrum provide different kinds of information as a result of the interaction.
The spectroscopic instrument represents - as its main part - a dispersion element (prism or grating with high separation capability of beam wavelengths). Usually this consists also of a slit, light collecting optics, and a detector (see Fig. 1.) called monochromator.

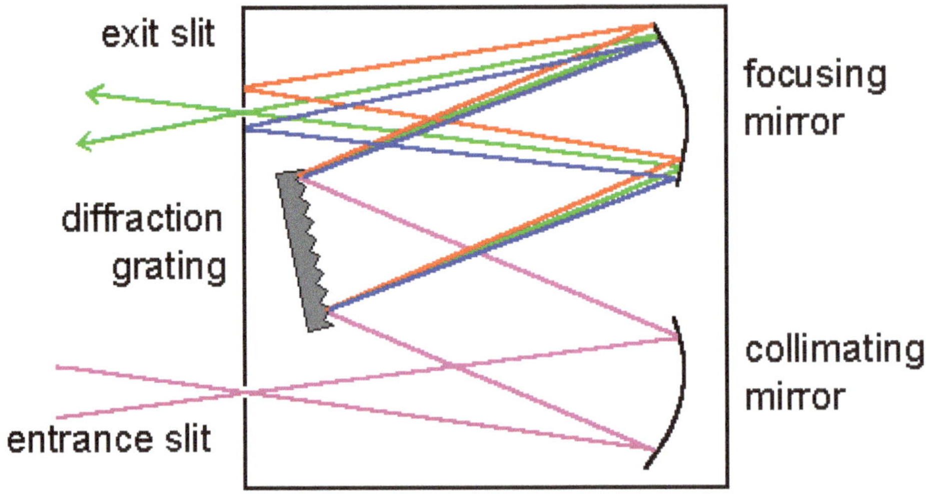

Fig. 1. A general scheme of a monochromator

Depending of the target material structure, composition, and the light-matter interaction type such as direct absorption, transmission, emission of scattering (type Rayleigh, Brillouin, Raman) one can summarize the classical spectroscopy methods as follow [5-11].

- Atomic spectroscopy:
    - absorption (AA)
    - emission (AES, OES)
    - fluorescence (AFS)
- Electron spectroscopy:
    - Auger electron spectroscopy (AES)
    - X-ray photoelectron spectroscopy (XPS, ESCA)
- Infrared spectroscopy
    - molecular spectroscopy
    - near infrared absorption spectroscopy (NIR)
    - ultraviolet and visible spectroscopy (UV-VIS)
- Nuclear and electron spectroscopy
    - electron paramagnetic resonance spectroscopy (EPR, ESR)
    - nuclear magnetic resonance spectroscopy (NMR)
- Fourier transform spectroscopy (FT)

- Laser spectroscopy
  - laser-induced fluorescence spectroscopy (LIF)
  - Raman spectroscopy (RAMAN)
  - resonance ionization spectroscopy (RIS)
  - X-ray and γ-ray spectroscopies
  - Mossbauer spectroscopy (MOSSBAUER)
  - Newton activation analysis spectroscopy (NAA)
  - X-ray fluorescence spectroscopy (XRF)
  - extended X-ray absorption fine structure (EXAFS)

and numerous other combined type spectroscopic methods [7-8].

6. Ultrafast spectroscopy

Ultrafast spectroscopy is based on using ultrashort laser pulses with pulse duration of ps-fs time region. This technique involves temporally short, therefore spectrally broad light pulses. These kinds of pulses are used to probe directly the dynamics of the system rather than the energy levels themselves. It is very important that the duration of the laser pulses must be shorter than the time scale of the dynamics that one wants to observe.

Taking into account the quantum mechanical considerations we reach the appearance of uncertainty principle e.g., time and energy resolution are related to each other through the Fourier transformation [7].

For Gaussian profile pulses the spectral bandwidth $\Delta\nu$ of the pulse and its temporal duration τ on the full width at the maximum level (FWHM) can be written as

$$\Delta\nu\tau \geq 2\ell n(2)/\pi \qquad (12)$$

e.g., if $\tau = 5$ fs we get $\Delta\nu = 8{,}8{\cdot}10^3$ Hz (2940 cm^{-1}).

7. Ultrafast laser excitation in materials

7.1 Impact of laser beam energy to a matter

Elementary excitations in solids show a complex nonequilibrium behavior. The fastest nonequilibrium processes occur on ultrafast time scales and strongly influence both optical properties and carrier transport. Among condensed phases metals and semiconductors represent an interesting class of practically important targets of nano and microelectronics. During these processes the electronic band structure, optical transition energies, carrier concentration, and phonon frequencies vary over a broad range leading to a variety of ultrafast phenomena. Moreover, the quantum confinement of wave functions in low-dimensional semiconductor nanostructures allows a systematic variation of material properties.

Optical spectroscopy with ultrashort pulses provides direct insight into these processes occurring on a time scale between about 10^{-14} and 10^{-10} s.

High amount of laser light energy can be deposited in a very small volume determined by the laser focal spot and penetration depth at a given wavelength. The electromagnetic incident wave will lead to photo-excitation of the electrons due to the large difference of electron and phonon heat capacities ($c_p >> c_e$). Therefore in the target material, especially in the case of metals and metal-based nanostructures - one creates a non-equilibrium electron distribution leaving the lattice temperature essentially unchanged ($T \sim 300$ K). The rise time of non-degenerate electron distribution creation is in the order of a few fs, thus we can say that high temperature non-equilibrium electron distribution has the same rise time as the laser pulse duration. Then over a time scale of a few fs the non-equilibrium electrons redistribute their energy among themselves. It takes place through e.g., electron - electron coulomb interactions resulting in a local equilibrium with temperature T_e, called the thermalized electron redistribution (with relaxation time τ_{e-e}).

The excited thermalized electron gas then transforms the energy through electron - phonon interactions within a relaxation time τ_{e-p}.

Energy transports between electron and phonon subsystems [12-16].

This mass energy transferred to phonon bath will be redistributed among phonons during the relaxation time τ_{p-p} leading to the equilibrium phonon temperature T_l.

Therefore we can consider the kinetic evolution of a photo-excited electron - phonon system.

- fast, involving the electron subsystem thermalization (at quite high temperatures) and electron-phonon scattering (at quite moderate temperatures). These are called non-thermal processes. The average time scale is about 1-500 fs.
- slow, involving the phonon-phonon scattering leading to heat conductions, thermal melting and probably ablation (called thermal processes). The average time scale in the case of metals is about 1 ps to a few ns.

In semiconductors and complex nanostructures, the relaxation processes are multistep ones and include different mechanism such as:

Below band gap excitations:

Transitions between electronic states may have different origins, such as:

Transitions from atoms or vacancies,

Transitions from impurity levels into the valence or conduction band continuum,

Transitions (indirect) between excited intraband levels,

Transitions due to the so-called free carrier absorption,

Inter-valence band transitions of holes, and intersubband transitions between valence and conduction subbands

Transitions in low-dimensional semiconductor nanostructures, e.g., quantum wells, wires and dots.

Indirect interband excitations: free carrier absorption due to the presence of free charges in both conduction and valence bands. This requires coupling to a third particle, e.g., a phonon or an impurity because of the conservation laws for wave number vectors.

Inter valence band transitions: due to dipole-allowed transitions of free holes from states in one valence band to states of higher energy in another valence band. For bulk semiconductors with a diamond-like, e.g., silicon and germanium, or zinc-blende lattice like most III-V semiconductors, inter-valence band absorption is dominated by transitions between the heavy hole (HH) and light hole (LH and split-off bands).

Intersubband transitions in quasi-two-dimensional nanostructures: They are characteristics for quantum wells or superlattices in which carrier motion is restricted to a quasi-two-dimensional semiconductor layer. Quantum confinement occurs in a situation where the length scale of the potential structure, i.e., the well width, is on the order of the de-Broglie wavelength of the carriers.

Dephasing of coherent polarizations

Resonant interaction of a coherent ultrashort pulse with a particular transition in the semiconductor creates both a coherent optical polarization between the optically coupled states and carriers (electrons or holes) from energetically lower to higher states in the same or a different band.

With time evaluation this well-defined phase relation is destroyed by a variety of scattering processes which change the relative phase of the wave function between the ground- and excited states. This phase relaxation or so-called dephasing process means a fast decay of the macroscopic polarization and results in a homogeneous broadening of the particular optical transition.

Therefore, the overall excitation-relaxation process could be characterized as it can be seen in Fig. 2.

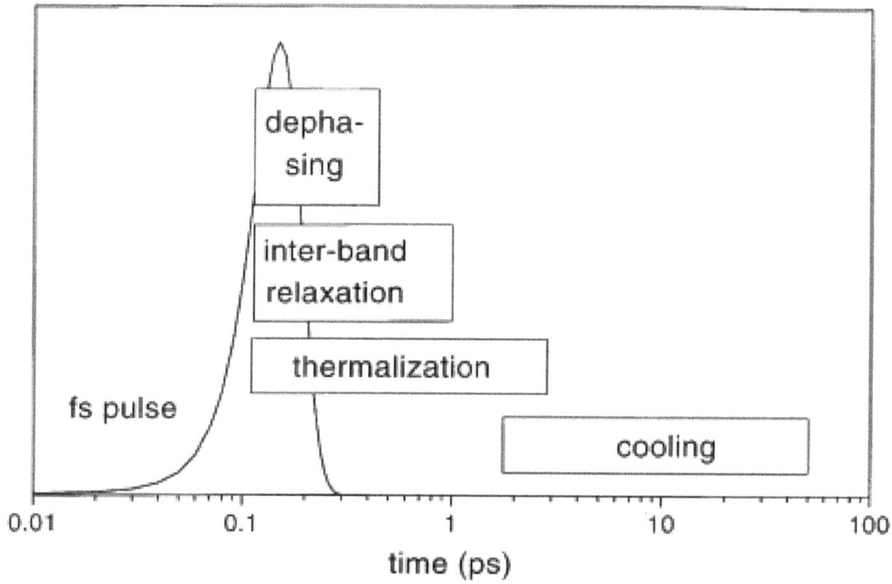

Fig. 2. The scheme of excitation-relaxation processes [12]

8. Measurements and instrumentations

As it had been mentioned before the ultrashort laser pulses provide an excellent tools to realize time-resolved experiments with which one can observe transient species in different chemical reactions and follow the dynamical behavior of physical-, chemical- and biological processes. Another important property is that with modest energy, the fs pulses can have huge peak powers. This also makes them suitable for many tasks that we would not normally think of as 'time resolved', including laser ablation of materials, multi-photon absorption (for imaging of biological materials), fragmentation (e.g., DNA into fragments that may be analyzed using mass spectrometry), the conversion to a range of new wavelengths using nonlinear techniques, e.g., infrared light to visible light conversion and 2-photon excited fluorescence, etc. Semiconductor processes and collisions in liquid phase materials are also in the range of a few hundreds of fs [18-20].

Direct measurements in fs region are not possible using electrical methods and other non-optical techniques. The use of specialized photodetectors such as streak cameras or avalanche photodiodes that can resolve picosecond or even 100s of femtoseconds transients in real-time, but are not able to resolve a necessary few fs events, therefore alternative detection techniques are required.

The techniques that are used most frequently are based on the auto- or cross-correlation of two beams of femtosecond pulses. If the target is a nonlinear crystal used for sum-frequency generation, this technique can be used to determine the shape and relative arrival time of two short pulses. If the sample consists absorbing materials normally one uses pump-probe experiments for temporal registrations of events [1,21-30]. Therefore, if we want to measure the dynamics of a fast event, we have to apply a faster tool to do it. Moreover, the use of a not as short as possible laser pulse can induce the shortening transient behavior [31]. The most commonly used scheme of a general pump and probe equipment is sketched in Fig 3. [24]

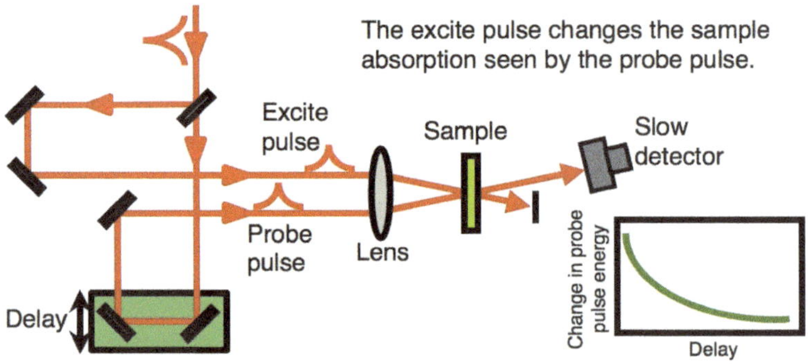

Fig. 3. Schematic of a general pump and probe equipment [24]

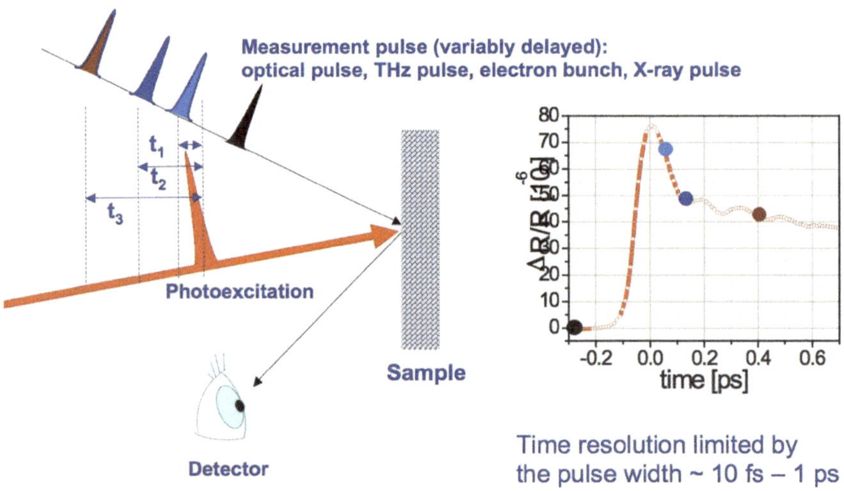

Fig. 4. Sketch of pump and probe for different sources

The delay in the probe arm is usually realized with an optical path enhancement done by a mirror system (Fig 4.).
As we can see the main laser beam is split with a mirror into 2 parts: pump beam with intensity of about 90% of the original and a probe beam of about 10 % of original. Both pulses are focused upon the target with their spatial overlapping. The delay is realized with variation of a beam path length compared to probe one. The weaker pulse in some of his characteristic (e.g., intensity, polarization, temporal duration) will be modified varying the delay *(Δt)*. This is the results of excitation in target material by the pump beam. Repeating the measurements by varying the time delay one determines the temporal dynamics of the excitation.
In some of more sophisticated measurements, one tries to use a focusing object as it can be seen in Fig 5.

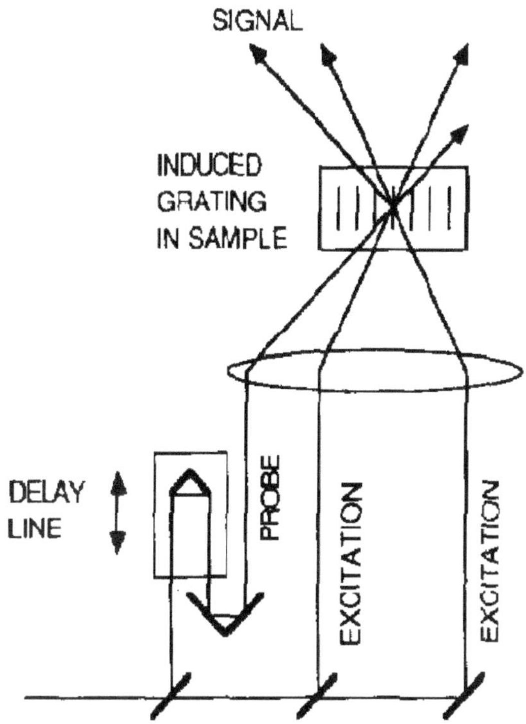

Fig. 5. Schematic of the transient grating experiment. Two excitation pulses are crossed in time and space in the sample. The resultant spatially periodic material excitation is probed by diffraction of a third, variably delayed beam [32].

Different variations of this technique can be used to determine dynamics of events in different fields, such as electron transport in solids, hetero- and nanostructures, induced spin dynamics by magnetic influences etc. Numerous applications had already been developed for chemistry, biology, and life sciences. The resolution achieved by pump and probe method nowadays reaches as hundreds of attoseconds.

9. Ultrafast X-ray spectroscopy

X-rays are very useful tools of modern science as well as solid state physics. The determination of the atomic structures became possible with achievement of coherent X-ray applications. In that frame one uses the static X-ray diffraction technique based on Bragg reflections.

However, the appearance of new pulsed coherent X-ray sources with extremely short pulse duration had opened a way for time dependent investigations. Femtosecond X-ray pulses enable atomic spatial (~0,1 mm) and high enough temporal resolution to observe the evolution of atomic configurations. In such a way one gets a direct dynamic structural picture [31-35].

Until now, a variety of methods have been developed to generate fs X-ray beams. For example, during the interaction of very high intensity laser pulses with material due to results of electron-atom interaction processes one yields to characteristic brehmstrahlung and line emission. The time duration of X-ray beam generated like as generating fs laser pulse duration, and the energies are in range of 10 eV ~ 1 MeV. Also high intensity coherent X-ray beams may be emerged from laser-produced plasma sources or laser-driven electron X-ray sources and synchrotron radiation induced sources [31, 33, 35].

To perform time-resolved measurements in the X-ray regime one can use suitable variants of pump and probe techniques like in optical region (Fig. 6.).

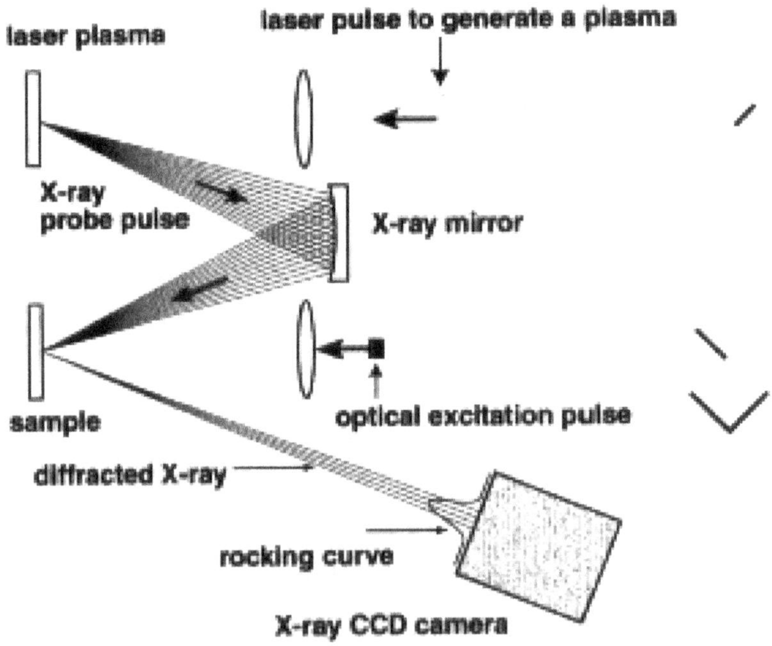

Fig. 6. Schematic of an optical pump- X-ray probe experiment [42].

One of the advantages of using X-ray beams for spectroscopic aims is the deeper penetration of coherent X-ray beam into the material if the wavelength is less or in the order of lattice

spacing. Under these conditions X-ray diffraction would be strongly dominated by the bulk crystal ignoring the damaged or melted subsurface layers.
In such a way, different X-ray spectroscopical techniques have been developed as X-ray absorption spectroscopy (XAS), extended X-ray absorption fine structure spectroscopy (EXAFS) absorption near edge spectroscopy (XANES) inelastic X-ray Raman scattering (XRS), and X-ray emission spectroscopy (XES).
As an example, we demonstrate the concept of XRS (Fig. 7.).

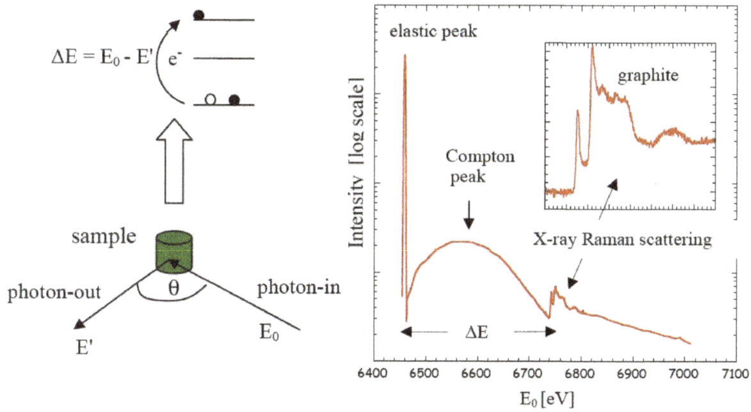

Fig. 7. Left: Concept of XRS. The energy transfer from an inelastically scattered photon results in the excitation of a core electron into an empty state. Right: Complete scattering spectrum from graphite. Intensity versus incident energy E_0 is plotted, analyzer energy E' is fixed at 6460 eV [41].

Concerning the instrumentation different kinds of wavelength dispersive devices are in utilizations for spectroscopical applications e.g., cylindrically curved analyzers and position sensitive detectors (PSD) (see Fig. 8)

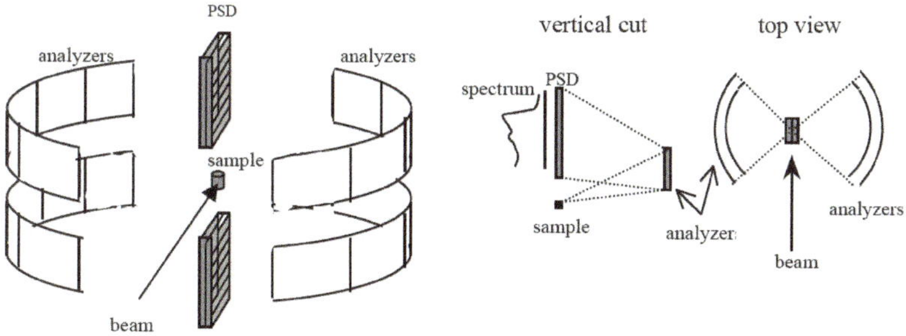

Fig. 8. Schematic setup four arrays of cylindrically curved crystals in sagital focusing mode. Scattering of a point source beam is analyzed at different energies (see vertical cut) resulting in a spectrum on the PSDs. For XRS the setup is rotated by 90° for scattering in the predominantly vertical plane [41].

As a sample of nice characteristic results of XANES/EXAFS we turn to Fig. 9.

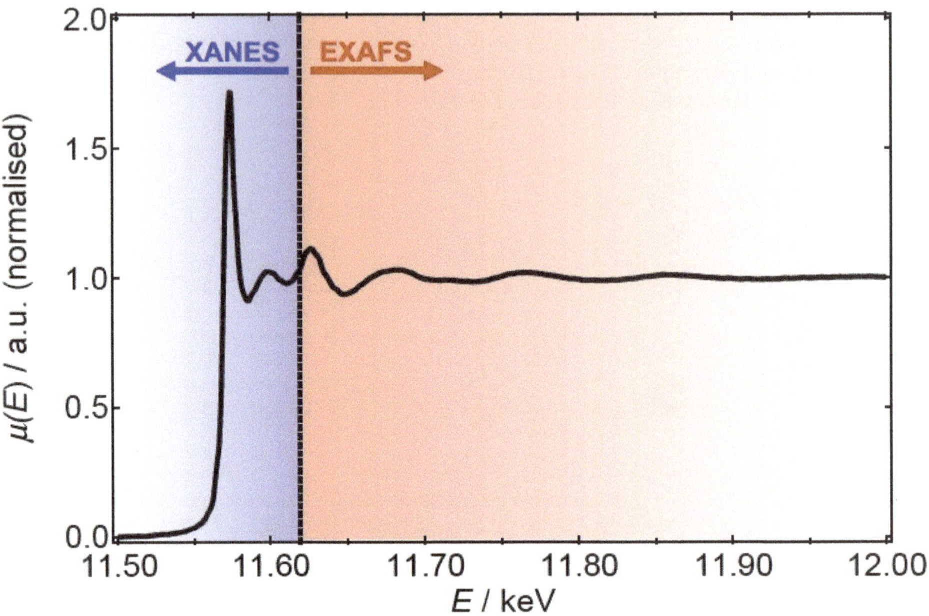

Fig. 9. XAS spectrum of a molecule (PtPOP) in solution illustrating the two regions: the low-energy XANES region up to ~50 eV above the IP and the high-energy EXAFS region >50 eV. The spectrum has been normalized.

10. Time resolved THz spectroscopy

In the optical wavelength scale the THz region makes a bridge between microwaves and infrared domains. This is located at about 10^{12} Hz, so called terahertz region. Because of the quite low phonon energies in this region, the terahertz spectroscopy mainly is devoted to carry investigations in the exploration of infraband/subband excitations (transition). The T-rays are harmless for the human body; therefore, one can find applications in basic medical research and security [34].

The materials used for generation of terahertz radiation by optical rectification can also be used for its detection by using the Pockels effect where certain crystalline materials become birefringent in the presence of an electric field. The birefringence caused by the electric field of a terahertz pulse leads to a change in the optical polarization of the detection pulse, proportional to the terahertz electric-field strength. With the help of polarizers and photodiodes, this polarization change can be measured.

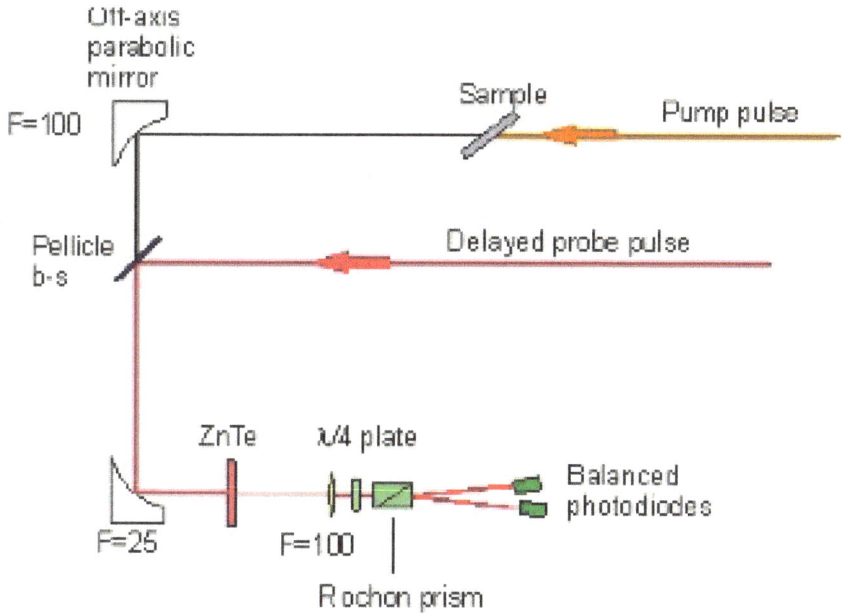

Fig. 10. Setup for the measuring of polarization change

A typical setup for transmission THz spectroscopy is shown in Fig. 11.

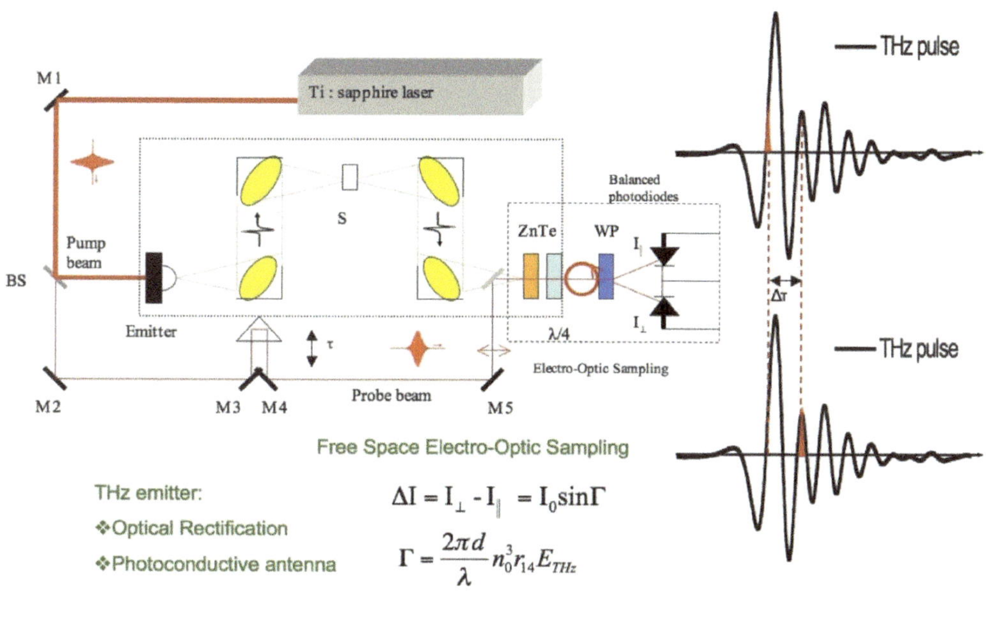

Fig. 11. A typical setup for transmission THz spectroscopy

As a nice example of THz transient conductivity spectrum we do example of Si measured and fitted by Drude model (Fig. 12) [37].

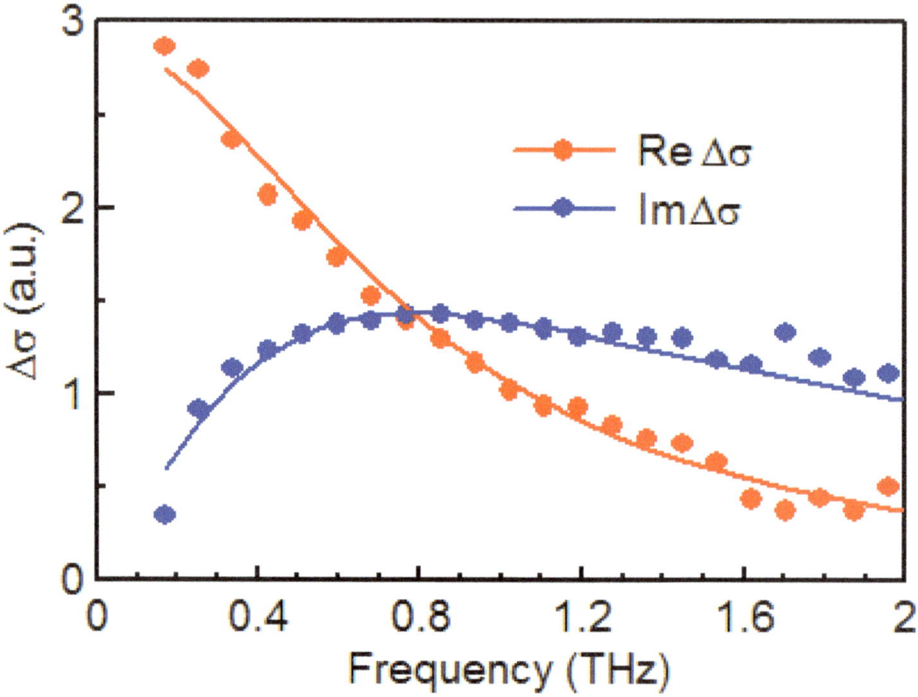

Fig. 12. Variation of conductivity in Si [37]

11. Ultrafast infrared spectroscopy

In the infrared spectral region (1-25 μm wavelengths) the consequences of Heisenberg uncertainty principle are especially significant, e.g., for a 200 fs pulse duration we get spectral bandwidth about 75 cm^{-1}.

For investigations of large and complex molecules the use of visible spectrum is not so convenient because of the overlapping features due to broad spectrum one can get only a few structural information. Therefore, to get useful information with high temporal resolution and sensitivity with proper reliability, we usually use combined (visible and IR) pump and probe variations [38].

As an example of a possible IR ultrashort arrangement, we show a scheme of experimental equipment devoted to examine charge-carrier dynamics in a polymer (PoV) (Fig. 13).

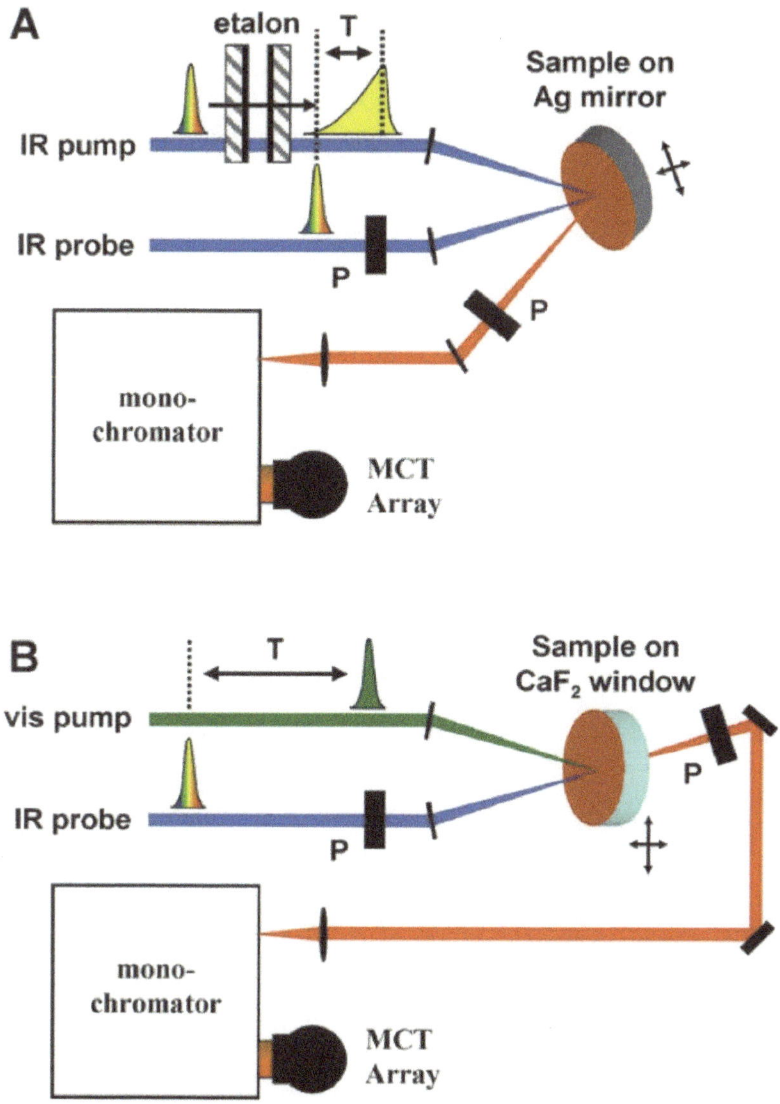

Fig. 13. Diagrams of the beam and sample geometries used for (A) 2D IR and (B) visible pump IR probe experiments. These geometries eliminate non-resonant signals that frequently interfere with ultrafast vibrational spectroscopic studies. The symbols represent: etalon – tunable Fabry-Perot interference filter, T – optical delay between the pump and probe pulses, P – polarizer, MCT Array – mercury cadmium telluride multichannel infrared array detector [33].

With the help of continuous wave IR source is used as a probe beam. Passing through the sample this probe beam is overlapped with fs pump beam. The time resolution is determined by applying a second fs visible pulse to gate the probe beam by up conversion. As a result of interaction we get a pulse with a sum frequency of IR and visible beams while the intensity will be related to the IR absorption.

The temporal delay changing of up conversion one can swap the dynamics of the event. Moreover due to such a up conversion linear detectors in visible are used with high sensitivity and resolution. The CW IR source usually have a very narrow line width, therefore the resolution depends only on homogeneous and inhomogeneous broadening caused by the sample itself.

There is another method avoiding the consequences of Heisenberg principle based on using a short IR pulse to be passed through the sample. After that this beam is dispersed in a monochromator to raise the frequency resolution [40].

In such a realization, the short pulse causes polarization field in the sample. As a polarization a coherent radiation emerges with the probe beam. Therefore, the resolution will be limited with the interaction time of the field with the sample (optical dephasing).

The ultrafast infrared spectroscopy proved to be an excellent tool to carry out structural and dynamical investigations in different areas of chemistry, biophysics, and organic chemistry, especially in the diagnostic of transient states. As an example of the dynamics of dissociation of ICN molecule is shown in Fig. 14.

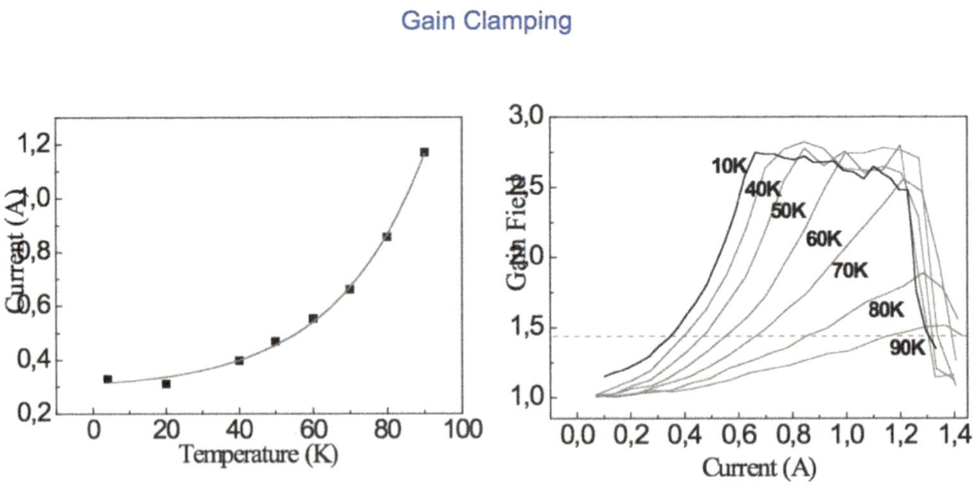

Fig. 14. Dissociation of ICN molecule

12. Some results of ultrashort spectroscopy methods' applications in different fields of material science

- Observations of ultrafast phase transitions

It is well known that VO_2 at moderate temperature T > 67 °C goes through a phase transformation from an insulating to a metallic phase [41-43].
The conductivity changes dramatically (~ about 10^5 fold) while the crystal structure changes from monoclinic to rutile.

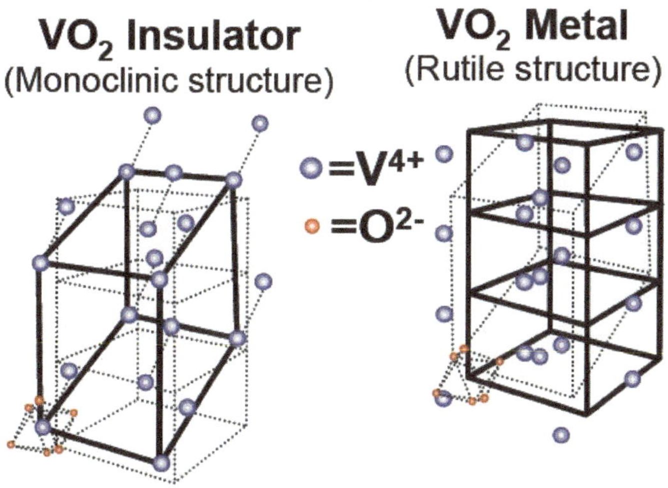

Fig. 15. The "structural bottleneck" in the transition arises from the time needed for the vanadium atoms to change from the monoclinic structure associated with the insulator to the rutile structure of the metal. [44].

The difference in a reflectivity of the structures is also dominant.

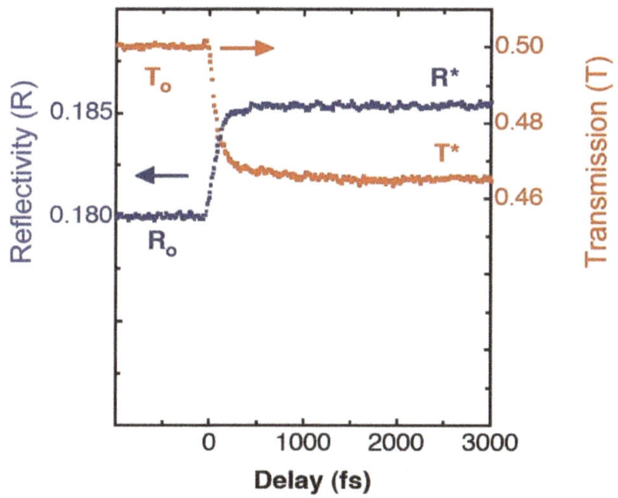

Fig. 16. The reflectivity of the metallic state is higher than that of the insulating state; use of an ultrafast probe pulse allows the measurement of the transition time with 15 fs resolution (1 fs = 10^{-15} second) [44].

The dynamics of the changes of electron correlation and so far reflectivities was studied with pump and probe spectroscopic methods; using ultrashort X-ray spectroscopy. The time transition limit for phase transformation from monoclinic to rutile was found about 80 fs [44].

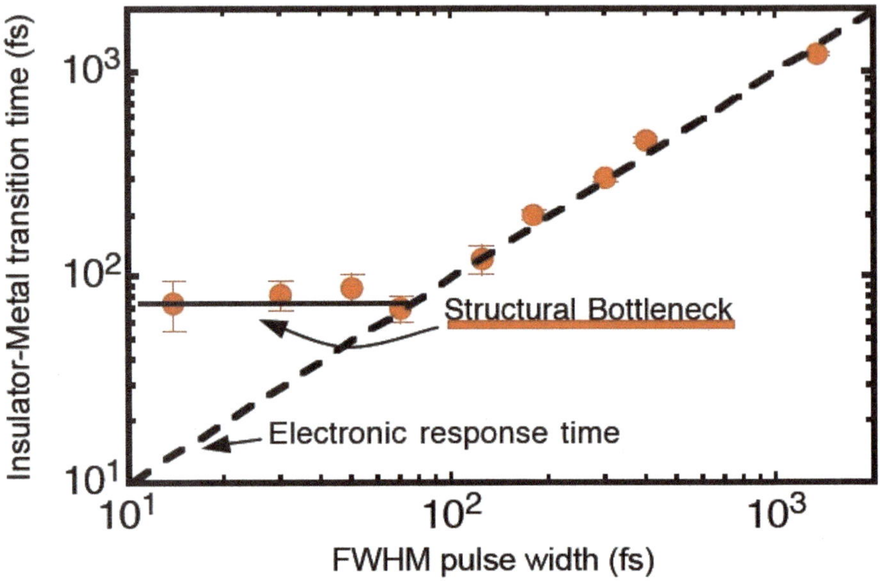

Fig. 17. The observed transition time for the insulator-metal transition (red data points) becomes progressively shorter with pulse duration down to 80 fs, at which point a fundamental limit is imposed by the material properties [44].

Direct band-to-band recombination in silicon nanocrystals and porous silicon
The silicon-bulk-crystal has interband optical transitions only with the participations of phonons (to evaluate the energy and momentum conservation laws). The reduction of size of silicon crystal e.g. nanosized or porous silicon show zero-phonon optical transitions which is promising for fabrication of light emitting devices for Si microelectronics. [44]
To demonstrate the existence of such a transition and to get data for dynamics of the different ultrafast spectroscopic methods had been used such as time resolved photoluminescence, pump and probe phonon echoes and for wave mixing.

13. References

[1] László Nánai, Zsolt Benkő, Renat R. Letfullin and Thomas F. George, Laser-Matter Interactions: Nanostructures, Fabrication and Characterization in T. F. George, D. Jelski, R. R. Letfullin and G. P. Zhang, Editors, Computational Studies of New Materials II: From Ultrafast Processes and Nanostructures to Optoelectronics, Energy Storage and Nanomedicine (World Scientific, Singapore, 2011), 540 pages.

[2] Trebino R., Frequency-resolved opticalgratings: The measurements of ultrashort pulses, Kluwer publisher 2002

[3] DeLong, K. W., R. Trebino, J. Hunter, and W. E. White, J. Opt. Soc. Amer. B, v. 11, p. 2206-2215 (1994).
[4] Nánai L., Szatmári S., Int. Conf. of Condensed. Matter. Mugla (Turkey) 26-28 May (2008) p. 27
[5] M. Dantus, P. Gross, Encyclopedia of Applied Physics 22, 431-456 (1998)
[6] Workman Jerry Jr., Springsteen Art, Applied Spectroscopy: A Compact Reference for Practitioners, Academic Press 1998.
[7] M. Cardona, R. Merlin, Light Scattering in Solids topics in Applied Physics, vol. 108 Springer 2007
[8] Rickard M. A., Pakoulev A. V., Mathew N. A., et al., J. Phys. Chem. A, 111, 1163-1166 (2007).
[9] Tranter G.E., Holmes J.L., Lindon J.C., Editors, Encyclopedia of Spectroscopy and Spectrometry, Volumes 1-3, Academic Press 2000
[10] Duckett S., Gilbert B., Foundations of Spectroscopy, Volumes 1-3, Oxford University Press 2000.
[11] Demtröder Wolfgang, Laser Spectroscopy, Volumes 1-3, 3rd Edition, Springer 2002.
[12] T. Elsaesser, M. Woerner, Physics Reports, 321 (1999) 253-305
[13] Cavalieri A. L. et al, Attosecond time-resolved spectroscopy at surfaces in Dynamics at Solid Surfaces and Interfaces, ed. By P. Hvroje and M. Wolf WILEY VCH 2010
[14] D. R. Brian, Science, 253/5078 p.1913
[15] Hörlein R., Thesis, Investigation of the XUV Emission from the Interaction of Intense Femtosecond Laser Pulses with Solid TargetsMünchen University 2008
[16] P.B. Allen, Phys. Rev. Lett. 59, 1460 (1987).
[17] M.I. Kaganov, I.M. Lifshitz, L.V.Tantarov, Zh. Exsp. Theor. Fiz. 31, 232 (1956) [Sov. Phys. JETP 4, 173 (1957)]
[18] Reid D. T., Measuring ultrafast laser pulses in Ultrafast photonics ed. By A. Miller, D. T. Reid. M Finlayson, Taylor and Francis 2004
[19] F. Krausz, Phys. World 14, 41 (2004)
[20] M. Schultze and al, Journal of electron spectroscopy and related phenomena (2011) doi:10.1016/j.elspec.2011.01003
[21] R.L. Samad et al., Ultrashort Laser Pulses Applications, IPEN-CNN/SP Brasil
[22] Krausz, F. & Ivanov, M. (2009). Attosecond physics. Rev. Mod. Phys., 81, 1, 163-234, ISSN: 0034-6861
[23] R. Paschotta, Encyclopedia of laser physics and technology Wiley VCH (WWW.virtualLibrary)
[24] Rick Trebino lectures at https://www.physics.gatech.edu/user/rick-trebino
[25] R. Hoffmann American Scientist, 87, 308-311 (1999)
[26] A. H Zewail et al, Journal of physical chemistry, 100:12701-12724. (1996)
[27] Reiter E. et al, PRL, 105 243902 (2010)
[28] Cavalieri, A. L. et al., Attosecond spectroscopy in condensed matter. Nature 449,1029 (2007).
[29] M. Dohle et al, Berichte der Bunsengesellschaft für physikalische Chemie Volume 99, Issue 3, pages 478–484, März 1995
[30] V. Schmidt, W. Husinsky*, G. Betz Applied Surface Science, 197–198 (2002) 145–155

[31] Diels J. C. W. Rudolph, Ultrafast laser pulse phenomena: fundamentals, techniques and applications on femtosecond time scale, Academic Press Burlington MA USA 2006

[31]Uwe Bergmann, LOI for proposal „Wavelength Dispersive Optics for Ultrafast X-ray Spectroscopy at LCLS" Stanford Synchrotron Radiation Laboratory Menlo Park, California 94025

[32] Renske Marjan van der Veen THÈSE NO 4905 (2010) Ultrafast X-ray and Optical Spectroscopy of Binuclear Molecular Complexes, Lausanne 2010.

[33] Pensack R.D et al. Phys. Chem. Chem. Phys 2009, 11, 2575 – 2591

[34] file:///F:/InTech/44_Carito_Time-resolved%20terahertz%20spectroscopy.htm

[35] Han-Kwang Nienhuys, and Villy Sundström, Appl. Phys. Lett. 87, 012101 (2005)

[36] Han-Kwang Nienhuys, and Villy Sundström, Phys. Rev. B 71, 235110 (2005)

[37] Hoffmann M C., Yeh K. L., Hebling J., Nelson K A., Optics Express 15(18) 11706-13 (2007)

[38] Jedju T. M. et al, Appl. Opt., 31 2684 (1992)

[39] Fecko J.C., Science 2003, Vol. 301 no. 5640 pp. 1698-1702

[40] Szekerle J.P. et al, J. Chem. Phys. 95, 5403 (1991)

[41] Berkeley Lab research Highlight by Schoenlein R.,"Ultrafast phase transitions" U. S. Dept of Energy Divison of Material Science

[42] Cavalieri A. et al, Phys rev B Rapid Comm., 70, 161102 (2004)

[43] Cavalieri A. et al, Phys Rev Lett., 95, 067405 (2005)

[44] Wen X., Ultrafast Spectroscopy of Semiconductor Nanostructures, PhD Thesis, Swinburne Univ. Melbourne, Australia, 2007

9

Time Resolved Spectroscopy with Femtosecond X-Ray Pulses

Enikoe Seres[1,2] and Christian Spielmann[1,2]
[1]*Institute of Optics and Quantumelectronics, Friedrich Schiller University Jena*
[2]*Helmholtz Institute Jena*
Germany

1. Introduction

Spectroscopy is of great interest for exploring the structure or composition of unknown matter. In 1813 Joseph von Fraunhofer discovered, independently from William Hyde Wollaston (1802), the dark lines in the spectrum of the sun, which has been later explained by the absorption of hydrogen. In 1859 Gustav Robert Kirchoff and Robert Wilhelm Bunsen discovered that the flame of gas burner changes its color, if different elements are heated up. In 1895 Wilhelm Conrad Röntgen (Röntgen, 1895) detected a new kind of radiation and named it X-ray radiation. In the first hundred years since their discovery, x-rays have played an important role in helping us understanding the structures of materials. Due to their short wavelength, x-rays are capable to resolve the structure of condensed matter but also the internal structure of atoms. The x-ray spectroscopic methods have been refined over the years. Nowadays physicists, chemists, biologists and material scientists rely on x-ray static structural analysis on a routine basis (Michette and Pfautsch, 1996).

In addition to the static structural information, transient structural information is required for a deeper understanding. Such dynamic processes include the breaking and formation of chemical bonds, protein motions, charge transfer, phase transitions and so on (Sundaram 2002; La-O-Vorakiat, 2009; Woerner 2010). In Figure 1, we summarize the main applications of the time resolved spectroscopy. Many of these problems have already been tackled by means of conventional optical pump/probe spectroscopy. Unfortunately, such optical measurements cannot be directly inverted to give the position of the atoms as a function of the times expect in very favorable cases. Unlike optical spectroscopy, x-ray diffraction (XRD) and x-ray absorption (XRA) do in principle provide direct ways to reconstruct the motion of atoms during dynamic processes. Thus, time-resolved XRD and XRA may serve as a more direct way to observe the microscopic processes by which biomolecules perform their tasks or to observe ultrafast processes in solid-state materials. An ideal x-ray system for time-resolved diffraction should have sufficient temporal and spatial resolution to resolve the dynamics of fast processes. It took almost one century since the discovery of x-rays to realize x-ray sources providing a temporal resolution which is sufficient to follow the atomic and molecular motion, or to monitor the dynamics of molecules such as rotation, vibration, dissociation, or to study the dynamics of electrons on their natural time scale. Atom motion takes place on a picosecond to femtosecond time scale, whereas attosecond resolution is necessary for revealing the electron dynamics (Bressler et al., 2004, Rousse et al., 2001). Now, in the 21th century, we are in the fortunate situation to routinely generate bright femtosecond and attosecond x-ray pulses. In this chapter we review the current progress of time resolved x-ray spectroscopy and present some of our recent results about x-ray source development and x-ray absorption studies of transient states of matter. This chapter is divided into the following sections:

1. Introduction
2. X-rays and x-ray absorption spectroscopy
3. Pulsed x-ray sources
4. High harmonic generation
5. Time-resolved X-ray spectroscopy

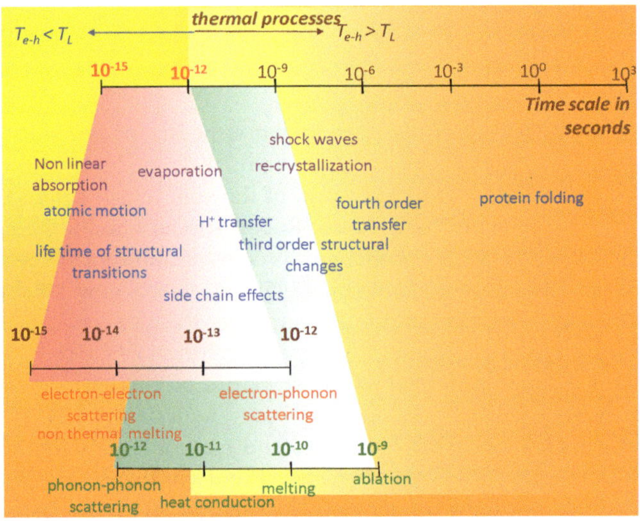

Fig. 1. Typical time scales for structural and electronic processes in solids.

2. X-ray radiation

X -ray radiation is a part of the electromagnetic radiation spectrum. The wavelength of X-rays, extends from 0.02 Å to 100 Å, and corresponds to the atomic and molecular length scales. X-ray radiation can be divided as follows:

- ultra-soft <1 keV (λ > 1 nm) and XUV
- soft 1 keV < E <10 keV as (0.1 nm - 1 nm),
- hard 10 keV < E < 100 keV (0.1 nm - 0.01 nm)
- ultra-hard E > 100 keV (λ < 0.01 nm)

The use of radiation in this wavelength range provides, direct information about the structure of matter (Agarwal 1991; Rose-Petruck et al., 1999; Bressler et al., 2002). Therefore, the development and application of X-ray sources for the structure determination is of great scientific interest. The first X-ray tube was realized by W. C. Röntgen. He discovered a new type of radiation arisen from the interaction of accelerated electrons with matter. The decelerated electron beam radiates a broad continuous spectrum with sharp characteristic lines on it. The continuous part is referred to as "bremsstrahlung" and its maximum frequency depends only on the applied voltage and is independent of the anode material. The frequency of the line radiation depends on the anode material. The impacting electrons excite electrons from core levels into the continuum. The created hole is filled by an electron from a higher shell and the excess energy is emitted as a photon with an energy corresponding to the energy difference of the two involved states. For an anode material with atomic number Z, Moseley law predicts the frequency of the line:

$$\nu = R_{\infty}c\left(\frac{M}{m+M}\right)Z^2\left(\frac{1}{n_f^2}-\frac{1}{n_i^2}\right) \tag{1}$$

Here m and M is the mass of the electron and the mass of the nucleus, n_f and n_i indicate the principal quantum number of initial and final states, R_∞ is the Rydberg constant.

2.1 X-ray spectroscopy methods

Both the bremsstrahlung and the characteristic radiation are used to investigate the structure of matter. Moreover, information about the atomic structure can be also obtained from the x-ray generation itself. The sample under investigation is illuminated with an electron or x-ray beam and its properties can be obtained by measuring the absorbed, diffracted, or scattered x-rays, the emitted x-ray fluorescence, or the ejected photoelectrons.
In the X-ray diffractometry (XRD) the X-ray beam interacts with the electron shells of the atoms fixed in a lattice. The diffraction pattern provides information about the atomic distances of the crystalline structure. At not too high photon energies, this scattering is elastic: there is no energy loss and then we also speak of coherent scattering (Rayleigh scattering). In this case, the wavelength of the scattered X-rays is the same as the original X-ray wavelength. When a core electron is ejected, then it may be followed by a recombination from higher occupied levels and a photon is generated. This effect is described and characterized by X-ray fluorescence (XRF). XRF is primarily only an element-specific method and allows not only qualitative but also quantitative analysis of the components contained in the sample volume. For special cases it can provide also information about the structure. X-ray spectroscopy for chemical analysis (ESCA) is based on the generation of photoelectrons. This technique is essentially limited to the surfaces. Typically, it is possibly to receive information from only two or three atomic layers, although the exciting X-rays penetrate much deeper.
X-ray absorption spectroscopy (XAS) provides information about the atomic structure and also about the atomic distances and chemical bonds. Matter can be characterized by the transmitted X-ray intensity I, if the sample is illuminated by an X-ray beam with intensity of I0:

$$I = I_0 e^{-\mu d}, \tag{2}$$

where d is the thickness of the material and μ is the linear attenuation coefficient. The incoming X-ray beam can excite core electrons to a higher unoccupied states or to the continuum. When the energy of the photons is increased, pronounced edges appears in the absorption spectrum, if the photon energy is high enough to excite electrons from a deeper core state (Figure 2) into the continuum.

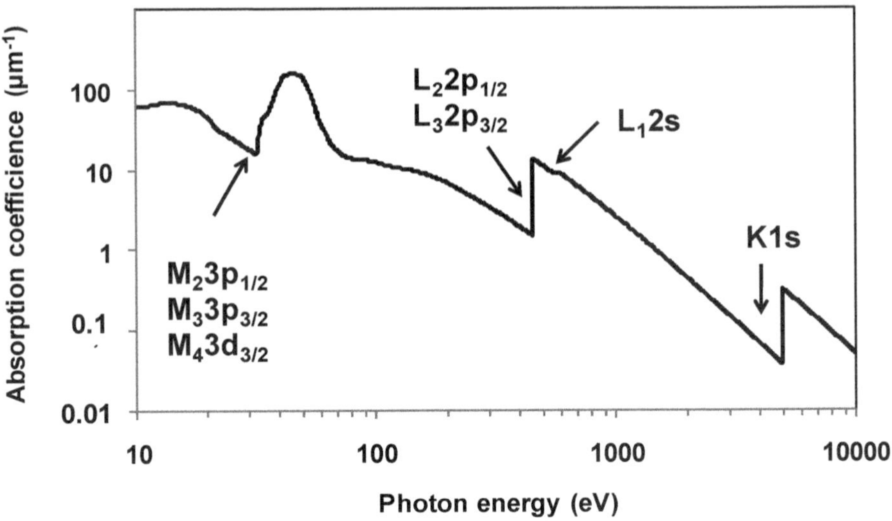

Fig. 2. X-ray absorption spectrum of Ti with K, L and M-edges.

A closer examination shows that only the K-edge consists of a simple jump. Near the L-absorption edge well-resolved jumps can be seen. The M-edges are not so well resolved.

2.1.1 Fine structure of the XAS signal

If the X-ray absorption spectrum is measured with high resolution, then a fine structure can be observed and resolved. The position and shape of the absorption edges is determined by the atom and is independent from environment of the atom, at least in a first approximation. The reason is that the x-ray absorption is related to the core levels, and only electrons from outer shell are involved in chemical bonds. However spectroscopy with sufficiently high resolution can detect an influence of the chemical bond on the energy and structure of the absorption edges. For determining the structure, i.e. getting information about the neighborhood of the atom of interest, we can rely on the following X-ray absorption methods:

- XAMES (X-Ray Absorption Main Edge Spectroscopy). The position of the absorption edge contains information about the electronic structure of atoms and the structure of the material. The pre-edge contains further information about the electron configurations and the symmetry around the absorbing atom. The measurements are made in a range from -10 eV to +10 eV around the absorption edge of the atom. The shift of the edge position is often referred as "chemical shift".
- XANES (X-Ray Absorption Near Edge Spectroscopy) provides information on the valence electrons and chemical bonds. It is necessary to record the signal in a range of 10 – 40 eV above the edge.
- EXAFS (Extended X-Ray Absorption Fine Structure) contains structural information, i.e. the distance to the neighboring atoms. The absorption spectrum is measured and evaluated in a range from 40 eV to 1000 eV beyond the absorption edge.

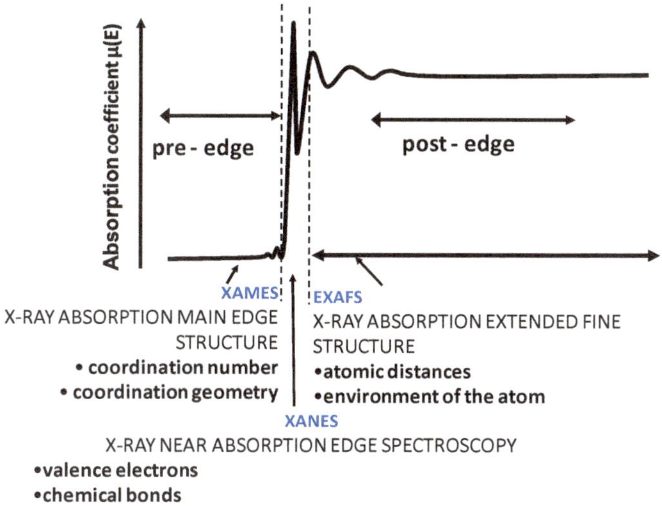

Fig. 3. Structure of the absorption edge can be divided into the ranges of XAMES, XANES and EXAFS.

2.1.2 EXAFS signal at the K-edge

In principle, structural information can be obtained with XANES or EXAFS. The measurement in a narrow range is easier, but XANES does not allow a simple determination of the distance to the next atoms from the experimental data. The basic principle of EXAFS is summarized in the following. The incident photons excite an electron from a core level into the continuum. The generated and outgoing photoelectron waves are elastically scattered by neighboring atoms. Quantum mechanically speaking, part of the wave function of the photoelectron is reflected by the neighboring atoms and the wave functions interfere. The modulation in the absorption spectrum depends on the path difference of the partial waves, and at a fixed atomic distance it is a function of the incident X-ray photon energy. (Stern 1974; Lytle et al., 1975; Lee et al, 1981; Rehr et al, 1992; Rehr & Albers 2000; Sipre, 2002.) The evaluation of a measured EXAFS spectrum requires the following steps. First, $\chi(E)$, the normalized EXAFS signal is calculated from the measured absorption spectrum:

$$\chi(E) = \frac{\mu(E) - \mu_0(E_0)}{\Delta\mu_0(E_0)}, \tag{3}$$

where $\mu(E)$ is the measured absorption coefficient, $\mu_0(E)$ is the smooth background function, or the absorption coefficient of the atom, and $\Delta\mu_0(E_0)$ is the measured jump of the absorption $\mu(E)$ at the edge and the energy of the absorption edge is E_0. Then the signal is converted into k-space: $k = \sqrt{\frac{2m_e}{\hbar^2}(h\nu - E_0)}$, where $h\nu$ is the incident energy of X-ray photons. The calculated signal $\chi(k)$ carries information about the distances and the type of neighboring atoms:

$$\chi(k)=\frac{1}{k}\sum_j \frac{N_j\left|f_j(k,\pi)\right|}{kr_j^2}\sin\left[2kr_j+\Psi_j(k)\right]\cdot e^{-2k^2\sigma_j^2}\, e^{-\frac{2r_j}{\lambda_j(k)}}$$

amplitude

wave number of the photo electrons $k=\sqrt{\frac{2m_e}{\hbar^2}(h\cdot\upsilon-E_0)}$

phase

Debye-Waller-factor

correction term

Here, N_j is the number of backscattering atoms at the same distance r_j from the absorber atom. f(k) and Ψ(k) corresponds to the amplitude and phase shift of the scattered wave, and σ^2 is the standard deviation of atomic distance. The factor $e^{-2k_j\sigma_j^2}$ also known as Debye-Waller factor is a measure for the smearing of the interference signal. Due to the spatial dispersion of the outgoing wave only backscattering from the next neighbors must be considered. Furthermore, the photoelectron must be scattered before the generated hole was filled. If these approximations are not valid a correction term of $e^{2r_j/\lambda_j(k)}$ is required. The size $\lambda_j(k)$ depicts the energy-dependent mean free path of photoelectrons. Phase shift impressed on the scattered electron wave is given

$$\Psi_j(k)=\Phi_j(k)+2\delta_l(k) \tag{4}$$

with $\Phi_j(k)$ phase shift through the backscattering from the jth neighboring atom and δ_l is the phase shift of the photoelectron in the potential of an atom in the lth shell. The amplitude of the backscattered wave is given:

$$f(k,\pi)=\left|f(k,\pi)\right|e^{i\Phi(k)}=\frac{1}{2ik}\sum_l(2l+1)\left(e^{2i\delta_l}-1\right)(-1)^l \tag{5}$$

With the help of the given formalism, the atomic distance can be precisely determined by EXAFS. Additionally it may be necessary to determine the composition of the nearest neighbors by other methods (Bzowski et al., 1993; Filiponi et al., 1989; Johnson et al.; 2003, D'Angelo et al., 1996, Bianconi et al., 1987, Farges et al., 1997; Faraci et al., 1997).

2.1.3 EXAFS signal at the L-edge

Previously it was assumed that the electron is generated by an excitation from the K-shell. However, EXAFS is also possible from e.g. the L-shell. For considering absorption from the L_1 shell, the same formalism as for the K-shall can be used. However for electrons from the $L_{2,3}$ shell the evaluation is more complicated. The reason is that the electron in the initial state is a p-state instead of a s-state. Because of the selection rule, the photoelectron generated can be a s or d-like electron. The difference is not only the energy but also the phase in the scattering is different and in addition, there is a superposition of two states. For EXAFS from the $L_{2,3}$ shell the signal can be calculated according to the following formula:

$$\chi(k)=\frac{1}{k}\sum_j\left\{\begin{array}{c}\frac{1}{2}\left(1+3\cos^2\Theta_j\right)M_{21}^2\sin\left[2kr_j+2\delta_2(k)+\Phi_j(k)\right]+\\+\frac{1}{2}M_{01}^2\sin\left[2kr_j+2\delta_0(k)+\Phi_j(k)\right]+\\+M_{01}M_{21}\left(1-3\cos^2\Theta_j\right)\sin\left[2kr_j+\delta_0(k)+\delta_2(k)+\Phi_j(k)\right]\end{array}\right\}\cdot \tag{6}$$

$$\cdot\left(M_{21}^2+\frac{1}{2}M_{01}^2\right)N_j\frac{\left|f_j(k,\pi)\right|e^{-2k^2\sigma_j^2}}{kr_j^2}e^{-\frac{2r_j}{\lambda_j(k)}}$$

where M_{01} and M_{02} are the elements of the radial dipole matrix between the l = 1 initial state and the l = 0 and l = 2 final state. The measured signal depends also on the orientation of the atoms. In the case of polycrystalline material, the third term of equation (6) disappears. Additionally the transition probabilities have been calculated and the p to d transition dominates over the p to s transition by a factor of 50, so the second term can be neglected. In a first approximation EXAFS spectra from the L-edge can be evaluated with the same simple approach as spectra above the K-edge.

3. X-ray sources

Synchrotrons represent the major source of powerful X-rays and will continue to play a dominant role for X-ray science in the foreseeable future. Nevertheless, a wide range of X-ray applications in science, technology and medicine would greatly benefit from i) X-ray pulse durations much shorter than routinely available from synchrotrons (few hundred picoseconds), ii) synchronizability of ultrashort pulses to other events, and iii) availability of useful fluxes from compact laboratory X-ray sources. Triggered by these needs, we witnessed in the last few years huge attempts towards the realization of such x-ray sources.
Advances in ultrashort-pulse high-power laser technology over the last decade (Perry and Mourou, 1994; Umstadter et al., 1998) triggered extensive research activity aiming at the development of compact, versatile laboratory X-ray sources in a number of laboratories for ultrafast as well as other applications. As a result, ultrashort-pulsed X-ray radiation became available from femtosecond-laser-produced plasmas (FLPP), (Murnane et al., 1991; Giulietti and Gizzi, 1998; and references therein).
These sources are now capable of converting up to several per cent of the driving laser pulse energy into incoherent X-rays emitted in a solid angle of 2π-4π and delivering pulses with durations down to the subpicosecond regime. FLPP sources matured to a point where a wide range of applications can be tackled all the way from the soft to the hard X-ray regime. Already demonstrated examples include time-resolved X-ray diffraction and absorption spectroscopy (Raksi et al., 1996; Rischel et al., 1997; Rose-Petruck et al., 1999) and medical radiology with improved contrast and resolution (Gordon III et al., 1995).
Many laboratory X-ray applications would greatly benefit from or rely on (spatially) coherent sources with high average and/or peak power. One of the major approaches to laboratory production of coherent X-rays is the development of X-ray lasers. Whereas short-wavelength lasing has been successfully demonstrated with compact, table-top setups using several promising schemes at $\lambda > 15$ nm in the XUV range (Rocca et al., 1994; Lemoff et al., 1995; Nickles et al., 1997; Korobkin et al., 1998), lasing at shorter (soft-X-ray) wavelengths

could only be achieved at large-scale facilities so far (Nagata et al., 1993; Da Silva et al., 1994; Zhang et al., 1997).

Another promising route to developing compact coherent X-ray sources is high-order harmonic generation (HHG) with ultrashort-pulse lasers (L'Huillier and Balcou, 1993; Macklin et al., 1993; Wahlström et al., 1993; Perry and Crane, 1993; Kondo et al., 1993). Extensive theoretical (Lewenstein et al., 1994; Antoine et al., 1996) and experimental (Salieres, 1995; Ditmire et al., 1995, 1996) research provided valuable insight into the microscopic (strongly-driven atomic dipole) and macroscopic (propagation effects, e.g. phase mismatch) phenomena relevant to HHG (for a review see Salieres et al., 1999). Recent investigations revealed that ultrashort drivers with pulse durations well below 100 fs (Zhou et al., 1996) can produce HH conversion efficiencies comparable to XUV lasers in the 50 - 20 nm range, $\eta_{conv} \approx 10^{-5} - 10^{-7}$, respectively (Sommerer et al., 1999; Constant et al., 1999).

Recently, few-cycle, sub-10 fs laser pulses produced HH radiation at 13-10 nm with efficiencies in the range of $\eta_{conv} \approx 10^{-7} - 10^{-8}$ and with pulse durations estimated as < 3 fs at a repetition rate of 1 kHz, resulting in the highest average and peak powers ever demonstrated from a coherent laboratory soft-X-ray ($E_{ph} \geq 100$ eV) source (Schnürer et al., 1999). Pulses in the 5-25 fs range have extended HHG even down to the water window, 2.3 - 4.4 nm (Spielmann et al., 1997; Schnürer et al., 1998). Theoretical investigations suggest that few-cycle-driven harmonic emission is confined temporally to a tiny fraction of the laser period in the cut-off region of the spectrum (Kan et al., 1997; Christov et al., 1997; Spielmann et al. 1998), resulting in a single XUV/X-ray burst of attosecond duration.

SOURCE	PHOTONS /PULSE/ 0.1% BW	PULSE WIDTH	ENERGY RANGE	REPETITION RATE
High Harmonic Generation		≥80 as	< 4 keV	kHz
Laser generated plasmas	10^2	~300 fs	≤ 3 keV	10 Hz- 1 kHz
Third generation synchrotrons	10^4-10^6	10-20 ps	0- 100 keV	≤ 500 MHz
Slicing scheme	10^1-10^3	~100 fs	0-100 keV	1-10 kHz
Short pulse photon source (SPPS)	10^8	~ 100 fs	Fixed energy	10 HZ

Table 1. Summary of the x-ray sources usable for time-resolved x-ray spectroscopy.

In spite of these advances, applications of coherent laboratory X-ray sources are at the beginning. E.g. the photon fluxes available from state-of-the-art harmonic sources are still low, and allow only selected experiments. However, increasing the power of the few-cycle-driven harmonic source by 1-2 orders of magnitude holds promise for opening up further intriguing application fields in science and technology and pushing the frontiers of physical sciences. The former include X-ray spectroscopy, X-ray microscopy, X-ray photoelectron spectroscopy and possibly X-ray interferometry all of which have had to rely on large-scale synchrotron facilities far. The unprecedented pulse durations and X-ray peak intensities that may become available from harmonic sources, opening the way to attosecond science.

4. High harmonic generation

High harmonics (HH) are generated by the interaction of intense linearly polarized laser pulses with atoms, molecules or atomic clusters, where photons are generated in the extreme ultraviolet spectral region and in the soft X-ray regime. The generation takes place in a gas jet with a focused, ultrashort, intense laser pulse. The relatively modest demands on the parameters of the laser pulses and the excellent temporal and spatial coherence of high harmonics pave the way for developing a brilliant compact short-wave radiation source. The rapid progress in high harmonics generation has made it possible for the first time to realize sophisticated inner-shell spectroscopy with a compact laboratory system.

These achievements are mainly based on the rapid development of ultrashort pulse lasers. With them, it is possible to build small, stable and high repetition laser systems. The intensity also plays an important role, since the maximum photon energy is directly proportional to it. Despite a number of advantages and excellent properties of HH sources they have the weakness of the relatively low conversion efficiencies of visible or near-IR laser light into XUV radiation. HH generation is a coherent process, so the signal grows quadratically with the propagation length. The fortunate length scaling is only applicable, if the generated short wavelength radiation remains in phase with the generating laser, which is termed as phase matching. If the condition for phase-matching is not fulfilled, the signal grows only along the often very short coherence length. Beside the linear (atomic) dispersion the major contribution to the phase mismatch is the plasma dipersion of free electrons. During the interaction of the laser pulse width the atoms, not only XUV photons are generated, but also lots of free electrons. The shorter the wavelength, the higher must be the laser intensity and the higher is the free electron density and larger the phase mismatch. So a major task in HHG is the development of phase-matching or quasi-phase matching schemas to extend the coherence length and hence the obtained photon flux. In this section we will briefly review the theoretical description of HHG and will present experimental results on approaches to extend the phase matching length.

To describe the process qualitatively, a bound electron can be released in the presence of a strong laser field by tunneling ionization, which can be only treated by applying quantum-mechanic. The subsequent motion of the free electron will be described by the classical equations of motions. The free electron first is accelerated away from its parent ion, and after changing the polarity of the field, it is accelerated toward the ion again. The returning electron can recombine and the excess energy is emitted as a photon with an energy corresponding to the binding energy plus the kinetic energy at the instant of recombination. The highest photon energy, the so called cut-off in the harmonic spectrum corresponds to the maximum energy and can be written:

$$N_c \hbar \omega_0 = W_b + 3.17 U_p \tag{7}$$

Where W_p is the ionization potential, N_c the number of the "cut-off" harmonics and $U_p = \frac{e^2 E_a^2}{4m\omega_0}$ is the ponderomotive potential, with the electron charge e and electron mass m, the frequency ω_0 and amplitude E_a of the laser field. This process is repeated in each optical cycle twice and the signal has a point symmetry compared to time zero. These two findings imply, that the spectrum will only contain the odd harmonics of the laser frequency. At very high intensities, the speed of the electron will be comparable to the speed of light. In this relativistic case, we must also consider the influence of the magnetic field on the motion of the electron. The resulting trajectories did not show a return to its parent ion, so no high harmonic signal will be generated.

However, for a more detailed description and explaining features like conversion efficiency and so on we must rely on a quantum mechanics. The nonlinear response of an atom to intense radiation can be divided into different intensity-dependent processes. At low and moderate intensity, if the external electric field is smaller than the static Coulomb field (perturbative nonlinear optics), the laser field modifies the atomic states: the energy levels are shifted by an energy proportional to electric field strength E_a which is known as Stark shift. The border between the region of the perturbative and nonpertubative nonlinear optics is determined by the following equations:

$$\frac{\chi^{(k+1)}E^{k+1}}{\chi^{(k)}E^{k}} \approx \frac{eE_a a_B}{\hbar\Delta} = \alpha_{bb} << 1\ (Bound-bound\ Transition) \tag{8}$$

$$\frac{1}{\gamma} = \frac{eE_a}{\omega_0\sqrt{2mW_b}} = \frac{eE_a a_B}{\hbar\omega_0} = \alpha_{bf} << 1\ (Bound-free\ Transition) \tag{9}$$

With e the elementary charge, m the mass of the electron, $W_b > \hbar\omega_0$ is the binding energy, $a_B = \frac{\hbar}{\sqrt{2mW_b}}$ the Bohr radius, E_a is the time-dependent amplitude of the linearly polarized radiation, and γ is a scaling parameter, the Keldysh parameter (Brabec et al., 2000). If the laser field is comparable to the binding Coulomb field ("Strong Field" area), the binding potential is deformed and a potential barrier appears. An initially bound electron can now tunnel through this barrier in a fraction of the laser - oscillation - cycle (T_0). Then, the electron follows adiabatically the variation of the optical field. It can move significantly away from its original position and acquires high kinetic energy (Brabec et al., 2000). The production of HH radiation takes place on the return of the electron to its parent ion.

The maximum energy E_c of the generated harmonic photons and the shape of the emitted spectrum critically depends on how and when the ionization takes place. The ionizing electric field is $E(t) = \tilde{E}_a(t)e^{-i\omega_0 t + i\phi_c} + c.c.$ where the electric field with amplitude E ($E_a = 2|\tilde{E}_a|$) and with a frequency of ω deforms the atomic binding potential. Through the formed barrier, the electron can tunnel with a frequency of $\omega_t = \frac{E}{\kappa} = \frac{2E}{E_0}\omega_0$, where $\gamma = \frac{\omega}{\omega_t} << 1$ (Ammosov 1986, Krainov 1997). The dipole apprioximation assumes a linearly polarized laser field, and requires the validity of the following assumptions:

- All bound states except the ground state are neglected.

- The laser intensity is lower than the saturation intensity, so the ground state is not depopulated.
- In the continuum the electron can be treated as free particles. The influence of the potential of the ion is negligible.

For very high intensities, however, the validity of the dipole approximation must be questioned and it is necessary to extend the theory with multi pole effect (Walser et al., 2000) to obtain an adequate description.

The emission of harmonic radiation from single atoms is determined by the $\frac{d^2}{dt^2}\langle\Psi|r|\Psi\rangle$ dipole acceleration arising as the solution of the Schrodinger equation. Here r is the vector of space and ψ is the wave function of the electron. Under the assumption of $\gamma < 1$, the quantum-mechanical expected value of the generation of high frequencies <D(t)> can be represented as a product of three atomic probability amplitudes:

$$\langle D(t)\rangle = \sum_{t_b} \frac{1}{\sqrt{i}} a_{ion}(t_b) a_{prop}(t_b, t) a_{rec}(t) \tag{10}$$

The electron can be released at the time t_b with the probability of $a_{ion}(t_b)$ by tunneling ionization. After that, it propagates in the laser field and returns back at time t with probability $a_{prop}(t_b,t)$ to the origin atom, where fall back to the ground state width the recombination probability of $a_{rec}(t)$ and creates a photon. The sum shows that there are several possible values for t_b. The resulting dipole emission spectrum is discrete and consists only of odd multiples of the fundamental laser frequency ω_0.

The conversion efficiency of high harmonic radiation is defined by the generation processes and the propagation processes. (Lewenstein et al., 1994; Scrinzi et al., 1999). Unfortunately, we do not have easily accessible experimental parameters to optimize the generation process, but there are several methods to enhance the signal by controlling the propagation. As for all coherent processes the signal grows only over the coherence length, i.e. as long as the condition for phase matching is fulfilled. In the following we discuss the major effects limiting the harmonic yield:

- Absorption of the radiation. During the propagation in the gas, the harmonic beam excites the core electrons and is re-absorbed. Although the absorption can be minimized by reducing the gas pressure or shortening the length of the gas jet, but it also reduces the conversion efficiency. It has been also shown that in the short wavelength range the absorption length is much longer than the coherence length and it is no longer the primary limitation. Its influence can be further reduced, by keeping the background pressure as low as possible in the vacuum system for generating and detecting HH radiation.
- Loss of phase matching. There are three effects that contribute to a phase mismatch. First, the phase shift between the HH pulse and the laser pulse by the dispersion of the free electrons background. Most of the freed electrons will not recombine, and remain as free electron background in the interaction region. The resulting coherence length is $L_{fe} = \frac{2\pi c\omega_0}{\omega_p^2(\tau)N}$, where $\omega_p = \sqrt{\frac{e^2 n_e}{m\varepsilon_0}}$ the plasma frequency, depending on the free-electron density n_e. Secondly, the curved wave front of the laser pulse (focused Gaussian beam)

produces a phase factor, which corresponds to a negative contribution to the refractive index. The corresponding coherence length is given by $L_{g,fs} = \frac{\pi z_0}{N} = \frac{\pi^2 w_0^2}{N\lambda_0}$, where z_0 and w_0 are the confocal parameter and the radius of the beam waist. This contribution is particularly large in the vicinity of the focus of the laser beam; therefore it is convenient to place the gas jet either before or after the focus, to minimize this contribution. Thirdly, the dispersion of the refractive index of the gas also causes dephasing. Since the refractive index is very close to 1 for the gas at the laser wavelength, and at the XUV wavelength it is approximately 1, this contribution is small and can be often neglected.

- Defocusing of the laser. During high harmonic generation, the laser pulse creates a free-electron density profile, which causes defocusing. This effect can be compensated with a spatially formed laser pulse profile. (Brabec et al., 2000).

After propagation over a distance longer than the coherence length, the generated x-rays destructively interfere with the already generated beam. Introducing a suitable long gap, where only propagation and phase shift happen, the signal can be enhanced in a subsequent region once again. Compared to the perfectly phase-matched case, quasi-phase-matching (QPM) leads to lower conversion efficiency, but for HH, QPM with its periodic structure is the only choice, because perfect phase-matching cannot be realized. Several implementations of this technique have been realized, including spatially modulated hollow fibers or using a gas jet sequence (Gibson et al., 2003, Paul et al., 2003, Zepf et al., 2007; Seres et al., 2007). Using a sequence of gas jet sequence, QPM has been successfully realized over a wide spectral range around 300 eV (Seres et al., 2007), see figure 4.

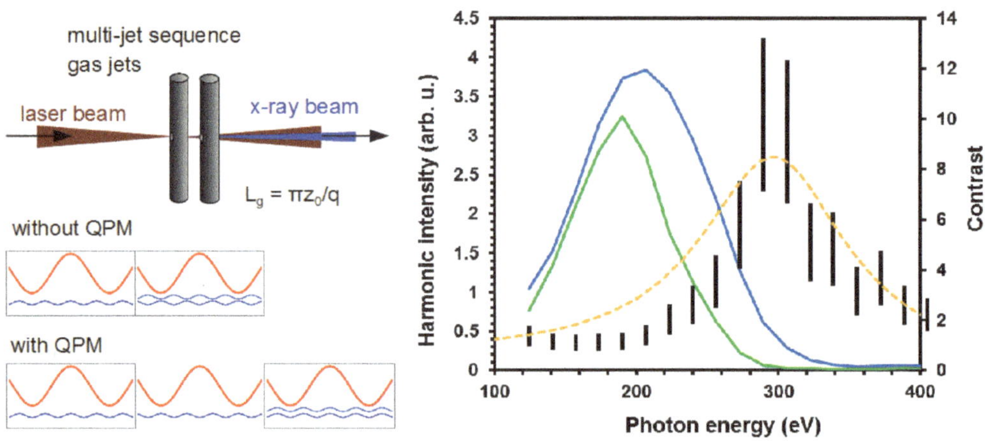

Fig. 4. Principe of the quasi-phase matching at 300 eV using two gas jets. When the two jets are merged (without QPM), the generated x-ray beams interfere destructively in the second jet and the photon yield after the second jet (green) is smaller than it would be after the first jet. When the two jets are moved to a suitable distance (with QPM), the generated x-rays in the second jet constructively interfere with the x-rays generated in the first jet. The photon yield (blue) get larger and the enhancement (black error bars) can be up to one order of magnitude with good agreement with the theory (orange dashed line).

Especially for radiation above 100 eV, the coherence length can be as short as a few microns representing a severe limitation for the obtainable flux. However, the conversion efficiency can be substantially increased by applying the method of nonadiabatic self-phase matching (NSPM) (Seres, 2005, 2006a). As mentioned above the free electron background strongly modifies the phase velocity for the laser beam. As a consequence the radiating atoms are no longer in phase, limiting the achievable flux. Recently it was shown, if the free electron density changes substantially within one optical cycle, the phase velocity of the laser beam will be strongly modified. Choosing the laser parameter right, i.e. the peak intensity should be one order of magnitude above the ionization threshold and the pulses should consist of only few cycles, this ionization induced phase shift cancels the free electron phase shift. This process is known as nonadiabatic self-phase matching (NSPM) (Tempea et al., 2000). With this method it was possible to extend the phase matching length by orders of magnitude even in the keV photon energy range and generate more than 2000th harmonics of Ti:sapphire laser wavelength of 800 nm (1.54 eV) with improved photon yield (Seres et al., 2004; Seres et al., 2005; Seres et al., 2006a). Figure 5 shows a typical spectrum generated in Ne gas with the 12-fs-long, 2.5-mJ pulses of a Ti:sapphire laser system (Seres et al., 2003).

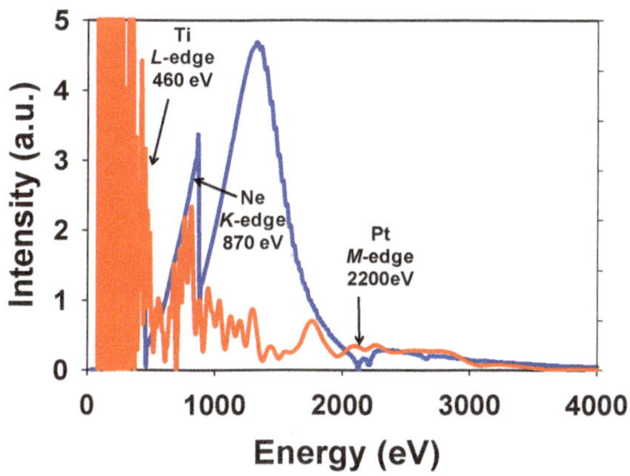

Fig. 5. Measured high harmonic spectrum (red) generated in Ne gas with the 12-fs-long, 2.5-mJ pulses of a Ti:sapphire laser system. The absorption edges of the used Ti foil, the used Ne gas and the Pt coating of the x-ray grating are well recognizable on both the measured spectrum and the calculated transmission curve (blue).

5. Time resolved x-ray spectroscopy

There is a broad spectrum of ultrafast processes in nature. For an experimental investigation we trigger the ultrafast process, i.e. we introduce an instantaneous structural change, with an ultrashort light pulse, the pump pulse. The characteristic of the material is probed by another light pulse with an adjustable delay in respect to the pump pulses. In an conventional pump-probe experiment both pulses are in the visible optical range. The development of reliable femtosecond solid-state laser brought new possibilities into time-resolved spectroscopy (Zewail 2000). For the first time it became possible, in principle to monitor the nuclear motion of molecules, crystal lattices and other out-of-equilibrium structures. However, usually it is very difficult to map the experimental observations to the structural dynamics. Therefore, experimental approaches are needed which can overcome the limitation of optical studies for structural determination, while the high temporal resolution of femtosecond lasers is maintained. Techniques such as X-ray diffraction (XRD) (Rousse 2001), X-ray absorption spectroscopy (Nakano 1999), or X-ray photoelectron spectroscopy deliver much more direct information about the structure. The key to the successful realization is the development of laser driven x-ray sources. Various schemes have been demonstrated but all of them rely on availability of state of the art femtosecond solid-state laser system. In this section we will describe our apparatus, which allowed for the first time time-resolved x-ray absorption spexctroscopy with femtosecond resolution.

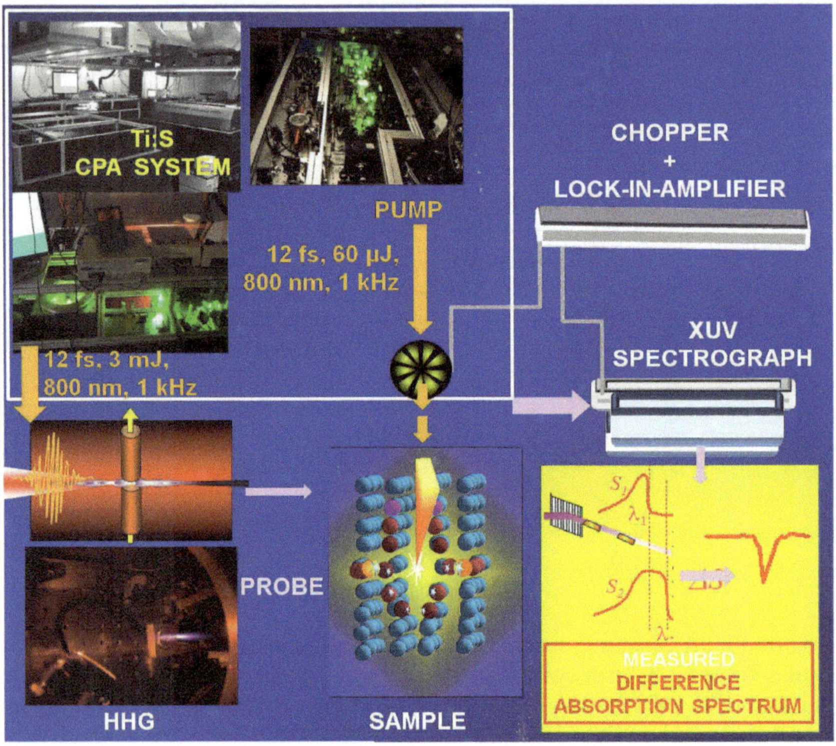

Fig. 6. Experimental setup for time-resolved x-ray absorption spectroscopy. A Ti:sapphire based CPA laser systems delivers energetic ultrahsort pulses. In a gas jet the laser pulses upconverted into the XUV range via high harmonic generation. The sample is pumped by a visible laser pulse. The structural changes are probed with a delayed XUV pulse. With a lock in amplifier and a spectrometer we can measure the XUV difference absorption spectrum, i.e. the change of the spectrum between the pumped and unpumped case.

Our pump-probe experimental setup is based on a state-of-the art multi pass Ti:sapphire CPA amplifier system. The front-end of the system is a mirror dispersion controlled Ti:sapphire oscillator delivering 10 fs, 5 nJ pulses. After a gentle stretching of the pulses to few picoseconds the pulses are amplified in a ten pass amplifier to amplifier to more than 1 mJ more than 1mJ at a repetition rate of 1 kHz. The pulses are compressed with double prism compressor to less than 30 fs and a final energy of 800 μJ. Such short pulses after compression are only possible by carefully compensating the residual high order dispersion with chirped mirrors and specially designed filters to counterfeit gain narrowing. Further spectral broadening is obtained with a hollow fiber filled with Ne. The output spectrum has a width supporting pulses in the order of 10 fs.

The phase and amplitude of the broadened pulses are shaped with an acousto-optic-modulator ("DAZZLER"), which proved later to be an indispensable tool for high harmonic optimization. The broadened and shaped pulses are launched into a second and third amplifier stage and finally compressed with triple-prism-compressor (Seres et al., 2003). After the third amplifier stage the compressed pulsesare 12 to 15 fs long and the pulse energy is 3 mJ. Adding an additional compressor stage consisting either of a filament (Seres et al., 2007) or a gas filled hollow-core fiber (Seres 2006a) and chirped mirrors we were able to compress the pulses down to 5–6 fs with energy in the range of 1–2 mJ. These laser pulses were up-converted in a noble gas jet into XUV via high harmonic generation. For maximizing the photon yield we applied quasi phase matching (Seres 2007) and NSPM (Tempea 2000, Seres 2004)

For the quasi-phase matching experiments we need intense pulses as short as possible. The output pulses of this system were spectrally broadened in an Ar filament and compressed with chirped mirrors to about 6 fs at pulse energies of 1.5 mJ. With a spherical mirror the pulses were focused into a sequence of two gas filled nickel tubes with a diameter of 0.2 mm. With a backing pressure of 80 mbar in He, the highest HH yield in the 400–600 eV spectral range was reached with a jet distance of 1.3 mm (Seres 2007).

For non-adiabatic technique, the 3 mJ, 12 fs laser pulses were strongly focused into a Nickel tube filled with Ne or He reaching an intensity of 2×10^{16} W/cm^2. The laser beam interacted with He atoms in 0.4-mm-long volume. The fast ionization gave the proper addition to the phase matching terms, so we could reach approximately the 2000th harmonics of the laser frequency corresponding to photon energies in the order of 3500 eV (Seres 2006a, 2006, Seres 2004). The broad spectrum of the x-ray pulses is well suited for static x-ray absorption spectroscopy. Different thin foils were inserted (200-nm-thick copper, 1-μm-thick aluminum, 300-nm-thick silicon) into the x-ray beam and the transmitted spectra were recorded with a scanning x-ray monochromator (248/310G, McPherson) equipped with a photoelectron multiplier Channeltron® 4715G. The absorption edges of Cu, Al, and Si at 0.94, 1.5, and 1.8 keV were clearly resolved, respectively. The shape of the measured spectra below and slightly above the absorption edge contains no information, which is of interest in our study.

5.1 Time resolved x-ray spectroscopy with HH radiation in material science

In this section we report on the generation of soft x-ray pulses via high harmonic generation and their first use for time resolved x-ray absorption spectroscopy (XAS) to investigate the structural dynamic of amorphous silicon with a temporal resolution of about 20 fs. To our knowledge this is the highest temporal resolution ever demonstrated in XAS (Seres E. 2008).

To tackle time-resolved XAS in the soft x-ray regime the light source must meet the following requirements: a) it must provide continuum radiation, b) it must provide ultrafast pulses, and c) it should have a sufficient photon flux (Bressler 2004). We realized it via high harmonic generation. HHG is a line radiation, and therefore of limited use for XAS. However, using very short driving laser pulses the line spectrum becomes continuous near the cut off (Brabec 2000). Due the generation process the XUV pulses are also always shorter

than the driving laser pulses. The short pulse duration and the excellent spatial and spectral characteristic make HHG based sources well suited for time-resolved XAS.
Our pump-probe experimental setup based on a single-stage Ti:sapphire CPA amplifier system. Most of the energy of the output beam was tightly focused with a broadband mirror with a focusing length of 150 mm into a Ne gas jet at an intensity of about 10^{15} W/cm^2. The XUV radiation hits our sample, which is a 100 nm thick silicon film, consisting of randomly oriented micro-crystallites (amorphous silicon a-Si). The transmitted beam was launched into a scanning grazing incidence monochromator the output of which was connected to a lock-in amplifier. The laser and gas jet parameters have been optimized to maximize the signal at around 100 eV, where we wanted to study dynamical structure modifications of silicon via changes of absorption near the L-edge. The signal has been safely above the noise level up to energies of about 500 eV opened the way to EXAFS (Rehr 2000). A small fraction of the output beam energy was delayed and focused onto the sample obtaining a pump fluence nearly two orders of magnitude below the damage threshold. Due to chopping of the pump beam and using a lock in amplifier it was now easily possible to detect changes of the transmitted spectra as small as a 10^{-4} .

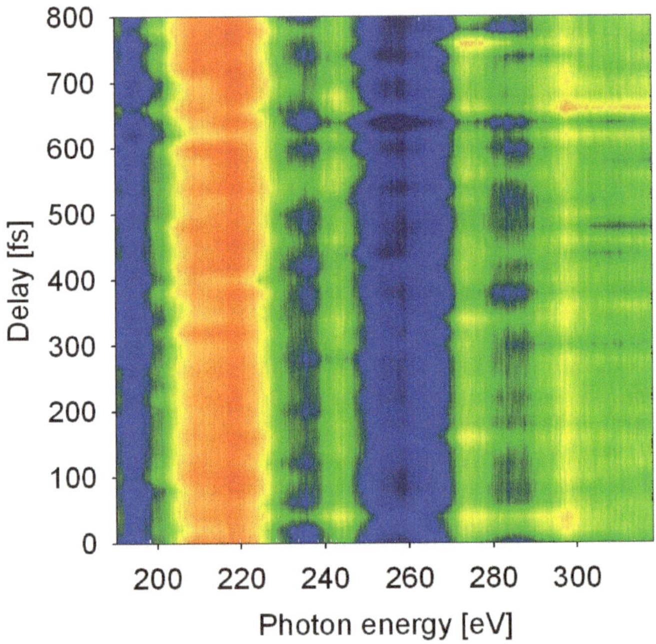

Fig. 7. Measured x-ray absorption difference spectra above the silicon L- edge after excitation with an ultrashort laser pulse. The delay has been varied in 20 fs steps in a range up to to 800 fs. From these data we can calculate via EXAFS the evolution of the interatomic distances.

To follow structural changes we opted for EXAFS (Oguri 2005, Brown 1999). From the measured absorption curves, the EXAFS signals $\chi(k)$ were calculated as a function of photoelectron wave number (k given in Å^{-1}) and from the measured absorption coefficient, the slowly varying contribution of the single atom absorption have been

eliminated with a high pass filter. After applying an amplitude window function and weighting the data with k^2, the atomic distance has been evaluated from the power density spectrum. The estimated interatomic distances are 2.20 ± 0.02 Å for Si which agrees very well with measurements at synchrotrons (Glover 2003). These measurements clearly demonstrate that it is possible generate HH signal up to several keV and the photon flux is sufficient for x-ray absorption spectroscopy. These results are the basis for our subsequent experiments, which made it possible to follow atomic motions in disorder materials with time resolved EXAFS.
The x-ray absorption spectra have been recorded in delay range of -300 fs to 800 fs in steps of 20 fs. The evaluation of the EXAFS signal gives the averaged atomic distances at the instant of probing. The calculated atomic distance shows a fast and slow oscillatory motion as a function of the delay. We fitted to the data the sum of two sinusoidal waves and found the best agreement for frequencies of 3.1 THz and 17 THz, respectively (Seres E 2007, Seres E 2008). As a check of the reliability of our evaluation and to minimize the influence of amplitude and phase noise we calculated the two dimensional FT of the whole data set. The evaluated spectrum shows again two distinct maxima at 3.6 THz and 17 THz. With two independent evaluation methods namely two-dimensional FT and least square fitting we identified two dominating oscillations at the same frequencies, making us confident that we have observed atomic oscillations. Moreover these frequencies agree very well with the predicted numbers for coherent phonons in Si after laser pulse excitation and have been published recently (Seres 2006, 2008, 2009).

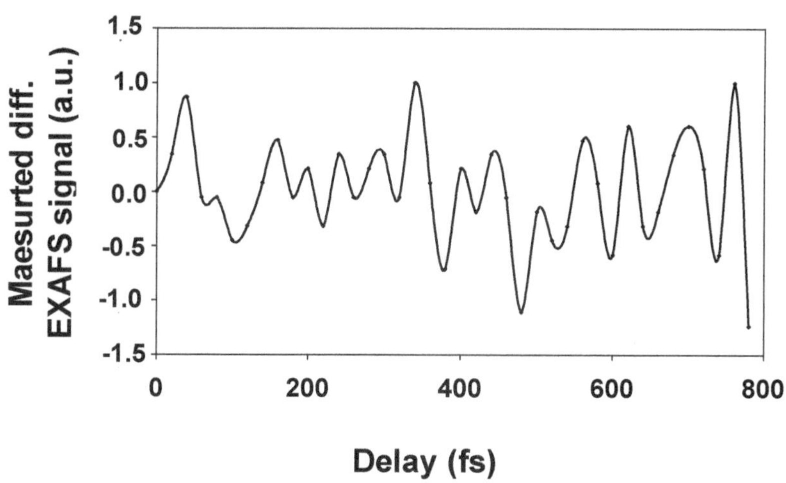

Fig. 8. The evaluated time resolved EXAFS signal at the Si L edge.

5.2 Changes in the atomic electronic levels, orbits

Most electronic devices have yet a limited temporal resolution, which prevent to detect fast electronic, magnetic changes on the atomic or molecular scale. But these changes come always together with the changing of the optical parameters in the x-ray regime. The absorption, reflection, or even the polarization of the x-ray light are sensitive to the electronic or magnetic states. Exciting a core electron from the valence level into the conduction level, the absorption cross-section is influenced by the changed density-of state (DOS) (Figure 9).

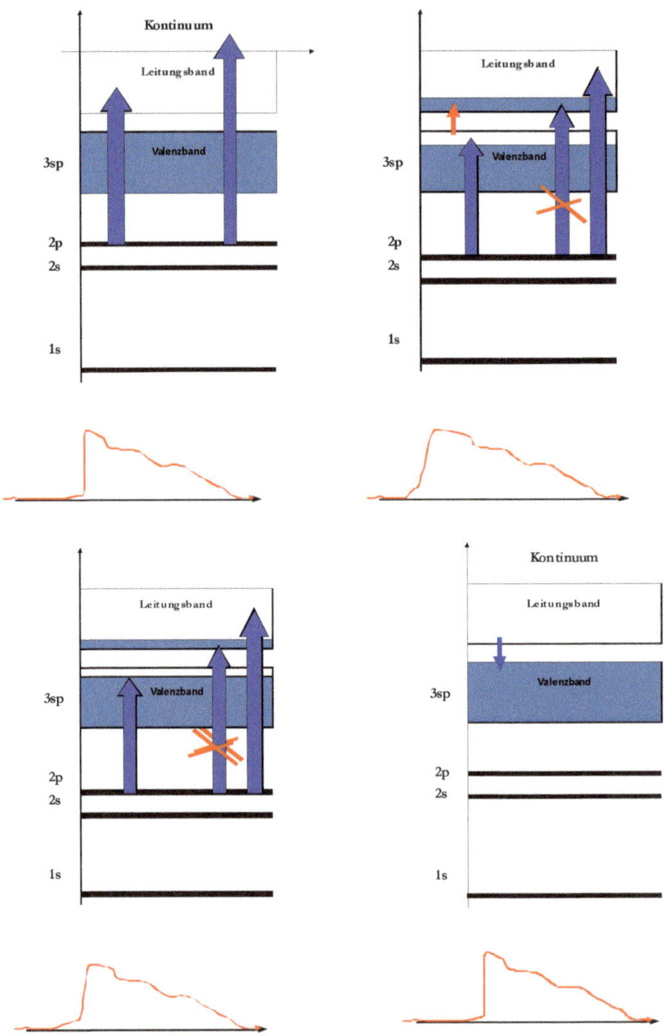

Fig. 9. The form of the absorption edge follows the changes of the electronic states produced by excitation.

A similar setup as for the time resolved EXAFS can record not only the atomic motion but the electronic changes after the short pump laser pulse. The laser pump light is absorbed through single and two photon absorption, creating a non-thermal electron distribution in the conduction band. The hot electron distribution thermalizes over different channels and time scales such as fast inter band thermalization via electron–phonon scattering and finally on a longer time-scale via recombination. The excitation of the electrons creates occupied states in the conduction band and unoccupied states in the valence band. These modifications have their signature also in the fine-structure of the soft-x-ray absorption spectrum (Nakano 1999).

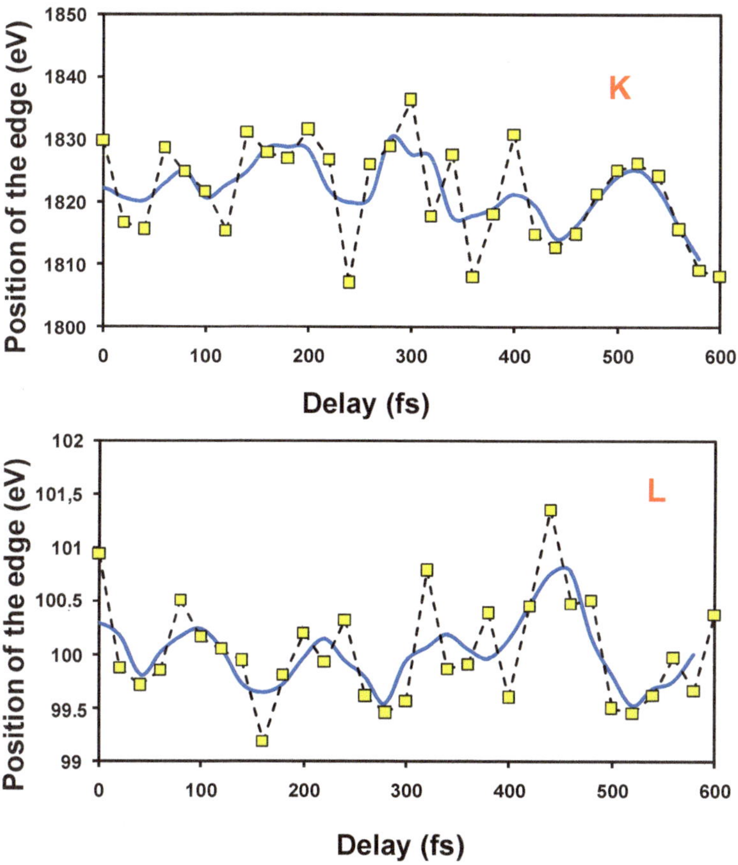

Fig. 10. Measured shift of the position of the (a) K- edge (1.8k eV) and (b) L-edge (100 eV) of silicon after excitation with an ultrahsort laser pulse (Seres 2009).

To investigate the carrier dynamic we recorded the difference absorption spectra in the vicinity of the L and K-edge of silicon as a function of the delay. The measured difference spectrum has been modified in very complex way and a full analysis would require a detailed knowledge about the exact band structure. To gain some information about the

relevant time constants we calculated the expectation value of the difference signal near the edge. The position of the L and K absorption edge of silicon oscillate with different amplitude and frequency as shown in Fig. 10. From a long-range measurement we have identified a fast and a slow time constant. The fast time constant is in the order of approx. 200 fs and the slow one in the order of 50 ps (Seres 2005). Comparing our findings with experiments based on conventional optical spectroscopy of silicon the fast time constant of about 200 fs is in reasonable agreement with the previously observed electron-phonon relaxation time in a-Si. The longer time constant of about 80ps corresponds to the carrier recombination time of electron and holes across the Si band gap (Sundram 2002).
Our pulsed XUV source delivering sub-20 fs pulses in an energy range up to 3 keV is bright enough for time-resolved XANES experiments. In a series of proof-of-principle experiments we have studied the electronic dynamics of the amorphous silicon bove the Si L edge and the C K edge, respectively. Further measurements are necessary to explore the underlying physics in full detail.

6. Acknowledgements

Financial support from the Deutsche Forschungsgemeinschaft (grant SE 1911/1-1 and TR 18, P10), the State of Thuringia (grant TMBWK B 715-08008) and the Friedrich Schiller University (grant ProChance 2009 A1) is gratefully acknowledged. We would like to express special thanks to all colleagues who have contributed to this work by supporting us with discussions and material.

7. References

Agarwal, B. K. (1991). *X-Ray Spectroscopy* Springer Verlag, Berlin

Ammosov, M. et al. (1986). Tunnel ionization of complex atoms and of atomic ions in an alternating electromagnetic field. *Sov. Phys. JETP* Vol. 64, pp. 1191-1194

Antoine, P. et al., (1996). Generation of attosecond pulses in macroscopic media. *Phys. Rev. A*, Vol. 56, No. 6 (December 1996), pp. 4960-4969

Bianconi, A et al., (1987). Multiple-scattering effects in the K-edge x-ray-absorption near-edge structure of crystalline and amorphous silicon. *Phys. Rev. B.* Vol. 36, No. 12, (October 1987) pp. 6426-6433

Brabec T. and Krausz, F. (2000). Intense few-cycle laser fields: Frontiers of nonlinear optics. *Rev. Mod. Phys.* Vol. 72, No. 2, (April 200), pp. 545-591

Bressler, Ch. et al., (2002). Towards structural dynamics in condensed chemical systems exploiting. *J. Chem. Phys.*, Vol. 116, No. 7, (February 2002), pp. 2295-2966

Bressler, Ch., Chergui M. (2004). Ultrafast X-ray Spectroscopy *Chem. Rev.*, Vol. 104, No. 4, pp. 1781-1812

Brown, F. L. H., et al. (1999). Ultrafast extended x-ray absorption fine structure (EXAFS)-theoretical considerations *J. Chem. Phys.*, Vol. 111, No. 14, (October 1999), pp. 6238-6246

Bzowski, S. A., & Sham T. K. (1993). Pd-Ti bimetallic: A study of the electronic structure using X-ray photoelectron spectroscopy and x-ray-absorption near-edge structure *Phys. Rev. B.* Vol. 48, No.11, (September 1993), pp. 7836-7840

Christov, I. P. et al., (1997). High-Harmonic Generation of Attosecond Pulses in the "Single-Cycle" Regime. *Phys. Rev. Lett.* 78, No. 7, (February 1997), pp. 1251-1254

Constant e., et al., (1999). Optimizing High Harmonic Generation in Absorbing Gases: Model and Experiment. *Phys. Rev. Lett.* Vol. 82, No. 8, (February 1999), pp. 1668-1671

D'Angelo, P. et al., (1996). Multi electron excitations at the L edges of barium in aqueous solution. *Phys. Rev. B.* Vol. 54, No. 17, (November 1996), pp. 12129-12138

Da Silva, L. B. et al., (1994). Generation of a 45-ps-duration 15.5-nm x-ray laser. *Opt. Lett.* Vol 19, No. 19, pp. 1532-1534

Ditmire, T. et al., (1995). Energy-yield and conversion-efficiency measurements of high-order harmonic radiation. *Phys. Rev. A*, Vol. 51, No. 2, (February 1995), pp. R902-R905

Faraci, G., et al., (1997). XANES of high-pressure Kr clusters in Be and Si. *Phys. Rev. B.*, Vol. 56, No. 19, (November 1995), pp. 12553-12559

Farges, F., et al.,(1997). Ti K-edge Xanes studies of Ti coordination and disorder in oxide compounds: Comparison between theory and experiment. *Phys. Rev. B.* Vol. 56, No. 4, (July 1997), pp. 1809-1819

Filiponi, A., et al., (1989). Structural investigations of a-Si and a-Si:H using X-ray-absorption spectroscopy at the Si K edge. *Phys. Rev. B.* Vol. 40, No. 14, (November 1989), pp.9636-9643

Gibson, E. A., et al. (2003). Coherent Soft X-ray Generation in the Water Window with Quasi-Phase Matching *Science,* Vol.302. No.3. (October 2003),8may 1995), pp. 95-98

Giulietti, D., & Gizzi, L. A. (1998). X-Ray Emission from Laser Produced Plasmas. *La Rivista del Nuovo Cimento,* Vol. 21, No.1, pp.1-103

Glover, C. J., Foran, G. J. & Ridgway, M.C. (2003). Structure of amorphous silicon investigated by EXAFS. Nucl. *Instr. And Meth. In Phys. Res. B,* Vol. 199, pp. 195-199

Gordon C. L. et al., (1995). Time-gated imaging with an ultrashort-pulse, laser-produced-plasma x-ray source, *Opt. Lett.,* Vol. *20*, No. 9, (Mai 1995), pp. 1056-1058

Johnson, S. L. et al. (2003). Properties of Liquid Silicon Observed by Time-Resolved X-Ray Absorption Spectroscopy. *Phys. Rev. Lett.,* Vol. 91, No. 15, (October 2003), pp. 157403

Kan, C. et al., (1997). Coherent XUV Generation from Gases Ionized by Several Cycle Optical Pulses. *Phys. Rev. Lett.* Vol. 79, No. 16, (October 1997), pp. 2971-2974

Kondo K. et al., (1993). Wavelength dependence of nonsequential double ionization in He. *Phys. Rev. A,* Vol. 48, No. 4, (October 1993), pp. R2531-R2533

Korobkin D., (1998). Soft X-Ray Amplification at 26.2 nm with 1-Hz Repetition Rate in a Table-Top System. *Phys. Rev. Lett.* Vol. 81, No. 8, (August 1998), pp. 1607-1610

Krainov, V. P. (1997).Ionization rates and energy-angular distributions at the barrier-suppression ionization of complex atoms and atomic ions *JOSA B* Vol. 14, No. 2, pp. 425-431

L'Huillier, A. & Balcou Ph. (1993). High-order harmonic generation in rare gases with a 1-ps 1053-nm laser. *Phys. Rev. Lett.* Vol. 70, No. 6, (February 1993), pp. 774-777

La-O-Vorakiat, C., et al., (2009). Ultrafast Demagnetization Dynamics at the M Edges of Magnetic Elements Observed Using a Tabletop High-Harmonic Soft X-Ray Source. *Phys. Rev. Lett.* Vol. 103, No. 25, (December 2009), pp. 257402

Lee, P. A. et al., (1981). Extended x-ray absorption fine structure - its strength and limitations as a structural tool. *Rev. Mod. Phys.* Vol.53, No. 4, (October 1981), pp. 769-806

Lemoff B. E. et al., (1995). Demonstration of a 10-Hz Femtosecond-Pulse-Driven XUV Laser at 41.8 nm in Xe IX. *Phys. Rev. Lett.* Vol. 74, No. 9, (February 1995) pp. 1574-1577

Lewenstein, M. ,et al. (1994). Theory of high-harmonic generation by low-frequency laser fields *Phys. Rev. A,* Vol. 49, No.3, (March 1994), pp. 2117-2132

Lytle, F. et al.,(1975). Theory of the extended x-ray-absorption fine structure II. Experimental Results and selected results. *Phys. Rev. B.* Vol. 11, No. 12, (March 1975), pp. 4825-4835

Macklin et al., (1993). High-order harmonic generation using intense femtosecond pulses. *Phys. Rev. Lett. Vol.* 70, No. 6, (February 1993), pp. 766-769

Michette A. and S. Pfautsch 1996, X-rays, the first hundred years, Wiley & Sons.

Murnane, M. M., et. al., (1991). Ultrafast X-ray Pulses from Laser-Produced Plasmas. *Science* Vol. 251,No. 4993, pp. 531-536.

Nagata Y. K., et al., (1993). Soft-x-ray amplification of the Lyman-α transition by optical-field-induced ionization. *Phys. Rev. Lett.* Vol. 71, No. 23, (December 1993), pp. 3774-3777

Nakano, H., et al., (1999). Time-resolved soft x-ray absorption spectroscopy of silicon using femtosecond laser plasma x rays. Appl. Phys. Lett. Vol. 75, No. 16, (October 1999), pp. 2350-2352

Nickles, P. V. et al., (1997). Short Pulse X-Ray Laser at 32.6 nm Based on Transient Gain in Ne-like Titanium. *Phys. Rev. Lett.* Vol. 78, No. 14, (April 1997) pp. 2748-2751

Oguri, K. et al.(2005). Transient observation of extended x-ray absorption fine structure in laser melted Si by using femtosecond laser-produced-plasma soft x ray. *Appl. Phys. Lett.* Vol. 87, (July 2005), pp. 011503

Paul, A. et al., (2003). Quasi-phase-matched generation of coherent extreme-ultraviolet light. *Nature* Vol. 421, (January 2003), pp. 51-54

Perry, M. D., & Crane J. K. (1993). High-order harmonic emission from mixed fields. *Phys. Rev. A,* Vol. 48, No. 6 (December 1993), pp. R4051-R4057

Raksi, F., et al.,(1996). Ultrafast x-ray absorption probing of a chemical reaction. *J. Chem. Phys.* Vol. 104, No. 15, (April 1996), pp. 6066-6069

Rehr J. J. & Albers R. C. (2000). Theoretical approaches to x-ray absorption fine structure. *Rev. Mod. Phys.* Vol. 72, No. 3, (July 2000), pp. 545-591

Rehr J. J. et al., (1992) High-Order-Multiple-Scattering Calculations of X-Ray-Absorption Fine Structure *Phys. Rev.Lett.,* Vol. 69, No. 23, (December 2002), pp. 3397-3400

Rischel, C., A., et al., (1997). Femtosecond time-resolved X-ray diffraction from laser-heated organic films . *Nature,* Vol. 390, (September 1997), pp. 490-492

Rocca, J. J. et al., (1994). Demonstration of a Discharge Pumped Table-Top Soft-X-Ray Laser, *Phys. Rev. Lett.* Vol. 73, No. 16, (October 1994), pp. 2192-2195

Rose-Petruck, C., et al., (1999). Picosecond-milliangstroem lattice dynamics measured by ultrafastX-ray diffraction. *Nature* Vol. 398,(25 March 1999), pp. 310-312

Rousse, A., et al. (2001). Femtosecond x-ray crystallography. *Rev. Mod. Phys.* Vol. 7317, No.1 pp. 17-31

Salieres, S. et al., (1995). Coherence Control of High-Order Harmonics. *Phys. Rev. Lett.* Vol. 74, No. 19, (Mai 1995), pp. 3776-3779

Salieres, S. et al., (1999). Study of the spatial and temporal coherence of high-order harmonics. *Adv. Atom. Mol. Opt. Phys.* Vol. 41, pp. 83-142

Schnürer, M. et al., (1999). Absorption-Limited Generation of Coherent Ultrashort Soft-X-Ray Pulses. *Phys. Rev. Lett.* Vol. 83, No. 4 (July 1999), pp. 722- 725

Schnürer, M., et al., (1998). Coherent 0.5keV x-ray emission from helium driven by a sub-10 fs laser. *Phys. Rev. Lett.* Vol. 80, No. 15, (April 1998) 3236.

Scrinzi, A., et al. (1999). Ionization above the Coulomb Barrier. *Phys. Rev. Lett.* Vol. 83, No.4, (July 1999), pp. 706-709

Seibert, M. M. et al., (2010). Femtosecond diffractive imaging of biological cells. *J. Phys. B,* Vol. 43, (September 2010), pp. 194015

Seres E. (2005) PhD Thesis, University of Würzburg

Seres, E., & Spielmann, C. (2007). Ultrafast soft x-ray absorption spectroscopy with sub-20-fs resolution. *Appl. Phys. Lett.* Vol. 91, No. 12, (September 2007), pp. 121919

Seres, E. et al. (2006). X-ray absorption spectroscopy in the keV range with laser generated high harmonic radiation. *Appl. Phys. Lett.* Vol. 89, (November 2006), pp. 181919

Seres, E. et al., (2003). Sub-10-fs, terawatt-scale Ti:sapphire laser system. *Opt. Lett.* Vol. 28, No. 19. (October 2003). Pp. 1832-1834

Seres, E. et al., (2004). Generation of Coherent Soft-X-Ray Radiation Extending Far Beyond the Titanium L Edge. Phys. Rev. Lett. Vol. 92, No. 16, (April 2004), pp. 163002

Seres, E. Seres, J. & Spielmann, Ch. (2009). Time resolved spectroscopy with femtosecond soft-x-ray pulses. Appl. Phys. A, Vol. 95, pp. 43-50

Seres, E., & Spielmann, Ch. (2008). Time-resolved optical pump X-ray absorption probe spectroscopy in the range up to 1 keV with 20 fs resolution. *J. Mod. Opt.* Vol. 55, No. 16, (September 2008), pp. 2643–2651

Seres, J. et al. (2006a). Generation of coherent keV x-rays with intense femtosecond laser pulses. *New. J. Phys.* Vol. 8 , (October 2006), pp. 251

Seres, J., et al. (2005). Source of coherent kiloelectronvolt X-rays. *Nature,* Vol., 433, (February 2005), pp. 596

Seres, J., et al.(2007). Coherent superposition of laser-driven soft-X-ray harmonics from successive sources. *Nature Phys.* Vol.3, (December 2007), pp. 878-883

Sipr, O., (2002). Spatial Distribution of photoelectrons participating in formation x-ray absorption spectra *Phys. Rev. B.* Vol. 65, No. 20, (Mai 2002), pp. 205115

Sommerer, G. rottke H., & Sandner W. (1999). Enhanced Efficiency in High-Order Harmonic Generation Using Sub-50-fs Laser Pulses. *Laser Phys.* Vol. 9, No. 1, pp. 430-432

Spielmann, Ch. et al, (1998). Ultrafast near-keV coherent x-ray sources pumped by sub-10fs lasers, IEEE JSTQE 4, 249

Spielmann, Chet al.,(1997). Generation of coherent x-ray in the water window using 5-fs laser pulses. *Science,* 278, 661.

Stern, E. A., (1974). Theory of the extended x-ray-absorption fine structure. *Phys. Rev. B.*Vol. 10, No. 8,(October 1974),pp. 3027-3037 (1974)

Sundaram, S. K. & Mazur, E. (2002). Inducing and probing non-thermal transitions in semiconductors using femtosecond laser pulses. *Nature Mater. Vol.* 1, (December 2002),pp. 217-224

Tempea, G., et al. (2000). Self-Phase-Matched High Harmonic Generation. *Phys. Rev. Lett.*, Vol.84, No. 19,(Mai 2000), pp. 4329-4332

Umstadter et al., (1998). Tabletop, Ultrahigh-Intensity Lasers: Dawn of Nonlinear Relativistic Optics.Opt. &Photon. News Vol. 9. Iss. 7.pp. 40-

Wahlstroem, C.-G., (1993). High-order harmonic generation in rare gases with an intense short-pulse laser. *Phys. Rev. A*, Vol. 48, No. 6, (December 1993), pp. 4709-4720

Walser, M. M., et al., (2000). High Harmonic Generation Beyond the Electric Dipole Approximation. *Phys. Rev. Lett.*, Vol. 85, No. 24, (December 2000), pp. 5082-5085

Woerner, H. J., et al. (2010). High-Harmonic Homodyne Detection of the Ultrafast Dissociation of Br_2 Molecules. *Phys. Rev. Lett.* Vol. 105, (September 2010), pp. 103002

Zepf, M. et al., (2007). Bright Quasi-Phase-Matched Soft-X-Ray Harmonic Radiation from Argon Ions. *Phys. Rev. Lett.* Vol. 999, (October 2007), pp. 143901

Zewail, A. H. (2000). Femtochemistry: Atomic-Scale Dynamics of the Chemical Bond. *J. of Phys. Chem.* A, Vol.104,(Mai 2000), pp. 5660-5694

Zhang, J. et al., (1997). A Saturated X-ray Laser Beam at 7 Nanometers. *Science* vol. 276, no. 5315, (Mai 1997) pp. 1097-1100

Zhou J. et al., (1996). Enhanced High-Harmonic Generation Using 25 fs Laser Pulses. *Phys. Rev. Lett.* Vol. 76, No.5, (January 1996), pp. 752-755

10

Interaction of Atom with Laser Pulses of Intra-Atomic Field Strength

A.V. Andreev, S.Yu. Stremoukhov and O.A.Shoutova
Physics Department, M.V. Lomonosov Moscow State University
Russia

1. Introduction

In spite of the twenty-year history, the effect of the high-order optical harmonic generation (HHG) is still under great interest of both experimentalists and theoreticians (see review (Ganeev, 2009)). The origin of this interest is manifold. From the practical point of view, the HHG is one of the effective mechanisms for producing coherent emission in broad range of electromagnetic wave spectrum. The plateau in the harmonic amplitude distribution in extreme ultraviolet (XUV) region affords grounds for development of subfemtosecond pulse formation methods (Paul, 2001). As a result, the new frontiers are opened up in science by extending the nonlinear optics and time-resolved spectroscopy to the XUV region (Papadogiannis, 2003) and pushing ultrafast science to the attosecond domain, enabling XUV spectroscopy and imaging of molecular orbitals (Itatani, 2004), surface dynamics (Tobey, 2007), and electron motion. Hence, the HHG effect opens up the new perspectives in attosecond science (Corkum, 2007; Sola, 2006).

At present days, the efficiency of conversion to high-order harmonics turns out to be insufficient for using them as real coherent short-wavelength radiation sources in biology, plasma diagnostics, medicine, microscopy, photolithography, etc. Hence, the search for ways of increasing the cut-off frequency and HHG efficiency in the XUV spectral range is still among the most topical problems of nonlinear optics.

The HHG effect was observed with the large number of periodic table elements having usually small and middle atomic numbers (Ganeev, 2009; 2005; 2007a;b; Redkin, 2010). As a rule, the interaction medium is gas jet, cell or laser plasma which is prepared by irradiation of metal surface by laser pre-pulse. The maximum harmonic order, or cut-off frequency (CF), obtained in plasma media to date varies from the sixties to seventies harmonics of fundamental frequency (Redkin, 2010; Suzuki, 2007). The highest-order harmonics (the 101st harmonic, λ =7,9 nm) have been obtained in manganese plasmas (Ganeev, 2007b). The efficiency of conversion in the plateau region amounts to $\sim 10^{-5}$ (Ganeev, 2007a). Recently, in experiments with silver plasma the CF values approximating seventies orders have been obtained (Ganeev, 2005). This value depends on both atomic levels structure and laser pulse parameters (intensity, energy and duration, envelope time-dependence, carrier-envelope phase). CF extension into higher frequencies band promises new possibilities in creating X-ray coherent sources and so the study of its behavior in dependence on media features and external laser pulse parameters is the problem of significant scientific and technical interest.

From the general point of view, it is clear that the emission spectrum is sensitive both to the spectral composition of the laser field and its polarization structure. At the beginning of the 1990s, the method of two-pulse and two-frequency HHG spectrum control was proposed in (Watanabe, 1994; Yin, 1992), and the idea of polarization control was seemingly first introduced in (Corkum, 1994). Recently, it has been shown that the use of the two-color schemes, where two pulses with fundamental and double frequencies (ω and 2ω) interact simultaneously with atomic or molecular media, enable to enhance significantly the power of THz emission (Bartel, 2005; Cook, 2000; Kress, 2004) in comparison with the single pulse schemes (Hamster, 1994; 1993; Sprangle, 2004). In the latter case the ultrashort pulses of the high intensity are usually used. Earlier theories suggest the basic mechanism of THz emission is based on the four-wave-mixing rectification (FWMR) process in laser induced plasma (Gorbunov, 1996; Sprangle, 2004). This phenomenological models, formulated in terms of plasma nonlinear susceptibilities, have been also used to interpret the results of experiments based on the two-color schemes.

Up to now, there have been developed a number of different theoretical models to describe the dynamics of atomic electron motion in strong laser field. These models are based on different approximations, which are usually valid in the restricted area of laser pulse field strength. The intra-atomic field strength, $E_{at} = e/a_B^2$ (where a_B is the Bohr radius), is used as unit for measure of field strength. The approximations are different for subatomic, $E_0 \ll E_{at}$, and overatomic, $E_0 \gg E_{at}$, fields. The unified theory applicable for both weak and strong laser fields has not yet been developed. Indeed, let us make some short review of the proposed theoretical approaches. The foundation of the atom ionization theory has been laid by the paper of Keldysh (Keldysh, 1965). In the frame of this theory it is assumed that the atom has one bound state and the wave functions of continuum are calculated in quasiclassical approximation. The Keldysh theory has got further development in the series of papers (Perelomov, 1967a;b; Popov, 1968). The theory developed in these papers is known as the Perelomov–Popov–Terent'ev (PPT) model. Similar approach based on the calculation of the matrix element of the transition from the initial bound state of a system belonging to the discrete spectrum to the final state described by the Volkov wave function was developed in (Parker, 1990). The approach proposed in (Faisal, 1973; Keldysh, 1965; Reiss, 1980) is usually called the Keldysh–Faisal–Reiss (KFR) approximation. The model known as the strong-field approximation (SFA) was proposed and developed in the series of papers by Reiss (Reiss, 1990; 1992; 1980). In contrast to the Keldysh's theory, this approximation does not use the saddle point method in calculating the matrix element of the transition from initial atomic state to ionized state of the Volkov continuum and, hence, it does not need an assumption that the photon energy is much smaller than electron binding energy. The Kramers-Henneberger model developed in (Kulander, 1991; Marte, 1991; Pont, 1990) is the model of dressed atom, the ionization potential of which decreases with laser field strengthening.

Among others there is a method of direct numerical solution of time-dependent Schrödinger equation (TDSE). The first numerical calculations for the case of hydrogen atom have been done in (Krause, 1992). Later this approach was successfully applied in studies of the one- and multi-photon ionization of the different hydrogen-like atoms (see e.g. (Rae, 1994)) and during this time undergo certain improvement getting more and more sophisticated (Bauer, 2006; Muller, 1999). With rapid progress in computer technique there appeared a conception, that this method is the most effective one in studies of light-atom interactions at strong laser field regime, which could serve a criterion of rightness for other theoretical approaches. But this seems to be a delusion. There are some reasons for that. Firstly, any computer simulation inevitably deals with a modeled atom. To avoid the singularity of the Coulomb potential, some empirical approximations for intra-atomic potential are usually used (see e.g. (Volkova,

2006; 2001; 2007) and references therein). Secondly, it is usually supposed that the wave functions of the continuum spectrum states are the Volkov wavefunctions. However, in the limit of zero field strength these wave functions tend to the free particle wave functions but not to the wave functions of a particle in the Coulomb field. Thirdly, it has been shown that the gauge choice dramatically change the results of numerical solutions of TDSE down to obvious contradiction with other models (Bauer, 2005).

In the frame of above models, the harmonics of laser radiation frequency result from a three-stage process that comprises the ionization of an atom, the electron acceleration in the electromagnetic field, and the subsequent recombination with ion and emission of harmonics. This process is periodically repeated every half cycle of the electromagnetic wave. The evolution of the ionized electron is usually described with the help of Volkov wave functions or classical electrodynamics equations. Notice, that the use of the Volkov wave functions is due to the variation of the spatial profile of the atomic continuous spectrum eigenfunctions in the presence of external field. However, the spatial profile of the atomic discrete spectrum eigenfunctions is also changed in the presence of the external electromagnetic field. This is the principle inconsistence of the above methods, because the wave functions of discrete (hydrogenic) and continuous (Volkov) spectra, which are used as basis for TDSE wave function expansion, do not compose the complete basis of the orthogonal functions.

The alternative approach in the theory of light-atom interaction has been proposed in (Andreev, 1999). This approach is based on the exact mathematical solutions of the boundary value problem for electron moving in superposition of centro-symmetric intra-atomic field and field of external electromagnetic wave. In the non-relativistic approximation the spectrum of eigenvalues for this problem coincides with that for free atom boundary value problem and eigenfunctions $\varphi_n(\vec{r},t)$ are related by $\varphi_n(\vec{r},t) = u_n(\vec{r})\exp\left(ie/\hbar c\vec{A}(t)\vec{r}\right)$ with the free atom eigenfuctions $u_n(\vec{r})$. The set of eigenfunctions $\varphi_n(\vec{r},t)$ provides the complete basis of the orthonormal functions which is used to calculate the matrix elements of quantum-mechanical operators. In the case of hydrogen-like atom or ion the dynamics of atomic electron over the states of both discrete and continuous spectra is calculated in consistent mathematical form. The main advantage of the proposed approach is its non-perturbative manner. The ratio of laser field strength E_0 to intra-atomic field strength $E_{at} = e/a_B^2$, where a_B is the Bohr radius, is not constrained by any conditions.

The paper is arranged as follows: Section 2 is devoted to the theory of eigensolutions of the boundary value problem for "atom in external field" and its applications to the general non-relativistic theory of light-atom interaction. The equations for atomic response are presented in Section 3. The results of computer simulations on the laser pulse interaction with silver atom are presented in Section 4. Finally, we summarize our findings in Section 5.

2. Basic principles of the theory

The traditional approach in the description of electromagnetic wave scattering by a single atom is based on the use of the eigenfunctions of free atom boundary value problem as the basis for wave function expansion. By solving the set of equations for probability amplitudes we can calculate the polarization of an ensemble of atoms. The polarization of atomic ensemble is the sum of the dipole moments of the individual atoms of ensemble. The eigenfunctions of the boundary value problem for an atom with the spherically symmetric intra-atomic potential have the following form

$$u_{nlm}(\vec{r}) = R_{nl}(r) Y_{lm}(\theta, \varphi), \tag{1}$$

where n is the principle quantum number, l and m are the angular momentum and its projection, respectively. The angular distribution of wave functions (1) possesses the spherical symmetry for the states with zero angular momentum and cylindrical symmetry for the states of non-zero angular momentum. Each state is characterized by the spatial parity $P = (-1)^l$. The matrix elements of dipole moment operator are defined by

$$\vec{d}_{nm} = e \langle n | \vec{r} | m \rangle . \tag{2}$$

Notice, that each of the states $|n\rangle$ is characterized by the three dimensional quantum number $n = (nlm)$. One can see from Eq.(2) that the dipole moment matrix elements take the non-zero values only for transitions between the states of opposite parity.
The angular dependency of the wave functions (1) is specified in the coordinate set of the individual atom configurational space. Indeed, the quantum mechanical average of the angular momentum in the states (1) reads as

$$\langle nlm | \vec{l} | nlm \rangle = \vec{n}_z \hbar m. \tag{3}$$

So, the direction of axis z in configurational space coincides with the direction of the average angular momentum of any individual atom. Under calculation of the matrix elements (2) it is usually assumed that the polarization vector of the linearly polarized wave is parallel to the axis z, i.e. $\vec{e}^{(0)} = \vec{n}_z$, and polarization vectors of circularly polarized wave lie in the perpendicular plane, i.e. $\vec{e}^{(\pm)} = (\vec{n}_x \pm \vec{n}_y) / \sqrt{2}$.
By expanding the wave functions of the time dependent Schrodinger equation (TDSE) into the series of eigenfunctions (1)

$$\psi(\vec{r}, t) = \sum_n a_n(t) u_n(\vec{r}) \tag{4}$$

and making summation over the physically small volume V_1, for the polarization of atomic ensemble we get

$$\vec{P}(\vec{r}_0, t) = \sum_{i \in V_1} \sum_{n,m} a_n^{(i)*}(t) a_m^{(i)}(t) \vec{d}_{nm}^{(i)}, \tag{5}$$

where $\vec{r}_0$ is the center-of-mass coordinate of the physically small volume V_1, and i is the summation index over the atoms located in this volume.
By summarizing the short sketch of the traditional methods based on the TDSE wave function expansion into the series of the free atom eigenfunctions we can conclude. Firstly, the atomic response is observed only in the process of inelastic electromagnetic wave scattering. Indeed, the matrix elements (2) have non-zero values only in the case when the initial and final states of transition have the opposite parity. As far as the eigenvalues corresponding to the eigenfunctions (1) depend on the angular momentum, l, hence, the atomic electron energies in the initial and final states of transitions are different. Secondly, the directions of dipole moments of all atoms in ensemble coincide. Indeed, the assumption on $\vec{e}^{(0)} \parallel \vec{n}_z^{(i)}$ and $\vec{e}^{(\pm)} \perp \vec{n}_z^{(i)}$ leads unambiguously to the following relationship $\vec{d}_{nm}^{(i)} - \vec{d}_{nm}$.
However, both these conclusions are in contradiction with the results of experimental measurements. Firstly, the dielectric permittivity of atomic gases does not equal unity even in spectral region of frequencies which are a few orders of magnitudes smaller than the

frequency of the nearest dipole allowed transition. In this case the population of the excited atomic states is negligibly small, and the atomic response is mainly due to the processes of electromagnetic wave elastic scattering. Secondly, the direction of electromagnetic wave polarization vector is strongly fixed at any spatial point of atomic ensemble. At the same time, the angular moments of different atoms of ensemble are chaotically directed. Hence, the assumption on $\left\langle \vec{l}_i \right\rangle \parallel \vec{e}$ could not be valid in principle.

Here we develop the theory of light-atom interaction, which is free of the above contradictions. The most principle innovation of the proposed theory is in the fact that the basis of eigenfunctions of the boundary value problem for "atom in the external field" is used to calculate the matrix elements of quantum mechanical operators. It was shown in (Andreev, 2010) that the bases of the "free atom" $u_n(\vec{r})$ and "atom in the external field" $\varphi_n(\vec{r},t)$ eigenfunctions are related by $\varphi_n(\vec{r},t) = u_n(\vec{r}) \exp\left[i\left(q/\hbar c\right)\vec{A}(t)\vec{r}\right]$. Each basis is the complete set of orthonormal functions, hence, any eigenfunction of any basis can be expanded into series of another basis eigenfunctions. For example, $u_n(\vec{r}) = \sum_m V_{nm}\varphi_m(\vec{r},t)$. By substituting this expansion into (4) we get for some individual atom

$$\vec{P}(t) = \sum_{n,m,p,q} a_n^*(t)\, a_m(t)\, V_{np}^{-1}(t)\, \vec{d}_{pq} V_{qm}(t). \tag{6}$$

By comparing the Eqs. (5) and (6) one can see the following principle difference between these equations. Firstly, the dipole moment matrix elements became time dependent

$$\vec{d}_{nm}(t) = \sum_{p,q} V_{np}^{-1}(t)\, \vec{d}_{pq} V_{qm}(t).$$

It is seen that the matrix elements $V_{nm}(t)$ play the crucial role in the frames of the developed approach. These matrix elements are the nonlinear functions of the electromagnetic field strength and they determine the temporal evolution of matrix elements of quantum mechanical operators. Secondly, and this is the most principle, the atomic polarization is time dependable even in the case when atom is in the ground state during the whole process of light-atom interaction. Indeed, let the state $n=0$ be the ground atomic state. Assuming in (6) $a_n(t) = \delta_{n0}$ we get

$$\vec{P}_0(t) = \sum_{p,q} V_{0p}^{-1}(t)\, \vec{d}_{pq} V_{q0}(t).$$

Thus we can see that an atom, being the whole time in the ground state, produces the response, the spectrum of which depends on temporal evolution of matrix elements $V_{nm}(t)$.

2.1 Boundary value problem for an atom in the external field

The boundary value problem for an atom with a spherically symmetric potential has the form

$$\left[\frac{\vec{p}^2}{2m} + U(r)\right] u_n(\vec{r}) = E_n u_n(\vec{r}). \tag{7}$$

Despite the explicit form of the spherical interatomic potential, the eigenfunction of this problem can be written as

$$u_{nlm}(\vec{r}) = R_{nl}(r)\, Y_{lm}(\theta,\varphi), \tag{8}$$

where $Y_{lm}(\theta,\varphi)$ are spherical functions and $R_{nl}(r)$ are radial functions determined by the given boundary conditions at $r=0$ and $r=\infty$.
Now we turn to the boundary value problem for an atom in the external field. Restricting our frames to the nonrelativistic approximation and taking $\vec{A}(\vec{r},t) \approx \vec{A}(t)$, we obtain the boundary value problem:

$$\left[\frac{1}{2m}\left(\vec{p}-\frac{q}{c}\vec{A}(t)\right)^2 + U(r)\right]\varphi_N(\vec{r},t) = E_N\varphi_N(\vec{r},t). \tag{9}$$

Substituting the wavefunction

$$\varphi_N(\vec{r},t) = u_n(\vec{r})\exp\left(i\frac{q}{\hbar c}\vec{A}(t)\vec{r}\right) \tag{10}$$

into (4), we obtain an expression duplicating (7), where

$$E_N = E_n. \tag{11}$$

Hence, we should conclude that the eigenvalues of the two problems (7) and (9) explicitly coincide and the eigenfunctions differ. As mentioned above, quantum number n represents a set of three quantum numbers $n=(nlm)$, which uniquely define the angular and radial functions within the $u_{nlm}(\vec{r})$. Due to the simplicity of expression (10), it is reasonable to put $N=(nlm)$. However, note that the angular and radial parts of wavefunctions $\varphi_N(\vec{r},t)$ are, in the general case, time-dependent features and therefore (nlm) no longer have the sense of quantum numbers, which are per se conservative values. Note that $|\varphi_N(\vec{r},t)|^2 = |u_n(\vec{r})|^2$, which means that the spatial distribution of probability does not change. This means that the energy of Coulomb interaction between the electron and atomic nucleus does not vary. Furthermore, the same can be stated about all spatial moments: $P_\alpha^{(nm)} = \int u_n^* x_\alpha u_m dV = \int \varphi_n^* x_\alpha \varphi_m dV$, $P_{\alpha\beta}^{(nm)} = \int u_n^* x_\alpha x_\beta u_m dV = \int \varphi_n^* x_\alpha x_\beta \varphi_m dV$, etc. This lies at the base of the equality of eigenvalues (11).

2.2 Relation between two sets of eigenfunctions

Set of functions $\varphi_n(\vec{r},t)$ also forms the complete orthonormal basis as the set $u_n(\vec{r})$. The orthonormality and completeness conditions for functions $\varphi_n(\vec{r},t)$ are

$$\begin{aligned}\int \varphi_n^*(\vec{r},t)\varphi_m(\vec{r},t)dV = \int u_n^*(\vec{r})u_m(\vec{r})dV = \delta_{nm}\\ \sum_n \varphi_n^*(\vec{r},t)\varphi_n(\vec{r}',t) = \exp\left[-i\frac{q}{\hbar c}\vec{A}(t)(\vec{r}-\vec{r}')\right]\sum_n u_n^*(\vec{r})u_n(\vec{r}') = \delta(\vec{r}-\vec{r}').\end{aligned} \tag{12}$$

Therefore, any function from one set can be represented as an expansion into series of eigenfunctions of the other:

$$\varphi_n(\vec{r},t) = \sum_m V_{nm}^{-1}u_m(\vec{r}), \quad u_n(\vec{r}) = \sum_m V_{nm}\varphi_m(\vec{r},t), \tag{13}$$

where transformation operator V, according to (10), takes the form

$$V = \exp\left(-i\frac{q}{\hbar c}\vec{A}(t)\vec{r}\right). \tag{14}$$

Introducing the three-dimensional form of indexes, we can rewrite (13) as

$$\varphi_{n_1l_1m_1}(\vec{r},t) = \sum_{n_2l_2m_2}\langle n_2l_2m_2|V^{-1}|n_1l_1m_1\rangle u_{n_2l_2m_2}(\vec{r}). \tag{15}$$

Using the well-known expansion for the exponent,

$$\exp\left(i\frac{q}{\hbar c}\vec{A}(t)\vec{r}\right) = 4\pi\sum_{l=0}^{\infty}\sum_{m=-l}^{l} i^l j_l\left(\frac{q}{\hbar c}A(t)r\right)Y_{lm}^*(\vec{e}(t))\,Y_{lm}(\vec{n}), \tag{16}$$

where $\vec{e}(t) = \vec{A}(t)/A(t)$ and $j_l(z)$ are spherical Bessel functions, for matrix elements on RHS of(15), we obtain the following expression:

$$\langle n_2l_2m_2|\,V^{-1}\,|n_1l_1m_1\rangle = 4\pi\sum_{l=0}^{\infty} i^l\int_0^{\infty} R_{n_2l_2}(r)\,j_l\left(\frac{q}{\hbar c}A(t)r\right)R_{n_1l_1}(r)\,r^2dr\cdot$$
$$\cdot\sum_{m=-l}^{l} Y_{lm}^*(\vec{e}(t))\int Y_{l_2m_2}^*(\theta,\varphi)\,Y_{lm}(\theta,\varphi)\,Y_{l_1m_1}(\theta,\varphi)\,do,$$

where $do = \sin\theta d\theta d\varphi$. Integration over angular variables can be performed analytically:

$$\int Y_{l_2m_2}^*(\theta,\varphi)\,Y_{lm}(\theta,\varphi)\,Y_{l_1m_1}(\theta,\varphi)\,do =$$
$$= (-1)^{m_2}\,i^{-l_2+l_1+l}\begin{pmatrix} l_2 & l & l_1 \\ -m_2 & m & m_1 \end{pmatrix}\begin{pmatrix} l_2 & l & l_1 \\ 0 & 0 & 0 \end{pmatrix}\cdot\sqrt{\frac{(2l+1)(2l_1+1)(2l_2+1)}{4\pi}} \tag{17}$$

so, the matrix element is modified as follows:

$$\langle n_2l_2m_2|\,V^{-1}\,|n_1l_1m_1\rangle = \sum_{l=|l_1-l_2|}^{l_1+l_2} Y_{lm}^*(\vec{e}(t))\,C(lm\,|l_2m_2,l_1m_1)\,\langle n_2l_2\,\|j_l\left(\frac{q}{\hbar c}A(t)r\right)\|\,n_1l_1\rangle, \tag{18}$$

where we have introduced the coefficients $C(lm\,|l_2m_2,l_1m_1)$ in the way seen from comparison (17) and (18) and the reduced radial matrix elements are

$$\langle n_2l_2\,\|j_l\left(\frac{q}{\hbar c}A(t)r\right)\|\,n_1l_1\rangle = \int_0^{\infty} R_{n_2l_2}(r)\,j_l\left(\frac{q}{\hbar c}A(t)r\right)R_{n_1l_1}(r)\,r^2dr. \tag{19}$$

Hence, the angular part of matrix elements V_{nm} is calculated analytically. The radial part can also be calculated in analytic form if we choose the set of hydrogenic eigenfunctions, which is only, known up to date, complete of three - dimensional boundary value problem.
Substituting (18) into (15), we finally obtain

$$\varphi_{n_1l_1m_1}(\vec{r},t) = \sum_{n_2l_2m_2}\sum_{l=|l_1-l_2|}^{l_1+l_2} R_{n_2l_2}(r)\,\langle n_2l_2\,\|j_l\|\,n_1l_1\rangle\,Y_{lm}^*(\vec{e})\,Y_{l_2m_2}(\vec{n})\cdot C(lm\,|l_2m_2,l_1m_1). \tag{20}$$

This expression reveals that the angular part of $\varphi_n(\vec{r},t)$ depends on both the electric field direction $\vec{e}$ and on the angular momentum of the atom $\langle\vec{l}\rangle$.
Wavefunctions $\varphi_n(\vec{r},t)$ take the form of axially symmetric function in two cases: (1) $l_1 = 0$, (2) $\vec{e}\parallel\vec{n}_z$.
In the first case, wavefunction (15) takes the form

$$\varphi_{n_1l_1=0}(\vec{r},t) = 2\sqrt{\pi}\sum_{n_2l_2}(-1)^{l_2}R_{n_2l_2}(r)\,\langle n_2l_2\,\|j_{l_2}\|\,n_10\rangle\sum_{m_2=-l_2}^{l_2} Y_{l_2m_2}^*(\vec{e})\,Y_{l_2m_2}(\vec{n}), \tag{21}$$

and if we use the relation

$$\sum_{m=-l}^{l} Y_{lm}^*(\vec{e})\,Y_{lm}(\vec{n}) = \frac{2l+1}{4\pi}P_l(\vec{e}\vec{n}),$$

we would see that function (15) is axially symmetric with the axis of symmetry, which coincides with the external electric field polarization.
In the second case, due to the relation

$$Y_{lm}\left(\vec{e}\right)|_{\vec{e}=\vec{n}_z} = i^l \sqrt{\frac{2l+1}{4\pi}} \delta_{m0},$$

we obtain

$$\varphi_{n_1 l_1 m_1}\left(\vec{r},t\right) = \sum_{n_2 l_2} \sum_{l=|l_1-l_2|}^{l_1+l_2} i^l \sqrt{\tfrac{2l+1}{4\pi}} R_{n_2 l_2}\left(r\right) \left\langle n_2 l_2 \left\| j_l \right\| n_1 l_1 \right\rangle Y_{l_2 m_1}\left(\vec{n}\right) C\left(l0 \left| l_2 m_1, l_1 m_1\right.\right). \quad (22)$$

Note, that wavefunction (15) is the superposition of wavefunctions $u_{n'l'm}\left(\vec{r}\right)$ with an identical value of the angular momentum projection on the external field direction. This testifies to the fact that only the angular momentum projection is a conservative value of problem (9) and the angular momentum itself is not.

2.3 Equations for the probability amplitudes

In the subrelativistic region of the electromagnetic field, the laser pulse interaction with a single atom is governed by the time-dependent Schrodinger equation (TDSE)

$$i\hbar \frac{\partial \psi}{\partial t} = \left[\frac{1}{2m} \left(\vec{p} - \frac{q}{c} \vec{A}\left(t\right) \right)^2 + U\left(r\right) \right] \psi. \quad (23)$$

The Hamiltonian matrix elements within sets of eigenfunctions (8) and (10) are, respectively,

$$\begin{aligned} \int u_n^*\left(\vec{r}\right) \left[\tfrac{1}{2m} \left(\vec{p} - \tfrac{q}{c} \vec{A} \right)^2 + U\left(r\right) \right] u_m\left(\vec{r}\right) dV &= \sum_p V_{np}^{-1}\left(t\right) E_p V_{pm}\left(t\right), \\ \int \varphi_n^*\left(\vec{r},t\right) \left[\tfrac{1}{2m} \left(\vec{p} - \tfrac{q}{c} \vec{A} \right)^2 + U\left(r\right) \right] \varphi_m\left(\vec{r},t\right) dV &= E_n \delta_{nm}. \end{aligned} \quad (24)$$

We see that the matrix elements within the set of eigenfunctions $\varphi_n\left(\vec{r},t\right)$ have rather plain structure, since they are the eigenfunctions of the Hamiltonian of Eq. (23). So, at first glance, we should find it more reasonable to expand the wavefunction of the TDSE into series of eigenfunctions of the boundary value problem (9). However, the presence of the time derivative on the left-hand side of Eq. (23) forces us to deal with the integral $\int \varphi_n^*\left(\vec{r},t\right) \partial \varphi_m\left(\vec{r},t\right) / \partial t dV$. And since the Hamiltonian of problem (9) is time-dependent, derivatives $\partial \varphi_n\left(\vec{r},t\right) / \partial t$ cannot be eigenfunctions of problem (9) and are therefore not orthogonal to functions φ_n.
At the same time, as mentioned above, the eigenfunctions of the problems (7) and (9) are related to each other in a simple way, so, we can expand wavefunction of the TDSE into series of free atom eigenfunctions

$$\psi\left(\vec{r},t\right) = \sum_n a_n\left(t\right) u_n\left(\vec{r}\right), \quad (25)$$

and then use the relation equation (13). Omitting some evident details, for the probability amplitudes $a_n\left(t\right)$ we can get the following set of equations

$$i\hbar \frac{da_n}{dt} = \sum_{m,k} V_{nk}^{-1} E_k V_{km} a_m. \quad (26)$$

We see that the matrix elements of operator V play the very important in the frame of developed theory. We now introduce the compound matrix elements:

$$M_{nm} = \sum_k V_{nk}^{-1} E_k V_{km}. \tag{27}$$

According to definition (14), operator V can be represented as an infinite series of the vector potential powers. If we restrict our consideration by the linear terms only, then in the weak field approximation we get

$$M_{nm} \approx \sum_k \left(1 + i\frac{q}{\hbar c}\vec{A}\vec{r} + \cdots\right)_{nk} E_k \left(1 - i\frac{q}{\hbar c}\vec{A}\vec{r} + \cdots\right)_{km}. \tag{28}$$

Retaining on the right-hand side only linear field terms, we obtain:

$$M_{nm} = E_n \delta_{nm} - i\omega_{nm}\frac{q}{c}\left(\vec{A}\vec{r}\right)_{nm} + \cdots .$$

In the case of $\vec{A}(t) = \vec{A}_0 \exp(-i\omega_0 t)$ and $\omega_0 \approx \omega_{nm}$, the relation between the field strength and the vector potential is

$$\vec{E}(t) = -\frac{1}{c}\frac{\partial \vec{A}}{\partial t} = \frac{i\omega_0}{c}\vec{A} \approx \frac{i\omega_{nm}}{c}\vec{A}.$$

Finally, for the compound matrix elements in the weak field approximation we get

$$M_{nm} \approx E_n \delta_{nm} - q\left(\vec{E}\vec{r}\right)_{nm} + \cdots .$$

So, one can see, that in the weak field approximation the set of equations (26) transforms to the set of equations for a two-level atom interacting with the field within the resonant electro–dipole approximation. Therefore, Eqs. (26) include not only resonant electro–dipole interaction (i.e. without the assumption $\omega_0 \approx \omega_{nm}$), but the non-resonant electro–dipole interaction, as well as interactions of higher multipolarity order, which originates from the terms in Eq. (27) with higher field power.

2.4 Basic peculiarities of the proposed theory

The general expression for the matrix elements V_{nm} is given by (18). Nevertheless, let us make some simple examples. These matrix elements have the most compact and simplest form when one of the states is $|ns\rangle$ state:

$$\langle ns| V^{-1} |n's\rangle = \langle ns \| j_0(z) \| n's\rangle, \tag{29}$$

$$\langle ns| V^{-1} |n'pm\rangle = i\sqrt{4\pi}\langle ns \| j_1(z) \| n'p\rangle Y_{1m}(\vec{e}), \tag{30}$$

$$\langle ns| V^{-1} |n'dm\rangle = -\sqrt{4\pi}\langle ns \| j_2(z) \| n'd\rangle Y_{2m}(\vec{e}). \tag{31}$$

where $z = \frac{q}{\hbar c}A(t)\,r$. In this case the dependency of matrix elements on the electromagnetic wave polarization vector is described by the spherical harmonics. The dependency on the field amplitude is associated with the reduced matrix elements $\langle n_2 l_2 \| j_l(z) \| n_1 l_1\rangle$. As we have mentioned above, in the case of hydrogenic wave functions the reduced matrix elements are calculated in the explicit analytic form. For example,

$$\langle 1s|\, j_0\, |1s\rangle = \frac{16}{(4+\mu^2)^2}, \quad \langle 2s|\, j_0\, |2s\rangle = \frac{1-3\mu^2+2\mu^4}{(1+\mu^2)^4},$$
$$\langle 2p, m=1|\, j_0\, |2p, m=1\rangle = \frac{1}{(1+\mu^2)^3}, \quad \langle 2p, m=0|\, j_0\, |2p, m=0\rangle = \frac{1-5\mu^2}{(1+\mu^2)^4},$$

$$\langle 1s|\, j_0\, |2s\rangle = \frac{256\sqrt{2}\mu^2}{(9+4\mu^2)^3}, \quad \langle 1s|\, j_1\, |2p\rangle = \frac{384\sqrt{2}\mu}{(9+4\mu^2)^3}, \quad \langle 1s|\, j_2\, |3d\rangle = \frac{6912\sqrt{6}\mu^2}{(16+9\mu^2)^4},$$

where

$$\mu(t) = \frac{qA(t)\, a_B}{\hbar c},$$

and a_B is the Bohr radius. By introducing the vector potential amplitude A_0, it is convenient to define the problem control parameter as follows

$$\mu_0 = \frac{qA_0 a_B}{\hbar c} = \frac{eE_0 a_B}{\hbar\omega} = \frac{2U_0}{\hbar\omega}\frac{E_0}{E_{at}}, \tag{32}$$

where $U_0 = Ry$ is the hydrogen atom ionization energy and $E_{at} = e/a_B^2$ is the strength of intra-atomic field.

It is seen that in the case of the hydrogen atom the parameter μ_0 is explicitly related with the adiabatic parameter of the Keldysh's ionization theory (Keldysh, 1965)

$$\gamma = \frac{\omega\sqrt{2mU_0}}{eE_0}.$$

Indeed, for the case of the hydrogen atom the parameter γ reads

$$\gamma = \frac{me\omega}{\hbar E_0} = \frac{\hbar\omega}{eE_0 a_B}.$$

Thus, the parameters μ_0 and γ are related by

$$\mu_0\gamma = 1.$$

Some reduced matrix elements for discrete – discrete transitions are shown graphically in Fig. 1a. The presented curves illustrate the main properties of these matrix elements. The diagonal matrix elements $\langle nl\, \| j_0 \|\, nl\rangle$ tend to unity at $\mu \to 0$, which is due to the eigenfunction normalization condition. The asymptotical behavior of the matrix elements at $\mu_0 << 1$ can be obtained from the asymptotical expansion of the Bessel function

$$\begin{aligned}\langle n_2 l_2 \,\| j_l(\mu(t)\rho) \|\, n_1 l_1\rangle = \mu^l \frac{\Gamma(1/2)}{2^{l+1}\Gamma(l+3/2)} \int_0^\infty R_{n_2 l_2}(\rho)\, R_{n_1 l_1}(\rho)\, \rho^{l+2} d\rho - \\ -\mu^{l+2} \frac{\Gamma(1/2)}{2^{l+3}\Gamma(l+5/2)} \int_0^\infty R_{n_2 l_2}(\rho)\, R_{n_1 l_1}(\rho)\, \rho^{l+4} d\rho + \cdots,\end{aligned} \tag{33}$$

where $\rho = r/a_B$. In accordance with the Wigner 3j symbol properties there are the following constraints: $|l_2 - l_1| \leqslant l \leqslant l_2 + l_1$. As a result, in the case of weak fields the matrix elements have the following asymptotics $\langle n_2 l_2 \,\| j_l \|\, n_1 l_1\rangle \sim \mu^{|l_2 - l_1|}$. On the other hand, in the region of the overatomic field strength, i.e. $\mu_0 >> 1$, the Bessel functions have the following

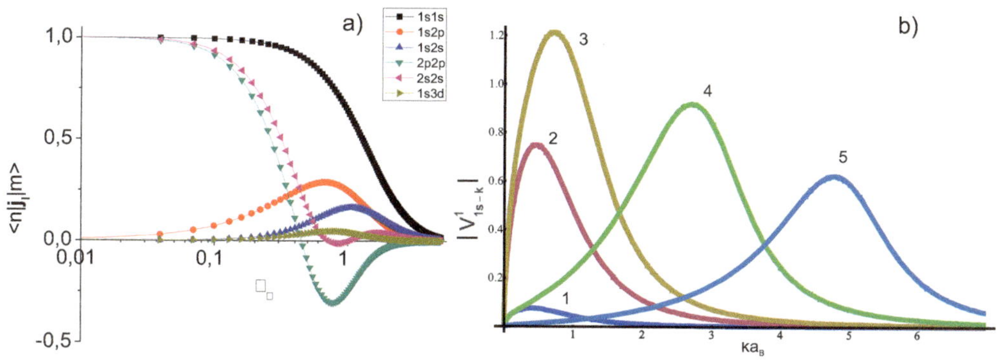

Fig. 1. Matrix elements for (a) discret-discret transitions as function of field strength, μ_0, and (b) discret-continuum transitions as function of photoelectron wave number, ka_B:
1. $V_{1s-k}(\mu_0 = 0.05, k)$, 2. $V_{1s-k}(\mu_0 = 0.5, k)$, 3. $V_{1s-k}(\mu_0 = 1, k)$, 4. $V_{1s-k}(\mu_0 = 3, k)$,
5. $V_{1s-k}(\mu_0 = 5, k)$

asymptotical behavior $j_l(\mu\rho) \approx \sin\left(\mu\rho - \frac{\pi l}{2}\right)/\mu\rho$. Hence, the matrix elements decrease with the field strength. Such behavior can be seen from Fig. 1a.

The matrix elements as function of photoelectron wave number for some discrete – continuum transitions are shown in Fig. 1b. One can see that the energy width of the populated continuum states increases with the increase of laser field strength.

Let us compare the common and distinctive features of equations (26) and equations for the probability amplitudes which follow from TDSE in electro-dipole approximation

$$i\hbar\frac{da_n}{dt} = \sum_m \left(E_n\delta_{nm} - \vec{E}\vec{d}_{nm}\right) a_m, \tag{34}$$

where E_n are the eigenvalues of the boundary value problem (7), δ_{nm} is the Kronecker delta function, and $\vec{d}_{nm}$ are the matrix elements of dipole moment operator. The most principle difference between the equations (26) and equations (34) is in the following.

It is seen that the diagonal elements of equations (34) coincide with the eigenvalues for "free atom" boundary value problem (7). At the same time, the diagonal elements of the Eq.(26) depend on the instantaneous laser field amplitude, because they are

$$M_{nn}(t) = \sum_k V_{nk}^{-1}(t)\, E_k V_{kn}(t). \tag{35}$$

Taking into account the above mentioned properties of the matrix elements V_{nm} it is seen that the diagonal elements (35) depend on the laser field amplitude. This dependency is due to the shifts of the effective energy of atom which in the presence of the external field is still in the "free atom" state (8). However, by comparing equations (8) and (10) we can see that state (8) is not an eigenstate of atom in the external field. Hence, its energy should depend on the field amplitude, and Eq.(35) describes the energy shifts, which are usually associated with the Stark shift, quadrupole splitting, etc. For the diagonal compound matrix elements $\langle nlm|M|nlm\rangle$ in the case of $|ns\rangle$ states we get

$$\langle ns|M|ns\rangle = \sum_{n'} \left[E_{n's}\left|\langle ns\|j_0(z)\|n's\rangle\right|^2 + 3E_{n'p}\left|\langle ns\|j_1(z)\|n'p\rangle\right|^2 + 5E_{n'd}\left|\langle ns\|j_2(z)\|n'd\rangle\right|^2\right]. \tag{36}$$

One can easily guess that the coefficients 3 and 5 appear due to the intermediate summation over the angular momentum projections of $|np\rangle$ and $|nd\rangle$ states. The diagonal matrix elements of $|npm\rangle$ states with $m = 0$ are

$$\begin{aligned}\langle npm| M |npm\rangle|_{m=0} = \sum_{n'} \Big\{ 3E_{n's} \langle np \| j_1(z) \| n's\rangle^2 \cos^2\theta + \\ + \tfrac{E_{n'p}}{2} \Big[2\langle np \| j_0(z) \| n'p\rangle^2 + 2\langle np \| j_0(z) \| n'p\rangle \langle n'p \| j_2(z) \| np\rangle + 5\langle np \| j_0(z) \| n'p\rangle^2 + \\ + 3\langle np \| j_2(z) \| n'p\rangle (2\langle n'p \| j_0(z) \| np\rangle + \langle n'p \| j_2(z) \| np\rangle)\cos 2\theta \Big] + \\ + \tfrac{3E_{n'd}}{10} \Big[7\langle np \| j_1(z) \| n'd\rangle^2 + 6\langle np \| j_1(z) \| n'd\rangle \langle n'd \| j_3(z) \| np\rangle + 12\langle np \| j_3(z) \| n'd\rangle^2 + \\ + \left(\langle np \| j_1(z) \| n'd\rangle^2 + 18\langle np \| j_1(z) \| n'd\rangle \langle n'd \| j_3(z) \| np\rangle + 6\langle np \| j_3(z) \| n'd\rangle^2 \right) \cos 2\theta \Big] \Big\}\end{aligned} \tag{37}$$

where $\theta = \arccos(\vec{n}_z \vec{e})$. In the case of $m = \pm 1$ the matrix elements are

$$\begin{aligned}\langle npm| M |npm\rangle|_{m=\pm 1} = \sum_{n'} \Big\{ \tfrac{3}{2} E_{n's} \langle np \| j_1(z) \| n's\rangle^2 \sin^2\theta + \\ + \tfrac{E_{n'p}}{4} \Big[4\langle np \| j_0(z) \| n'p\rangle^2 - 2\langle np \| j_0(z) \| n'p\rangle \langle n'p \| j_2(z) \| np\rangle + 7\langle np \| j_0(z) \| n'p\rangle^2 - \\ - 3\langle np \| j_2(z) \| n'p\rangle (2\langle n'p \| j_0(z) \| np\rangle + \langle n'p \| j_2(z) \| np\rangle)\cos 2\theta \Big] - \\ - \tfrac{3E_{n'd}}{10} \Big[-13\langle np \| j_1(z) \| n'd\rangle^2 + 6\langle np \| j_1(z) \| n'd\rangle \langle n'd \| j_3(z) \| np\rangle - 18\langle np \| j_3(z) \| n'd\rangle^2 + \\ + \left(\langle np \| j_1(z) \| n'd\rangle^2 + 18\langle np \| j_1(z) \| n'd\rangle \langle n'd \| j_3(z) \| np\rangle + 6\langle np \| j_3(z) \| n'd\rangle^2 \right) \cos 2\theta \Big] \Big\}\end{aligned} \tag{38}$$

One can see from Eqs. (36) - (38) that the effective energies of states are the non-linear functions of laser field strength, which can be expanded into the infinite series of the even powers of field. The effective energies of $|ns\rangle$ states do not depend on the orientation of the electromagnetic field polarization vector. It is quite evident because the spatial distribution of wave functions of $|ns\rangle$ states is spherically symmetric. At the same time the effective energies of $|npm\rangle$ states depend on the polarization vector orientation, because the wave function spatial distribution for the states with $l > 0$ possesses only cylindrical symmetry with respect to direction of angular momentum $\langle \vec{l} \rangle$. Hence, if the vectors $\langle \vec{l} \rangle$ and $\vec{e}$ are non-collinear the energy of the atomic electron interaction with the electromagnetic wave depends on the mutual orientation of these two vectors.

Notice, if it is assumed that atom during the whole process of light-atom interaction is unpolarized, i.e. $a_{nlm} = a_{nl}$, then the effective energy reads as

$$\langle nl| M |nl\rangle = \frac{1}{2l+1} \sum_{m=-l}^{l} \langle nlm| M |nlm\rangle.$$

In this case the diagonal compound matrix element $\langle np| M |np\rangle$ is

$$\begin{aligned}\langle np| M |np\rangle = \tfrac{1}{3} \sum_{m=-1}^{1} \langle npm| M |npm\rangle = \\ = \sum_{n'} \Big[E_{n's} \langle np \| j_1(z) \| n's\rangle^2 + E_{n'p} \left(\langle np \| j_0(z) \| n'p\rangle^2 + 2\langle np \| j_2(z) \| n'p\rangle^2 \right) + \\ + E_{n'd} \left(2\langle np \| j_1(z) \| n'd\rangle^2 + 3\langle np \| j_3(z) \| n'd\rangle^2 \right) \Big].\end{aligned} \tag{39}$$

One can see that the effective energy of $|np\rangle$ states averaged over the angular momentum projections does not depend on the polarization vector orientation. It should be noted that the approximation $a_{nlm} = a_{nl}$ can have some sense only in the case when atom interacts with the unpolarized light.

So, one can see that there is a cardinal difference between the diagonal elements of equations (26) and (34). The compound matrix elements M_{nn} approach E_n only in the limit of the very weak fields. It is evident that the non-diagonal matrix elements of these two sets of equations differ most principally. The non-diagonal compound matrix elements for transitions $|ns\rangle \to |n's\rangle$ are

$$\begin{gathered}\langle ns|\, M\, |n's\rangle = \sum_{n''} [E_{n''s} \langle ns \,\|j_0 (z)\|\, n''s\rangle \langle n''s \,\|j_0 (z)\|\, n's\rangle + \\ +3E_{n''p} \langle ns \,\|j_1 (z)\|\, n''p\rangle \langle n''p \,\|j_1 (z)\|\, n's\rangle + 5E_{n''d} \langle ns \,\|j_2 (z)\|\, n''d\rangle \langle n''d \,\|j_2 (z)\|\, n's\rangle] .\end{gathered} \tag{40}$$

For transitions $|ns\rangle \to |n'pm\rangle$ they are

$$\begin{gathered}\langle ns|\, M\, |n'pm\rangle|_{m=0} = \\ = i\sqrt{3}\cos\theta \sum_{n''} [E_{n''s} \langle ns \,\|j_0 (z)\|\, n''s\rangle \langle n''s \,\|j_1 (z)\|\, n'p\rangle + \\ +E_{n''p} \langle ns \,\|j_1 (z)\|\, n''p\rangle (\langle n''p \,\|j_0 (z)\|\, n'p\rangle + 2 \langle n''p \,\|j_2 (z)\|\, n'p\rangle) \\ +E_{n''d} \langle ns \,\|j_2 (z)\|\, n''d\rangle (2 \langle n''d \,\|j_1 (z)\|\, n'p\rangle + 3 \langle n''d \,\|j_3 (z)\|\, n'p\rangle)] ,\end{gathered} \tag{41}$$

$$\begin{gathered}\langle ns|\, M\, |n'pm\rangle|_{m=\pm1} = \\ = \mp i\sqrt{\tfrac{3}{2}} \sin\theta e^{\pm i\phi} \sum_{n''} [E_{n''s} \langle ns \,\|j_0 (z)\|\, n''s\rangle \langle n''s \,\|j_1 (z)\|\, n'p\rangle + \\ +E_{n''p} \langle ns \,\|j_1 (z)\|\, n''p\rangle (\langle n''p \,\|j_0 (z)\|\, n'p\rangle + 2 \langle n''p \,\|j_2 (z)\|\, n'p\rangle) + \\ +E_{n''d} \langle ns \,\|j_2 (z)\|\, n''d\rangle (2 \langle n''d \,\|j_1 (z)\|\, n'p\rangle + 3 \langle n''d \,\|j_3 (z)\|\, n'p\rangle)] .\end{gathered} \tag{42}$$

It is seen that in contrast to the dipole selection rules the states $|ns\rangle$ and $|n'pm = \pm1\rangle$ are coupled if the vectors $\left\langle \vec{l} \right\rangle$ and $\vec{e}$ are non-collinear.

Additionally, the states of the same parity are also coupled. For example, for transitions $|ns\rangle \to |n'dm\rangle$ we have

$$\begin{gathered}\langle ns|\, M\, |n'dm\rangle|_{m=0} = \\ = -\frac{1}{28\sqrt{5}} (1 + 3\cos 2\theta) \sum_{n''} [35E_{n''s} \langle ns \,\|j_0 (z)\|\, n''s\rangle \langle n''s \,\|j_2 (z)\|\, n'd\rangle + \\ +E_{n''p} \langle ns \,\|j_1 (z)\|\, n''p\rangle (42 \langle n''p \,\|j_1 (z)\|\, n'd\rangle + 63 \langle n''p \,\|j_3 (z)\|\, n'd\rangle) \\ +E_{n''d} \langle ns \,\|j_2 (z)\|\, n''d\rangle (35 \langle n''d \,\|j_0 (z)\|\, n'd\rangle + 50 \langle n''d \,\|j_2 (z)\|\, n'd\rangle + 90 \langle n''d \,\|j_4 (z)\|\, n'd\rangle)] ,\end{gathered} \tag{43}$$

$$\begin{gathered}\langle ns|\, M\, |n'dm\rangle|_{m=\pm1} = \\ = \pm\frac{1}{14}\sqrt{\frac{3}{10}} \sin 2\theta e^{\pm i\phi} \sum_{n''} [35E_{n''s} \langle ns \,\|j_0 (z)\|\, n''s\rangle \langle n''s \,\|j_2 (z)\|\, n'd\rangle + \\ +E_{n''p} \langle ns \,\|j_1 (z)\|\, n''p\rangle (42 \langle n''p \,\|j_1 (z)\|\, n'd\rangle + 63 \langle n''p \,\|j_3 (z)\|\, n'd\rangle) + \\ +E_{n''d} \langle ns \,\|j_2 (z)\|\, n''d\rangle (35 \langle n''d \,\|j_0 (z)\|\, n'd\rangle + 50 \langle n''d \,\|j_2 (z)\|\, n'd\rangle + 90 \langle n''d \,\|j_4 (z)\|\, n'd\rangle)] ,\end{gathered} \tag{44}$$

$$\begin{gathered}\langle ns|\, M\, |n'dm\rangle|_{m=\pm2} = \\ = -\frac{1}{14}\sqrt{\frac{3}{10}} \sin^2\theta e^{\pm i2\phi} \sum_{n''} [35E_{n''s} \langle ns \,\|j_0 (z)\|\, n''s\rangle \langle n''s \,\|j_2 (z)\|\, n'd\rangle + \\ +E_{n''p} \langle ns \,\|j_1 (z)\|\, n''p\rangle (42 \langle n''p \,\|j_1 (z)\|\, n'd\rangle + 63 \langle n''p \,\|j_3 (z)\|\, n'd\rangle) + \\ +E_{n''d} \langle ns \,\|j_2 (z)\|\, n''d\rangle (35 \langle n''d \,\|j_0 (z)\|\, n'd\rangle + 50 \langle n''d \,\|j_2 (z)\|\, n'd\rangle + 90 \langle n''d \,\|j_4 (z)\|\, n'd\rangle)] .\end{gathered} \tag{45}$$

Thus, the selection rules which determine the limits of summation on the right-hand-side of equations (26) and (34) are drastically different for these two sets of equations.

In conclusion of this subsection, let us notice some additional remarkable properties of matrix elements V_{nm}, which will be very important for future analysis. Firstly, the polarization vector $\vec{e}(t)$ has been defined as follows: $\vec{A}(t) = \vec{e}(t)\,A(t)$. In the case of linearly polarized wave we have $\vec{e}(t) = \vec{e}_0$ and field amplitude $A(t)$ varies in time. However, in the case of circularly polarized wave we have the opposite situations: $A(t) = A_0$ and $\vec{e}(t) = \vec{n}_x \cos(\omega t) + \vec{n}_y \sin(\omega t)$. Taking this into account it is absolutely clear that the width of nonlinear atomic response spectrum in the case of linearly polarized wave exceeds significantly that for circularly polarized wave. Indeed, the reduced matrix elements are the nonlinear functions of $A(t)$. Hence, in the case of linearly polarized wave the matrix elements $\langle n_2 l_2 \| j_l(z) \| n_1 l_1 \rangle$ will include a lot of harmonics of laser pulse carrier frequency. Contrary, in the case of circularly polarized wave the reduced matrix elements do not depend on time and the nonlinear response can be only associated with the angular part of matrix elements V_{nm}. Secondly, the normalization and orthogonality properties (12) result in the following equation

$$S_n = \sum_{m=0}^{\infty} |V_{nm}|^2 = 1,$$

where summation over m is made over the whole spectrum of atomic states. The spectrum of any atom includes the infinite number of discrete spectrum states and the uncountable number of continuum spectrum states. Hence, to solve the set of equations for probability amplitudes we should restrict ourselves by some finite number of the most important states. Let us introduce the following sum

$$S_n^{(N)} = \sum_{m=0}^{N} |V_{nm}|^2. \tag{46}$$

The numerical value of this sum depends on field amplitude and it does not exactly equal unity in whole range of the field strength variation. At the same time the sum $S_n^{(N)}$ can serve as a measure of completeness of the finite basis of eigenfunctions. Fig. 2 shows the sum (46) as a function of field strength for the case of hydrogen atom. One can see that the basis of the eigenfunctions consisting of 6 low-lying discrete and 280 continuum states can be considered as complete basis in the following range of laser field strength: $0 < \mu_0 \leqslant 1$. It should be noted, that the further increase in the number of states will not result in sufficient increase of calculation accuracy.

3. Atom response field

3.1 Polarization features of the response field

In the far-field range the spectrum of vector potential of atomic response field is given by

$$\vec{A}_r(\vec{r},\omega) = \frac{\exp(ikr)}{rc} \int \vec{j}(\vec{r}',\omega) \exp\left(-i\vec{k}\vec{r}'\right) dV',$$

where $\vec{j}(\vec{r},\omega)$ is the spectrum of atomic current density, which is defined by the well known expression (Landau, 1981)

$$\vec{j}(\vec{r},t) = \frac{q}{2m}\left[\psi^* \cdot \left(\vec{p} - \frac{q}{c}\vec{A}\right)\psi + \left(\left(\vec{p} - \frac{q}{c}\vec{A}\right)\psi\right)^* \cdot \psi\right]. \tag{47}$$

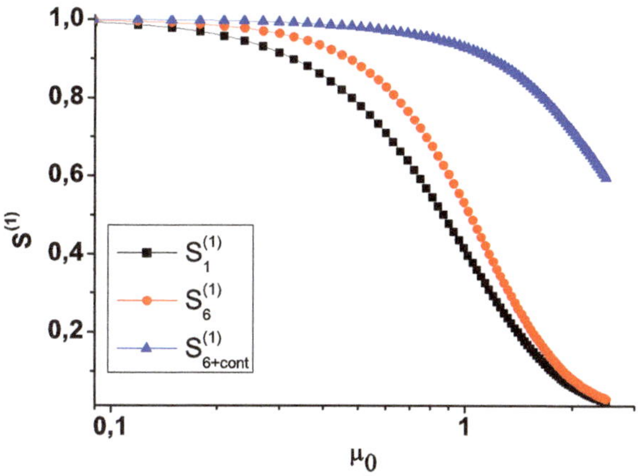

Fig. 2. The sum $S_n^{(N)}$ as a function of parameter μ_0 for the different finite sets of hydrogen atom eigenfunctions

In its turn the spectral density of response field intensity is given by

$$\frac{dI}{d\omega} = \frac{c}{4}\left|B_r(\vec{r},\omega)^2\right| r^2 do = \frac{\omega^2}{4c^3}\left|\int \left[\vec{j}(\vec{r}',\omega)\vec{n}\right]\exp\left(-i\vec{k}\vec{r}'\right)dV'\right|^2 do. \tag{48}$$

In the case when the long-wave approximation, $ka_0 << 1$ (where a_0 is the amplitude of electron oscillation in external field), holds for any harmonic, the intensity spectrum of single atom response is given by

$$\frac{dI}{d\omega} = \frac{\omega^2}{4c^3}\left|\left[\vec{J}(\omega)\vec{n}\right]\right|^2 do, \tag{49}$$

where

$$\vec{J}(\omega) = \int \vec{j}(\vec{r},\omega)\,dV. \tag{50}$$

However, if we deal with the spatially distributed ensemble of atoms, the equation (50) takes the form

$$\vec{J}\left(\vec{k},\omega\right) = \sum_{i=1}^{N}\vec{J}_i(\omega)\exp\left[i\left(\vec{k}\vec{r}_i - \omega t_i\right)\right], \tag{51}$$

where $\vec{r}_i$ is the coordinate of i-th atom of ensemble. Let $\vec{k}_0$ is the wave vector of the incident electromagnetic wave. Then the phase of the field in the position of i-th atom is $\omega_0 t - \vec{k}_0\vec{r}_i = \omega_0(t - t_i)$, where $t_i = \vec{k}_0\vec{r}_i/\omega_0$. By substituting t_i into the equation (51) we get

$$\vec{J}\left(\vec{k},\omega\right) = \sum_{i=1}^{N}\vec{J}_i(\omega)\exp\left[i\frac{\omega}{c}\left(\vec{n}(\omega) - \vec{n}_0(\omega_0)\right)\vec{r}_i\right],$$

where $\vec{n}(\omega) = \vec{k}c/\omega$ and $\vec{n}_0(\omega_0) = \vec{k}_0 c/\omega_0$. If the atoms of ensemble are identical it is convenient to introduce the form factor $f\left(\vec{k},\vec{k}_0\right)$ which is defined by

$$\vec{J}\left(\vec{k},\omega\right) = \vec{J}(\omega)\sum_{i=1}^{N}\exp\left[i\frac{\omega}{c}\left(\vec{n}(\omega) - \vec{n}_0(\omega_0)\right)\vec{r}_i\right] = \vec{J}(\omega) f\left(\vec{k},\vec{k}_0\right). \tag{52}$$

It is seen that the form factor takes into account the retardation effects, which depend on the dispersive properties of atomic media and the geometry of the radiative volume. Let us illustrate the difference between the frequency-angular spectra of single atom and spatially distributed ensemble of atoms. In the case of Gaussian incident beam the density of responding atoms in illuminating area can be approximated as follows

$$\frac{N}{V}(\vec{r}) = \frac{N_0}{V}\frac{\theta(z+L/2-\theta z-L/2)}{2}\exp\left(-\frac{\rho^2}{\rho_0^2}\right),$$

where $\theta(z)$ is the unit step function, ρ_0 is the beam radius, and L is the length of gas volume in direction of the laser pulse propagation. We assume that the z axis is directed along the direction of the wave vector $\vec{k}_0$ of the incident wave, and the wave vector of the response field is defined by $\vec{k} = \{k\sin\theta\cos\varphi, k\sin\theta\sin\varphi, k\cos\theta\}$. In this case the form factor $f(\theta,\omega)$ defined by Eq. (52) takes the following form

$$f(\theta,\omega) = \frac{\sin\left[kL(n(\omega)\cos\theta - n_0(\omega_0))/2\right]}{kL(n(\omega)\cos\theta - n_0(\omega_0))/2}\exp\left[-\frac{1}{4}(k\rho_0 n(\omega)\sin\theta)^2\right],$$

It is seen from the last equation that if we neglect the atomic media dispersive properties, i.e. if we assume $|\vec{n}(\omega)| = |\vec{n}_0(\omega_0)| = 1$, then the intensity of all harmonics reaches the maximum at $\theta = 0$. However, if we take into account the dispersive properties of the atomic ensemble then we get the conical emission. The intensity of different harmonics reaches maximum at different angles θ_N with respect to the direction of the laser pulse wave vector. It is well known that the refraction index $n(\omega)$ is varied significantly in the visible and UV ranges and approaches unity in XUV spectral range. So, it is supposed that the frequency-angular spectrum of emission is strongly varied in the visible and UV ranges and becomes more rigid in XUV and x-ray region.

The matrix elements of the generalized momentum operator $\vec{P} = \vec{p} - q\vec{A}/c$ in sets of eigenfunctions for problems (7) and (9) appear as

$$\int u_n^*(\vec{r})\left(\vec{p} - \frac{q}{c}\vec{A}\right)u_m(\vec{r})\,dV = \sum_{k,p} V_{nk}^{-1}(t)\,\vec{p}_{kp}V_{pm}(t), \tag{53}$$

$$\int \varphi_n^*(\vec{r},t)\left(\vec{p} - \tfrac{q}{c}\vec{A}\right)\varphi_m(\vec{r},t)\,dV = \int u_n^*(\vec{r})\,\vec{p}\,u_m(\vec{r})\,dV = \vec{p}_{nm},$$

respectively. Substituting (25) into (47) with the use of (53) for the total current of atomic electrons, we obtain

$$\vec{J}(t) = \frac{q}{m}\sum_{n,m,p,q} a_n^*(t)\,a_m(t)\,V_{np}^{-1}(t)\,\vec{p}_{pq}V_{qm}(t). \tag{54}$$

Note that Eq. (53) manifests the relation between the matrix elements of generalized momentum and the matrix elements of momentum in the basis of free atom eigenfunctions $\vec{p}_{nm} = -i\hbar\int u_n^*(\vec{r})\,\vec{\nabla}u_m(\vec{r})\,dV$. In contrast to the generalized momentum matrix elements, the matrix elements of momentum operator $\vec{p}$ are directly related with the coordinate matrix elements $\vec{p}_{nm} = im\omega_{nm}\vec{r}_{nm}$. So, finally, we get

$$\vec{J}(t) = i\sum_{n,m,p,q} a_n^*(t)\,a_m(t)\,\omega_{pq}V_{np}^{-1}(t)\,\vec{d}_{pq}V_{qm}(t). \tag{55}$$

We should recall that the probability amplitudes $a_n(t)$ are the coefficients of wavefunction expansion into series of free atom eigenfunctions, so all summation indexes are three-dimensional values $n = (n, l, m)$. Using these 3D representation for the partial matrix elements of expression (55), we get

$$\langle n_1 l_1 m_1 | \vec{J} | n_2 l_2 m_2 \rangle = i \sum_{n_3 l_3 m_3} \sum_{n_4 l_4 m_4} (\omega_{n_3 l_3} - \omega_{n_4 l_4}) \langle n_1 l_1 m_1 | V^{-1} | n_3 l_3 m_3 \rangle \langle n_3 l_3 m_3 | \vec{d} | n_4 l_4 m_4 \rangle \langle n_4 l_4 m_4 | V | n_2 l_2 m_2 \rangle, \tag{56}$$

where $\omega_{nl} = E_{nl}/\hbar$ and E_{nl} are the energy eigenvalues for the free atom problem. Using the angular momentum summation rules, for the dipole operator $\vec{d}$ we obtain

$$\langle n_3 l_3 m_3 | \vec{d} | n_4 l_4 m_4 \rangle = q \langle n_3 l_3 \| r \| n_4 l_4 \rangle (-1)^{m_3} i^{l_4 - l_3} \sqrt{(2l_3+1)(2l_4+1)} \cdot \sum_{m=-1}^{1} \vec{n}^{(m)} \begin{pmatrix} l_3 & 1 & l_4 \\ -m_3 & m & m_4 \end{pmatrix} \begin{pmatrix} l_3 & 1 & l_4 \\ 0 & 0 & 0 \end{pmatrix},$$

here

$$\vec{n}^{(0)} = -i\vec{n}_z, \quad \vec{n}^{(\pm)} = \pm i \frac{\vec{n}_x \mp i\vec{n}_y}{\sqrt{2}}$$

The matrix elements of the operator V have been calculated above. So, Eq. (56) can be transformed as

$$\langle n_1 l_1 m_1 | \vec{J} | n_2 l_2 m_2 \rangle = 4\pi q i^{l_2 - l_1 + 1} \sqrt{(2l_1+1)(2l_2+1)} \sum_{n_3 l_3} \sum_{n_4 l_4} (\omega_{n_3 l_3} - \omega_{n_4 l_4}) (2l_3+1)(2l_4+1) \cdot \sum_{l=|l_1-l_3|}^{l_1+l_3} \sum_{l'=|l_2-l_4|}^{l_2+l_4} \sqrt{(2l+1)(2l'+1)} \langle n_1 l_1 \| j_l \| n_3 l_3 \rangle \langle n_3 l_3 \| r \| n_4 l_4 \rangle \langle n_4 l_4 \| j_{l'} \| n_2 l_2 \rangle \cdot \begin{pmatrix} l_1 & l & l_3 \\ 0 & 0 & 0 \end{pmatrix} \begin{pmatrix} l_3 & 1 & l_4 \\ 0 & 0 & 0 \end{pmatrix} \begin{pmatrix} l_4 & l' & l_2 \\ 0 & 0 & 0 \end{pmatrix} \cdot \sum_{m=-1}^{+1} \sum_{m_3=-l_3}^{l_3} (-1)^{l'-m_2+m_3} Y_{l(m_3-m_1)}(\vec{e}) \vec{n}^{(m)} Y_{l'(m_3-m_2-m)}(\vec{e}) \cdot \begin{pmatrix} l_1 & l & l_3 \\ -m_1 & m_1-m_3 & m_3 \end{pmatrix} \begin{pmatrix} l_3 & 1 & l_4 \\ -m_3 & m & m_3-m \end{pmatrix} \begin{pmatrix} l_4 & l' & l_2 \\ m_3-m & m_2-m_3+m & -m_2 \end{pmatrix}, \tag{57}$$

where, as previously, $j_l = j_l(qA(t)r/\hbar c)$.

Thus, the mathematical formalism of this section allows us to calculate the angular-frequency spectrum (AFS) of the atomic response field in the case of arbitrary mutual orientation of the angular momentum of the atom and the external field polarization if it is linearly polarized, as well as for arbitrary state of its polarization. As follows from (57), the polarization of AFS components depends on both angular momentum direction and polarization vector of the incident field.

3.2 Atom response at subatomic laser field strength

We now comprehend the main characteristics of the response field in the subatomic range at $\mu << 1$. Matrix elements of dipole momentum are nonzero for the states with angular momentums shifted to unity: $l' = l \pm 1$. As mentioned above, the diagonal reduced matrix elements $\langle n_1 l_1 \| j_0 \| n_1 l_1 \rangle$ are the even functions of the field amplitude and at $\mu << 1$ take values close to unity. Reduced matrix elements $\langle n_1 l_1 \| j_1 \| n_3 (l_1 \pm 1) \rangle$ are the odd functions and at $\mu \to 0$ vary as the first power of the field $\langle n_1 l_1 \| j_1 \| n_3 (l_1 \pm 1) \rangle \sim \mu$. All other matrix elements have noticeably smaller values because they vary as higher powers of the field $\langle n_1 l_1 \| j_1 \| n_3 l_1 \rangle \sim \mu^2$ when $n_3 \neq n_1$, and $\langle n_1 l_1 \| j_1 \| n_3 (l_1 \pm k) \rangle \sim \mu^k$.

We now proceed with the analysis of diagonal matrix elements of the atomic current. Taking into account the above considerations for the most valuable contributions in the subatomic range, we obtain

$$\begin{gathered}\langle n_1 l_1 m_1 | \vec{J} | n_1 l_1 m_1 \rangle = \sum_n k(n_1 l_1, nl) \langle n_1 l_1 \| j_0 \| n_1 l_1 \rangle \langle n_1 l_1 \| r \| nl \rangle \langle nl \| j_1 \| n_1 l_1 \rangle \cdot \\ \sum_{m=-1}^{+1} \vec{n}^{(m)} Y_{1m}(\vec{e}) \sum_{m'=-1}^{+1} \langle l_1 m_1 | Y_{1m}(\vec{n}) | lm' \rangle \langle lm' | Y^*_{1m}(\vec{n}) | l_1 m_1 \rangle,\end{gathered} \tag{58}$$

where $l = l_1 \pm 1$. For brevity, we have introduced the new term $k(n_1 l_1, nl)$, the explicit expression of which can be found from comparison of (58) and (57). Performing summation over m', for the angular dependence of the matrix elements (58), we obtain

$$\vec{F}_{l_1 m_1}(\vec{n}, \vec{e}) = \sum_{m=-1}^{+1} \vec{n}^{(m)} Y_{1m}(\vec{e}) C_m(l_1 m_1), \tag{59}$$

where

$$\left\{ \begin{array}{c} C_1(l_1 m_1) \\ C_0(l_1 m_1) \\ C_{-1}(l_1 m_1) \end{array} \right\} = \frac{1}{(2l_1+1)(2l_1+2)(2l_1+3)} \cdot \left\{ \begin{array}{c} (l_1 - m_1 + 1)(l_1 - m_1 + 2) \\ 2(l_1 + m_1 + 1)(l_1 - m_1 + 1) \\ (l_1 + m_1 + 1)(l_1 + m_1 + 2) \end{array} \right\}. \tag{60}$$

Specifically,

$$C_m(l_1 = 0) = \frac{1}{3}, \qquad \sum_{m_1=-l_1}^{l_1} C_m(l_1 m_1) = \frac{1}{3}.$$

Thus, from equations (59), (60) it follows that the direction of atomic current coincides with the external field polarization vector only in the case of atom with zero angular momentum $l_1 = 0$, or non-polarized ensemble of atoms, i.e. when the sublevels of atomic energy structure are degenerated: $a_{n_1 l_1 m_1}(t) = a_{n_1 l_1}(t)$. Indeed, in these cases, we obtain

$$\sqrt{12\pi} \sum_{m_1} \vec{F}_{l_1 m_1}(\vec{n}, \vec{e}) = \sqrt{\frac{4\pi}{3}} \sum_{m=-1}^{+1} \vec{n}^{(m)} Y_{1m}(\vec{e}) = \vec{e}. \tag{61}$$

At the same time, Eq. (58) shows that in general case the direction of atomic current diagonal matrix elements depend on both external field and angular momentum directions. In principle, because the eigenvalues of both "free atom" and "atom in the field" boundary value problems depend on the two quantum numbers (nl) only, the matrix elements $\langle n_1 l_1 m_1 | \vec{J} | n_1 l_1 m_1' \rangle$ could be treated as diagonal. However, in the external field, the shifts appear between the sublevels of different m. Hence, the latter approximation can be valid only in the case when these shifts are negligible. Thus, in general case the atomic current diagonal matrix elements posses the tensor structure and depend on both external field and angular momentum directions.

Now, we turn to analysis of the nondiagonal matrix elements of atomic current. Taking into account the properties of reduced matrix elements $\langle n'l' \| j_l \| n''l'' \rangle$ at subatomic field strength, we find that the main contribution comes from the term

$$\begin{gathered}\langle n_1 l_1 m_1 | \vec{J} | n_2 l_2 m_2 \rangle = k(n_1 l_1, n_2 l_2) \langle n_1 l_1 \| j_0 \| n_1 l_1 \rangle \langle n_1 l_1 \| r \| n_2 l_2 \rangle \cdot \\ \cdot \langle n_2 l_2 \| j_0 \| n_2 l_2 \rangle \sum_{m=-1}^{+1} \vec{n}^{(m)} \langle l_1 m_1 | Y_{1m}(\vec{n}) | l_2 m_2 \rangle,\end{gathered} \tag{62}$$

where $l_2 = l_1 \pm 1$, and coefficient $k(n_1l_1, n_2l_2)$ has been introduced above. It is seen that in contrast to diagonal matrix elements, these depend exclusively on the angular momentum direction and not on the external field direction:

$$\vec{F}_{l_1m_1,l_2m_2}(\vec{n}) = \sum_{m=-1}^{+1} \vec{n}^{(m)} \langle l_1m_1|\, Y_{1m}(\vec{n})\, |(l_1 \pm 1)\, m_2\rangle. \tag{63}$$

In subatomic fields, as follows from (27), matrix element $\langle nl\, \|j_0\|\, nl\rangle \approx 1$; therefore,

$$\langle n_1l_1m_1|\, \vec{J}\, |n_2l_2m_2\rangle = k(n_1l_1, n_2l_2)\, \langle n_1l_1m_1|\, \vec{r}\, |n_2l_2m_2\rangle\,. \tag{64}$$

Hence, the selection rules for nondiagonal current matrix elements $l_1 \to l_2 = l_1 \pm 1$ agree with the dipole selection rules governed by the angular momentum of the atom, and the directions of matrix elements $\langle n_1l_1m_1|\, \vec{J}\, |n_2l_2m_2\rangle$ in nonpolarized media are completely chaotic.
The population amplitudes in the subatomic range are calculated using perturbation theory if there are no resonances between the pulse carrier frequency and the frequencies of atomic transitions. Under the assumption that $|a_{N_0}(t)| \approx 1$, Eqs. (26) yield

$$\begin{aligned} a_{N_0}(t) &= a_{N_0} \exp\left[-i\Phi_0(t)\right], \\ a_{N_1}(t) &= a_{N_0} \exp\left[-i\Phi_1(t)\right] \cdot \int_{-\infty}^{t} M_{N_1N_0}(t') \exp\left[i\left(\Phi_1(t') - \Phi_0(t')\right)\right] dt', \end{aligned} \tag{65}$$

where $\Phi_i(t) = \frac{1}{\hbar} \int_{-\infty}^{t} M_{N_iN_i}(t')\, dt'$. As mentioned above, in the subatomic range, the matrix elements between neighboring states $l_0 \to l_1 = l_0 \pm 1$ make the maximal contribution. For such transitions, Eq. (35) takes the form

$$\begin{aligned} M_{N_1N_0}(t) &= \sqrt{12\pi}\, i^{l_0 - l_1} (-1)^{m_0} \sqrt{(2l_1+1)(2l_0+1)}\, \langle n_1l_1\, \|j_1\|\, n_0l_0\rangle \\ &\left(E_{n_1l_1} \langle n_1l_1\, \|j_0\|\, n_1l_1\rangle + E_{n_0l_0} \langle n_0l_0\, \|j_0\|\, n_0l_0\rangle\right) \begin{pmatrix} l_1 & 1 & l_0 \\ -m_1 & m_1 - m_0 & m_0 \end{pmatrix} \begin{pmatrix} l_1 & 1 & l_0 \\ 0 & 0 & 0 \end{pmatrix} Y_{1(m_0 - m_1)}(\vec{e})\,. \end{aligned} \tag{66}$$

In particular, for diagonal matrix elements, we obtain

$$M_{N_1N_1}(t) = E_{n_1l_1} \left(\langle n_1l_1\, \|j_0\|\, n_1l_1\rangle\right)^2. \tag{67}$$

So, for the nondiagonal term of the total atomic current, we finally obtain

$$\begin{aligned} \vec{J}_{N_1N_0}(t) &= a^*_{N_1}(t)\, a_{N_0}(t)\, \langle n_1l_1m_1|\, \vec{J}\, |n_0l_0m_0\rangle = |a_{N_0}|^2 k_1(n_1l_1, n_0l_0)\, \langle n_1l_1\, \|j_0(t)\|\, n_1l_1\rangle \\ &\langle n_1l_1\, \|r\|\, n_0l_0\rangle \langle n_0l_0\, \|j_0(t)\|\, n_0l_0\rangle \exp\left[i\left(\varphi_1(t) - \varphi_0(t)\right)\right] \\ &\int_{-\infty}^{t} dt' \langle n_1l_1\, \|j_1(t')\|\, n_0l_0\rangle \left(E_{n_1l_1} \langle n_1l_1\, \|j_0(t')\|\, n_1l_1\rangle + E_{n_0l_0} \langle n_0l_0\, \|j_0(t')\|\, n_0l_0\rangle\right) \cdot \\ &\cdot \exp\left[-i\left(\varphi_1(t') - \varphi_0(t')\right)\right] \sum_{m=-1}^{+1} \vec{n}^{(m)} Y_{1m}(\vec{e}(t'))\, C_m(l_1m_1, l_0m_0), \end{aligned} \tag{68}$$

where

$$C_m(l_1m_1, l_0m_0) = \begin{pmatrix} l_1 & 1 & l_0 \\ -m_1 & m & m_0 \end{pmatrix}^2.$$

For example, in the case when $l_1 = l_0 + 1$, for the coefficients $C_m(l_1 m_1, l_0 m_0)$ with the use of relation $m_1 = m_0 + m$, we obtain

$$\left\{\begin{array}{c} C_1(l_0 m_0) \\ C_0(l_0 m_0) \\ C_{-1}(l_0 m_0) \end{array}\right\} = \frac{1}{(2l_0+1)(2l_0+2)(2l_0+3)} \cdot \left\{\begin{array}{c} (l_0+m_0+1)(l_0+m_0+2) \\ 2(l_0-m_0+1)(l_0+m_0+1) \\ (l_0-m_0+1)(l_0-m_0+2) \end{array}\right\}.$$

3.3 Atomic response in the case of impact of two linearly polarized pulses

We now examine the situation when an atom interacts with the superposition of two linearly polarized pulses with arbitrary directions of polarization vectors $\vec{e}_1$ and $\vec{e}_2$:

$$\vec{A}(t) = \vec{e}_1 A_1(t) + \vec{e}_2 A_2(t), \tag{69}$$

where $A_{1,2}(t)$ are the magnitudes of vector potentials of laser field components, which can be represented as an envelope with harmonic stuffing:

$$A_{1,2}(t) = f_{1,2}(t)\cos(\omega_{1,2}t + \varphi_{1,2}),$$

here $\omega_{1,2}$ are the carrier frequencies, $f_{1,2}(t)$ are the pulse envelopes, and $\varphi_{1,2}$ are the phase shifts. Apparently, the spatial direction of the superposed field changes during the pulse propagation:

$$\vec{A}(t) = \vec{e}_1 A_1(t) + \vec{e}_2 A_2(t) = \vec{e}(t) A(t).$$

To clarify our interpretation, we turn to a certain special case without loss of generality. We assume that vector $\vec{e}_1$ is directed along the z axis in the laboratory system of coordinates and vector $\vec{e}_2$ lies in the (y,z) plane and makes an angle θ_0 with the z axis. Then, the vectors $\vec{A}_{1,2}(t)$ look like

$$\vec{A}_1(t) = \{0, 0, A_1(t)\}, \quad \vec{A}_2(t) = \{0, A_2(t)\sin\theta_0, A_2(t)\cos\theta_0\}.$$

Therefore, vector $\vec{A}(t)$ should always lie in the (y,z) plane and make an angle $\theta(t)$ with the z axis, which is found to be

$$\theta(t) = \arctan\frac{A_2(t)\sin\theta_0}{A_1(t) + A_2(t)\cos\theta_0}. \tag{70}$$

In addition, the vector $\vec{e}(t)$ and magnitude of vector $\vec{A}(t)$ are determined as follows

$$\vec{e}(t) = \{0, \sin\theta(t), \cos\theta(t)\}, \tag{71}$$

$$A(t) = A_1(t)\cos\theta(t) + A_2(t)\cos(\theta_0 - \theta(t)). \tag{72}$$

It is seen that angle $\theta(t)$ does not vary in time exclusively under the circumstances of equality of amplitudes $A_1(t) = A_2(t)$ or collinear geometry $\theta_0 = \pi n$.

Hence, the problem of polychromatic field interaction with an ensemble of atoms requires special examination even in the simplest case, when the ensemble consists of a single atom, because the direction of the total polarization vector is generally time-dependent. Equation (70) shows that in noncollinear geometry, this effect remains even in the case of identical carrier frequencies ($\omega_1 = \omega_2$) and pulse envelopes, but with some delay between them ($A_2(t) = A_1(t - t_0)$). In the linear regime, the response to a superposition of fields appears to be a superposition of responses to each component, but the nonlinear interaction makes the whole picture rather more complicated.

4. Numerical research

The application of the developed method to the study of atomic response specific features in laser fields of near-atomic strength is presented below.
In computer simulations we shall assume that the laser pulse has the Gaussian temporal profile

$$\vec{A} = \vec{e} A_0 \exp\left(-\frac{(t-t_0)^2}{\tau_0^2}\right) \sin(\omega t), \tag{73}$$

where ω is the laser pulse carrier frequency and τ_0 is the pulse temporal width. The delay time t_0 does not play any significant role and we have chosen it to place pulse peak at the center of integration interval.

4.1 Selection rules

In Section 2 we have already discussed the principle differences between the set of equations for probability amplitudes, which follow from the Hamiltonian in electro-dipole approximation (34), and equations (26) of the proposed theory. Indeed, the angular momentum selection rules of equations (26) are not restricted by condition $\Delta l = \pm 1$ of the electro - dipole approximation theory. For example, this difference will manifest itself in the angular distribution of photoelectron emission.
Let us consider the process of hydrogen atom ionization by electromagnetic wave with the carrier frequency of $\hbar\omega = 15.11 eV$. Because the quantum energy exceeds the binding energy of $1s$ electron, then in this case we have the one-quantum ionization. As a result the energy-level diagram can be approximated by the following way. We take into account the ground state and continuum spectrum states with the angular momentum values lying in the interval $l = 0 - 5$. The energy interval of continuum spectrum states which should be taken into account is determined by the energy dependency of reduced matrix elements (19). This dependency is non monotonic and reach maximum at certain value of ionized electron wave number, $E_{max} = \hbar^2 {k_{max}}^2 / 2m$. The matrix element $\langle 1s|V|k_{max} l\rangle$ as a function of field strength for transitions from the ground state to the continuum spectrum states of different angular momentum $l = 0 - 5$ are shown in Fig. 3. It is seen that the dimensionless parameter μ_0 is varied in the interval $\mu_0 = (0.01 - 5)$, it corresponds to the variation of the laser pulse intensity in the interval $I = 2.16 \cdot 10^{12} - 5.4 \cdot 10^{17}$ W/cm^2.
It is seen, that in the region of subatomic field strength the magnitude of matrix elements $\langle 0|V|k_{max}, l = 1\rangle$ exceeds the magnitude of all other elements (this region is marked as "I" in Fig.3). Such ratio of matrix elements of different multipolarity transitions indicates that in this region of laser field amplitude the selection rules, associated with the traditional electro-dipole allowed transitions, play the dominant role. However, when the field amplitude approach the near-atomic field strength the magnitude of matrix elements for $\Delta l = 2$ transition becomes initially equal and then exceeds the magnitude of $\Delta l = 1$ transition (the region II). Further increase of the laser field amplitude results in the successive increase of magnitude of matrix elements for transitions $\Delta l = 3$, $\Delta l = 4$, etc. Hence, the electro-dipole selection rules are violated in the region of over-atomic field strength. In the region II the most probable transition become the transition corresponding to the selection rule of $\Delta l = 2$. In the region III there is no any preferred transition.
As we have mentioned above, the profiles of angular spectra of photoelectron emission corresponding to different Δl transitions are different. The results of computer simulations have shown that in the region of subatomic field strength the angular distributions are

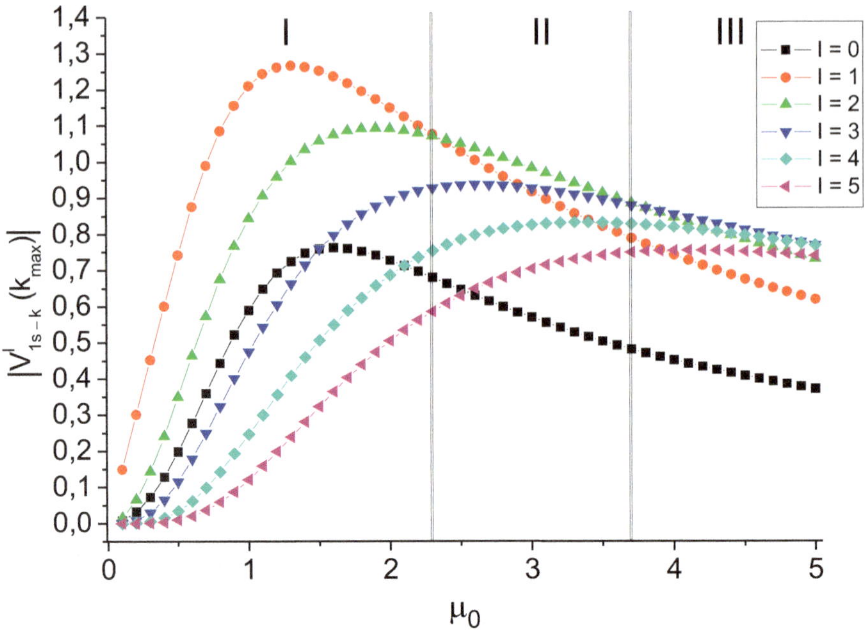

Fig. 3. Maximum value of matrix elements as a function of parameter μ_0. Angular momentum l runs from 0 to 5.

described by the Legendre function of the first order. Such distributions correspond to the electro-dipole transitions. With the increase of the pulse intensity the additional lobes in angular distribution of low energy photoelectrons appear. The angular distribution of the high energy photoelectrons remains approximately invariable. Fig. 4 shows the angular distributions of photoelectron emission for the case of laser pulse of field strength $\mu_0 = 5$ ($I = 5.4 \cdot 10^{17}$ W/cm^2). The angular distributions are averaged in time over the laser pulse duration. It is seen that the multilobe directional pattern for low energy photoelectrons is transformed to the unidirectional angular distribution for high energy photoelectrons. In the case of intense ultrashort laser pulses the angular spectra demonstrate the specific feature consisting in the asymmetry of emission in direction along (0°) and opposite (180°) to polarization vector. We suppose that this asymmetry is due to the fact that $A(t)$ given by (73) is odd function of time. From the general point of view, it is evident that the most energetic photoelectrons arise in time interval near the maximum of the instantaneous laser field strength. For the laser pulse of time profile (73) the maximum of the field strength magnitude corresponds to its negative value. This is in agreement with the curve (c) in the Fig. 4. It is also evident that the width of the energy region, where the selection rules differ from the electro-dipole one, grows with the increase of laser pulse intensity.

4.2 Ionization probability as a function of laser field amplitude

As we have mentioned in Introduction the first consistent theory of atom ionization has been proposed by Keldysh (Keldysh, 1965). Keldysh's theory is based on the model atom having only one bound state. If it is assumed that the wave functions of continuum spectrum states can be calculated in quasiclassical approximation then the probability of ionization reads as

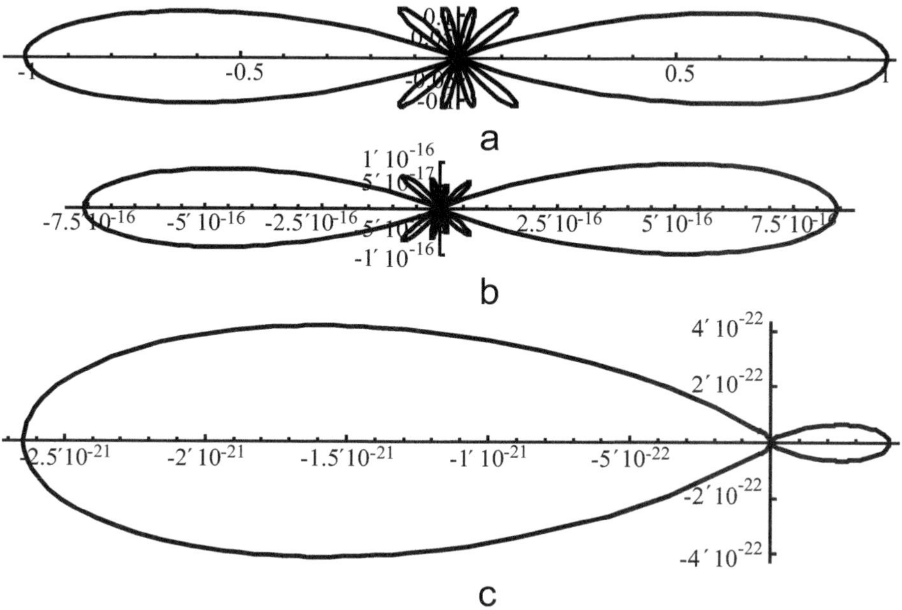

Fig. 4. Angular distributions of photoelectrons with $(a)\, k = 0.05$, $(b)\, k = 0.95$, $(c)\, k = 4.55$ at the external field amplitude $\mu_0 = 5$

$$w_{Keld}(\mu_0) \sim \exp\left[-2Im\left[\int_{\tau}^{\tau_0} \mu_0^2 \sin^2(\frac{\hbar\omega}{E_i}\tau')d\tau' + \tau_0\right]\right], \tag{74}$$

where E_i is the binding energy of electron, and ω is the carrier frequency of electromagnetic field. According to Eq. (74) the probability of ionization increases monotonically with the field strength in subatomic region, and it is saturated in the region of the over-atomic field strength. Here, we shall demonstrate that if we take into account the multi-level structure of the discrete spectrum states then the ionization probability ceases to be a monotonic function of field strength. The integral probability of ionization is defined as the total population of the continuum spectrum states after the termination of the laser pulse action

$$w_{ion} = \sum_{k,l,m} |a_{k,lm}(t \to \infty)|^2. \tag{75}$$

To specify the process under consideration let us turn on to the silver atom interaction with the pulses of *Ti:Sapphire* laser at wavelength 800 *nm*. The spectrum of the discrete and continuum states of silver is infinite, as for any other atom in nature. Neither analytical nor numerical research allows us to take into account all of them. Therefore, we need for a mathematical criterion for the selection of states making a substantial contribution to the process of light–atom interaction. This becomes extremely important because we have gone beyond the approximations of perturbation theory and, hence, the resonant transitions cease to play any significant role in the dynamics of level populations. As we have mentioned in Sec.2 the role of such criterion play the sum $S_n^{(N)}$.

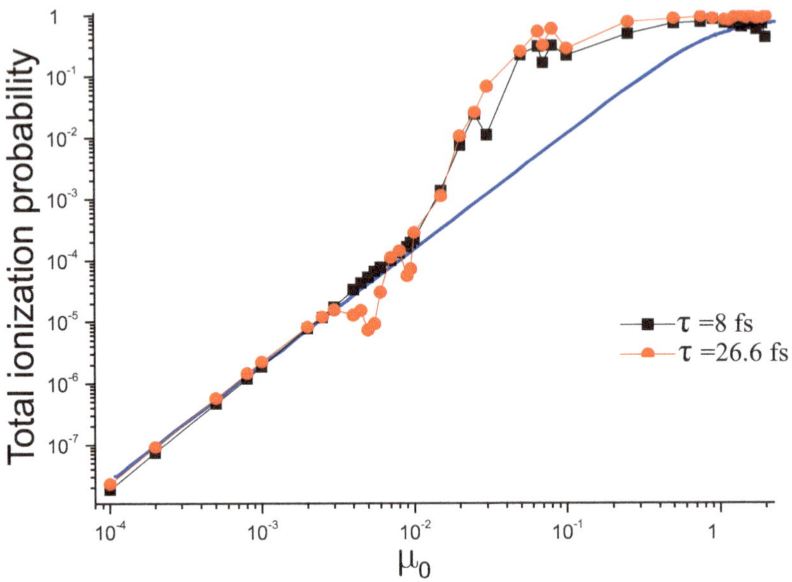

Fig. 5. Total ionization probability as a function of laser field amplitude μ_0 ($\lambda = 800$nm)

Fig. 5 shows the total ionization probability as a function of laser field strength for the case of atomic silver ionization by *Ti : Sapphire* laser pulses of duration: $\tau = 3T$ (squares) and $10T$ (circles). Solid line depicts ionization probability dependence calculated on the base of Keldysh model (Keldysh, 1965), in frame of which it is supposed that atom has only one bound state. From this figure we see, that in the region of substantially subatomic fields ($\mu_0 \leq 10^{-3}$) ionization rate actually does not depend on pulse duration and coincide with the curve, predicted by Keldysh model. At the same time in near-atomic field $10^{-3} \leq \mu_0 \leq 4 \cdot 10^{-1}$ this dependence demonstrate some new features: it ceases to be monotonic function; it depends not only on field strength, but also on pulse duration, i.e. on pulse energy. One more peculiarity is that ionization probability gets not only larger magnitudes than the Keldysh curve has, but also smaller. In the region $\mu_0 \geq 4 \cdot 10^{-1}$ ionization probability saturates.

4.2.1 Subatomic fields

Fig. 6 shows the population of discrete and continuum spectrum states after the termination of laser pulse as a function of field strength for the two values of the *Ti : Sapphire* laser pulse temporal width: $\tau = 10T$ (a,b) and $3T$ (c,d). As it is seen from fig. 5, for the case of pulse duration of $\tau = 3T$ the ionization probability is well-fitted by Keldysh curve till the values $\varphi \leq 10^{-2}$. In this region, as fig.6 c shows, the population of excited states rises monotonically with the laser field strength. The total population of all excited states is less than tenth part of percent and the population of $5p$ level exceeds populations of higher levels. For the case of $\tau = 10T$ the region of agreement with Keldysh formula extends to $\mu_0 \leq 2.5 \cdot 10^{-3}$. In this case we also observe monotonic growth of discrete state populations. The distribution

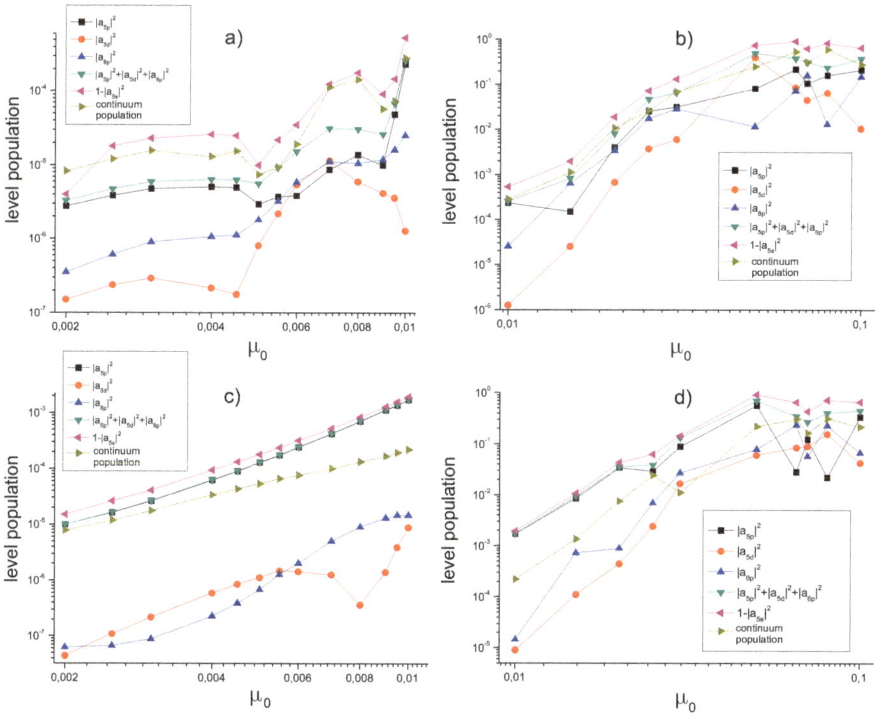

Fig. 6. Post-pulse population of silver atom energy states as a function of laser field strength for pulse temporal width $\tau = 10T$ (a,b) and $3T$ (c,d) ($\lambda = 800$nm)

of population of discrete spectrum states is "quasiequilibrium", i.e. the population of upper states exceeds lower ones. At the same time, the total population of excited states is still less than hundredth of percent.

Summarizing, we can say, that a good agreement with Keldysh model is observed when the ionization from ground state is dominating process, while the population of excited states is small and increases monotonically with the laser field strength.

4.2.2 Ionization stabilization

At pulse duration $\tau = 10T$ and field strength $\mu_0 \geq 2.5 \cdot 10^{-3}$ the ionization probability gets the values less, than predicted by Keldysh model. Such decrease in ionization probability is usually called by ionization stabilization effect (see, e.g. (Popov, 2004)). Fig. 6 a provides a clear explanation of this effect. Indeed, one can see, that in this region the population of all discrete states is saturated. This is conditioned by the fact that the rate of population is determined not only by the transitions from ground state, but also by recombination transitions. As it is seen from fig.6 a, at $\mu_0 \geq 2.5 \cdot 10^{-3}$ the rates of these processes first equalize, and then at $\mu_0 \approx 5 \cdot 10^{-3}$ the rate of recombination processes starts to prevail. This dominating effect expresses in the fact that at $\varphi \geq 5 \cdot 10^{-3}$ the populations of excited states are equalized, and the ground state population decreases, which means that the population distribution more and more decline from "quasiequilibrium" one. Addressing to fig. 1a, one

can see, that the reason of this declination consists in nonlinearity of the M_{nm} dependence on field strength. Notice another specific feature of the process. By comparing figs. 6a and 6c we can see that the level populations depend not only on field strength, but also on pulse duration i.e. on pulse energy. We can also see that the temporal dynamics of population distribution in multilevel atom results not only in the ionization stabilization but also in suppression of ionization rate with external laser field growth.
Note, that figures represented above demonstrate that the model of atom with one or even a few discrete states is insufficient for the consistent interpretation of yielded data, because the distribution of level populations changes non-monotonically. This non-monotonic character of the dynamics is conditioned by the competition between three kind of the processes: ionization, recombination, and interlevel transitions within the discrete band. The final populations depend essentially on ratio between the rates of these three processes.

4.2.3 Enhanced ionization

At field strength $\mu_0 \geq 10^{-2}$ the ionization probability starts to exceed values, predicted by Keldysh curve. This phenomenon has quite obvious explanation associated with the temporal dynamics of population of discrete spectrum states. Figs. 6b and 6d represent post-pulse population dependence as a function of field strength. It is seen that total population of excited discrete states exceeds 10% at field strength $\mu_0 = 3 \cdot 10^{-2}$, which means that contribution from this states becomes noticeable. In spite of the fact that in the case $\tau = 10T$ the integral population of the continuum spectrum states is close to the integral population of discrete spectrum states, the energy density of continuum state population, $d|a|^2/dE$, is small, because the energy width of photoionization cross section (see, fig.1b) increases rapidly in the region of near-atomic field strength. As a result, in this region of field strength the processes of ionization starting from the excited bound states become to play the dominating role. In spite of the fact that the rate of ionization exceeds the value predicted by Keldysh's model this enhancement can be explained in the frame of the Keldysh theory. Indeed, the parameter γ of Keldysh's theory depends on the ratio $N_0 = U_0/\hbar\omega$, which determines how much photons need to overcome the ionization threshold U_0. It is evident that N_0 decreases with the decrease of binding energy U_0. Hence, the ionization from the excited states became dominant.

4.2.4 Ionization rate saturation

In fig. 5 one can see, that at $\mu_0 > 0.1$ the ionization probability starts to saturate and that is to be associated with its approaching to unity. Such kind of the dependence is predictable from general point of view and the fact that our numerical results agree with this evident circumstance and nowhere exceeds unity verify the fact that our theory is non-perturbative (Andreev, 2009; 1999; 2006; 2007; 2008; 2010), i.e. it is not restricted by the small values of ratio E/E_{at}. Furthermore, in full agreement with properties of matrix elements showed in fig. 1a, the atomic silver ionization probability under action of laser pulse of duration $\tau = 3T$ starts to fall at field strength $E > E_{at}$.

4.3 Photoemission spectrum of atomic response

We now analyze the photoemission spectra of atomic silver response. The photoemission spectra for some values of the laser field strength are shown in Fig 7. In the region of substantially subatomic field strength ($\mu_0 < 10^{-4}$) the spectrum of response include only the fundamental frequency of incident pulse. Hence, in this case the response is linear. The odd harmonics appear in spectrum with the field strengthening and their amplitudes rapidly fall with the harmonic number. At the field strength $\mu_0 \sim 10^{-3}$ we see quantitative changes in the spectrum profile: its width grows and the plateau with pronounced cut-off frequency arises. Further field strengthening leads to the strengthening of these tendencies: spectrum width considerably grows and cut-off frequency gets more contrast ($\mu_0 = 1.2 \cdot 10^{-2}$).

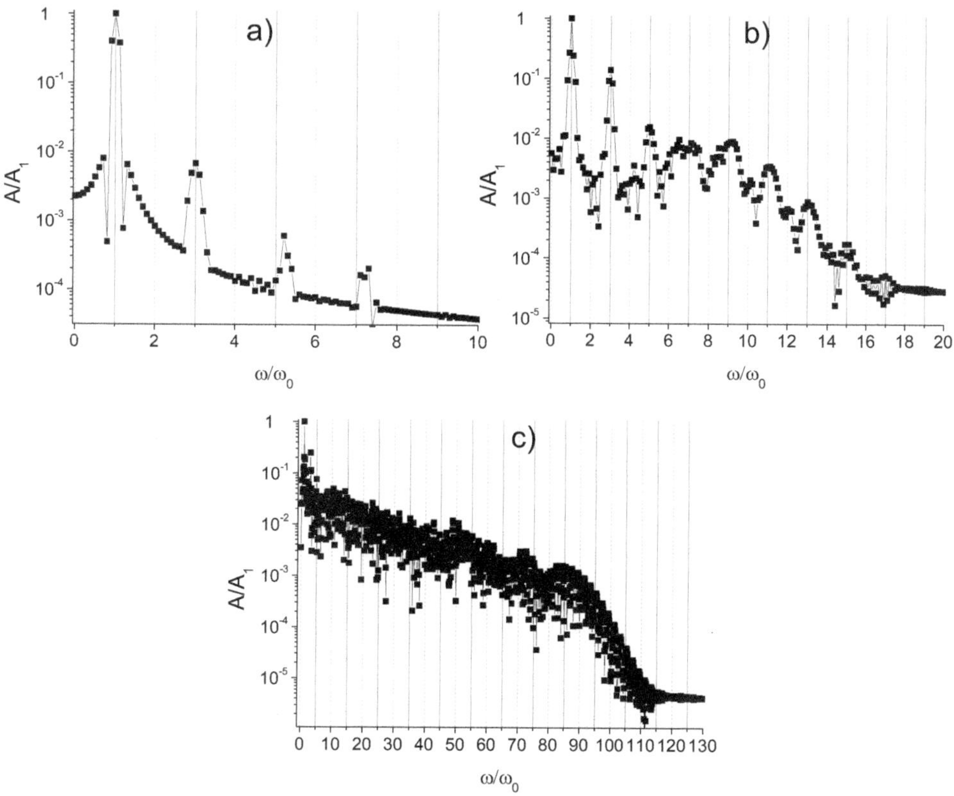

Fig. 7. The photoemission spectra generated in silver at $\mu_0 = 0.0045$ (a), $\mu_0 = 0.025$ (b), $\mu_0 = 0.08$ (c)

The dependence of cut-off frequency on the laser field strength is shown in fig. 8. One can see that in the weak field range ($\mu_0 < 0.1$) there is the quadratic growth of cut-off frequency with the field strength. However, at laser pulse intensity $I > 10^{14}\, W/cm^2$ the CF is saturated, i.e. it ceases to be intensity dependent. The reason of such behavior is quite obvious if we take into account that the probability of electron ionization approaches unity in this region of pulse intensity (see fig. 5). It means that the atomic electron is mostly localized in the continuum spectrum states and it does not collide with its parent ion. At the same time, as far as the time profile (73) of laser pulse has relatively soft slope in front of pulse, hence, the harmonics are effectively generated here. So, the results of computer modeling show that the most probable reason of the cutoff frequency saturation is the total ionization of the irradiated atom.
Some interpretation of such modification of response spectra can be done with the help of analytic solutions (65) obtained in perturbation theory approximation. To avoid the overcomplicated mathematical expressions for definition of CF let us concentrate on the

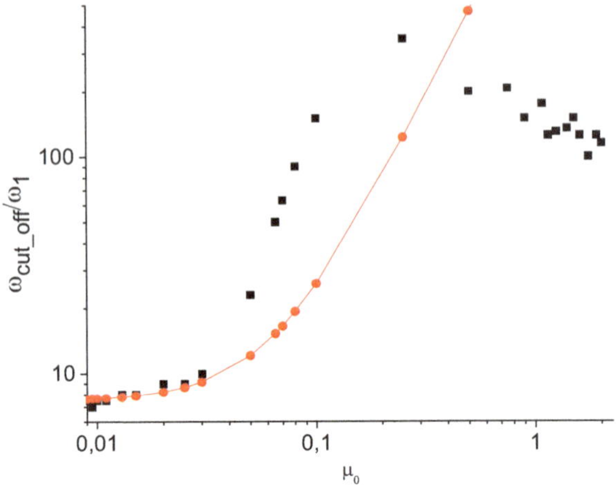

Fig. 8. Cut-off frequency as a function of laser pulse intensity. The solid red curve is the approximated quadratic dependence

hydrogen atom. In subatomic region of the laser field strength we can restrict ourselves by account of the two lowest discrete states and quadratic approximation for the compound matrix elements

$$M_{2p}(t) \approx E_{2p}\left(1 - 10\mu^2(t)\right), \quad M_{1s}(t) \approx E_{1s}\left(1 - \mu^2(t)\right). \tag{76}$$

Here we have signed $M_n = M_{nn}$ for brevity. Assuming that incident pulse profile has the form (73) and executing time integration, we find the expression for $\Phi_n(t)$

$$\begin{gathered}\Phi_{2p}(t) - \omega_{2p}t = \\ = -\omega_{2p}\sqrt{\frac{\pi}{2}}\frac{5\mu_0^2\tau_0}{4}\left[2\mathrm{erf}\left(\frac{\sqrt{2}t}{\tau_0}\right) + i\exp\left(-\frac{\omega_0^2\tau_0^2}{2}\right)\left(\mathrm{erfi}\left(\frac{-2it+\omega_0\tau_0^2}{\tau_0}\right) - \mathrm{erfi}\left(\frac{2it+\omega_0\tau_0^2}{\tau_0}\right)\right)\right]\end{gathered} \tag{77}$$

$$\begin{gathered}\Phi_{1s}(t) - \omega_{1s}t = \\ = -\omega_{1s}\sqrt{\frac{\pi}{2}}\frac{\mu_0^2\tau_0}{8}\left[2\mathrm{erf}\left(\frac{\sqrt{2}t}{\tau_0}\right) + i\exp\left(-\frac{\omega_0^2\tau_0^2}{2}\right)\left(\mathrm{erfi}\left(\frac{-2it+\omega_0\tau_0^2}{\tau_0}\right) - \mathrm{erfi}\left(\frac{2it+\omega_0\tau_0^2}{\tau_0}\right)\right)\right],\end{gathered} \tag{78}$$

where $\mathrm{erf}(z)$ is the error function and $\mathrm{erfi}(z) = \mathrm{erf}(iz)/i$. The first term in square brackets describes energy shift governed by pulse profile and the second one shows oscillations with double carrier frequency. The second term equals zero, when external pulse has rectangular profile, and is negligibly small, when external pulse has Gaussian profile and its duration satisfies the condition $\omega_0\tau_0 >> 1$. Then in subatomic region ($\mu_0 << 1$) the phases $\Phi_i(t)$ are finally approximated by the following expressions

$$\Phi_{2p}(t) = \omega_{2p}\left[t - \frac{5}{2}\sqrt{\frac{\pi}{2}}\mu_0^2\tau_0 \tanh\left(\frac{\sqrt{3}t}{\tau_0}\right)\right],$$

$$\Phi_{1s}(t) = \omega_{1s}\left[t - \frac{1}{4}\sqrt{\frac{\pi}{2}}\mu_0^2\tau_0 \tanh\left(\frac{\sqrt{3}t}{\tau_0}\right)\right].$$

Therefore, during the laser pulse action the level energy spacing depends on time and the instantaneous frequency reads as

$$\Omega(t) = \frac{d}{dt}\left(\Phi_{2p}(t) - \Phi_{1s}(t)\right) = \omega_{2p} - \omega_{1s} - \sqrt{\frac{3\pi}{2}}\mu_0^2\left(\frac{5}{2}\omega_{2p} - \frac{1}{4}\omega_{1s}\right)\cosh^{-2}\left(\frac{\sqrt{3}t}{\tau_0}\right).$$

It is seen, that the magnitude of energy shift is

$$\Delta\omega = \sqrt{\frac{3\pi}{2}}\mu_0^2\left(\frac{5}{2}\omega_{2p} - \frac{1}{4}\omega_{1s}\right). \tag{79}$$

Energy shift $\Delta E = \hbar\,|\Delta\omega|$ in hydrogen-like atom for transition $1s \to 2p$ is found then to be

$$\Delta E = \sqrt{\frac{3\pi}{2}}\frac{3}{4}U_p,$$

where U_p is ponderomotive potential, given by well-known expression

$$U_p = \frac{e^2E_0^2}{4m\omega^2}.$$

In the previous section we have discussed the difference between the diagonal ($n = m$) and non-diagonal ($n \neq m$) partial matrix elements of atomic current. As we have seen, the diagonal elements result in sequence of odd harmonics. At the same time, the shape of spectra resulted from the non-diagonal elements depends significantly on the laser pulse spectral width. In the case of a few–cycle laser pulses this part of the integral spectrum takes the shape of quasi-continuum plateau, spread till double energy shift of Eq. (6), instead of a sequence of distinct harmonics. Hence, in the subatomic region of field strength the cutoff energy, E_c, can be estimated as

$$E_c = \frac{3}{2}\sqrt{\frac{3\pi}{2}}U_p = 3.26\cdot U_p.$$

It is seen that the obtained equation coincides approximately with the well known semi-empirical equation $E_c = U_0 + 3.17\cdot U_p$.
Let us remind that Eq. (79) is true only in subatomic region at $\mu_0 \ll 1$, when the approximation (76) for compound matrix elements is legal. In Sec.2.4 we have shown that the reduced matrix elements fall with field strength in the region $\mu_0 \geq 1$. It is this property of matrix elements that provides the mathematical explanation of cut-off frequency saturation in over-atomic fields.
The similar calculations can be easily made for the case of silver atom. However, the appropriate equations have the very cumbersome form, therefore we shall not bring them here. The approximated analytical dependence calculated for silver atom is shown in Fig. 8 by solid line with circles. In the region of the laser pulse intensity of $(10^{11} \leq I \leq 10^{12})\ W/cm^2$ the results of computer calculations are in good agreement with the approximate equation. This agreement is due to the following. The dependency described by solid line has been derived under account of only two discrete states, namely $5s$ and $5p$. On the other hand, in this region of pulse intensity the population of $5p$ state exceeds significantly the population of other excited states of discrete spectrum, as it has been shown in previous section. Under further increase in the laser pulse intensity the quadratic curve raises faster than the numerically calculated curve. In this region of pulse intensity the population of other discrete spectrum excited states becomes comparable and, due to the recombination process, even exceeds the population of $5p$ state. Hence, the above approximation ceases to be valid.

4.4 Atom interaction with two-color laser field

Let us now turn to the problem of atom interaction with two-color laser field. As a practical example of the problem we shall consider the THz emission in argon gas, because there is number of available data of experimental measurements. The incident laser field constitutes of the fundamental frequency and second harmonic of $Ti : Sapphire$ laser. The main goal of research is to study the modification of photoemission spectrum under variation of laser field parameters (field amplitudes, pulse durations, delay times, and angles between polarization vectors of components). We have mentioned in Introduction that the traditional interpretation of experiments on THz emission in two-color laser fields follows the theoretical description of THz emission in monochromatic laser field, which is usually based on the four-wave-mixing rectification (FWMR) process in laser produced plasma. Here, we are going to show that in multi-color fields the atomic nonlinearities take the dominant role and may prevail over the plasma nonlinearities. Therefore, in the computer simulations we shall assume that the laser pulse intensity is below the ionization threshold intensity. In accordance with this assumption we shall take into account the thirteen low-lying states of argon atom. The degree of completeness of the chosen set of eigenfunctions is determined by the sum $S_n^{(N)}$ (see Eq. (46)). So, our calculations show that in the region of the laser field strength $\mu_0 < 0.1$ the chosen set of eigenstates is practically complete. Note, that the energy difference between highest and lowest state of the modeled atom amounts to 96.5% of ionization energy of a real argon atom. The condition $\mu_0 < 0.1$ means that the laser pulse intensity is limited by the value $I < 6.77 \cdot 10^{12}$ W/cm^2).

Figure 9 shows some typical spectra of the atomic response in the case of two pulses with the following parameters: $\mu_{01} = 0.1, \mu_{02} = 0.0316, \tau_1 = 120fs, \tau_2 = 85fs$, and a delay time of 100 fs. Figure 9a corresponds to the collinear geometry, and Fig. 9b, to the orthogonal geometry. In the first case ($\theta = 0$), the atomic response includes even as well as odd harmonics of the external field, and in the second case ($\theta = \pi/2$), only odd harmonics. This indicates that the response symmetry features, which are strictly related to the character of the response anisotropy, are sensitive to the angle between the polarization vectors of the pulses. It should be noted that the component of the atomic response field with polarization collinear to the polarization vector of the incident field at the fundamental frequency is shown only in Fig. 9.

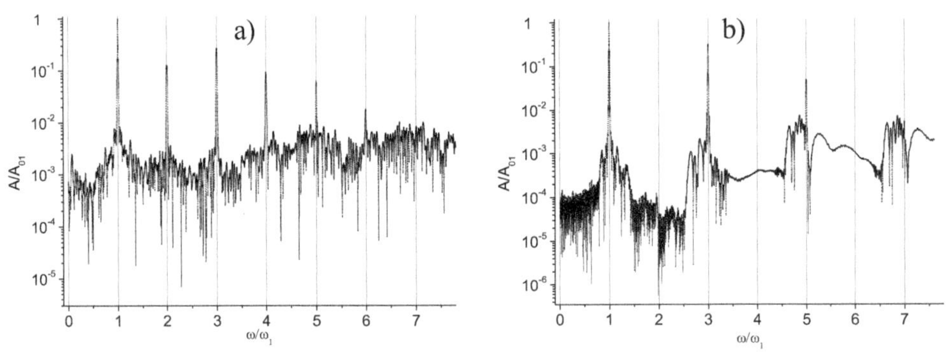

Fig. 9. Atomic response spectra: (a) collinear geometry, (b) orthogonal geometry

Figure 10 provides more detailed information on the low-frequency (THz) part of the spectra, shown in Fig 9. The enlargement of this detail of the spectrum immediately reveals that the shape of the THz signal strongly depends on the angle between polarizations. In order to study this dependence more accurately, we varied the angle within the range $[0, \pi]$. Figure 11 represents the corresponding dependence of THz-signal output for laser pulses with parameters $\mu_{01} = 0.1, \mu_{02} = 0.1, \tau_1 = \tau_2 = 4.25fs$ and a delay of time 0 fs (a) or 13.33 fs (b). The THz-signal output is determined as the total signal recorded at a frequency of 1 THz. The most remarkable feature of this curve is its nonmonotonic behavior.

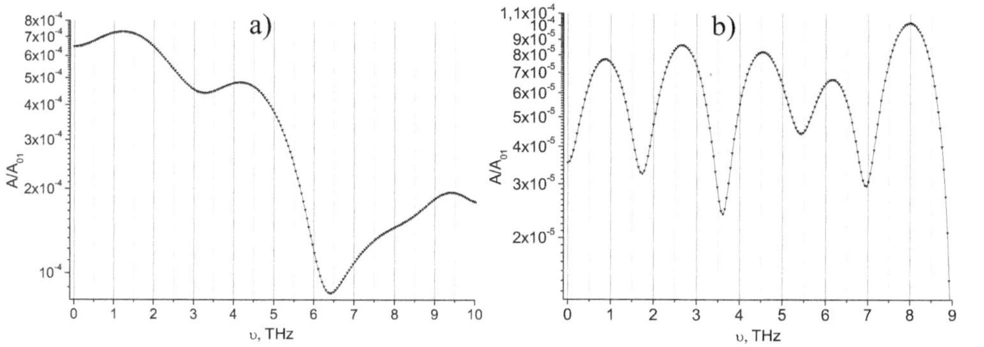

Fig. 10. THz part of atomic response spectra: (a) collinear geometry, (b) orthogonal geometry

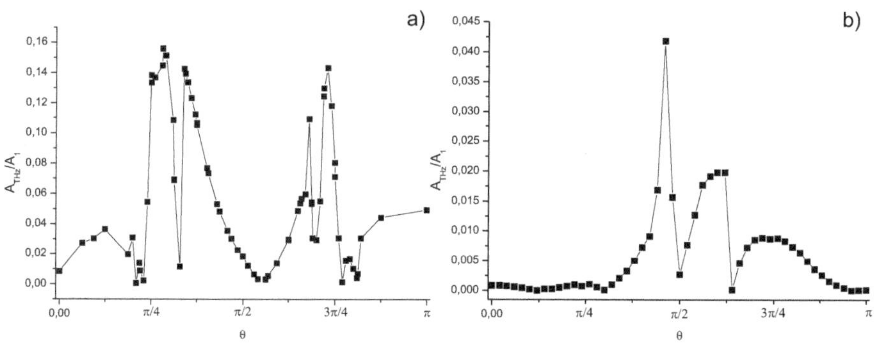

Fig. 11. THz response field amplitude as a function of angle between the polarization vectors of laser field components at frequencies ω and 2ω for $\mu_{01} = 0.1, \mu_{02} = 0.1, \tau_1 = \tau_2 = 4.25fs$ and for the delay times: 0 fs (a) 13.33 fs (b)

It is seen that a slight variation in angle may lead to a considerable change in the efficiency of THz-signal generation. Such behavior is consistent with the above discussion, and its origin is illustrated by the Fig. 12, where the trajectory described by the end of vector $\vec{A}(t)$ (see Eq.(69)) is shown. It is seen that the polarization state of the laser field depends significantly on the temporal profiles of constituents and the integral field could not be described in terms of linear, circular, or elliptic polarization.

The spectra of atomic response in the two-color laser fields depend on a number of parameters of constituent fields, which includes the amplitudes, temporal widths, delay times, and

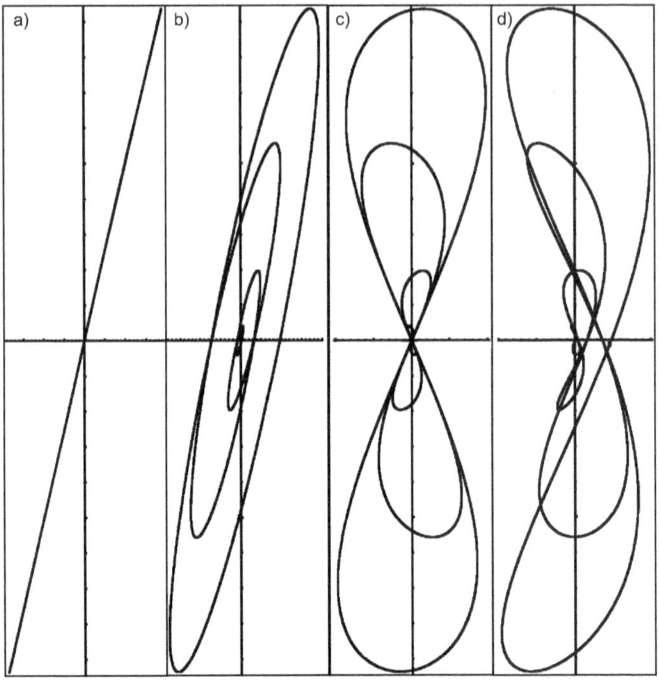

Fig. 12. Trajectory outlined by the end of integral field vector $\vec{A}(t)$ in the plane (y,z) (see (69)) for $A_{01} = 1$, $A_{02} = 0.2$, $\theta_0 = \pi/2$ and $\omega_1 = \omega_2, t_{01} = t_{02} = 0$ (a); $\omega_1 = \omega_2$, $t_{01} = 0, \omega_2 t_{02} = 0.5$ (b); $\omega_1 = 2\omega_2, t_{01} = t_{02} = 0$ (c); $\omega_1 = 2\omega_2, t_{01} = 0, \omega_2 t_{02} = 0.5$ (d)

mutual orientation of the fields at frequencies ω and 2ω. Fig. 13 illustrates the transformations of the response spectrum profile under variations of these parameters.

Figure 14 shows the angular dependence of the ninth harmonic of the fundamental frequency, which corresponds to XUV emission. By comparing Figs. 11 and 14, we can easily see the common features of the response in the long- (THz) and short- (XUV) wavelength regions. The power of emission in long and short wavelength parts of spectrum is small, if the pulses at frequencies ω and 2ω, have the same temporal profile. However, we can significantly enhance the generation efficiency by delaying the second-harmonic pulse. It should be noted again that the component of response field with the polarization collinear to polarization vector of laser field at fundamental frequency has been shown in above figures.

The dependency of the THz emission power on mutual orientation of polarization vectors of waves at frequencies ω and 2ω have been experimentally studied in recent work (Kim, 2008). The results of comparison of dependencies shown in fig.14 with the experimental data are presented in fig. 15. One can see that there is a good agreement between the results of computer simulations and in-situ measurements.

Thus, the results of computer simulations clearly demonstrate that variation in the mutual polarizations and temporal profiles of two-color field pulses is an effective tool for modifying the nonlinear atomic response spectra in a strongly controllable way.

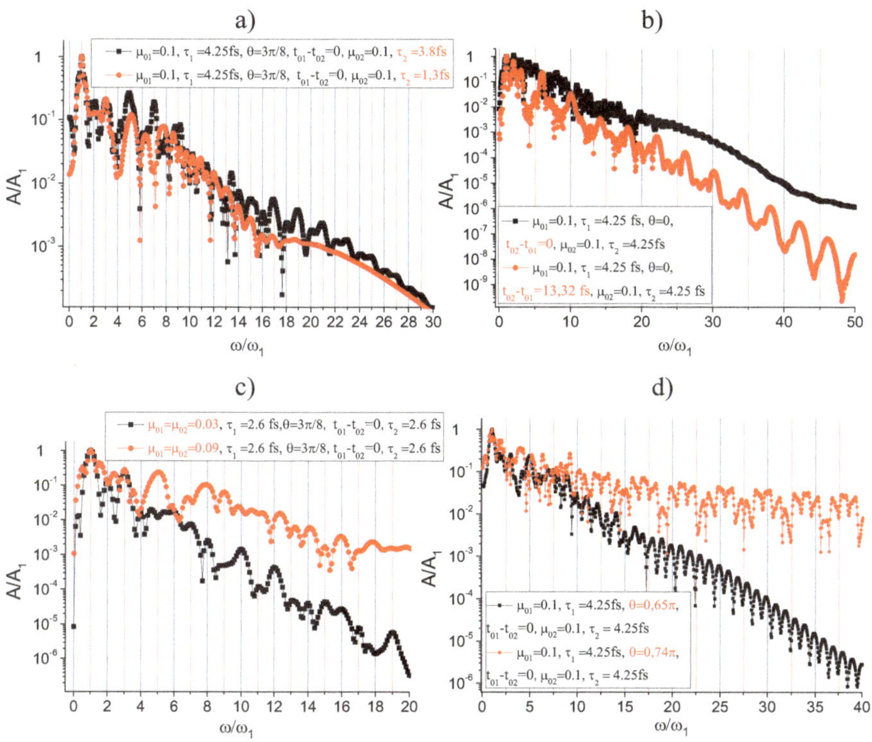

Fig. 13. Spectra of argon atom response in two-color field: (a) $\mu_{01} = \mu_{02} = 0.1, \tau_1 = 4.25fs, t_{01} - t_{02} = 0, \theta = 3\pi/8, \tau_2 = 3.8fs$ (squares) and $\tau_2 = 1.3fs$ (circles); (b) $\mu_{01} = \mu_{02} = 0.1, \tau_1 = \tau_2 = 4.25fs, , \theta = 0, t_{01} - t_{02} = 0$ (squares) and $t_{01} - t_{02} = 13.32fs$ (circles); (c) $\tau_1 = \tau_2 = 2.6fs, t_{01} - t_{02} = 0, \theta = 3\pi/8, \mu_{01} = \mu_{02} = 0.03$ (squares) and $\mu_{01} = \mu_{02} = 0.09$ (circles); (d) $\mu_{01} = \mu_{02} = 0.1, \tau_1 = \tau_2 = 4.25fs, t_{01} - t_{02} = 0, \theta = 0.65\pi$ (squares) and $\theta = 0.74\pi$ (circles)

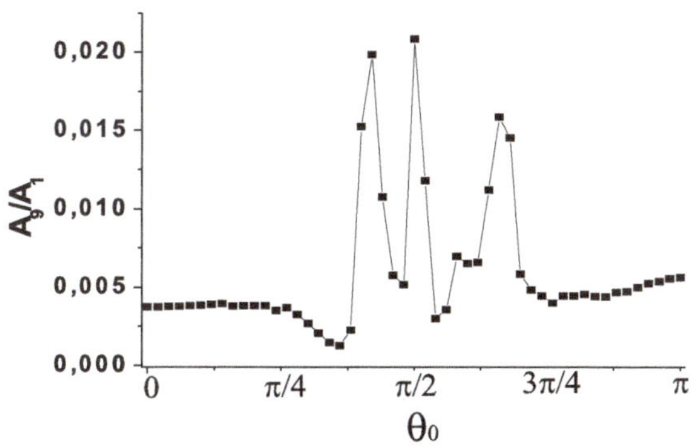

Fig. 14. Ninth harmonic field amplitude as a function of angle between the polarization vectors of laser field components at frequencies ω and 2ω

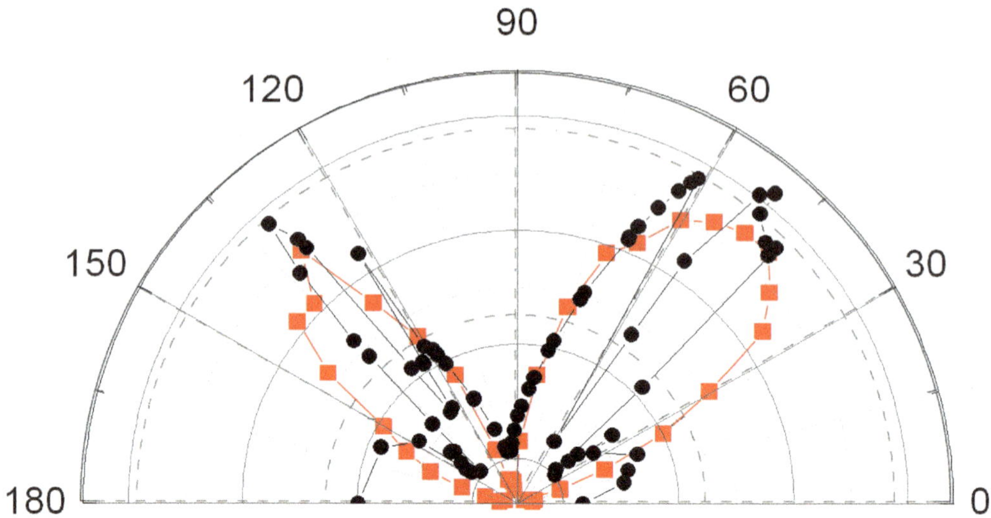

Fig. 15. THz emission power under variation of the 2ω polarization angle: experiment (squares) and computer simulations (circles)

5. Conclusions

The new theoretical approach in the theory of an atom interaction with the intense laser pulses has been developed. The proposed approach is based on the exact analytical solutions of the boundary value problem for "atom in the external field". The obtained solutions form the complete basis of the orthonormal eigenfunctions including both the discrete and continuous spectrum states. The spectrum of eigenvalues of "atom in the external field" problem coincides exactly with that for free atom boundary value problem. The eigenfunction bases of these two boundary value problems are the one-to-one sets. The transformation matrix is $\langle nlm| \exp\left(i\frac{q}{\hbar c}\vec{A}(t)\vec{r}\right)|n'l'm'\rangle$, where n is the principle quantum number, l and m are the angular momentum and its projection. For any atom with the spherically symmetric intra-atomic potential the angular part of transformation matrix is calculated in explicit analytic form; the radial part is also calculated in analytic form for the case of hydrogenic radial functions. The developed approach is non-perturbative one, because it is free of some constraints on the ratio of laser to intra-atomic field strengths. This is the most principle benefit of the developed approach. Notice, that as far as the theory is based on the Schrodinger equation we can speak here on the non-relativistic interactions. The generalization of the proposed approach for the region of relativistic field strength is given in book (Andreev, 2009). The developed theory has been applied to study the specific features of non-linear atomic response in the case of the laser pulses of near-atomic field strength and multi-color laser fields. The presented results of the mathematical modeling provide the interpretation of a number of phenomena, which has been observed experimentally and closely related to near-atomic field strength of laser pulses. (i) The violation of the electro-dipole selection rule. This phenomenon is due to the nonlinear dependency of compound matrix elements M_{nm} on the laser field strength. As it has been shown these matrix elements are linear functions of laser field amplitude only in region of subatomic field strength. The growth of the field amplitude results in the dominating role of high order terms with respect to the degrees of field amplitude. (ii) The stabilization of ionization, enhanced ionization, and saturation of ionization rate. The nature of these phenomena is at least twofold. Firstly, the account of multilevel structure of bound atomic states affects drastically on the ionization probability dependency on the laser field strength. This is due a number of different reasons. The integral rate of level population is a sum of the rates of ionization, recombination, and interlevel transitions among the atomic discrete spectrum. Any of these rates is nonlinear function of the field amplitude. The energy width of continuum populated states is also nonlinearly depends on the field amplitude. The stabilization ionization process occurs when the recombination process becomes dominating. The enhanced ionization is due to the ionization from the excited bound states and it occurs when the population of excited discrete spectrum states becomes appreciable. The saturation of ionization probability is due to the total single ionization of an atom. (iii) The cut-off frequency saturation. The good agreement of the results of computer simulations on the silver atom photoemission spectra and experimentally measured spectra shows that the cut-off frequency saturation is due to the total single ionization. (iv) The developed theory, supported by results of computer simulations, shows that in multicolor laser fields the variation of mutual polarization of field constituents provides the most effective method of photoemission spectrum control both in short and long wavelength regions. The results of calculations of THz emission power as a function of mutual orientation of two-color field polarization vectors show that in multicolor fields the atomic response is due mainly by atomic but not plasma nonlinearities. The significant enhancement in the intensity of high harmonics under variation of mutual orientation of two-color field polarization vectors is of great practical interest as a source of intensive x-ray emission and for development of subfemtosecond pulse formation methods.

6. References

Andreev, A. V. (2009). *Relativistic Quantum Mechanics*, Fizmatlit.

Andreev, A. V. (1999). Interaction of an atom with superstrong laser fields, *Journal of Experimental and Theoretical Physics* 89(3): 421–427.

Andreev, A.V. & Shoutova, O. (2006). Single hydrogen like atom ionization by ultrastrong laser field: non-perturbative approach, *Physics Letters A* 350(3-4): 309–314.

Andreev, A.V.; Shoutova, O. . S. S. Y. (2007). Ionization of a single hydrogen-like atom by laser pulses of near-atomic strength, *Laser Physics* 17(4): 496–507.

Andreev, A.V.; Stremoukhov, S. . S. O. A. (2008). Atom in electromagnetic field of near-atomic strength, *Journal of Russian Laser Research* 29(3): 203–218.

Andreev, A. V.; Stremoukhov, S. Yu. & Shoutova, O. A.(2010). Ionization of a multilevel atom by ultrashort laser pulses, *Zh. Eksp. Teor. Fiz.* 111(6): 936Ű948.

Bartel, T.; Gaa, P.; Reimann, K.; Woerner, M. & Elsaesser, T.(2005). Generation of single-cycle thz transients with high electric-field amplitudes, *Optics Letters* 30(20): 2805–2807.

Bauer, D. & Koval, P. (2006). Qprop: A schrodinger-solver for intense laser-atom interaction, *Comp. Phys. Comm.* 174(5): 396–421.

Bauer, D.; Milosevic, D. B. & Becker, W.(2005). Strong-field approximation for intense-laserŰatom processes: The choice of gauge, *Phys. Rev. A* 72: 023415.

Cook, D. J. & Hochstrasser, R. M. (2000). Intense terahertz pulses by four-wave rectification in air, *Optics Letters* 25(16): 1210–1212.

Corkum, P. B. & Krausz, F. (2007). Attosecond science, *Nature Physics* 3: 381–387.

Corkum, P. B.; Burnett, N. H. & Ivanov, M. Y.(1994). Subfemtosecond pulses, *Optics Letters* 19 (22): 1870–1872.

Faisal, F. H. M. (1973). Multiple absorption of laser photons by atoms, *J. Phys. B: At. Mol. Phys.* 6: L89.

Ganeev, R. (2009). Generation of high-order harmonics of high-power lasers in plasmas produced under irradiation of solid target surfaces by a prepulse, *Phys. Usp* 52: 55–77.

Ganeev, R. A.; Baba, M.; Suzuki, M. & Kuroda, H.(2005). High-order harmonic generation from silver plasma, *Phys. Lett. A* 339: 103–109.

Ganeev, R. A.; Elouga Bom, L. B.; Kieffer, J.-C.; Suzuki, M.; Kuroda, H. & Ozaki, T.(2007a). Demonstration of the 101st harmonic generated from a laser-produced manganese plasma, *Phys. Rev. A* 76: 023831.

Ganeev, R.A.; Singhal, H.; Naik, P.A.; Chakravarty, U.; Arora, V.; Chakera, J.A.; Khan, R.A.; Raghuramaiah, M.; Kumbhare, S.R.; Kushwaha, R.P. & Gupta, P.D. (2007b). Optimization of the high-order harmonics generated from silver plasma, *Applied Physics B: Lasers and Optics* 87(2): 243–247.

Gorbunov, L. M. & Frolov, A. A. (1996). Emission of low-frequency electromagnetic waves by a short laser pulse in stratified rarefied plasma, *JETP* 83(5): 967.

Hamster, H.; Sullivan, A.; Gordon, S.& Falcone, R. W.(1994). Short-pulse terahertz radiation from high-intensity-laser-produced plasmas, *Phys. Rev. E* 49: 671Ű677.

Hamster, H.; Sullivan, A.; Gordon, S.; White, W. & Falcone, R. W.(1993). Subpicosecond, electromagnetic pulses from intense laser-plasma interaction, *Phys. Rev. Lett.* 71: 2725Ű2728.

Itatani, J.; Levesque, J.; Zeidler, D.; Niikura, H.; Pepin, H.; Kieffer, J. C.; Corkum, P. B. & Villeneuve, D. M. (2004). Tomographic imaging of molecular orbitals, *Nature (London)* 432: 867Ű871.

Keldysh, L.V. (1965). Ionization in the field of a strong electromagnetic wave, *Sov. Phys. JETP* 20: 1307.

Kim,K.Y.; Taylor,A.J.; Glownia,A.J. & Rodriguez, G.(2008). Coherent control of terahertz supercontinuum generation in ultrafast laser-gas interactions, *Nature photonics* 2: 605–609.

Krause, J. L.; Schafer, K. J. & Kulander, K. C.(1992). High-order harmonic generation from atoms and ions in the high intensity regime, *Phys. Rev. Lett.* 68: 3535Ű3538.

Kress, M.; Loffler, T.; Eden S.; Thomson, M. & Roskos, H. G.(2004). Terahertz-pulse generation by photoionization of air with laser pulses composed of both fundamental and second-harmonic waves, *Optics Letters* 29(10): 1120–1122.

Kulander, K. C.; Schafer, K. J. & Krause, J. L.(1991). Dynamic stabilization of hydrogen in an intense, high-frequency, pulsed laser field, *Phys. Rev. Lett.* 66: 2601Ű2604.

Landau, L. D. & Lifshits, E. M. (1981). *Quantum Mechanics, 3rd ed.*, Elsevier Science.

Marte, P. & Zoller, P. (1991). Hydrogen in intense laser fields: Radiative close-coupling equations and quantum-defect parametrization, *Phys. Rev. A* 43: 1512Ű1522.

Muller, H. (1999). An efficient propagation scheme for the timedependent schrodinger equation in the velocity gauge, *Laser Phys.* 9: 138.

Papadogiannis, A.; Nikolopoulos, L.A.A.; Charalambidis, D.; Tzallas, P.;Tsakiris, G. & Witte, K.(2003). Two xuv-photon ionization of he through a superposition of higher harmonics, *Phys. Rev. Lett* 90: 133902.

Parker, J. & Stroud, C. R. (1990). Population trapping in short-pulse laser ionization, *Phys. Rev A* 41(3): 1602.

Paul, P. M.; Toma, E. S.; Breger, P.; Mullot, G.; Aug, F.; Balcou, Ph.; Muller, H. G. & Agostini, P. (2001). Observation of a train of attosecond pulses from high harmonic generation, *Science* 292: 1689.

Perelomov, A. M. & Popov, V. S. (1967a). Ionization of atoms in alternating electric field. iii, *Sov. Phys. JETP* 25: 336–343.

Perelomov, A.M.; Popov, V.S. & Terent'ev, M.V. (1967b). Ionization of atoms in an alternating electric field, *Sov. Phys. JETP* 24: 207.

Pont, M. & Gavrila, M. (1990). Stabilization of atomic hydrogen in superintense, high-frequency laser fields of circular polarization, *Phys. Rev. Lett.* 65: 2362Ű2365.

Popov, V. S. (2004). Tunnel and multiphoton ionization of atoms and ions in a strong laser field (keldysh theory), *Phys. Usp.* 47: 47 855Ű885.

Popov, V. S.; Kuznetsov, V. P. & Perelomov, A. M.(1968). *Sov. Phys. JETP* 26: 222.

Rae, S.C.; Chen, X. & Burnett, K.(1994). Saturation of harmonic generation in one- and three-dimensional atoms, *Phys. Rev. A* 50: 1946–1949.

Redkin, P. V. & Ganeev, R. A. (2010). Simulation of resonant high-order harmonic generation in a three-dimensional fullerenelike system by means of a multiconfigurational time-dependent hartree-fock approach, *Phys. Rev. A* 81: 063825.

Reiss, H. R. (1990). Complete keldysh theory and its limiting cases, *Physical Review A* 42 (3): 1476–1486.

Reiss, H. R. (1992). Theoretical methods in quantum optics: S-matrix and keldysh techniques for strong-field. processes, *Prog. Quantum Electron* 16: 1.

Reiss, H. R. (1980). Effect of an intense electromagnetic field on a weakly bound system, *Phys. Rev. A* 22: 1786Ű1813.

Sola, I. J.; Melvel, E.; Elouga, L.; Constant, E.; Strelkov, V.; Poletto, L.; Villoresi, P.; Benedetti, E.; Caumes, J.-P.; Stagira, S.; Vozzi, C.; Sansone, G. & Nisoli, M.(2006). Controlling attosecond electron dynamics by phase-stabilized polarization gating, *Nature Physics* 2: 319 – 322.

Sprangle, P.; Peano, J. R.; Hafizi, B. & Kapetanakos, C. A.(2004). Ultrashort laser pulses and electromagnetic pulse generation in air and on dielectric surfaces, *Phys. Rev. E* 69: 066415.

Suzuki, M.; Baba, M.; Kuroda, H.; Ganeev, R. A. & Ozaki, T.(2007). Intense exact resonance enhancement of single-high-harmonic from an antimony ion by using ti:sapphire laser at 37 nm, *Optics Express* 15(3): 1161–1166.

Tobey, R. I.; Siemens, M. E.; Cohen, O.; Murnane, M. M. & Kapteyn, H. C.(2007). Ultrafast extreme ultraviolet holography: Dynamic monitoring of surface deformation, *Opt. Lett.* 32: 286.

Volkova, E. A.; Gridchin, V. V.; Popov, A. M. & Tikhonova, O. V.(2006). Tunneling ionization of a hydrogen atom in short and ultrashort laser pulses, *Journal of Experimental and Theoretical Physics* 102(1): 40–52.

Volkova, E. A.; Popov, A. M. & Tikhonova, O. V. (2001). Ionization and stabilization of a three-dimensional system with a short-range potential in a strong laser field, *Journal of Experimental and Theoretical Physics* 93(6): 1155–1162.

Volkova, E. A.; Popov, A. M.; Tikhonov, M. A. & Tikhonova, O. V.(2007). Atom under an intense laser pulse: Stabilization effect and strong-field approximation, *Journal of Experimental and Theoretical Physics* 105(3): 526–534.

Watanabe, S.; Kondo, K.; Nabekawa, Y.; Sagisaka, A. & Kobayashi, Y.(1994). Two-color phase control in tunneling ionization and harmonic generation by a strong laser field and its third harmonic, *Phys. Rev. Lett.* 73: 2692Ű2695.

Yin, Y.-Y.; Chen, C.; Elliott, D. S. & Smith, A. V.(1992). Asymmetric photoelectron angular distributions from interfering photoionization processes, *Phys. Rev. Lett.* 69: 2353Ű2356.

Part 2

Time Resolved Laser Spectroscopy and Coherent Control Techniques

11

Quantum Key Distribution

Philip Chan[1], Itzel Lucio-Martínez[2], Xiaofan Mo[2] and Wolfgang Tittel[2]
[1]*Institute for Quantum Information Science, and Department of Electrical and Computer Engineering, University of Calgary*
[2]*Institute for Quantum Information Science, and Department of Physics and Astronomy, University of Calgary*
Canada

1. Introduction

This chapter describes the application of lasers, specifically diode lasers, in the area of quantum key distribution (QKD). First, we motivate the distribution of cryptographic keys based on quantum physical properties of light, give a brief introduction to QKD assuming the reader has no or very little knowledge about cryptography, and briefly present the state-of-the-art of QKD. In the second half of the chapter we describe, as an example of a real-world QKD system, the system deployed between the University of Calgary and SAIT Polytechnic. We conclude the chapter with a brief discussion of quantum networks and future steps.

2. Motivation

The importance of communication networks has increased significantly in the last three decades. As an ever-growing fraction of our daily activities (including e-mail and e-banking) now depends on communication over public channels such as optical fibers, the security of the exchange of sensitive information also became an issue of high importance. Of special concern are credit card numbers, personal health records and business-related information, just to give some examples.

The problem of guaranteeing security is solved by encrypting the sensitive information (referred to as the *plaintext* and assumed to be in its binary form) before transmission. During encryption, the plaintext is processed using a *key* and a certain algorithm (the *cipher* or *cryptosystem*). The encrypted message is referred to as the *cryptogram* (or *ciphertext*), and the sender is typically denoted *Alice*. The cryptogram is sent through the communication channel to the receiver, *Bob*. If intercepted during transmission, it should be incomprehensible to an eavesdropper, typically called *Eve*. The encrypted message becomes meaningful only once it is decrypted. This requires a *secret key*, which must be known only to the legitimate receiver and, depending on the cryptosystem, the sender of the message.

Different types of ciphers exist; they can be divided into two groups: symmetric and asymmetric ciphers.

- *Asymmetric* ciphers use two different keys: a *public key* with which anyone can encrypt a message and a *private key* that belongs to the receiver of the message. Only the private key allows decrypting the message.

- *Symmetric* ciphers use the same secret key for encryption and decryption.

Obviously, the encrypted message must not reveal any information about the plaintext. Hence, the secrecy of the encrypted message transmitted over a public channel relies on the secrecy of the key used for decryption. The security of a cryptosystem is generally assessed in terms of the time required to break it; two general categories exist:

- *Computational security* assumes the eavesdropper has limited computational power, and relies on assumptions about the difficulty to solve a certain mathematical problem. An example of a cryptosystem that provides computational security is RSA (named after the inventors Ronald Rivest, Adi Shamir and Leonard Adleman)[1]; its security relies on the difficulty to factorize a large number into its primes. In the factoring problem, the number of computational steps required increases exponentially with the number of bits used to represent the number to be factorized (Nielsen & Chuang (2004)). This is generally believed not to be computable for sufficiently large numbers. For instance, to factorize a 768-bit number, the best known classical algorithm has been estimated in 2010 to require on the order of 1500 years if a single-core, high-end processor is used (Kleinjung et al. (2010)). Furthermore, the factorization of a 1024-bit number, the current standard for RSA, is believed by the same authors to be 1000 times harder. This should suffice to safeguard any information one may want to encrypt.

  However, the difficulty to factorize large numbers on a classical computer (a computer whose operation can be described using classical physics) is not proven, and less time-consuming algorithms may exist, or a large computer cluster may be used. Referring again to the example of a 768-bit number given above, this number has actually been factorized in 2010 using many hundreds of computers over a period of only two years (Kleinjung et al. (2010)). Furthermore, the researchers estimated that it is not unreasonable to expect that a 1024-bit RSA key can be factored within the next decade. Probably even more threatening, it is known that a quantum computer can factorize large numbers efficiently (i.e. in polynomial time – exponentially faster than the best known classical algorithm) by means of Shor's algorithm (Nielsen & Chuang (2004)). While these threats to RSA encryption do not yet exist (or rather, are currently not known to exist), today's eavesdropper could simply copy the encrypted information, wait for algorithmic or technological advances, and then decrypt the message efficiently.

  We thus have to ask: when is it necessary to research alternative methods to safeguard information in transit? To answer this question, let us assume that the information has to remain encrypted for x years. Furthermore, let y be the time needed to retool the current secure-communication infrastructure. Hence, if disruptive technology appears in z years, where $z < x + y$, then a secret that has been encoded before the encryption infrastructure was improved becomes unprotected before the end of its lifetime. There is thus clear need to investigate and implement new encryption systems long before the old ones are broken (Stebila et al. (2010)).

- *Information-theoretic security* relies only on information-theoretic arguments. In particular, the security of the encrypted message does not depend on any assumptions about the computational power of an eavesdropper. The *one-time pad* satisfies this stringent condition. In the one-time pad the message is encrypted by combining it bit by bit with a

[1] RSA encryption was independently invented by Clifford Cocks four years earlier. Yet, his discovery was classified top secret by British Intelligence and was only revealed in 1997.

Cryptosystems		
Type	*Example*	*Security*
Asymmetric	RSA	Computational security
Symmetric	One-time pad	Information theoretic security

Table 1. Examples for two different cryptosystems

random binary key, see figure 1. The key must be used only once, must be as long as the message that is to be encrypted, and, obviously, must only be known to the sender and the receiver. Then, regardless of the available time, an eavesdropper will never obtain the message from the ciphertext.

An important problem when using the one-time pad is thus the distribution of the secret key. This is generally accomplished using trusted couriers, a cumbersome solution that restricts its wide implementation and leaves a security loophole in the overall encryption procedure. Indeed, the key itself does not reveal if a non-trustworthy courier has duplicated it. This raises the question of whether alternative key distribution mechanisms exist that operate over standard communication channels and allow the detection of eavesdropping.

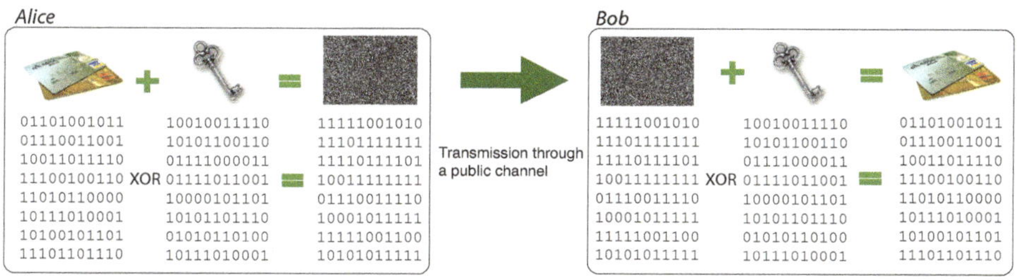

Fig. 1. The one-time pad. Alice encodes her message with a random binary key using the XOR operation (i.e. 0+0=0, 0+1=1, 1+0=1, and 1+1=0). The encrypted message is then sent over a public channel to Bob. He decodes it with the same key, again using the XOR operation, and thereby obtains Alice's original message.

3. Quantum key distribution

Quantum key distribution (QKD) (Gisin et al. (2002); Scarani et al. (2009)) takes advantage of the peculiar properties of individual photons to provide two parties with arbitrarily long secret keys, provided a short key for authentication purposes is initially available. Hence, a more appropriate name for QKD would be quantum key growing. Yet, as is common use, we will refer to this procedure as QKD. For QKD, information is encoded into one degree of freedom of photons (e.g. their polarization state), while all other degrees of freedom (phase, wavelength, etc.) must not contain any information. Each information-carrying photon then becomes a *quantum bit (qubit)*[2]. Due to their quantum nature, it is impossible for an eavesdropper to extract information about the encoded quantum states during transmission

[2] Encoding information into so-called continuous quantum variables (Scarani et al. (2009)) is possible as well but will not be discussed here.

without altering them, which can be detected by Alice and Bob. Ignoring for the moment loopholes arising from imperfect implementations (we will discuss this problem later), the combination of QKD with one-time pad encoding provides information-theoretic secure communication[3] that withstands any technological and algorithmic advances – even the development of the quantum computer.

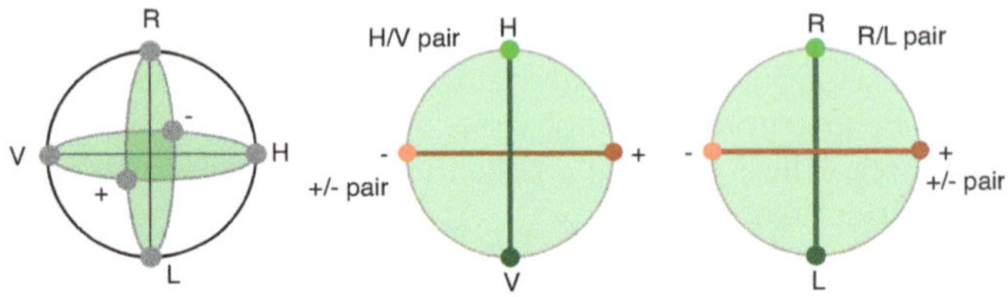

Fig. 2. The Poincaré sphere (left-hand picture), and two examples of polarization states that are suitable for the BB84 protocol (center and right-hand picture). The first example comprises the $|H\rangle$, $|V\rangle$, $|+\rangle$ and $|-\rangle$ linear polarization states, the second one comprises the $|R\rangle$, $|L\rangle$ circular polarization states and the $|+\rangle$ and $|-\rangle$ linear polarization states. Note that the states belonging to each example are arranged equally on two great circles around the Poincaré sphere.

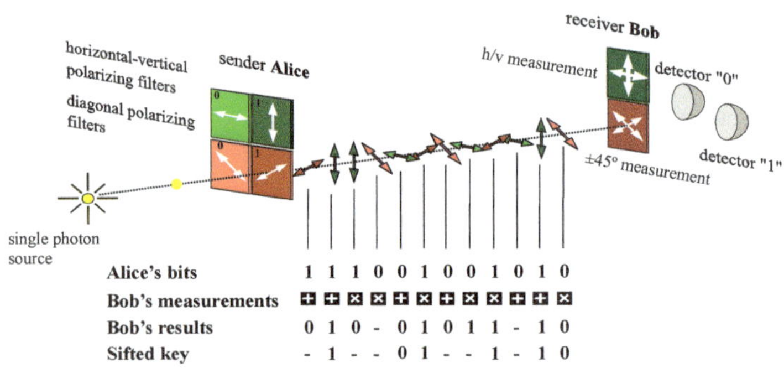

Fig. 3. Illustration of the BB84 protocol.

For QKD, Alice and Bob share two communication channels: a quantum channel that is used to transmit qubits, and a classical (standard) channel to send classical messages. The quantum channel is generally an optical fiber or a free-space link connecting Alice and Bob, while the classical channel may be the Internet.

3.1 The BB84 protocol

Quantum key distribution was proposed in 1984 by Charles Bennett and Gilles Brassard (Bennett & Brassard (1984)). The first QKD protocol, generally referred to as the BB84 or four-state protocol, uses four different quantum states that form two pairs, chosen such that

[3] This property is sometimes also referred to as *unconditional security*, a technical term that refers to the fact that the security is not based on mathematical assumptions. This meaning must not be confused with *secure without any conditions*.

the difference (overlap) between any two states, one from each pair, is the same. A specific bit value is assigned to each state. For example, if the linear polarization states horizontal ($|H\rangle$), vertical ($|V\rangle$), +45° ($|+\rangle$) and -45° ($|-\rangle$) are used, $|H\rangle$ and $|V\rangle$ form one pair, $|+\rangle$ and $|-\rangle$ form the other pair, then $|H\rangle$ and $|-\rangle$ correspond to bit 0, while $|V\rangle$ and $|+\rangle$ correspond to bit 1, see figure 2[4]. For each photon Alice sends to Bob, she randomly selects one out of these four states. Bob, before receiving it, randomly decides whether to make a measurement that allows discriminating horizontally from vertically polarized photons, or +45° from -45° polarized photons. It is important to note that it is impossible to make a measurement that allows distinguishing between all four possible states (Nielsen & Chuang (2004)). Hence, whenever Bob picks the measurement that allows him to distinguish between the state that Alice sent and the orthogonal state, his measurement result indicates with certainty which state was originally prepared by Alice. Identifying states with bits, as described above, then obviously results in equal bits at Alice's and Bob's. Conversely, in the case where Bob choses to do the 'wrong' measurement, the measurement result as well as the associated bit value will be uncorrelated with the state sent by Alice.

Let us use the example of Alice sending state $|H\rangle$ to make this concept more concrete: If Bob makes the 'correct' measurement (i.e. the measurement that distinguishes between $|H\rangle$ and $|V\rangle$), he will find the result $|H\rangle$. Hence, Alice and Bob have sent and received the same bit value: 0. However, if Bob choses the measurement that distinguishes between $|+\rangle$ and $|-\rangle$, he will randomly obtain one of these two possible outcomes, and hence randomly obtain the bit value 0 or 1. From this point on all information exchange between Alice and Bob is performed over the classical channel, and the remaining steps in the protocol are known under the name of *classical post-processing*. These procedures are explained below.

Assume now that an eavesdropper intercepts photons during transmission from Alice to Bob. She (the eavesdropper) may duplicate each photon, keep one copy to measure it and obtain a bit value, and send the other copy to Bob. Fortunately, one of these steps is not allowed by quantum theory: the *no-cloning theorem* states that it is impossible to make a perfect copy of a photon in an unknown quantum state (Nielsen & Chuang (2004)). In fact, both 'duplicates' will only approximately resemble the original. Hence, the eavesdropper's attempt to eavesdrop inevitably leads to a modification of the state of the photon sent on to Bob, which results in errors in the bits that should be perfectly correlated and thereby reveals eavesdropping[5].

4. Classical post-processing

Classical post-processing in QKD consists of four operations, as shown in Figure 4.

4.1 Sifting

Alice and Bob start by performing key sifting: Bob communicates which photons he detected, and which measurements he performed in these cases. However, he does not specify the results he obtained. Alice and Bob only keep those bits from their *raw keys* (i.e. the list of bits specifying the quantum states Alice sent and Bob detected, see figure 3) where Bob detected a photon, and that resulted from correct measurements (as defined above). All other bits are discarded. The result of this step is called the *sifted key*. Ideally, the sifted key would be error

[4] We use the usual ket-notation to denote quantum states, e.g. $|\psi\rangle$ denotes the quantum state ψ.

[5] Note that other, more sophisticated eavesdropping strategies exist. However, regardless of the strategy, an eavesdropper gaining information about the photon states inevitably introduces errors at Bob's.

free, but in practice, no communication system is perfect and thus a small error rate, generally referred to as the *quantum bit error rate* (QBER), always remains. In addition, errors introduced by eavesdropping may also be present in the sifted key.

4.2 Error-correction

The next step is to perform error correction: Alice sends Bob additional information that allows him to generate an *error-corrected key* that is identical to Alice's. Furthermore, this procedure yields the QBER. Error correction in QKD is similar to error correction in classical communications, with the sifted key in QKD being analogous to the transmitted message. However, there is one important difference: rather than combining information that allows correcting errors directly into the message to be transmitted, this information is sent after key sifting is complete as it is only at this point that the message to be corrected is known. Furthermore, we need not worry about protecting the information for error correction against transmission errors as we can use existing protocols that provide error-free communication. With this in mind, the Cascade error correction protocol was originally designed for QKD (Bennett et al. (1992)). It requires many rounds of back-and-forth communication between Alice and Bob, which limits the maximum key rates that can be handled, due to communications delays. More recently, Low-Density Parity-Check codes (Gallager (1962); MacKay & Neal (1997)) have been adapted from classical communication protocols – they are capable of handling larger key rates (Pearson (2004)).

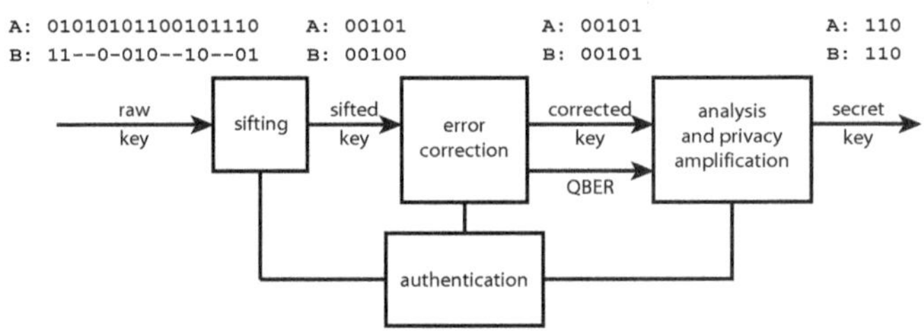

Fig. 4. Post-processing in QKD. Example keys for Alice (A) and Bob (B) are shown. Only key bits resulting from correct measurements and photon detections are kept during sifting, yielding the sifted key. Remaining errors are corrected to form the error-corrected key. The eavesdropper's information about the error-corrected key is removed, yielding the secret key. Authentication is required for all steps.

4.3 Privacy amplification

Next, the amount of information that an eavesdropper may have obtained is estimated. The analysis considers several factors:

- First, the maximum amount of information Eve may have gained from measuring photons in transit, I_t, is evaluated from the error rate introduced into the sifted key (Gottesman et al. (2004)). Generally, all errors are attributed to eavesdropping as this is the worst-case scenario. If QKD is implemented using extremely faint laser pulses (containing on average less than one photon) instead of single photons, this procedure has to take into account the possibility of a so-called photon number splitting (PNS) attack (Brassard et al. (2000)). This attack exploits the fact that faint laser pulses sometimes contain more than one photon. It is generally thwarted by adding decoy states – qubits encoded into faint pulses with different mean photon number – to the signals used to establish the sifted key (Scarani et al. (2009)). This allows determining the information gained by the eavesdropper from PNS attacks, and makes implementations using faint laser pulses comparable to those using (much more difficult to generate) single photons.

- Second, the eavesdropper gains additional information by monitoring the error-correction procedure. The amount, I_{ec}, is directly given by the number of bits exchanged.
- Third, as we will discuss in more detail later, an implementation of QKD using imperfect devices can impact the security of the protocol. Sometimes, it is possible to quantify the information leakage I_{leak} (Lamas-Linares & Kurtsiefer (2007)).

Using these three contributions, the length of the secret key to be distilled from the corrected key is computed. Assuming for simplicity that Alice has used true single photons[6], the secret key per error-corrected bit amounts to

$$I_{secret} = 1 - I_t - I_{ec} - I_{leak}. \quad (1)$$

Privacy amplification, implemented by both Alice and Bob, then maps the error-corrected keys onto shorter secret keys in a way that knowledge of many bits in the corrected key is required to calculate any bit of the secret key. This step is typically performed by multiplication of the corrected key (expressed as a bit vector) with a suitably chosen binary Toeplitz matrix. This removes the eavesdropper's information about the key.

4.4 Authentication

An important consideration in all post-processing steps is to ensure that Alice and Bob are in fact going through this process with each other. Otherwise, an eavesdropper could simply block all quantum and classical communication between Alice and Bob and perform QKD with Alice while taking on Bob's role and vice versa. This is known as a man-in-the-middle attack, and would allow Eve to establish different secret keys with both Alice and Bob. She could then intercept a secret message, for example, being sent from Alice to Bob, decrypt it using the key she shares with Alice, read it, and then encrypt it again using the key she shares with Bob before forwarding it to him. She could thus read the entire message. To avoid man-in-the middle attacks, the classical channel has to be authenticated. In other words, Alice and Bob have to identify each message they send as originating from themselves. This can be achieved using a protocol known as Wegman-Carter authentication (Wegman & Carter (1981)). Hence, while the eavesdropper can listen to the conversation during post-processing (i.e. the classical channel is not required to be secure), she cannot modify or replace it. Authentication requires a short initial key, which is consumed during the first round of QKD. For subsequent rounds, it is replaced by some of the key generated during the key expansion[7].

5. Security loopholes in QKD

As introduced above, eavesdropping qubits encoded into individual photons during transmission is revealed through the observed error rate, regardless of the strategy. However,

[6] This formula is easily adapted to the case of faint laser pulses. However, for pedagogical reasons, we only give this simple expression.

[7] Note that the security of QKD is not compromised if the initial key is revealed after the first round of QKD. This property therefore allows, for instance, the use of a computational secure cryptosystem for the distribution of this initial key, provided it features sufficient short-term security.

loopholes in the actual implementation of QKD may exist and can be exploited for attacks that are not reflected in the QBER. This was already noted in the very first implementation of QKD, where Charles Bennett realized that the noise emitted by the QKD system rendered the key only secure against an eavesdropper who happened to be deaf.
Quantum hacking has become an important research field during the past five years, and various attacks have been proposed and experimentally studied. The most important ones are briefly introduced below.

- The already mentioned *photon number splitting attack* takes advantage of the fact that faint laser pulses sometimes contain more than one photon (Brassard et al. (2000)). This opens a security loophole when using the original BB84 protocol. Interestingly, this loophole can be closed by a small modification of the protocol, i.e. the addition of decoy states (Scarani et al. (2009)).
- In the *Trojan-horse attack* (Vakhitov et al. (2001)), Eve exploits the fact that every optical element reflects some of the incident light. It is then possible to analyze the status of optical components such as phase (polarization) modulators by reflecting short pulses of light from them, yielding for instance information about the qubit state that is being generated. This technique is called reflectometry and is well known to optical engineers. Counter measures include active monitoring of light at the input of Alice and Bob, and, for Alice, an optical isolator.
- QKD systems rely on single photon detectors. In the *detector blinding attack*, an eavesdropper exploits that these detectors can be prevented from detecting photons, and then forced to announce detections at will using various mechanisms (Makarov (2011)). In the *time-shift attack* (Lamas-Linares & Kurtsiefer (2007); Qi et al. (2007)) the eavesdropper exploits a possible detection efficiency mismatch between two detectors in the time domain. In this case, controlling the arrival time of each photon at Bob's device allows the eavesdropper to modify the probabilities for certain detectors to detect a given photon, and thereby yields information about the key. Counter measures against attacks exploiting vulnerabilities of single photon detectors, often combining hardware and protocol modifications, have already been proposed and investigated (Makarov (2011); Yuan et al. (2010)).

It is of utmost importance to critically assess vulnerabilities of QKD systems and devise counter measures, either of theoretical nature or on the technological level, to remove the threat to security. Yet, even in the case of remaining potential loopholes, one should not underestimate that the security of QKD depends on the technological capabilities of the adversary at the time of the key exchange, in contrast to complexity-based cryptosystems that generate ciphertexts that can be recorded and decoded later. This point is important for secrets that should remain secure over many years.

6. State-of-the-art

Quantum key distribution is the most mature application in the field of quantum information processing. For a few years, QKD systems have been commercially available (idQuantique (2011); MagiQ (2011); Quintessence (2011)) but research still continues to progress both theoretically and experimentally. Comparing different QKD systems is not a simple task, and trying to identify the 'best' system in a field that still evolves quickly is quite a pointless effort. Furthermore, it is unclear what figure of merit one should use. However, to give some idea about what is currently achievable, let us briefly summarize recent results that have advanced practical QKD in terms of maximum distance and key rate. Please note that, while these systems deliver secret keys if eavesdropping is restricted to measuring qubits in transit[8], we make no claim concerning their robustness against security loopholes such as those discussed above.

QKD systems differ in terms of the type of quantum channel used (fiber or free space), the degree of freedom utilized to encode qubits (e.g. polarization or phase), and the nature of the quantum effect exploited (qubit states encoded into faint laser pulses, or so-called entangled qubits (Tittel & Weihs (2001))). Furthermore, they may employ single photon detectors based on avalanche photo diodes or on superconductors. As the state-of-the-art (in terms of distance and key rate) is only weakly system-dependent, we will not distinguish between different implementations.

Distances over which QKD systems operate can, in the best case, exceed $\sim$200 km (Liu (2010); Schmitt-Manderbach et al. (2007); Stucki et al. (2009)). Due to the high channel loss experienced, the secret key rates are typically limited to $\sim$10 bps. On the other hand, QKD systems have been demonstrated to deliver secret key rates up to $\sim$1 Mpbs (Dixon et al. (2010); Restelli et al. (2009)). Obviously, the distances are reduced compared to those mentioned previously; the current maximum is 50 km. It is likely that these two benchmarks will not be improved significantly over the next few years, the only exception possibly being QKD over a free-space link between a ground station and a (very distant) satellite (Ursin et al. (2009)).

An important concern in QKD systems, in particular in those clocked at high rates, is the generation of random numbers. For instance, in faint pulse based QKD, the security relies on Alice generating randomly selected qubit states encoded into laser pulses with average photon number chosen randomly from a small set. Furthermore, Bob has to randomly select which measurement to perform[9] and some randomness is required for privacy amplification. Hence, true random number generators, possibly exploiting randomness of certain quantum effects, are needed. A lot of progress has been made over the past years; quantum random number generators delivering random numbers at 16 Mbps are commercially available (idQuantique (2011)), and 50 Mbps rates have meanwhile been achieved in an academic effort (Fürst et al. (2010)). Yet, further improvement is required for QKD systems clocked at Gbps rates. Another possibility is the use of physical (non-quantum) RNGs for which Gbps rates have been reported (Honjo et al. (2009)).

7. Our QKD system - a case study

The optical part of a QKD system consists of subsystems for signal generation, modulation, transmission, demodulation and detection, see figure 5. Figure 6 shows pictures of the sender and receiver of the QKD system that is currently being developed in our group (Lucio-Martinez et al. (2009)). Laser pulses generated by a standard telecommunication laser diode are attenuated to faint pulses and are used as carriers for qubits. Each of these laser pulses are modulated to a random polarization state (i.e. a qubit) and a random intensity

---

8 Note that some QKD protocols still lack an unconditional security proof, i.e. a proof that takes into account any eavesdropping allowed by the laws of quantum physics. More precisely, some protocols have so far only been shown to resist attacks on individual photons, as opposed to attacking all photons coherently (jointly).

9 Often, the latter condition is satisfied by a beamsplitter that randomly reflects, or transmits, photons to different measurement devices; as we describe below, our implementation is an example for this very simple approach.

level to implement the BB84 protocol supplemented with decoy states[10]. The pulses then pass through a standard, 12 km long telecommunication fiber and arrive at the receiver of the QKD system. On Bob's side, each qubit is measured using one out of two, randomly selected devices (see figure 3). The measurement results are post-processed, as described above, resulting in a shared secret key. The details of each subsystem are given below.

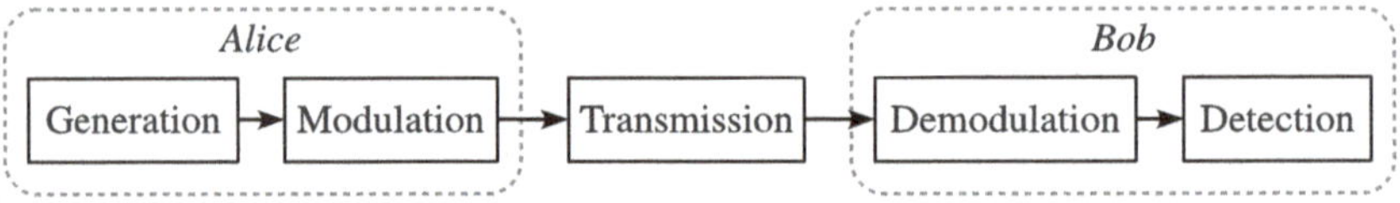

Fig. 5. Diagram of the optical part of a QKD system, where Alice and Bob denote the sender and the receiver, respectively.

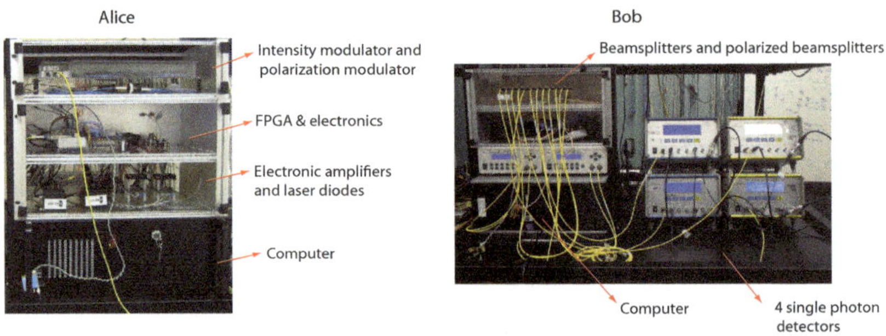

Fig. 6. The QKD system currently being developed at the Quantum Communication and Cryptography (QC2) Lab at the University of Calgary. The left-hand picture shows Alice (located at SAIT Polytechnic), the right-hand one depicts Bob (located at the University of Calgary).

7.1 Generation

Figure 7 depicts the schematics of Alice's part of the QKD system. All fiber-optical components are polarization maintaining unless stated otherwise. We assume that the polarization beam splitter (PBS) transmits horizontal, and reflects vertical polarization.

The laser diode LD_Q is driven by a homemade laser diode driver that is under the control of a field-programmable gate-array (FPGA)-based circuit. When the circuit sends a short digital signal to the driver, a 500 ps long, horizontally polarized laser pulse is generated. It features a central wavelength of 1548.07 nm, and a spectral width of 0.214 nm (full-width at half-maximum). The ratio between the power levels of a laser pulse and the background is around 100 to 1. An optical attenuator (ATT) reduces the energy of the laser pulse down to single-photon level; the background is reduced accordingly.

[10] Our QKD system currently employs a software-based pseudo-random number generator. While being acceptable in the current development phase, true random numbers, as described above, are needed in a system that is used to encode actual secrets.

7.2 Modulation

The horizontally polarized laser pulse passes through the PBS, enters the subsequent polarization-maintaining fiber with its polarization aligned along one of the fiber's principal axes (slow, let's say), and arrives at an element denoted R_{45}. The axes of the polarization maintaining fibers on both sides of R_{45} are rotated by 45° with respect to each other. This rotation leads to a decomposition of the incoming laser pulse into two orthogonally polarized components with equal intensity, but randomly varying phase difference. The two components are polarized along the two polarization-maintaining axes of the second fiber, and arrive parallel to the slow (S) and the fast (F) axes of the subsequent $LiNbO_3$ phase modulator (PM). This modulator introduces a random and a controllable phase shift (ϕ_{in}) to one of the two components (slow, let's say), and another random phase shift to the component parallel to the other axis. Next, the Faraday mirror reflects the input light. Due to the Faraday effect, the polarization of the input light and that of the reflected light are orthogonal to each other, i.e. the component previously traveling along the slow axis of the modulator now travels along the fast one and vice versa. Hence, provided the random phase shifts introduced by the PM during the first passage of light have not changed before the second passage (a correct assumption given the small time difference of a few nanoseconds), the random phase shifts are equally present in both polarization components, and the phase difference is entirely due to the controllable phases (ϕ_{in} and ϕ_{out}) that are applied during the two passages. When arriving again at the PBS, the polarization of the reflected laser pulse can thus be expressed by the Jones vector

$$J_{out} = \begin{bmatrix} -i \sin \Delta\phi \\ \cos \Delta\phi \end{bmatrix}, \tag{2}$$

where $\Delta\phi = \frac{1}{2}(\phi_{out} - \phi_{in})$. The details of the calculation can be found in (Lucio-Martinez et al. (2009)). Equation 2 shows that the polarization of the reflected laser pulse varies as a function of $\Delta\phi$, which is determined by the modulation voltages applied to the phase modulator.

The horizontally polarized component of the laser pulse passes again through the PBS and is absorbed by the attenuator. The vertical polarization component is reflected by the PBS and is subsequently used to encode a qubit. By changing the polarization of the laser pulse, its intensity (or rather the average number of photons in the laser pulse) is modulated to a high or low value, as required by the decoy-state protocol. Furthermore, so-called vacuum states (pulses without photons) are generated by not triggering the laser diode.

Due to the use of a Faraday mirror, this intensity modulator has the following advantages compared to a standard modulator based on a Mach-Zehnder interferometer:

- The modulator is insensitive to changes in the environment, such as temperature and mechanical stress in the fiber. As an example, we have observed a variation of the output intensity of less than ±1.5% over 12 hours.
- The modulator is insensitive to polarization mode dispersion, which makes it suitable for use with a light source with large spectral width. As a second example, we routinely observe 20 dB power extinction ratio.

The faint laser pulses are then transmitted to the second, equally built polarization modulator. Depending on the value of the phase difference $\Delta\phi$, we generate +45° ($\Delta\phi = 0$), −45° ($\Delta\phi = \frac{\pi}{2}$), left-hand circular ($\Delta\phi = -\frac{\pi}{4}$) or right-hand circular ($\Delta\phi = \frac{\pi}{4}$) polarization. This is equivalent to the example given in the text above, as can be seen by comparing the center and right-hand parts of figure 7. We note that the second modulator comprises an optical circulator (CIR) instead of a PBS. Hence, both polarization components will exit the modulator, which then acts as a polarization modulator instead of an intensity modulator.

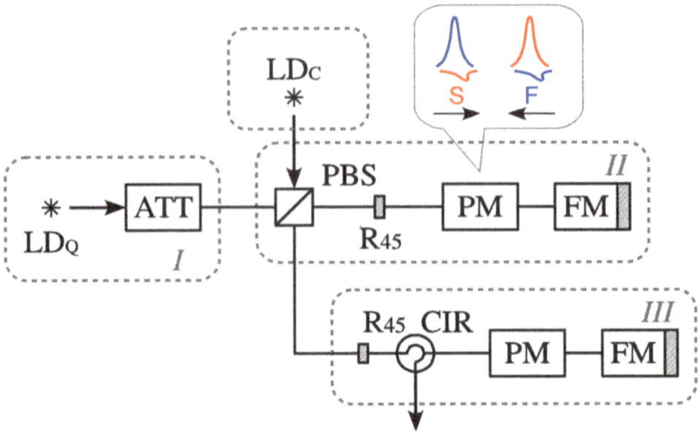

Fig. 7. Schematics of Alice's system, which consists of sub-systems for signal generation (*I*), intensity modulation (*II*) and polarization modulation (*III*).

7.3 Transmission, demodulation and detection

The generated qubits are transmitted from Alice to Bob using a dedicated (dark) fiber, and then demodulated and detected using appropriate measurement devices. Before going into detail, we will briefly introduce the concept of quantum frames, which play an important role for selecting and maintaining quantum channels suitable for QKD.

The idea of *quantum frames* is inspired by the Ethernet protocol. A quantum frame consists of alternating sequences of high-intensity pulses (the classical control frame, providing a platform to include classical control information into quantum communication) and qubits encoded into faint laser pulses (the quantum data), see figure 8. Adding classical control frames (also referred to as data headers) allows for a variety of tasks related to establishing a link for QKD (e.g. in a network environment, which will be discussed below), and maintaining its properties:

- *Routing:* To allow all-optical routing of quantum data in a network, the classical control frames include information about the sender and the receiver. This information can be read using standard detectors, which, in turn, can activate optical switches to route entire quantum frames along specific optical paths. Work on quantum networks will be discussed briefly in section 8.
- *Compatibility*: In a future quantum network, it is likely that different types of QKD systems co-exist. They may vary in the degree of freedom chosen to encode quantum information into photons, or the protocol employed, which impacts on the way post-processing (in particular privacy amplification) is done. This information can be included into the classical control frames, and allows Bob to take appropriate action.
- *Polarization compensation*: Unfortunately, a fiber that maintains all polarization states required for the implementation of the BB84 protocol does not exist[11]. Hence, Bob will receive photons in states that differ from those sent by Alice. Moreover, the polarization change during transmission through the link varies with time (e.g. due to temperature fluctuations) rather than being constant. A feedback mechanism that tracks and compensates the change introduced by the link is thus required. We exploit

[11] We point out that the so-called polarization-maintaining fibers only maintain polarization for two, orthogonally polarized polarization states, which is insufficient for QKD.

strong pulses of light in specific polarization states to solve this problem. More precisely, the classical control frames contain information for Bob, which allows him to actively compensate for these polarization changes.

- *Clock Synchronization*: One of the challenges in QKD, especially in high-rate QKD, is clock synchronization between the two parties. This is needed to associate the generation with the corresponding detection of qubits. Information for clock synchronization is included into the classical control frames, allowing for periodic synchronization of Alice's and Bob's clocks.

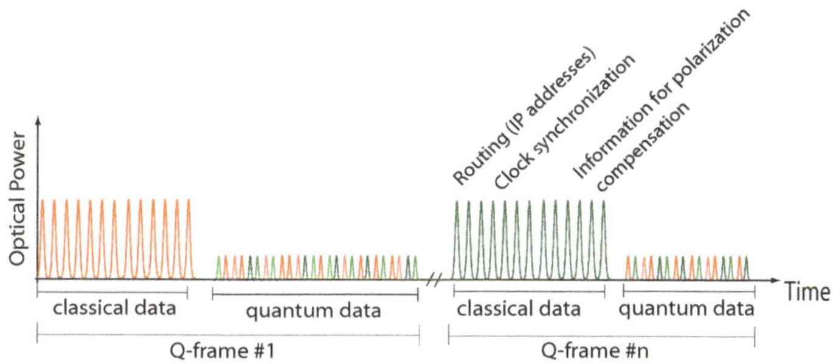

Fig. 8. Structure of a quantum frame (not to scale). Classical information (high intensity pulses) are time-multiplexed with quantum bits (faint laser pulses). Each color in the quantum data indicates one of the four different polarization states required in the BB84 protocol.

The laser pulses constituting the classical control frames are generated by the laser diode LD_C in figure 7: Horizontally polarized laser pulses pass through the PBS and are modulated to a specific polarization by the polarization modulator. On Bob's side, as shown in figure 9, 10% of the optical power is reflected towards a standard photodiode (PD), which detects the data header of the quantum frame to provide timing information for Bob's control circuit and thereby allow for clock synchronization.

Ninety percent of the optical power is transmitted through a 10/90 beam splitter (BS_1) and is then equally divided by a 50/50 beam splitter (BS_2). The outputs of BS_2 are connected to two polarization measurement devices. Each device consists of a polarization controller (PC_1, PC_2), a polarization beam splitter (PBS_1, PBS_2) and two single photon detectors (SPD_{1a}, SPD_{2a} and SPD_{1b}, SPD_{2b}).

PC_1 ensures that $-45°$ polarized classical data, and hence qubits, emitted at Alice's arrive horizontally polarized at PBS_1 and will be detected by SPD_{1a}. Similarly, PC_2 is set up such that right circular polarized classical data and qubits emitted at Alice's always impinge horizontally polarized on PBS_2, and will thus be detected by SPD_{2a}. This directly implies that qubits prepared with $+45°$ or left circular polarization arrive vertically polarized on PBS_1 or PBS_2, respectively, and will thus be detected by SPD_{1b} and SPD_{2b}. Hence, the two sets of PC, PBS and two SPDs allow compensation of unwanted polarization transformations in the quantum channel, and appropriate measurements of qubit states.

At the end of each classical data header, the control circuit disables the polarization controller as well as the clock synchronization and starts gating single photon detectors based on InGaAs/InP avalanche photo-diodes operated in Geiger mode. The outputs of these detectors produce the raw key at Bob's, which is transmitted to a personal computer for classical post-processing.

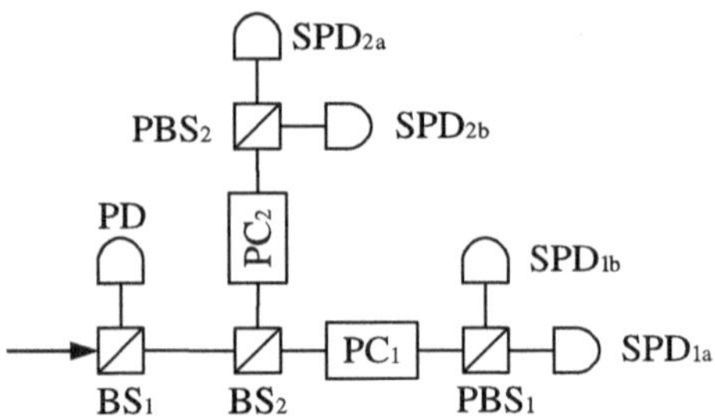

Fig. 9. Schematics of Bob's system, which consists of demodulation and detection subsystems.

7.4 QKD performance

As described above, each deployed optical fiber introduces polarization changes to transmitted states of light (both quantum and classical) that vary over time. As shown in figure 10a, these variations are rapid during day-time (when the sun is out and the fiber heats up despite running through underground conduits) and are less pronounced during the night. However, even in the worst case, the polarization transformation is stable on a timescale of seconds. Due to the feedback system described above, our QKD system is able to perform continuously during $\sim$30 hours: as depicted in figure 10b, the QBER remains at $\sim$3% (a typical value for a QKD setup operating over $\sim$10 km optical fiber), independent on the time of the day.

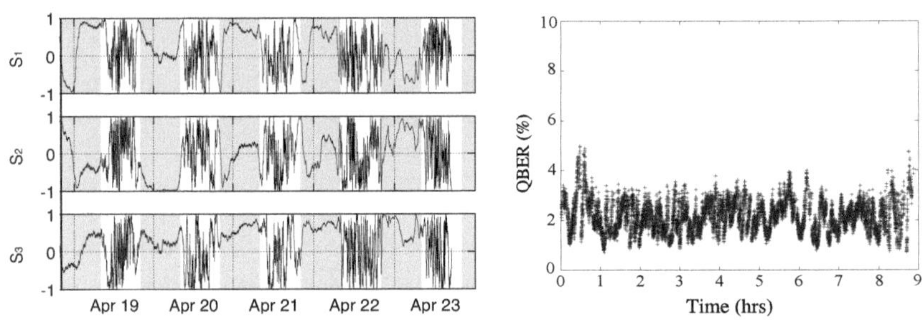

(a) Polarization variation (b) QBER as a function of time

Fig. 10. a) Plot of the polarization state of originally +45° polarized classical control frames arriving at Bob's as a function of time. The states are parametrized in terms of their Stokes vectors S_1-S_3. A correlation between the time of the day and the polarization variation can be seen. b) QBER as a function of time, measured over a period of 9 hours.

Our QKD system currently features a raw key rate[12] of $\sim$0.5 KHz, which is sufficient to provide cryptographic keys for encoding using the Advanced Encryption Standard (AES)[13], but too small to provide keys for one-time pad encoding in real-time. The raw key rate is determined by the repetition rate at which qubits are produced or the single-photon detectors can be gated (whatever is smaller), the loss in the channel between Alice to Bob ($\sim$6.5dB), the

[12] The raw key rate is defined as the rate with which photons are detected at Bob's.

[13] AES is a widely used symmetric cipher. It uses a short key to encrypt large amount of data, and therefore does not provide information-theoretic security.

fraction of time that is used for qubit generation and transmission (~10%; the remaining time is used for classical post-processing, which is currently done sequentially, and polarization compensation), loss in the optical components at Bob's (~3dB), and the quantum efficiency (i.e. the probability to detect a photon) of the single-photon detectors (~10%). The sifted key rate is ~0.25 KHz leading to a secret key rate of ~50 bps for an average of 0.5 photons per faint laser pulse.

The performance can be improved by several orders of magnitude by employing high-rate single photon detectors[14], and parallel implementation of classical post-processing using dedicated hardware.

8. Quantum networks

As described above, the no-cloning theorem prevents an eavesdropper from copying quantum data, but excludes at the same time broadcasting of identical quantum keys to several legitimate users. While QKD systems can thus operate only in point-to-point (P2P) fashion, it still benefits from being implemented in networks. This is due to the possibility to connect different users in a flexible and efficient way. Various types of quantum networks have been considered; the differences are determined by the available (or assumed) technology.

8.1 Trusted-node quantum networks

A trusted-node QKD network is composed of dedicated QKD links, each one connecting two neighboring locations or *nodes*. Secret keys for encoding messages (henceforth referred to as the message-encryption-key, MEK) are distributed between arbitrary (non-neighboring) nodes using a chain of QKD links. More precisely, the MEK is encrypted using the one-time pad and a key-encryption-key established with the neighboring node by means of QKD. The MEK is sent to the next node, attached with an authentication tag. Upon reception, the authentication tag is verified, and the MEK is decrypted. The process repeats until the MEK reaches its final destination. A potential drawback is that the security of the distribution of the MEK is only ensured if all intermediate nodes between the sender and receiver can be trusted as they posses full information about the MEK. However, on the other hand, the distribution of the MEK is not limited in distance as any distance can always be covered using many short-distance QKD links.

An example of a trusted-node quantum network is the *SECOQC Network*, implemented in October 2008 in Vienna, Austria (Peev et al. (2009)). This network comprised six nodes, connected by different QKD systems (or platforms). One of the main results of the deployment of this network was the development of an interface between a QKD system and existing (non-quantum) information and communication technology (ICT) systems. Also as a result of this project, an Industry Specification Group (ISG) of the European Telecommunications Standards Institute (ETSI) for QKD was put in operation to create universally accepted QKD standards (Langer & Lenhart (2009)). Another example is the *Tokyo QKD Network*, inaugurated in October 2010, also comprising six trusted-nodes connected by P2P QKD links forming the quantum backbone. The Tokyo Network includes a key management server for centralized management of the key life cycle and to determine the secure paths between two distant nodes (Sasaki et al. (2011)).

---

[14] The current state-of-the-art allows gate rates of 2 GHz – an improvement of a factor of 1000 compared to commercially available detectors.

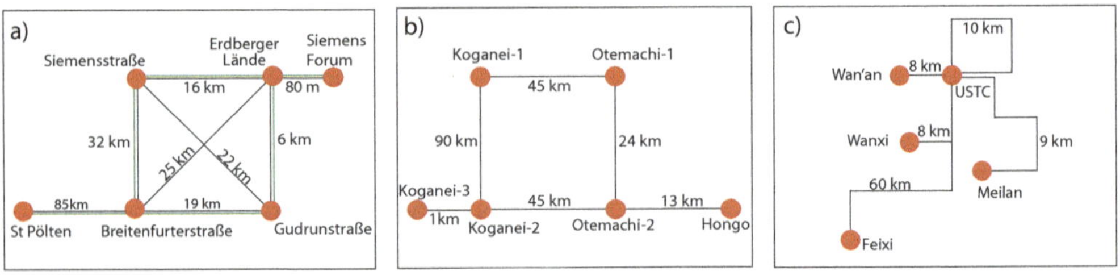

Fig. 11. Examples for different quantum network topologies (all based on trusted nodes). a) SECOQC Network showing quantum channels (black) and classical channels (green) between the nodes. b) Tokyo Network. c) Hefei Network showing a star-type network.

8.2 Optically configurable quantum networks

An optically configurable quantum network utilizes passive or active optical devices (e.g. beam splitters, optical switches, wavelength multiplexers, etc.) in the quantum channel to allow QKD between multiple pairs of users. The benefit of this kind of network, in contrast to the previous kind, is that the nodes between the two parties that establish a secret key do not have to be trusted. Configurable quantum networks can be implemented with current technology. However, the distance over which a secret key can be established is limited to $\sim$100 km, restricting its use to local or metropolitan areas.

The first investigations towards optically configurable quantum networks were pursued within the framework of the *DARPA Quantum Network* (Elliot et al. (2005)) which was operational from 2004 to approximately 2008 in Massachusetts, USA. This network consisted of 8 nodes and employed various QKD platforms; 4 of these nodes were connected via active optical switches. A second example is the quantum network located in Hefei (Anhui), China. This hybrid network combines trusted relays and all-pass optical switches, allowing interconnection among all 5 nodes (Chen et al. (2010)).

8.3 Fully quantum-enabled networks

Fully quantum-enabled networks require technologies such as entanglement swapping, entanglement purification, quantum error correction and quantum memories (Kimble (2008)). These networks are not limited by the distance barrier imposed to optically configurable networks, and do not require trust in nodes, as opposed to trusted-node networks. However, the technology required for fully quantum-enabled networks is not yet mature, even though all basic building blocks have meanwhile been demonstrated in proof-of-concept experiments (Lvovsky et al. (2009); Sangouard et al. (2011)).

9. The future

To conclude this chapter, let us briefly address a few directions in which QKD is likely to evolve in the near future.

First, QKD systems will be improved to deliver secret keys at Mbps rates, as has already been demonstrated in the QKD system developed by Toshiba. Progress beyond this rate can be expected to be slow, as many constituents, such as single photon detectors, face technical limitations that would require expensive and time-consuming engineering at the component level.

Next, the critical assessment of QKD implementations with respect to security loopholes will continue, and a lot of work will be devoted to the elimination of attacks through the development of better technology and the improvement of security proofs and protocols in the sense of making them more applicable to real-world devices. Key words here are *squashing models* (Lütkenhaus (2000)) and *device-independent protocols* (Masanes et al. (2011)). Squashing models allow the use of qubit-based security proofs for non-qubit-based implementations. Device-independent protocols go beyond squashing and attempt to remove not only all assumptions about the nature of the quantum systems used to transmit quantum information, but also those about the classical devices used by Alice and Bob.

Furthermore, going beyond QKD, other quantum cryptography protocols that provide benefits compared to their classical analogs will receive more attention. Examples include quantum coin flipping (Berlin et al. (2009)) or quantum private database queries (Jakobi et al. (2011)). An interesting problem here is to devise protocols that can tolerate loss and errors, which are both present in real-world implementations.

Finally, the integration of QKD into networks will be improved, and technologies required for breaking the distance barrier by means of quantum repeaters and for fully quantum-enabled networks will be further advanced. This challenge is huge and immensely interesting, but progress over the past 5 years has been surprisingly rapid, and we believe that a fully quantum-enabled network will eventually be built.

10. Acknowledgements

The authors thank V. Kiselyov for technical support and S. Hosier for all the help with setting up the QKD system. This work is supported by General Dynamics Canada, Alberta's Informatics Circle of Research Excellence (iCORE, now part of Alberta Innovates Technology Futures), the National Science and Engineering Research Council of Canada (NSERC), QuantumWorks, Canada Foundation for Innovation (CFI), Alberta Advanced Education and Technology (AET), and the Mexican Consejo Nacional de Ciencia y Tecnología (CONACYT).

11. References

Bennett, C. & Brassard, G. (1984). Quantum cryptography: Public key distribution and coin tossing, *Proceedings of IEEE International Conference on Computers Systems and Signal Processing* pp. 175–179.

Bennett, C. H., Bessette, F., Brassard, G., Salvail, L. & Smolin, J. (1992). Experimental quantum cryptography, *Journal of Cryptology* 5(3): 253–265.

Berlin, G., Brassard, G., Bussières, F. & Godbout, N. (2009). Fair loss-tolerant quantum coin flipping, *Physical Review A* 80(6):062321.

Brassard, G., Lütkenhaus, N., Mor, T. & Sanders, B. (2000). Limitation on practical quantum cryptography, *Physical Review Letters* 85: 1330–1333.

Chen, T., Wang, J., Liang, H., Liu, W.-Y., Liu, Y., Jiang, X., Wang, Y., Wan, X., Cai, W.-Q., Ju, L., Chen, L.-K., Wang, L.-J., Gao, Y., Chen, K., Peng, C.-Z., Chen, Z.-B. & Pan, J.-W. (2010). Metropolitan all-pass and inter-city quantum communication network, *Optics Express* 18(26): 27217–27225.

Dixon, A., Yuan, Z. L., Dynes, J., Sharpe, A. W. & Shields, A. (2010). Continuous operation of high bit rate quantum key distribution, *Applied Physics Letters* 96: 161102–161105.

Elliot, C., Colvin, A., Pearson, D., Pikalo, O., Schlafer, J. & Yeh, H. (2005). Current status of the DARPA quantum network, *Proceedings of SPIE* 5815: 138–149.

Fürst, M., Weier, H., Nauerth, S., Marangon, D. G., Kurtsiefer, C. & Weinfurter, H. (2010). High speed optical quantum random number generation, *Optics Express* 18(12): 13029–13037.

Gallager, R. (1962). Low-density parity-check codes, *IRE Transactions on Information Theory* 8(1): 21–28.

Gisin, N., Ribordy, G., Tittel, W. & Zbinden, H. (2002). Quantum cryptography, *Review Modern Physics* 74: 145–195.

Gottesman, D., Lo, H., Lütkenhaus, N. & Preskill, J. (2004). Security of quantum key distribution with imperfect devices, *Quantum Information and Computation* 4: 325.

Honjo, T., Uchida, A., Amano, K., Hirano, K., Someya, H., Okumura, H., Yoshimura, K., Davis, P. & Tokura, Y. (2009). Differential-phase-shift quantum key disitrbution experiment using fast physical random bit generator with chaotic semiconductor lasers, *Optics Express* 17(11): 9053–9061.

idQuantique (2011). `www.idQuantique.com`.

Jakobi, M., Simon, C., Gisin, N., Bancal, J.-D., Branciard, C., Walenta, N. & Zbinden, H. (2011). Practical private database queries based on a quantum-key-distribution protocol, *Physical Review A* 83(2): 022301.

Kimble, H. J. (2008). The quantum internet, *Nature* 453: 1023–1030.

Kleinjung, T., Aoki, K., Franke, J., Lenstra, A., Thomé, E., Bos, J., Gaudry, P., Kruppa, A., Montgomery, P., Osvik, D., te Riele, H., Timofeev, A. & Zimmermann, P. (2010). Factorization of a 768-bit RSA modulus, *in* T. Rabin (ed.), *Advances in Cryptology – CRYPTO 2010*, Vol. 6223 of *Lecture Notes in Computer Science*, Springer Berlin / Heidelberg, pp. 333–350.

Lamas-Linares, A. & Kurtsiefer, C. (2007). Breaking a quantum key distribution system through a timing side channel, *Optics Express* 15(15): 9388–9393.

Langer, T. & Lenhart, G. (2009). Standarization of quantum key distribution and the ETSI standarization initiative ISG-QKD, *New Journal of Physics* 11(5): 055051.

Liu, Y. (2010). Decoy-state quantum key distribution with polarized photons over 200 km, *Optics Express* 18(8): 8587.

Lucio-Martinez, I., Chan, P., Mo, X.-F., Hosier, S. & Tittel, W. (2009). Proof-of-concept of real world quantum key distribution with quantum frames, *New Journal of Physics* 11(9): 095001.

Lütkenhaus, N. (2000). Security against individual attacks for realistic quantum key distribution, *Physical Review A* 61: 052304.

Lvovsky, A. I., Sanders, B. C. & Tittel, W. (2009). Optical quantum memory, *Nature Photonics* 3: 706–714.

MacKay, D. & Neal, R. (1997). Near Shannon limit performance of low density parity check codes, *Electronics Letters* 33(6): 457–458.

MagiQ (2011). `http://www.magiqtech.com/MagiQ/Home.html`.

Makarov (2011). Secure gated detection scheme for quantum cryptography, *Physical Review A* 83(3): 032306.

Masanes, L., Pironio, S. & Acín, A. (2011). Secure device-independent quantum key distribution with causally independent measurement devices, *Nat. Commun.* 2(238).

Nielsen, M. A. & Chuang, I. L. (2004). *Quantum Computation and Quantum Information (Cambridge Series on Information and the Natural Sciences)*, 1 edn, Cambridge University Press.

Pearson, D. (2004). High-speed QKD reconciliation using forward error correction, *Proc. 7th International Conference on Quantum Communication, Measurement and Computing (QCMC)*, Vol. 734, pp. 299–302.

Peev, M., Pacher, C., Alléaume, R., Barreiro, C., Bouda, J., Boxleitner, W., Debuisschert, T., Diamanti, E., Dianati, M., Dynes, J. F., Fasel, S., Fossier, S., Fürst, M., Gautier, J.-D., Gay, O., Gisin, N., Grangier, P., Happe, A., Hasani, Y., Hentschel, M., Hübel, H., Länger, G. H. T., Legré, M., Lieger, R., Lodewyck, J., Lorünser, T., Lütkenhaus, N., Marhold, A., Matyus, T., Maurhart, O., Monat, L., Nauerth, S., Page, J.-B., Poppe, A., Querasser, E., Ribordy, G., Robyr, S., Salvail, L., Sharpe, A. W., Shields, A. J., Stucki, D., Suda, M., Tamas, C., Themel, T., Thew, R. T., Thoma, Y., Treiber, A., Trinkler, P., Tualle-Brouri, R., Vannel, F., Walenta, N., Weier, H., Weinfurter, H., Wimberger, I., Yuan, Z. L., Zbinden, H. & Zeilinger, A. (2009). The SECOQC quantum key distribution network in Vienna, *New Journal of Physics* 11: 075001.

Qi, B., Zhao, Y., Ma, X., Lo, H.-K. & Qian, L. (2007). QKD Quantum key distribution with dual detectors, *Physical Review A* 75(5): 052304–052314.

Quintessence (2011). `http://www.quintessencelabs.com/`.

Restelli, A., Bienfang, J. C., Mink, A. & Clark, C. W. (2009). Quanutm key distribution at GHz transmission rates, *Proceedings of SPIE* 7236(72360L): 1–7.

Sangouard, N., Simon, C., de Riedmatten, H. & Gisin, N. (2011). Quantum repeaters based on atomic ensembles and linear optics, *Review Modern Physics* 83(1): 33–80.

Sasaki, M., Fujiwara, M., Ishizuka, H., Klaus, W., Wakui, K., Takeoka, M., Tanaka, A., Yoshino, K., Nambu, Y., Takahashi, S., Tajima, A., Tomita, A., Domeki, T., Hasegawa, T., Sakai, Y., Kobayashi, H., Asai, T., Shimizu, K., Tokura, T., Tsurumaru, T., Matsui, M., Honjo, T., Tamaki, K., Takesue, H., Tokura, Y., Dynes, J. F., Dixon, A. R., Sharpe, A. W., Yuan, Z. L., Shields, A. J., Uchikoga, S., Legré, M., Robyr, S., Trinkler, P., Monat, L., J.-B., Ribordy, G., Poppe, A., Allacher, A., Maurhart, O., Länger, T., Peev, M. & Zeilinger, A. (2011). Field test of quantum key distribution in the Tokyo QKD network, *Optics Express* 19(11): 10387–10409.

Scarani, V., Bechmann-Pasquinucci, H., Cerf, N. J., Dušek, M., Lütkenhaus, N. & Peev, M. (2009). The security of practical quantum key distribution, *Review Modern Physics* 81: 1301.

Schmitt-Manderbach, T., Weier, H., Furst, M., Ursin, R., Tiefenbacher, F., Sheidl, T., Perdigues, J., Sodnik, Z., Kurtsiefer, C., Rarity, J. G., Zeilinger, A. & Weinfurter, H. (2007). Experimental demonstration of free-space decoy-state quantum key distribution over 144 km, *Physical Review Letters* 98(1): 010504.

Stebila, D., Mosca, M. & Lütkenhaus, N. (2010). The case for quantum key distribution, *Lecture Notes of the Institute for Computer Sciences, Social Informatics and Telecommunications Engineering* 36: 283–296.

Stucki, D., Walenta, N., Vannel, F., Thew, R. T., Gisin, N., Zbinden, H., Gray, S., Towery, C. R. & Ten, S. (2009). High rate, long-distance quantum key distribution over 250 km of ultra low loss fibers, *New Journal of Physics* 11: 075003.

Tittel, W. & Weihs, G. (2001). Photonic entanglement for fundamental tests and quantum communication, *Quantum Information and Computation* 1(2): 3–56.

Ursin, R., Jennewein, T., Kofler, J., Perdigues, J., Cacciapuoti, L., de Matos, C., Aspelmeyer, M., Valencia, A., Scheidl, T., Acin, A., Barbieri, C., Bianco, G., Brukner, C., Capmany, J., Cova, S., Giggenbach, D., Leeb, W., Hadfield, R., Laflamme, R., Lütkenhaus, N., Milburn, G., Peev, M., Ralph, T., Rarity, J., Renner, R., Samain, E., Solomos, N., Tittel,

W., Torres, J., Toyoshima, M., Ortigosa-Blanch, A., Pruneri, V., Villoresi, P., Walmsley, I., Weihs, G., Weinfurter, H., Zukowski, M. & Zeilinger, A. (2009). Space-quest, experiments with quantum entanglement in space, *Europhysics News* 40(3): 26–29.

Vakhitov, A., Makarov, V. & Hjelme, D. (2001). Large pulse attack as a method of conventional optical eavesdropping in quantum cryptography, *Journal of Modern Optics* 48: 2023–2038.

Wegman, M. & Carter, J. L. (1981). New hash function and their use in authentication and set equality, *Journal of Computer and System Sciences* 22(3): 265–279.

Yuan, Z. L., Dynes, J. & Shields, A. J. (2010). Avoiding the blinding attack in qkd, *Nature Photonics* 4: 800–801.

12

Generation and Detection of Mesoscopic Pulsed States of Light for Quantum Information

Alessia Allevi[1] and Maria Bondani[2]
[1]*Dipartimento di Fisica e Matematica - Università degli Studi dell'Insubria*
U.d.R. Como - Consorzio Nazionale Interuniversitario per le Scienze fisiche della Materia
[2]*Istituto di Fotonica e Nanotecnologie - Consiglio Nazionale delle Ricerche*
U.d.R. Como - Consorzio Nazionale Interuniversitario per le Scienze fisiche della Materia
Italy

1. Introduction

Generation and measurement of quantum states of the electromagnetic field represent a hot and widely discussed topic in the Physics community, since optical states are useful not only in fundamental experiments but also in several applications in the fields of Quantum Optics and Quantum Information (Bouwmeester et al., 2000). In fact, optical radiation is endowed with relevant characteristics as it travels at maximum possible speed, it is almost unaffected by the interaction with the environment, which in many cases represents an unavoidable decoherence source, and it can be integrated in optoelectronic circuits to implement quantum computing protocols (Sansoni et al., 2010). The possibility of using optical fields in applicative protocols depends on the capability of generating and manipulating optical states that are robust with respect to losses (i.e. that contain a sizeable number of photons) and that can be produced and measured at high rate. For all these reasons, mesoscopic pulsed optical states containing few tens of photons per pulse seem to be the ideal candidates for applications to quantum communication protocols in which each pulse must be addressed individually. Moreover, the problem of generating suitable pulsed states must be considered together with the question of their measurement and characterization. In fact, nowadays we have many types of pulsed-light sources at our disposal, each of them characterized by different pulse durations (from few fs up to few ns), different pulse-repetition rates (from few Hz to 100 MHz) and a wide range of energies per pulse (from few nJ to several J). However, not all these pulsed light can be easily measured. In the mesoscopic intensity regime, for example, detectors endowed with photon-number resolution are required so as to determine shot-by-shot photon numbers or, at least, to characterize some features of the optical states. In fact, full characterization of the optical state, obtained *e.g.* by evaluating its Wigner function, can be achieved with optical homodyne tomography (OHT) (Raymer, 1997), a technique that over the years has proved its effectiveness in reconstructing both classical and quantum states, even if further optimizations are needed in the case of pulsed fields (Zavatta et al., 2006).
Direct detection schemes are an alternative to OHT in all the cases in which knowing the statistics of detected photons is sufficient to characterize the signal state, as demonstrated in the pioneering work of Arecchi (Arecchi, 1965). Obviously, the implementation of a direct detection scheme requires a proper choice of the detector, the development of a calibration

strategy and a robust reconstruction method independent of any a-priori knowledge of the state under investigation. For example, valuable results have been obtained by means of an ON/OFF method assisted by a maximum likelihood algorithm (Zambra et al., 2005).
To this aim, several different strategies are being followed either to improve the traditional detectors or to build new ones (Haderka et al., 2010). In fact, the mature silicon-based technologies have encouraged the optimization of both photomultipliers and avalanche photodiodes, by enhancing their quantum efficiency (*QE*) and reducing their dark-count rate. More recently, a new type of detector, usually called hybrid photodetector (HPD), as it combines a photocathode with a diode structure operated below the breakdown threshold, has been developed and successfully tested (Bondani et al., 2009a). As in this case the amplification process takes part in one step, the excess noise is small enough to allow photon-number resolution (up to 6 detected photons). The *QE*, mainly given by the photocathode, can reach good values (up to 0.5 in the visible range), whereas the dark-count rate can be neglected. As an alternative, the accomplishment of photon-counting capability can be obtained by splitting the light to be measured either in space or in time prior to detection so that at most one photon at a time hits the detector sensitive area. Among them, we mention the visible light photon counter (VLPC) (Kim et al., 1999); the fiber-loop detector, which is a time-multiplexed detector based on one (Rehacek et al., 2003) or more (Fitch et al., 2003) single-photon avalanche diodes (SPADs); the Silicon photomultiplier (SiPM), that is constituted by a matrix of SPADs with a common output (Akindinov et al., 1997). Due to their composite structure, SiPMs have a good photon-counting capability, though presenting large dark-count rate (600 kHz) and cross-talk probability (Afek et al., 2009; Ramilli et al., 2010). Finally, two types of CCD cameras can be used to detect light in the mesoscopic domain, namely the intensified CCD (iCCD) (Haderka et al., 2005) and the electron-multiplied CCD (EM-CCD) (Blanchet et al., 2008).
In the last decade, a tremendous progress has been achieved in the field of superconductors so that new types of detectors have been developed, such as the transition-edge sensor (TES) (Lita et al., 2008) and the superconducting nanowires (Gol'tsman et al., 2001). Despite having a good *QE*, these detectors must operate at cryogenic temperatures and thus they are rather cumbersome. As of today, the ideal detector has yet to appear and the optimal choice is application specific.

2. Theory

We present the procedure for analyzing the output of a detector measuring light pulses in order to recover light statistics (Andreoni & Bondani, 2009a). With reference to Fig. 1, we consider a pulsed light field having photon-number distribution, P_n. The field is delivered to the detector by suitable optics and the overall losses of the detection chain are summarized by a single factor η, which represents the overall photon-detection efficiency of the apparatus. Since $\eta < 1$, the distribution of the number of detected photons, P_m, and the photon-number

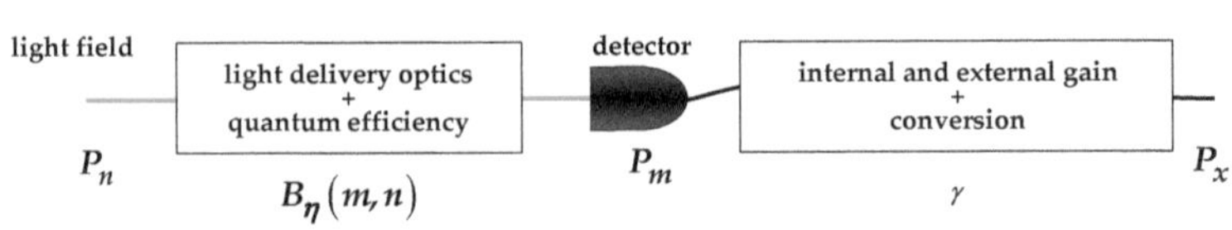

Fig. 1. Scheme for light detection: η is the overall detection efficiency and γ is the overall gain.

distribution P_n are linked by

$$P_m = \sum_{n=m}^{+\infty} B_\eta(m,n) P_n = \sum_{n=m}^{+\infty} \binom{n}{m} \eta^m (1-\eta)^{n-m} P_n\,, \tag{1}$$

from which we obtain the relation between the first moments of photons and detected-photon distributions: $\langle m \rangle = \eta \langle n \rangle$, $\langle m^2 \rangle = \eta^2 \langle n^2 \rangle + \eta(1-\eta)\langle n \rangle$ (Bondani et al., 2009a).
We consider a detector operated within its range of linearity on pulsed light states. The output pulses of the detector are amplified and integrated (see Fig. 1) over a time-gate synchronous with the detector output pulse and slightly longer than its time duration (typically few nanoseconds). The output of the integrator is then sampled and digitized. The final output number x is stored in the memory of a computer to be processed offline. The obtained pulse-height spectrum normalized to its integral can be interpreted as the probability density P_x of the variable x. The zero of the x-scale is set at the value that is equal to the mean of the experimental P_x distribution obtained in the absence of light. The overall conversion process of the m detected photons into a x-value can be characterized by a single conversion factor γ, which is a stochastic variable distributed according to p_γ and having mean value $\overline{\gamma}$ and variance σ^2.
As the detection events giving different m values are mutually exclusive, P_x can be written as

$$P_x = P_{m=0} p_\gamma^{(0)} + P_{m=1} p_\gamma + P_{m=2}(p_\gamma * p_\gamma) + \ldots + P_{m=k}(p_\gamma * p_\gamma * \ldots * p_\gamma)_k\,, \tag{2}$$

where $p_\gamma^{(0)}$ is the probability density of the x values in the absence of light, whose mean value is zero by definition, and the $*$ symbols indicate convolution products. Note that we allow $p_\gamma^{(0)}$ to be different from p_γ, which accounts for the fact that in photoemissive detectors the anodic charge distribution for dark counts is different from that of the single-electron response (SER).
The output values $x^{(k)}$ recorded when k photons are detected is given by $x^{(k)} = \sum_{i=1}^{k} \gamma_i$ where all γ_i are distributed according to p_γ. We can thus write for the cumulants (Mandel & Wolf, 1995) $\kappa_r^{(\sum_{i=1}^{k}\gamma_i)} = \sum_{i=1}^{k} \kappa_r^{(\gamma_i)}$, in which $\kappa_1^{(\gamma_i)} = \overline{\gamma}$, $\kappa_2^{(\gamma_i)} = \sigma^2$, $\kappa_3^{(\gamma_i)} = \widetilde{\mu}_3$, $\kappa_4^{(\gamma_i)} = \widetilde{\mu}_4 - 3\sigma^4$, $\kappa_5^{(\gamma_i)} = \widetilde{\mu}_5 - 10\sigma^2\widetilde{\mu}_3$ and $\widetilde{\mu}_r$ are the values assumed for the central moments of p_γ. In the case of $r = 1$ we get

$$\langle x \rangle = \overline{\gamma} \sum_{k=0}^{+\infty} k P_{m=k} = \langle m \rangle \overline{\gamma}\,. \tag{3}$$

We can now relate the central moments $\mu_r(x) = \langle (x - \langle x \rangle)^r \rangle$ of the experimental P_x to those of the unknown probability density P_m, $\mu_r(m) = \langle (m - \langle m \rangle)^r \rangle$

$$\mu_r(x) = \sum_{k=0}^{+\infty} P_{m=k} \int_{-\infty}^{+\infty} (x - \langle x \rangle)^r (p_\gamma * p_\gamma * \ldots * p_\gamma)_k dx = \sum_{k=0}^{+\infty} P_{m=k} \mu_r(x^{(k)})\,, \tag{4}$$

where $\mu_r(x^{(k)}) \equiv \int_{-\infty}^{+\infty} (x - \langle x \rangle)^r (p_\gamma * p_\gamma * \ldots * p_\gamma)_k dx$. For $r = 1$ Equation 4 obviously vanishes. For $r \geq 2$ we insert the binomial expansion $(x - \langle x \rangle)^r = \sum_{j=0}^{r} \binom{r}{j} x^j (-\langle x \rangle)^{r-j}$ and use Equation 3 to find

$$\mu_r(x) = \sum_{j=0}^{r} \binom{r}{j} (-\langle m \rangle \overline{\gamma})^{r-j} \sum_{k=0}^{+\infty} P_{m=k} \mu'_j(x^{(k)})\,, \tag{5}$$

in which the "prime" distinguishes the moments from the central moments. It can be shown (Andreoni & Bondani, 2009a) that the moments $\mu'_j(x^{(k)})$ are polynomials whose terms contain the number of detected photons and the cumulants of the probability distribution p_γ. The results for $r = 2$ and $r = 3$ are

$$\frac{\mu_2(x)}{\langle x\rangle} = \bar{\gamma}\left(\frac{\mu_2(m)}{\langle m\rangle} + \frac{\sigma^2}{\bar{\gamma}^2}\right) \tag{6}$$

$$\frac{\mu_3(x)}{\langle x\rangle} = \bar{\gamma}^2\left(\frac{\mu_3(m)}{\langle m\rangle} + 3\frac{\mu_2(m)}{\langle m\rangle}\frac{\sigma^2}{\bar{\gamma}^2} + \frac{\tilde{\mu}_3}{\bar{\gamma}^3}\right). \tag{7}$$

If p_γ is narrow enough so that $\sigma^2/\overline{\gamma}^2 \to 0$ only one term of the polynomials survives and $\mu'_j(x^{(k)})$ reduces to $\mu'_j(x^{(k)}) \cong k^j(\kappa_1^{(\gamma_i)})^j = k^j\overline{\gamma}^j$, which we substitute into Equation 5 to obtain $\mu_r(x) = \overline{\gamma}^r\mu_r(m)$ (Andreoni & Bondani, 2009a) that we conveniently rewrite as

$$\frac{\mu_r(x)}{\langle x\rangle} = \overline{\gamma}^{r-1}\frac{\mu_r(m)}{\langle m\rangle}. \tag{8}$$

The link between $\mu_r(x)$ and $\mu_r(m)$ in Equation 8, which holds when σ is sufficiently smaller than $\overline{\gamma}$, shows that the simple knowledge of $\overline{\gamma}$ allows retrieving P_m from P_x, since Equation 8 holds for moments of any order and guarantees that the distribution that is obtained by binning the values x data into bins of width $\overline{\gamma}$ coincides with P_m.

The result in Equation 8 allows adopting a self-consisting procedure to determine $\overline{\gamma}$ without calibrating both detector and signal processing electronics, but simply using the measurements performed to determine P_x for the light field under investigation and without any previous knowledge of its statistics (Bondani et al., 2009a;b;c). In fact, by taking into account the properties derived from Equation 1, we can calculate

$$F_m \equiv \frac{\mu_2(m)}{\langle m\rangle} = \frac{\eta^2\mu_2(n) + \eta(1-\eta)\langle n\rangle}{\eta\langle n\rangle} = \eta F_n + 1 - \eta = \eta Q + 1, \tag{9}$$

in which $F_n = \mu_2(n)/\langle n\rangle$ is the Fano factor and $Q = (\mu_2(n) - \langle n\rangle)/\langle n\rangle$ is the Mandel parameter of the light entering the experimental apparatus in Fig. 1 and containing $\langle n\rangle$ photons in the measure time T_M (Mandel & Wolf, 1995). In analogy to light, we define a sort of Fano factor for detected photons, $F_m = \mu_2(m)/\langle m\rangle$, and for outputs, $F_x = \mu_2(x)/\langle x\rangle$. Substituting Equation 9 into Equation 8 with $r = 2$ and using Equation 3 yields

$$F_x = \frac{\mu_2(x)}{\langle x\rangle} = \frac{Q}{\langle n\rangle}\langle x\rangle + \overline{\gamma}\left(1 + \frac{\sigma^2}{\bar{\gamma}^2}\right). \tag{10}$$

Equation 10 is the key result of our method that allows determining $\overline{\gamma}$ by a self-consistent procedure when $\sigma^2/\overline{\gamma}^2 \ll 1$, that is when Equation 8 holds. In fact, since measurements performed on the same input light with different η-values, thus producing different $\langle x\rangle$-values, would leave the factor $Q/\langle n\rangle$ unaltered, the plots of F_x as a function of $\langle x\rangle$ should exhibit a linear dependence with slope $Q/\langle n\rangle$ with $\overline{\gamma}$ as the intercept. The value of the slope is significant of the statistics of the number of photons P_n, being it zero for coherent light, positive for light with super-poissonian variance and negative for light with sub-poissonian variance (Mandel & Wolf, 1995). In the following Sections we will show the retrieval of the slope values for several different light states (Bondani et al., 2009a;b;c).

3. Validation of the theory

We tested the analysis method by measuring a coherent light with a number of linear detectors endowed with different features. In all the measurements presented here, the coherent light was provided by the second-harmonic output of a Nd:YLF mode-locked laser amplified at 5 kHz (High Q Laser Production, Austria) producing pulses of ~5.4 ps duration at 523 nm wavelength. To minimize the collection of stray light, we used a multimode fiber to deliver the light to the detectors. Unless otherwise indicated, the output current pulses of the detectors were suitably gate-integrated by SR250 modules (Stanford Research Systems, CA) over integration gates shorter than 500 ns and sampled to produce a voltage which was digitized and recorded at each shot to produce the output x. The variation of η required to implement the analysis was obtained by inserting a variable neutral-density filter or a polarizer in the beam path. The linearity of the detectors allows the determination of the actual value of η.
For a coherent field we expect a Poissonian photon-number distribution

$$P_n = \frac{\langle n \rangle^n}{n!} \exp(-\langle n \rangle) \,, \tag{11}$$

for which mean value and variance are $\langle n \rangle$ and $Q = 0$. By applying Equation 1 to Equation 11 we obtain $P_m = \langle m \rangle^m / m! \exp(-\langle m \rangle)$, that has mean value $\langle m \rangle = \eta \langle n \rangle$, which means that the statistics is still Poissonian. We remark that this property of invariance of the statistical distribution under Bernoullian convolution has a more general application and holds in particular for all the states considered in this Chapter. For a coherent light we have

$$F_x = \frac{\mu_2(x)}{\langle x \rangle} = \overline{\gamma} \,, \tag{12}$$

independent of the mean value $\langle x \rangle$. The value of $\overline{\gamma}$ obtained from the experimental values of F_x is then used to convert the output x into a number of detected photons m and to re-bin the data to obtain the experimental probability distribution of detected photons $P_{m,\mathrm{exp}}$. The quality of the reconstructed distributions can be estimated by their fidelity f with the expected theoretical distributions P_m (Jozsa, 1994)

$$f = \sum_{m=1}^{\infty} \sqrt{P_{m,\mathrm{exp}} P_m} \,. \tag{13}$$

3.1 Photon-resolving photomultiplier

The first detector we tested is the Quantacon Burle 8850 photomultiplier ($\mathrm{PMT_{Burle}}$, Burle Industries Inc.). This photomultiplier is based on a semitransparent bialkali photocathode (maximum $QE = 0.24$ at ~400 nm) and a first dynode with a high-gain followed by 11 dynodes. The $\mathrm{PMT_{Burle}}$ detector is endowed with partial photon resolving capability, *i.e.* it gives partially distinct outputs for different numbers of photons detected within the pulse-response time-duration. The detector has 200 counts/s of dark noise, that give a negligible contribution in the gate time. The after-pulse probability has no effect on the detection since the light source is operated at 500 Hz. In Fig. 2(a) we show some typical pulse-height spectra of the $\mathrm{PMT_{Burle}}$ at different light intensities. We note that the pulse-height spectra lose the peak structure as soon as the field intensity becomes mesoscopic but remain linear in the mean values over a wide range of intensities.

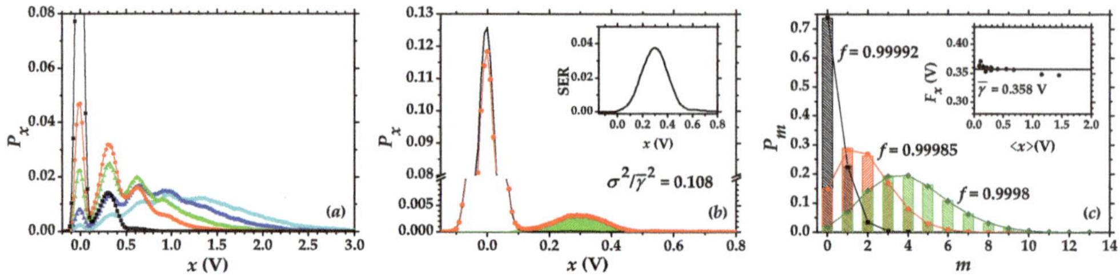

Fig. 2. (*a*) Typical pulse-height spectra measured with the $\mathrm{PMT}_{\mathrm{Burle}}$ at different light intensities. (*b*) Estimation of the single-electron response (SER) of the detector by subtraction of the response to the dark from the response to a very weak light. Note the interruption in the vertical axis. Inset: normalized SER. (*c*) Inset: Plot of the Fano factor, F_x, for the output voltages as a function of the mean voltage $\langle x \rangle$ for coherent states having different intensities (dots) and linear fit (line). Main panel: Reconstructed distribution, $P_{m,\mathrm{exp}}$, for coherent light reconstructed from some of the data sets used to obtain the calibration in the Inset (bars). Line+symbols: theoretical curves, P_m, from Equation (11).

To evaluate the applicability of the theory described above to the detector, we introduce a simple analysis of its response to obtain an estimation of the SER. We consider the normalized pulse-height spectrum P_x produced by very low illumination, virtually in the so-called "single-photon regime" (red line+dots in Fig. 2(*b*)) and subtract from it a fraction of the normalized pulse-height spectrum obtained in the dark (black line). We interpret the resulting curve obtained by varying the fraction until we get a single peak (green area) as p_γ, *i.e.* the SER of the system. The inset shows the normalized SER together with the evaluation of $\sigma^2/\overline{\gamma}^2 \sim 0.108$, a value that suggests the applicability of our analysis method.

In Fig. 2(*c*) we show the application of the analysis technique to the recorded data. In the inset of the figure we plot the Fano factor of the outputs (dots) along with the linear fit, that gives $\overline{\gamma} = (0.358 \pm 0.002)$ V, while in the main panel we show the experimental distributions $P_{m,\mathrm{exp}}$ (bars) reconstructed by using $\overline{\gamma}$ along with the theoretical curves P_m evaluated at the measured mean values. The values of the fidelity reported in the figure indicate the good quality of the reconstructions.

3.2 Hybrid photodetector

The second detector we tested is the H8236-40 hybrid photodetector (HPD, Hamamatsu, Japan). The light-sensitive element of the detector is a GaAsP(Cs) photocathode (maximum $QE = 0.40$ at $\sim$550 nm). Instead of a series of dynodes, the detector incorporates a semiconductor element into an evacuated electron tube. The photoelectrons emitted by the photocathode are accelerated and strike the semiconductor undergoing multiplication. Since the fluctuation in the multiplication process is very low, the HPD displays high electron resolution and excellent stability. The HPD has negligible dark-count rate and after-pulse probability. In Fig. 3(*a*) some typical pulse-height spectra taken at different light intensities are shown. We note that the HPD has a partial peak resolution, but again the peak structure disappears as soon as the field intensity becomes mesoscopic, even if the detector remains linear in the mean values over a wide range of intensities. The estimation of the SER of the HPD (see Fig. 3(*b*)) gives a value of $\sigma^2/\overline{\gamma}^2 \sim 0.126$, that is still small enough to apply the theory. The results for the reconstructed distributions are shown in Fig. 3(*c*): in the inset we

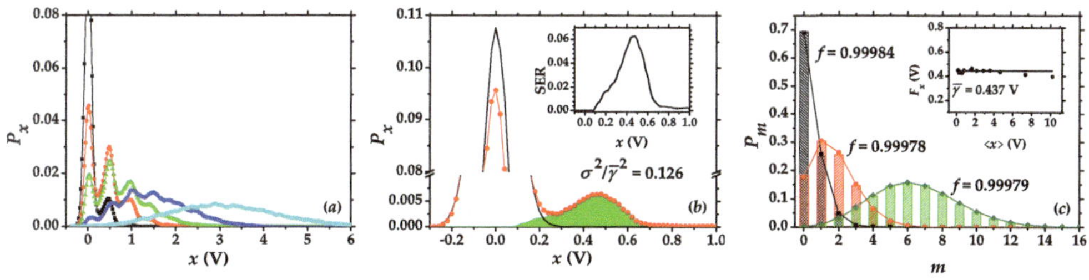

Fig. 3. Same as in Fig. 2 for HPD.

show the experimental results for the Fano factor of the output F_x (dots) and the linear fit that gives $\overline{\gamma} = (0.437 \pm 0.005)$ V, while in the main panel we show the reconstructed distributions and the theoretical curves. The fidelity is very high.

3.3 Single-photon-counting photomultiplier

We now check the theoretical description with a H5773P photomultiplier ($\text{PMT}_{\text{single}}$, Hamamatsu, Japan) that has no photon-number resolving power but that can be used for single-photon counting in that its output for zero detected photons can be discriminated from SER. According to the specifications, for light pulses of duration well below its time response (typical rise time: 0.78 ns) delivered at kHz repetition rate as in our experiments, the $\text{PMT}_{\text{single}}$ should saturate for m-values of few units. The $\text{PMT}_{\text{single}}$ current output was processed by an amplifier (ZFL 1000NL, Mini-Circuits) and integrated by a charge digitizer (V792, CAEN) with a 12-bit resolution over a temporal gate (40 ns) synchronized by means of the Q-switching signal of the laser amplifier. The final x outputs obtained for each ensemble of 30000 light pulses were then recorded by a computer.

The pulse height spectra in Fig. 4(a) at increasing light intensity show that the PMT has no photon-number resolving power, being just able to distinguish between $m = 0$ and $m > 0$. From the estimation of the SER in Fig. 4(b), we see that the response of the $\text{PMT}_{\text{single}}$ has

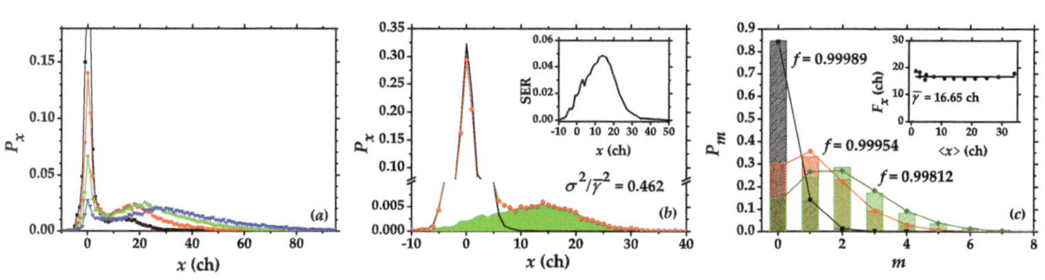

Fig. 4. Same as in Fig. 2 for $\text{PMT}_{\text{single}}$ detector.

$\sigma^2/\overline{\gamma}^2 \sim 0.462$, a value that could still be low enough to apply the theory.

In the inset of Fig. 4(c), we report the value of the Fano factor of the output $\overline{\gamma} = (16.7 \pm 0.3)$ ch. According to the main panel of Fig. 4(c), we are able to reconstruct detected photon distributions P_m with high fidelity for mean values slightly above $\langle m \rangle = 2$, that is when the probability of detecting photons, in the case of Poissonian statistics, remains non-negligible up to $m = 7$. We remark that this result is achieved by using a PMT that can hardly distinguish between $m = 0$ and $m = 1$.

3.4 Proportional photomultiplier

Finally, we want to demonstrate that the condition $\sigma^2/\overline{\gamma}^2 \ll 1$ is actually necessary for a detector to be used to reconstruct the photon-number statistics by measuring shot-by-shot photon numbers. To this aim we used a 931B photomultiplier (PMT, Hamamatsu, Japan) that is endowed with a very poor photon-number resolution (see Fig. 5(a)). Although the detector

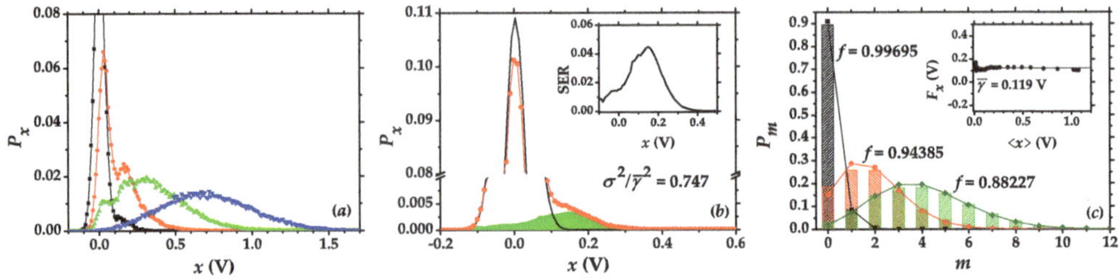

Fig. 5. Same as in Fig. 2 for PMT_{prop} detector.

does not resolve even the first terms in Equation (2) as separate peaks, we try to recover its SER (see Fig. 5(b)). The obtained value of $\sigma^2/\overline{\gamma}^2 \sim 0.747$ confirms the expected non-applicability of the analysis method.

If we apply our procedure to the measurements performed on coherent light, we obtain the results displayed in Fig. 5(c): the value of γ obtained from the fit of the Fano factor (see inset) produces reconstructed statistical distributions having a limited fidelity with the expected theoretical distributions evaluated at the experimental mean values (see main panel).

4. Applications

We exploited the capability of the HPD described above, assisted by the analysis technique, for some different applications. First of all we retrieved the P_m distributions for light states having different photon-number statistics. Secondly, we used the measured statistics to reconstruct the Wigner function of some classical optical states. Third, we demonstrated that we can measure shot-by-shot photon-number correlations between bipartite states and finally we used correlations to produce conditional non-Gaussian states.

4.1 Statistics

We produced a number of "artificial" field states by mixing, at a beam splitter, different fields generated by a pulsed laser source (Bondani et al., 2009c). Here we show the results obtained for two of such states, the displaced thermal state and the phase-averaged coherent state, which are relevant for the applications described in Section 4.2. As displayed in Fig. 6(a), both these states were obtained by mixing two fields at a beam splitter (BS). To obtain the displaced thermal state, we produced a single-mode thermal state containing $n_{\rm th}$ mean photons by inserting a rotating ground glass plate (D) and selecting a single speckle by a pin-hole (PH). Then we mixed the single-mode thermal field with a coherent field containing $|\alpha|^2$ mean photons at the BS, thus implementing the displacement operation. The resulting state is

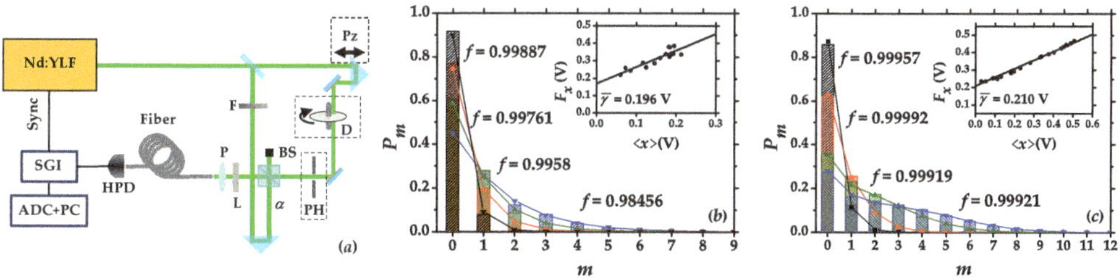

Fig. 6. (*a*) Experimental setup for the generation and measurement of either a displaced single-mode thermal state or a phase-averaged coherent state. Dashed boxes: elements inserted to obtain the different states. (*b*) Displaced single-mode thermal state. Inset: Fano factor, F_x, as a function of $\langle x \rangle$. Main panel: reconstructed detected-photon distributions for some of the data sets used to calculate the Fano factor (bars) and theoretical curves (lines+symbols). (*c*) Same as (*b*) for phase-averaged coherent state.

described by the following photon-number distribution

$$P_n = \frac{n_{\text{th}}^n}{(n_{\text{th}}+1)^{n+1}} \exp\left(-\frac{|\alpha|^2}{n_{\text{th}}+1}\right) L_n\left(-\frac{|\alpha|^2}{n_{\text{th}}\,(n_{\text{th}}+1)}\right), \tag{14}$$

in which L_n is the Laguerre polynomial L_n^γ for $\gamma = 0$. We have $\langle n \rangle = n_{\text{th}} + |\alpha|^2$, $\mu_2(n) = n_{\text{th}} + |\alpha|^2 + n_{\text{th}}\,(n_{\text{th}} + 2|\alpha|^2)$ and hence, according to Equation 10, we get $F_x = K\langle x \rangle + \overline{\gamma}$, with $K = n_{\text{th}}\,(n_{\text{th}} + 2|\alpha|^2)\,/\,(n_{\text{th}} + |\alpha|^2)^2$.

In Fig. 6(*b*) we plot the experimental results for a number of reconstructed distributions (bars) along with the theoretical distributions calculated from Equation 14. In the inset, we show the values of F_x as a function of $\langle x \rangle$.

The phase-averaged coherent state was obtained by superimposing two coherent fields whose relative phase was randomized from shot to shot. We obtained phase randomization by changing the path of one of the two fields at a frequency of ~100 Hz with a piezoelectric device, Pz, covering a travel length of 1.28 μm (see Fig. 6(*a*)). If the two coherent states contain $|\alpha|^2$ and $|\alpha_0|^2$ mean photons, the photon-number distribution is (Zambra et al., 2007)

$$P_n = \frac{A^n e^{-A}}{2\pi n!} \sum_{k=0}^{n} \binom{n}{k} \left(-\frac{B}{A}\right)^k \frac{\Gamma(1/2 + k/2)\,\Gamma(1/2)}{\Gamma(1 + k/2)}\, {}_1F_2\left[\left\{\frac{1}{2} + \frac{k}{2}\right\}, \left\{\frac{1}{2}, 1 + \frac{k}{2}\right\}, \frac{B^2}{4}\right] \tag{15}$$

where $A = |\alpha|^2 + |\alpha_0|^2$, $B = 2|\alpha||\alpha_0|$ and ${}_1F_2(a,b,z)$ is the generalized hypergeometric function. We get $\langle n \rangle = |\alpha|^2 + |\alpha_0|^2$, $\mu_2(n) = \langle n \rangle\,(K'\langle n \rangle + 1)$, with $K' \equiv 2|\alpha|^2|\alpha_0|^2/(|\alpha|^2 + |\alpha_0|^2)^2$ and $F_x = K'\langle x \rangle + \overline{\gamma}$.

In Fig. 6(*c*) we plot the experimental results for a number of reconstructed distributions (bars) along with the theoretical distributions calculated from Equation 15 and the values of the Fano factor for outputs (inset).

4.2 Wigner function

The capability of reliably reconstructing generic light states can be exploited to directly reconstruct the Wigner function of a state (Banaszek & Wodkiewicz, 1996; Banaszek et al., 1999; Cahill & Glauber, 1969; Wallentowitz & Vogel, 1996), with a technique alternative to OHT. The method consists in detecting the light exiting a beam-splitter that mixes the signal with a variable coherent probe field (see Fig. 6(a)). The Wigner function of the state can be written as

$$W(\alpha) = \frac{2}{\pi} \sum_{n=0}^{\infty} (-1)^n P_{n,\alpha} \,, \tag{16}$$

where $P_{n,\alpha}$ is the photon-number distribution of the field displaced by α. To obtain the photon-number distribution, reconstruction/inversion methods must be implemented (Allevi et al., 2009a) that require a known and sufficiently high photon-detection efficiency (Kiss et al., 1995). If, instead of $P_{n,\alpha}$, we consider $P_{m,\beta}$, the distribution of detected photons, where $\beta = \sqrt{\eta}\alpha$ is the detected amplitude of the displacement field, we obtain the Wigner function in the presence of losses (Banaszek & Wodkiewicz, 1996)

$$\overline{W}(\beta) = \frac{2}{\pi(1-\eta)} \int d^2\beta' e^{-\frac{2}{1-\eta}|\beta-\beta'|^2} W(\beta'/\sqrt{\eta}) = \sqrt{2}\pi \sum_{m=0}^{\infty} (-1)^m P_{m,\beta} \tag{17}$$

where $W(\beta/\sqrt{\eta}) = W(\alpha)$ is the Wigner function for photons. It can be demonstrated that,

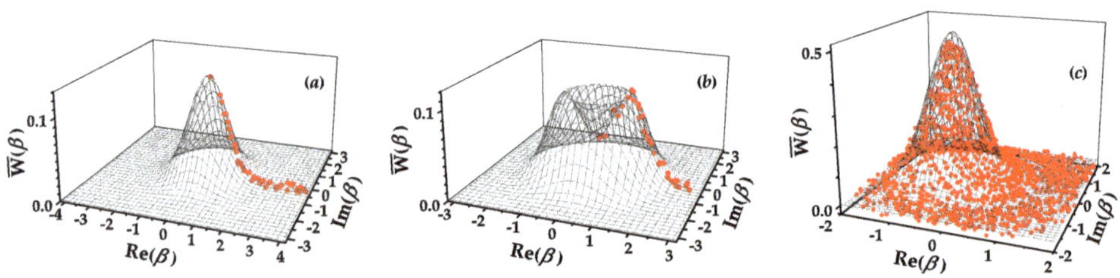

Fig. 7. Reconstructed Wigner functions (dots) superimposed to their theoretical surfaces for (a): single-mode thermal state, (b): phase-averaged coherent state and (b) coherent state.

for classical states such as non-squeezed Gaussian states and their linear superpositions, the integral in Equation (17) preserves the functional form of the Wigner distribution, so that the information directly extracted from detected photons is enough to characterize the state.

We used the experimental setups in Fig. 6(a) but with a frequency-doubled Q-switched Nd:YAG laser at 15 kHz repetition rate (Quanta System) delivering pulses at 532 nm of $\sim$ 200 ns duration as the laser source. The beam was spatially filtered and split into two parts serving as signal and probe fields. The probe was left Poissonian while that of the signal was modified in order to get the different states to be measured. The probe, whose intensity was modified by a variable neutral density filter, was then mixed with the signal at the PBS.

We begin by considering the reconstruction of the Wigner function of the two states considered in Section 4.1, the single-mode thermal state and the phase-averaged coherent state. Since both these states are phase-insensitive, their Wigner function is symmetric with respect to the origin and thus it is not necessary to control the phase of the probe field. In Fig. 7 we plot the reconstructed Wigner function for a single-mode thermal field (a) and for the phase-averaged coherent field (b). The dots in the figure are the experimental data and the surface is the theoretical Wigner function evaluated at the experimental parameters including some corrections that account for the mode-mismatching between probe and signal (Bondani et al., 2009b).

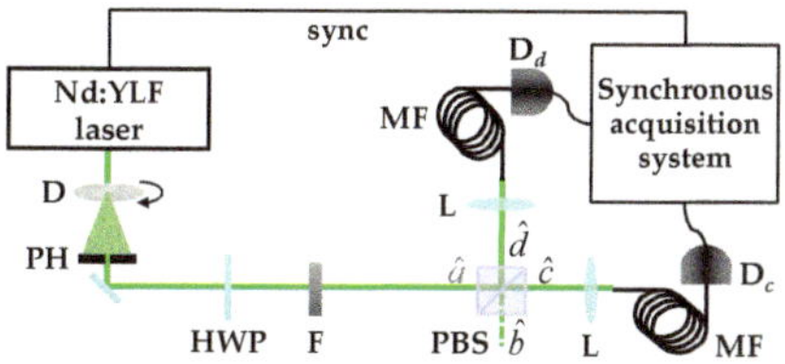

Fig. 8. Scheme of the experimental setup for the measurement of shot-by-shot correlations in single-mode thermal fields.

As an example of reconstruction of the Wigner function of a phase-sensitive state, we consider a coherent state. To perform the measurements we varied the phase of the probe field by means of a piezoelectric movement and evaluated the actual values of the phase by exploiting the linearity properties of the HPD (Bondani et al., 2010). In Figure 7(c) we plot the resulting Wigner function.

4.3 Correlations

Till now, we have demonstrated that HPDs are endowed with the capability of reconstructing the correct statistics of detected photons provided their outputs are analyzed in the proper way (Bondani et al., 2009a;b;c). Now we exploit the fact that, in principle, our analysis allows us to determine the number of photons detected at each laser shot correctly enough to reveal shot-by-shot photon-number correlations between the components of a bipartite state. We tested measurement apparatus and procedure by measuring the correlations in a classically-correlated bipartite state, namely a single-mode thermal field divided by a beam-splitter. The two HPDs used in these measurements were model R10467U-40 (Hamamatsu, Japan). According to the experimental setup sketched in Fig. 8, we produced a single-mode thermal field as described in Section 4.1. The selected mode was then sent to a polarizing beam-splitter (PBS) whose outputs were collected by two multimode fibers and delivered to the detectors. To obtain a fine tuning of the beam-splitter transmittance we inserted a half-wave plate (HWP) between the PH and the PBS. The light pulses were then delivered to the detectors ($D_{c,d}$) through two multimode fibers (MF). Single-shot detector outputs were then amplified, integrated, digitized and recorded. We measured 30000 subsequent shots at a number of different intensity values set by a variable neutral density filter (F). In Fig. 9(a) we plot the joint probability distributions obtained by re-scaling the outputs of the detectors by the proper value of $\overline{\gamma}$ without rebinning the data.

Referring to Fig. 8, the field operators at the outputs of the beam splitter ($\hat{c}$ and $\hat{d}$) are linked to the input ones ($\hat{a}$ and $\hat{b}$) by $\hat{c} = \sqrt{\tau}\hat{a} - \sqrt{1-\tau}\hat{b}$ and $\hat{d} = \sqrt{1-\tau}\hat{a} + \sqrt{\tau}\hat{b}$, τ being the beam splitter transmittance. If $\hat{b}$ is in the vacuum, the moments of the bipartite state at the output are given by $\langle \hat{n}_c \rangle = \tau \langle \hat{n}_a \rangle$, $\sigma^2_{n_c} = \tau \left[\tau \sigma^2_{n_a} + (1-\tau)\langle \hat{n}_a \rangle \right]$, $\langle \hat{n}_d \rangle = (1-\tau)\langle \hat{n}_a \rangle$, $\sigma^2_{n_d} = (1-\tau)\left[(1-\tau)\sigma^2_{n_a} + \tau \langle \hat{n}_a \rangle \right]$ and $\langle \hat{n}_c \, \hat{n}_d \rangle = \tau(1-\tau)\left(\langle \hat{n}^2_a \rangle - \langle \hat{n}_a \rangle \right)$, from which we obtain the expression for the normalized correlation coefficient:

$$\Gamma \equiv \frac{\langle (n_c - \langle \hat{n}_c \rangle)\,(n_d - \langle \hat{n}_d \rangle) \rangle}{\sigma^2_{n_c} \sigma^2_{n_d}} = \frac{\sqrt{\tau(1-\tau)}\left[\sigma^2_{n_a} - \langle \hat{n}_a \rangle \right]}{\sqrt{\left[\tau \sigma^2_{n_a} + (1-\tau)\langle \hat{n}_a \rangle \right] \left[(1-\tau)\,\sigma^2_{n_a} + \tau \langle \hat{n}_a \rangle \right]}} . \tag{18}$$

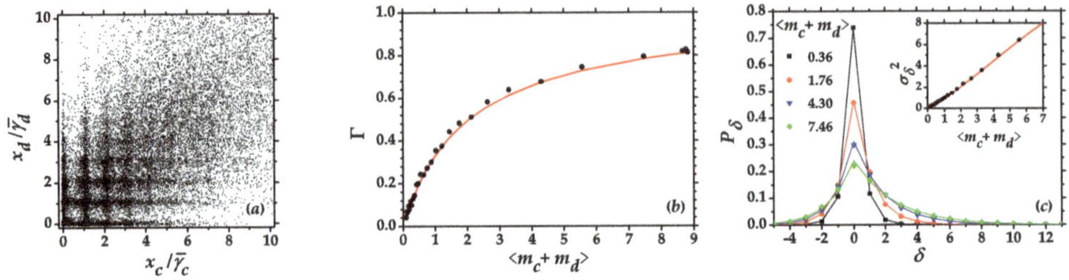

Fig. 9. (*a*) Experimental joint probability distribution. (*b*) Shot-by-shot correlation, Γ, as a function of the total mean value. (*c*) Experimental results for P_δ (symbols) and theoretical predictions (full lines). Inset: measured variance σ_δ^2 of P_δ as a function of the total number of detected photons (dots) and theoretical prediction (full line).

The single-mode thermal state at the input port *a* of the PBS can be written as

$$\nu_{n_{\text{th}}} = \sum \frac{n_{\text{th}}^n}{(1+n_{\text{th}})^{n+1}} |n\rangle\langle n| = \sum_n P_n |n\rangle\langle n| \tag{19}$$

and obviously $\langle \hat{n}_a \rangle = n_{\text{th}}$. Equation 18 becomes

$$\Gamma = \frac{\sqrt{\tau(1-\tau)n_{\text{th}}^2}}{\sqrt{(\tau n_{\text{th}}+1)[(1-\tau)n_{\text{th}}+1]}} = \frac{\sqrt{\langle n_c \rangle \langle n_d \rangle}}{\sqrt{(\langle n_c \rangle + 1)(\langle n_d \rangle + 1)}}, \tag{20}$$

where $\langle n_c \rangle = \langle \hat{n}_c \rangle$ and $\langle n_d \rangle = \langle \hat{n}_d \rangle$. If the overall detection efficiencies are $\eta_{c,d}$, the expression in Equation (20) remains the same upon substituting the mean number of photons $\langle n_{c,d} \rangle$ with that of detected photons $\langle m_{c,d} \rangle = \eta_{c,d} \langle n_{c,d} \rangle$ (Agliati et al., 2005). Figure 9(*b*) shows the results for shot-by-shot correlation measured as a function of the total mean value of the output detected light $\langle m_c + m_d \rangle$. Full line represent the theoretical curve evaluated according to Equation (20) by inserting the experimental mean values. The results obtained with HPDs perfectly superimpose to the theory.

Relevant applications of correlated bipartite states require the determination of the nature of the correlations. A quite standard way to discriminate between classical and nonclassical states (Agliati et al., 2005) is to study the statistics of the difference, P_δ, in the number of photons detected shot-by-shot at the outputs of the beam splitter. Figure 9(*c*) displays the experimental results for P_δ evaluated by subtracting the number of detected photons measured shot-by-shot together with the theoretical values evaluated according to Ref. (Agliati et al., 2005). The inset of the figure displays the measured variance σ_δ^2 as a function of the total number of detected photons, as compared to the theoretical predictions $\sigma_\delta^2 = (\langle m_c \rangle - \langle m_d \rangle)^2 + \langle m_c \rangle + \langle m_d \rangle$ (Agliati et al., 2005). The correct evaluation of σ_δ^2 is crucial for the estimation of the noise reduction in a bipartite state $R = \sigma_\delta^2/(\langle m_c \rangle + \langle m_d \rangle)$, which is a means to discriminate classical from nonclassical states. The results superimpose to the theory.

4.4 Conditional states

Nonclassical optical states are a crucial ingredient for fundamental tests of quantum mechanics and represent a resource for quantum communication and high-precision measurements. Besides squeezing, nonclassical effects are generally observed in connection with non-Gaussian states of light. It has been demonstrated that non-Gaussian states and operations represent a valuable resource for Quantum Information (Genoni et al., 2010): in fact, they have been studied in connection with entanglement distillation (Aoki et al., 2009; Dong et al., 2008; Eisert et al., 2002; Fiurasfiek, 2002), teleportation (Cochrane et al., 2002; Olivares et al., 2003; Opatrny et al., 2000), cloning (Cerf et al., 2005) and quantum storage (Casagrande et al., 2007). The simplest way to generate a non-Gaussian optical state starting from a Gaussian one consists in subtracting photons from it (Ourjoumtsev et al., 2007; Parigi et al., 2009), an operation that can be implemented by inserting a beam splitter in the optical path of the original state, detecting the number of photons at one output port and selecting the output of the other port only if a certain condition on the number of detected photons is satisfied. When short-pulsed fields are used, the challenging part of this scheme is the shot-by-shot measurement of sizeable numbers of photons.

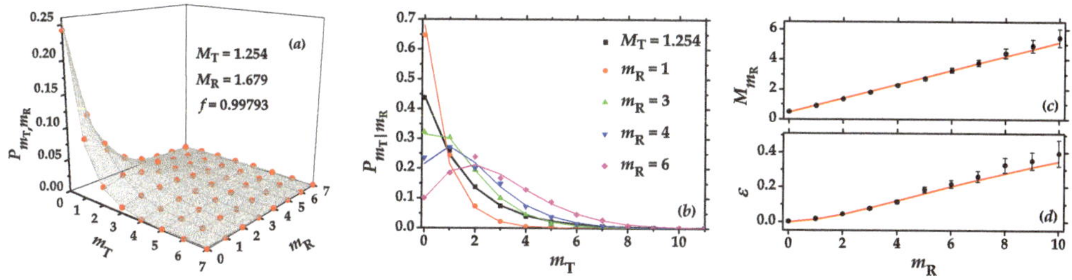

Fig. 10. (*a*) Joint detected-photon distribution: the experimental data are plotted as dots and the theoretical surface as grey mesh. (*b*) Reconstructed distributions of different conditional states (colored symbols) obtained starting from the same unconditioned one (black squares). The theoretical curves are plotted as lines according to the same choice of colors. (*c*) Mean number of detected photons of the conditional states and (*d*) Lower bound of non-Gaussianity measure, plotted as functions of the conditioning value (experimental data (dots) and theoretical curve (solid line)).

Here we present the production and characterization of non-Gaussian states of light starting from both classically and quantum correlated optical fields and exploiting the characteristics of the HPDs. First of all, we report the results obtained by using single-mode thermal fields as input states (Allevi et al., 2010b). Though photon-subtracted thermal states are not directly involved in any quantum information protocol, they are suitable probes to investigate the performances of our particular scheme. In fact, thermal states are Gaussian states diagonal in the photon-number basis, thus, the knowledge of their photon statistics fully characterizes them and their conditional non-Gaussian counterparts, which are still diagonal. Thanks to this property, we can give a complete analytical description of the behavior of the setup, including the actual expressions of the conditional states, and we can verify, with very high accuracy and control, the agreement of the theoretical expectations with the experimental results. We refer to the experimental setup sketched in Fig. 8, in which the input state at the PBS is the single-mode thermal state $\nu_{n_{\rm th}}$ and the transmitted and reflected fields are described by the operators $\hat{c}$ and $\hat{d}$, respectively. The measurements performed in the reflected beam irreversibly modify the states measured in the transmitted arm and, in particular, make them non-Gaussian. As an example, here we show the results obtained by choosing a set of measurements with $\langle m_c \rangle = \mathrm{M_T} \sim 1.254$. The joint probability $P_{m_{\rm T},m_{\rm R}}$ of measuring $m_{\rm R}$ photons in the reflected arm and $m_{\rm T}$ photons in the transmitted one is plotted in Fig. 10(*a*) as dots together with the theoretical surface to which it perfectly superimposes. Starting from the theoretical joint probability, we can calculate the expected photon-number distribution of the states obtained by performing different conditional measurements in the reflected arm. As it is shown in panel (*b*) of Fig. 10, the larger the conditioning value, the more different the statistics of the conditional state (lines + symbols) is from that of the incoming one (black squares + line). The accordance with the theoretical expectation, which can be quantified by calculating the fidelity (see f values in the figure), is rather satisfactory. In Fig. 10(*c*), we also plot the behavior of the mean number of photons of the conditional states as a function of the different conditioning values m_{R}. Finally, to deeply characterize the output conditional states we define the non-Gaussianity measure as

$$\delta = S[\nu_{N_{n_{\mathrm{R}}}}] + \sum_{n} P_n \log P_n \tag{21}$$

where $N_{n_{\mathrm{R}}}$ is the mean photon number of the conditional states, and $S[\nu_{n_{\mathrm{th}}}] = n_{\mathrm{th}} \log(1 + 1/n_{\mathrm{th}}) + \log(1 + n_{\mathrm{th}})$ is the entropy of the single-mode thermal state $\nu_{n_{\mathrm{th}}}$ (see Equation 19). However, due to the inefficient detection, we cannot reconstruct the actual photon-number distribution P_n, but only the detected-photon-number distribution $P_{m_{\mathrm{T}},m_{\mathrm{R}}}$. Remembering that for thermal states the statistics of detected photons is still thermal, we evaluate the quantity

$$\varepsilon = S[\nu_{M_{m_{\mathrm{R}}}}] + \sum_{m_{\mathrm{T}}} P_{m_{\mathrm{T}},m_{\mathrm{R}}} \log P_{m_{\mathrm{T}},m_{\mathrm{R}}} \leq \delta. \tag{22}$$

The last inequality follows from the fact that the inefficient detection may be described by a Gaussian lossy channel that does not increase the non-Gaussianity, followed by an ideal (*i.e.* unit quantum efficiency) detection. The quantity ε turns out to be a lower bound for the actual non-Gaussianity, that is, significant values of ε correspond to more markedly non-Gaussian states. The behavior of ε as a function of the conditioning value m_{R} predicted by the theory (line) is well reproduced by the experimental data (dots) shown in Fig. 10(*d*).

As a second application, we address multimode conditional measurements on the quantum correlated optical states produced by a pulsed multimode spontaneous parametric down-conversion (Allevi et al., 2010c).

The experimental setup is sketched in Fig. 11(*a*). The pump light source was the fourth harmonics of a Nd:YLF ps-pulsed laser (High-Q Laser Production, Austria) produced by a non-collinear sum-frequency generation. The pump was then sent into a β-BaB_2O_4 (BBO) crystal (Kaston, China, cut angle 48.84 deg, 4 mm length) to produce pairwise entanglement at 523.5 nm, in order to match the maximum quantum efficiency ($QE = 0.5$) of the HPDs. The UV stray light was cut-off by a filter and by two harmonic separators. Signal and idler were selected by two pin-holes (300 μm diameter, located at 1 m from BBO2). Notice that the number of temporal modes, which is evaluated from the marginal detected-photon number distribution, cannot be reduced at will. The only way to reduce the number of modes is to select a single spatial mode, which involves the challenging matching of the collection areas in signal and idler. The possible mismatch between the collection areas results in an effective detection efficiency, reduced in comparison to the nominal efficiency of the detectors, which can be estimated through the level of noise reduction $R = 1 - \eta$ (see Fig. 11(*c*)) exhibited by the two beams (Bondani et al., 2007). For our beams we obtained, without noise subtraction, $\eta \sim 0.06$. The light passing the pin-holes was coupled to two multimode optical fibers and delivered to the detectors. Each experimental run was performed on 50000 subsequent laser

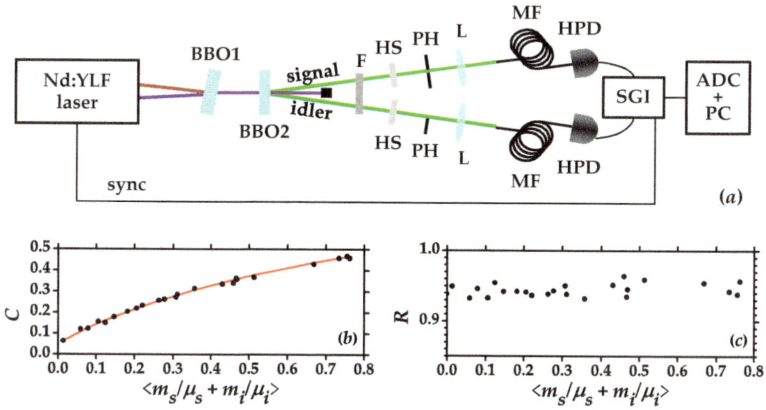

Fig. 11. (a) Schematic diagram of the experimental setup. BBO1 and BBO2: nonlinear crystals; F: cut-off filter; HS: harmonic separators; PH: pin-hole apertures; L: lenses; MF: multimode optical fibers; HPDs: amplified hybrid photodetectors; SGI: synchronous gated integrator; ADC+PC: digitizing PC board. (b) Detected-photon correlation coefficient and (c): shot-noise reduction, plotted as functions of the mean number of detected photons per mode (experimental data (dots) and theoretical curve (solid line)).

shots at fixed values of the pump intensity. First of all, we characterized the entangled states by measuring the correlation coefficient

$$C = \frac{\sqrt{(M_s/\mu_s)(M_i/\mu_i)} + \eta}{(M_s/\mu_s + 1)(M_i/\mu_i + 1)}, \tag{23}$$

where $M_{s,i}$ is the mean number of detected photons and $\mu_{s,i}$ is the number of modes in signal and idler. In Fig. 11(b) we plot C as a function of the total mean number of detected photons per mode: the experimental results are in perfect accordance with theory.

For what concerns the generation of non-Gaussian states, we notice that the intrinsic multimode nature of our entangled states makes the theoretical description of the photon-subtracted states non trivial. Moreover, a quantification of the amount of non-Gaussianity is not possible, as all the existing descriptions refer to single-mode states (Genoni & Paris, 2010). The experimental joint detected-photon distribution $P_{m_\mathrm{s},m_\mathrm{i}}$, plotted in Fig. 12($a$), is well superimposed to the theoretical curve derived from the multimode description of the process. Starting from its expression we calculated the photon distribution of the conditional states obtained by choosing the values of the measured photons on the idler beam, m_i. Panel (b) of Fig. 12 displays the detected-photons distributions of the conditional states obtained by choosing a definite number of detected photons on the idler ($m_\mathrm{i} = 10$ and $m_\mathrm{i} = 15$). We notice that the results are in excellent agreement with the theory and, despite the small value of effective quantum efficiency, the conditioning power of the measurement is appreciable. This is clearly illustrated by the behavior of the mean values of the distributions, which is reported in panel (c) of Fig. 12 as a function of the conditioning value: the experimental data are again in agreement with the theoretical prediction.

The conditioning protocol is thus effective even though the results are somehow limited by the specific characteristics of the original twin-beam state and in particular by its multimode

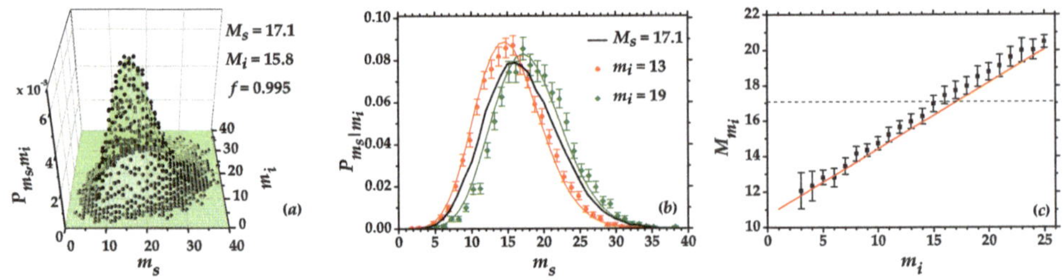

Fig. 12. (*a*) Joint detected-photon distribution: the experimental data are plotted as dots and the theoretical surface as grey mesh. (*b*) Reconstructed distributions of two different conditional states (symbols) obtained starting from the same unconditioned one (black line). The theoretical curves are plotted as lines according to the same choice of colors. (*c*) Mean number of detected photons of the conditional states as a function of the conditioning value: experimental data (dots) and theoretical curve (solid line). The dashed line corresponds to the mean value of the unconditioned state.

nature. The results could be improved by decreasing the number of temporal modes, either by reducing the pump duration or by changing the nonlinear crystal: work is in progress to follow both these lines.

5. Conclusions

We have presented the generation and characterization of pulsed optical states in the mesoscopic photon-number regime. In particular, the good results obtained for the reconstruction of detected-photon distributions and Wigner functions, for the measurement of shot-by-shot photon-number correlations and for the implementation of multiple photon-subtraction by means of HPDs open the way to the exploitation of these detectors for several applications in the fields of Quantum Optics and Quantum Information. Their main power is represented by the possibility to provide all the information contained in each laser shot, which is a required feature for the implementation of quantum communication protocols. The satisfactory experimental results obtained till now with classical states encourage us to explore more exotic classical states, such as the non-Gaussian ones, that could be useful as probe fields in many application schemes.

Nevertheless, beside the many advantages, we are conscious that the HPDs present an intrinsic deviation from ideality in the single-electron response. This limitation seems to be irrelevant for classical states, but becomes critical in case of detection of nonclassical states, as the uncertainty in the determination of the photon peaks corresponds to a reduction of the overall quantum efficiency. In order to minimize this problem, we could try to increase the gain of the detection chain by operating the HPDs at the maximum voltage below the breakdown threshold and at the minimum temperature allowed, and to refine the analysis of the detector outputs to include this effect in the description of its response.

6. References

Afek, I.; Natan, A.; Ambar, O., & Silberberg, Y. (2009). Quantum state measurements using multipixel photon detectors. *Phys. Rev. A.*, Vol. 79, No. 4, April 2009, 043830, ISSN 1050-2947.

Agliati, A.; Bondani, M.; Andreoni, A.; De Cillis, G., & Paris, M. G. A. (2005). Quantum and classical correlations of intense beams of light investigated via joint photodetection. *J. Opt. B Quantum Semiclass. Opt.*, Vol. 7, No. 12, November 2005, S652–S663, ISSN 1464-4266.

Akindinov, A. V.; Martemianov, A. N.; Polozov, P. A.; Golovin, V. M., & Grigoriev, E. A. (1997). New results on MRS APDs. *Nucl. Instr. and Meth. in Phys. Res. A*, Vol. 387, No. 1-2, March 1997, 231–234, ISSN 0168-9002.

Allevi, A.; Andreoni; Bondani, M.; Brida, G.; Genovese, M; Gramegna, M.; Olivares, S.; Paris, M. G. A.; Traina, P., & Zambra, G. (2009). State reconstruction by on/off measurement. *Phys. Rev. A*, Vol. 80, No. 2, August 2009, 022114, ISSN 1050-2947.

Allevi, A.; Bondani, M., & Andreoni, A. (2010). Photon-number correlations by photon-number resolving detectors. *Opt. Lett.*, Vol. 35, No. 10, May 2010, 1707–1709, ISSN 0146-9592.

Allevi, A.; Bondani, M.; Andreoni, A.; Genoni, M. G., & Olivares, S. (2010). Reliable source of conditional states from single-mode pulsed thermal fields by multiple-photon subtraction. *Phys. Rev. A*, Vol. 82, No. 1, July 2010, 013816, ISSN 1050-2947.

Allevi, A.; Andreoni, A.; Bondani, M.; Beduini, F. A.; Genoni, M. G.; Olivares, S., & Paris, M. G. A. (2010). Conditional measurements on multimode pairwise entangled states from spontaneous parametric downconversion. *Europhys. Lett.*, Vol. 92, No. 2, October 2010, 20007, ISSN 0295-5075.

Andreoni, A. & Bondani, M. (2009). Photon statistics in the macroscopic realm measured without photon counters. *Phys. Rev. A*, Vol. 80, No. 1, July 2009, 013819, ISSN 1050-2947.

Andreoni, A.; Allevi, A., & Bondani, M. (2009). Mesoscopic photon-number statistics measured with a non-counting PMT. *J. Russ. Laser Res.*, Vol. 30, No. 5, September 2009, 418–426, ISSN 1071-2836.

Aoki, T.; Takahashi, G.; Kajiya, T.; Yoshikawa, J.; Braunstein, S. L.; van Loock, P., & Furusawa, A. (2009). Quantum error correction beyond qubits. *Nat. Phys.*, Vol. 5, No. 8, August 2009, 541–546, ISSN 1745- 2473.

Arecchi, F. T. (1965). Measurement of the Statistical Distribution of Gaussian and Laser Sources. *Phys. Rev. Lett.*, Vol. 15, No. 24, December 1965, 912–916, ISSN 0031-9007.

Banaszek, K. & Wódkiewicz, K. (1996). Direct Probing of Quantum Phase Space by Photon Counting. *Phys. Rev. Lett.*, Vol. 76, No. 23, June 1996, 4344–4347, ISSN 0031-9007.

Banaszek, K.; Radzewicz, C.; Wódkiewicz, K., & Krasin'ski, J. S. (1999). Direct measurement of the Wigner function by photon counting. *Phys. Rev. A*, Vol. 60, No. 1, July 1999, 674–677, ISSN 1050-2947.

Blanchet, J.-L.; Devaux, F.; Furfaro, L., & Lantz, E. (2008). Measurement of Sub-Shot-Noise Correlations of Spatial Fluctuations in the Photon-Counting Regime. *Phys. Rev. Lett.*, Vol. 101, No. 23, December 2008, 233604, ISSN 0031-9007.

Bondani, M.; Allevi, A.; Zambra, G.; Paris, M. G. A., & Andreoni, A. (2007). Sub-shot-noise photon-number correlation in a mesoscopic twin beam of light. *Phys. Rev. A*, Vol. 76, No. 1, July 2007, 013833, ISSN 1050-2947.

Bondani, M.; Allevi, A.; Agliati, A., & Andreoni, A. (2009). Self-consistent characterization of light statistics. *J. Mod. Opt.*, Vol. 56, No. 2, January 2009, 226–231, ISSN 0950-0340 print/ISSN 1362-3044 online.

Bondani, M.; Allevi, A., & Andreoni, A. (2009). Wigner function of pulsed fields by direct detection. *Opt. Lett.*, Vol. 34, No. 9, May 2009, 1444–1446, ISSN 0146-9592.

Bondani, M.; Allevi, A., & Andreoni, A. (2009). Light Statistics by Non-Calibrated Linear Photodetectors. *Adv. Sci. Lett.*, Vol. 2, No. 4, December 2009, 463–468, ISSN 1936-6612.

Bondani, M.; Allevi, A., & Andreoni, A. (2010). Self-consistent phase determination for Wigner function reconstruction. *J. Opt. Soc. Am. B*, Vol. 27, No. 2, February 2010, 333–337, ISSN 0740-3224.

Bouwmeester, D.; Ekert, A., & Zeilinger, A. (2000). The Physics of Quantum Information, Springer-Verlag, Berlin.

Cahill, K. E. & Glauber, R. J. (1969). Density operators and quasiprobability distributions. *Phys. Rev.*, Vol. 177, No. 5, January 1969, 1882–1902.

Casagrande, F.; Lulli, A., & Paris, M. G. A. (2007). Improving the entanglement transfer from continuous-variable systems to localized qubits using non-Gaussian states. *Phys. Rev. A*, Vol. 75, No. 3, March 2007, 032336, ISSN 1050-2947.

Cerf, N. J.; Krüger, O.; Navez, P.; Werner, R. F., & Wolf, M. M. (2005). Non-Gaussian Cloning of Quantum Coherent States is Optimal. *Phys. Rev. Lett.*, Vol. 95, No. 7, August 2005, 070501, ISSN 0031-9007.

Cochrane, P. T.; Ralph, T. C., & Milburn, G. J. (2002). Teleportation improvement by conditional measurements on the two-mode squeezed vacuum. *Phys. Rev. A*, Vol. 65, No. 6, May 2002, 062306, ISSN 1050-2947.

Dong, R.; Lassen, M.; Heersink, J.; Marquardt, C.; Filip, R.; Leuchs, G., & Andersen, U. L. (2008). Experimental entanglement distillation of mesoscopic quantum states. *Nat. Phys.*, Vol. 4, No. 12, December 2008, 919–923, ISSN 1745- 2473.

Eisert, J.; Scheel, S., & Plenio, M. B. (2002). Distilling Gaussian states with Gaussian operations is impossible. *Phys. Rev. Lett.*, Vol. 89, No. 13, September 2002, 137903, ISSN 0031-9007.

Fitch, M. J.; Jacobs, B. C.; Pittman, T. B., & Franson, J. D. (2003). Photon-number resolution using time-multiplexed single-photon detectors. *Phys. Rev. A*, Vol. 68, No. 4, October 2003, 043814, ISSN 1050-2947.

Fiurášfiek, J. (2002). Gaussian transformations and distillation of entangled Gaussian states. *Phys. Rev. Lett.*, Vol. 89, No. 13, September 2002, 137904, ISSN 0031-9007.

Genoni, M. G.; Beduini, F. A.; Allevi, A.; Bondani, M.; Olivares, S., & Paris, M. G. A. (2010). Non-Gaussian states by conditional measurements. *Phys. Scr.*, Vol. 2010, No. T140, September 2010, 014007, ISSN 0031-8949.

Genoni, M. G. & Paris, M. G. A. (2010). Quantifying non-Gaussianity for quantum information. *Phys. Rev. A*, Vol. 82, No. 5, November 2010, 052341, ISSN 1050-2947.

Gol'tsman, G. N.; Okunev, O.; Chulkova, G.; Lipatov, A.; Semenov, A.; Smirnov, K.; Voronov, B.; Dzardanov, B.; Williams, C., & Sobolewski, R. (2001). Picosecond superconducting single-photon optical detector. *Appl. Phys. Lett.*, Vol. 79, No. 6, August 2001, 705–707, ISSN 0003-6951.

Gramegna, M.; Genovese, M; Brida, G.; Bondani, M.; Zambra, G.; Andreoni, A.; Rossi, A., & Paris, M. G. A. (2005). Measuring the photon distribution by ON/OFF photodectors. *Laser Physics*, Vol. 16, No. 2, February 2006, 385–392, ISSN 1054-660X.

Haderka, O.; Peřina Jr., J.; Hamar, M., & Peřina, J. (2005). Direct measurement and reconstruction of nonclassical features of twin beams generated in spontaneous parametric down-conversion. *Phys. Rev. A*, Vol. 71, No. 3, March 2005, 033815, ISSN 1050-2947.

Haderka, O.; Peřina Jr., J.; Hamar, M.; Michálek, V.; Černoch, A., & Soubusta, J. (2010). Photon-number resolving detectors. *Proceedings of SPIE 7746, 17th Slovak-Czech-Polish*

Optical Conference on Wave and Quantum Aspects of Contemporary Optics, pp. 774603, ISBN 9780819482365, Liptovsky Jan, Slovakia, September 2010, SPIE Publications.

Jozsa, R. (1994). Fidelity for Mixed Quantum States. *J. Mod. Opt.*, Vol. 41, No. 12, December 1994, 2315–2323, ISSN 0950-0340 print/ISSN 1362-3044 online.

Kim, J.; Takeuchi, S.; Yamamoto, Y., & Hogue, H. H. (1999). Multiphoton detection using visible light photon counter. *Appl. Phys. Lett.*, Vol. 74, No. 7, February 1999, 902–904, ISSN 0003-6951.

Kiss, T.; Herzog, U., & Leonhardt, U. (1995). Compensation of losses in photodetection and in quantum-state measurements. *Phys. Rev. A*, Vol. 52, No. 3, September 1995, 2433–2435, ISSN 1050-2947.

Lita, A. E.; Miller, A. J., & Nam, S. W. (2008).Counting near-infrared single-photons with 95% efficiency. *Opt. Express*, Vol. 16, No. 5, March 2008, 3032–3040, ISSN 1094-4087.

Mandel, L. & Wolf, E. (1995). *Optical Coherence and Quantum Optics*, Cambrige University Press, ISBN 0521417112, New York.

Martienssen, W & Spiller, E. (1966). Fluctuation Measurements in Mixed Light Fields. *Phys. Rev.*, Vol. 145, No. 1, May 1966, 285–287.

Olivares, S.; Paris, M. G. A., & Bonifacio, R. (2003). Teleportation improvement by inconclusive photon subtraction. *Phys. Rev. A*, Vol. 67, No. 3, March 2003, 032314, ISSN 1050-2947.

Opatrný, T.; Kurizki, G., & Welsch, D.-G. (2000). Improvement on teleportation of continuous variables by photon subtraction via conditional measurement. *Phys. Rev. A*, Vol. 61, No. 3, February 2000, 032302, ISSN 1050-2947.

Ourjoumtsev, A.; Dantan, A.; Tualle-Brouri, R., & Grangier, P. (2007). Increasing Entanglement between Gaussian States by Coherent Photon Subtraction. *Phys. Rev. Lett.*, Vol. 98, No. 3, January 2007, 030502, ISSN 0031-9007.

Parigi, V.; Zavatta, A., & Bellini, M. (2009). Implementation of single-photon creation and annihilation operators: experimental issues in their application to thermal states of light. *J. Phys. B*, Vol. 42, No. 11, June 2009, 114005, ISSN 0953-4075.

Ramilli, M.; Allevi, A.; Chmill, V.; Bondani, M.; Caccia, M., & Andreoni, A. (2010). Photon-number statistics with Silicon photomultipliers. *J. Opt. Soc. Am. B*, Vol. 27, No. 5, May 2010, 852–862, ISSN 0740-3224.

Raymer, M. G. (1997). Measuring the quantum mechanical wave function. *Contemp. Phys.*, Vol. 38, No. 5, September 1997, 343–355, ISSN 1366-5812 (electronic)/0010-7514 (paper).

Řeháček, J.; Hradil, Z.; Haderka, O., Peřina Jr., J.; & Hamar, M. (2003). Multiple-photon resolving fiber-loop detector. *Phys. Rev. A*, Vol. 67, No. 6, June 2003, 061801(R), ISSN 1050-2947.

Sansoni, L.; Sciarrino, F.; Vallone, G.; Mataloni, P.; Crespi, A.; Ramponi, R., & Osellame, R. (2010). Polarization entangled state measurement on a chip. *Phys. Rev. Lett.*, Vol. 105, No. 20, November 2010, 200503, ISSN 0031-9007.

Wallentowitz, S. & Vogel, W. (1996). Unbalanced homodyning for quantum state measurements. *Phys. Rev. A*, Vol. 53, No. 6, June 1996, 4528–4533, ISSN 1050-2947.

Zambra, G.; Bondani, M.; Spinelli, A. S.; Paleari, F., & Andreoni, A. (2004). Counting photoelectrons in the response of a photomultiplier tube to single picosecond light pulses. *Rev. Sci. Instrum.*, Vol. 75, No. 8, August 2004, 2762–2765, ISSN 0034-6748 print/1089-7623 online.

Zambra, G.; Andreoni, A.; Bondani, M.; Gramegna, M.; Genovese, M; Brida, G.; Rossi, A., & Paris, M. G. A. (2005). Experimental reconstruction of photon statistics without photoncounting. *Phys. Rev. Lett.*, Vol. 95, No. 6, August 2005, 063602, ISSN 0031-9007.

Zambra, G.; Allevi, A.; Andreoni, A.; Bondani, M., & Paris, M. G. A. (2007). Nontrivial photon statistics with low resolution-threshold photon counters. *Int. J. Quant. Inf.*, Vol. 5, No. 1-2, February-April 2007, 305–309, ISSN 0219-7499.

Zavatta, A.; Viciani, S., & Bellini, M. (2006). Non-classical field characterization by high-frequency, time-domain quantum homodyne tomography. *Laser Phys. Lett.*, Vol. 3, No. 1, January 2006, 3–16, ISSN 1612-2011print/1612-202X online.

Ultrafast Photoelectric Effect in Oxide Single Crystals and Films

Hao Ni and Kun Zhao
Laboratory of Optic Sensing and Detecting Technology,
China University of Petroleum, Beijing
People's Republic of China

1. Introduction

In recent years there has been an increasing demand for detecting the ultrafast laser pulse to apply in environmental monitoring such as solar ultraviolet (UV) radiation monitoring, ozone-hole sensing, flame detection for a fire alarm, and secure space-to-space communication and many research groups focus their attention on the detector materials and its various kinds of detecting performance parameter. Main efforts are currently directed to wide band gap semiconductors, such as III-V nitrides,[1] silicon carbide,[2] zinc oxide,[3] and diamond,[4] which are a much more attractive choice for selective UV detection. A response of 8 ns rise time and 1.4 μs fall time has been achieved in UV photoconductive detector based on epitaxial $Mg_{0.34}Zn_{0.66}O$ thin films.[5] Highly quality GaN devices exhibit 1.4-ps-wide electrical transients under illumination by 100 fs duration and 360 nm wavelength laser pulses.[6] In general, thermal detectors such as bolometers are independent of the wavelength with a major disadvantage of the slow response time, which limits their applications especially where the fast measurement is required. In the past few years, there have been active studies of the fast photoresponse of superconductor materials such as $YBa_2Cu_3O_{7-\delta}$ thin films irradiated by visible and infrared laser pulses, and the transport measurements of photo-induced voltages have probed the time scale from picosecond to nanosecond.[7-10] However, these devices require a complicated fabrication process and high-cost manufacturing.

In this paper, a brief review of our recent research results on oxide single crystals and films, such as tilted perovskite oxides,[11-13] polar crystals[14] and manganite films,[15,16] is presented. Some experimental results are introduced and then, theoretical investigations about interesting experimental phenomena are presented.

2. Ultrafast photovoltaic effects in oxide single crystals

2.1 Ultrafast photovoltaic effects in tilted perovskite oxide single crystals

Perovskite oxides, as one of the leading research topics, have been intensively studied due to their intrinsic properties such as insulating, ferroelectric, superconducting, colossal magnetoresistance, optical properties and so on. Some research groups focused their interests on the photoelectric effect of perovskite oxide materials, and the research results showed that perovskite oxides have the potential and foreground as the photodetector materials in different wavelengths.

Strontium titanate oxide $SrTiO_3$ (STO) is a promising wide band gap material and appears to be excellent in the field of electronic industry. Over the years, STO has been mainly used for dielectric devices,[17–19] whereas very little work has been devoted to optical applications.[20,21] In agreement with its band gap of 3.2 eV, STO absorbs lights with a wave-length of less than 390 nm, presenting a high transparency in the visible and infrared wavelength range and selectively sensitive to a UV light.

As-supplied STO (001) single crystals with the purity of 99.99% are mirror double polished. The geometry of the sample is 3 mm × 10 mm with the thickness of 0.5 mm. The (001) plane is tilted as an angle of α with respect to the surface. The angles ranged from 0 to 45°. As shown in Fig. 1, the tilting of the c axis was confirmed by x-ray diffraction (XRD) measurement with the usual θ-2θ scan. To satisfy the Bragg's diffraction geometry, the [001] axis was aligned carefully and the offset point was set as ω. Here, ω was α or α-45° to detect (00*l*) or (*l*0*l*) peaks, respectively.

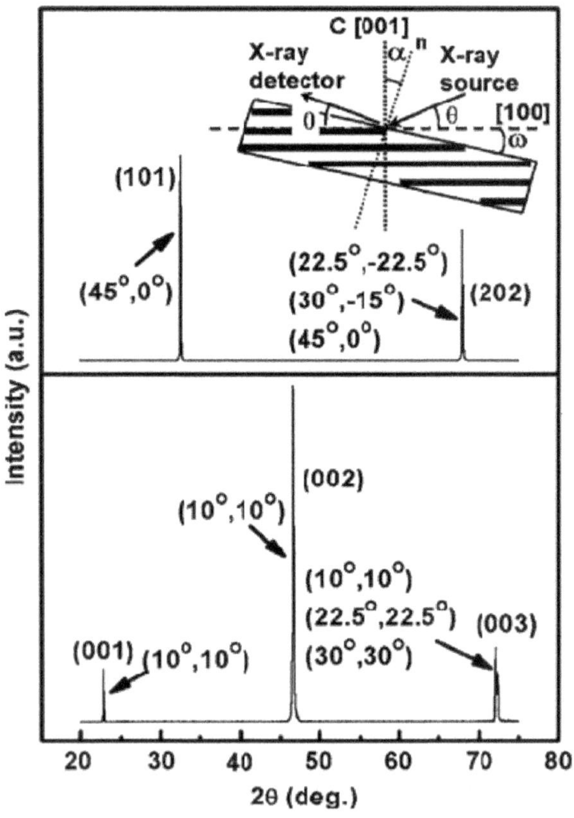

Fig. 1. X-ray diffraction patterns of tilted STO crystals. The inset shows the configuration of θ-2θ scan. To get different diffraction peaks, (00*l*), (101) and (202), ω was adjusted as shown in the angle arrays (α,ω) for different α. [11]

For the photovoltaic measurements, two indium electrodes separated by 1 mm were prepared on the laser-irradiated surface of the STO, and were always kept in the dark to prevent the generation of any electrical contact photovoltage. The third harmonic of an actively passively mode-locked Nd:yttrium–aluminum–garnet (Nd:YAG) laser was used as the light source, operating at a wavelength of 355 nm (3.49 eV photon energy) with 25 ps duration at a 1 Hz repetition rate and an energy density of 0.05 mJ mm^{-2} . Thus the on-sample energy is 0.15 mJ. The photovoltaic signals were monitored with a Tektronix sampling oscilloscope (2.5 GHz bandwidth) terminated into 1 MΩ.
The pulsed photovoltage was observed when the tilted STO surface was irradiated directly by a laser pulse. Figure 2(a) presents a typical transient open-circuit photovoltage of a 15° tilting STO single crystal. The peak photovoltage V_L^P is 52 mV, the 10%~90% rise time (RT) is 80 ns, and the full width at half maximum (FWHM) is 2.8 μs. To reduce the influence of the circuit in the measurement and the long tail of the decay time due to the RC effect in the photovoltaic signal, a 0.5 Ω load resistance was connected in the two electrodes and parallel with the STO crystal. As shown in Fig. 2(b), the RT is dramatically reduced to about 130 ps, which is limited by the oscilloscope, and the FWHM is also reduced to be about 230 ps. The peak photocurrent I_L^P is 4.7 mA, here, $I_L = V_L/0.5\ \Omega$.

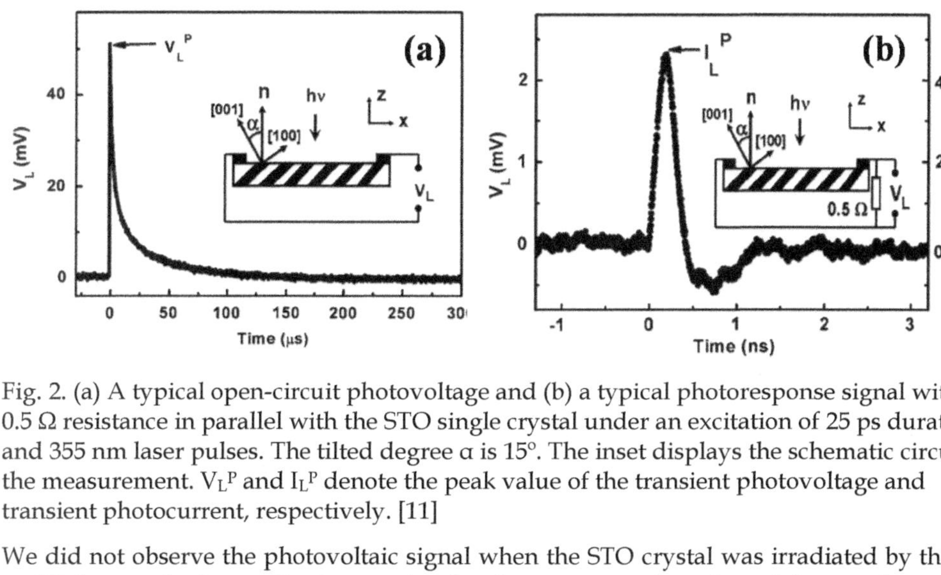

Fig. 2. (a) A typical open-circuit photovoltage and (b) a typical photoresponse signal with a 0.5 Ω resistance in parallel with the STO single crystal under an excitation of 25 ps duration and 355 nm laser pulses. The tilted degree α is 15°. The inset displays the schematic circuit of the measurement. V_L^P and I_L^P denote the peak value of the transient photovoltage and transient photocurrent, respectively. [11]

We did not observe the photovoltaic signal when the STO crystal was irradiated by the 532 or 632.8 nm pulse lasers. To characterize the photo-generated carriers of the samples used in present study, we determined the optical absorption spectrum from the optical transmission measurement using a SpectraPro500i spectrophometer. Figure 3 shows the UV-visible absorption spectrum of STO samples as a function of the wavelength. The sharp absorption edge is at ~385 nm, which is in agreement with the optical band gap of STO and the experimental results, indicating that photovoltaic signals cannot have been observed when the samples are irradiated by lasers with the photon energies smaller than the band gap of STO. This property demonstrates that the production of the photongenerated carriers plays a crucial role in the process of laser-induced voltage in this system.

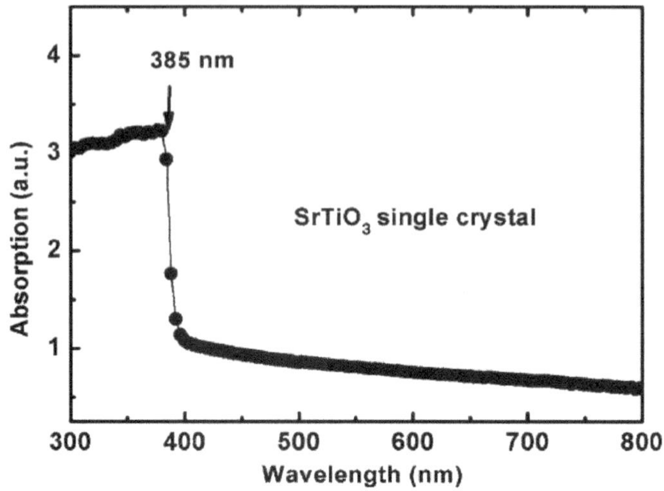

Fig. 3. Spectral absorption of STO single crystal with a thickness of 0.5 mm. [11]

To understand the mechanism of the photoelectric effect, we illuminate the STO crystal through the reverse face which is opposite to the side where leads are attached. The signal polarity reversal is obvious, and the response is essentially mirror images of each other. In addition, no photovoltage was generated along the nontilting [010] direction. These facts can be considered to support the suggestion that this phenomenon is due to a transverse thermoelectric effect.[22] In this case, the voltage can be represented by

$$E_x = l(S_{[100]} - S_{[001]})\sin(2\alpha)(dT/dz)/2. \tag{1}$$

Here l is the illumination length, $S_{[100]}$ and $S_{[001]}$ are the Seebeck coefficients along the [100] and [001] axis, and dT/dz is the temperature gradient in the direction of irradiation (the surface normal direction of STO crystal).

Furthermore, we chose several STO samples with different tilting angles (1°, 7°, 10°, 15°, 20°, 30°, 40°, 45°) and measured V_L^P and I_L^P as the function of α. The experimental procedures are the same as shown in Figs. 2(a) and 2(b). Results are plotted in Fig. 4. The signal amplitude increases with α until 20°, and then decreases. At α = 45°, the detected signal, similar to that at α = 0°, is basically a random noise due to the electrical-magnetical emission by the operation of the excimer laser. The maximum of the signal height, a photovoltage of 63 mV and a photocurrent of 67 mA, appears at α = 20°. It has been well known that each element of Seebeck tensor is proportional to the corresponding electrical resistivity of the system, and the electrical resistivity along each axis is reverse proportional to the transient carrier density correspondingly. In this way, Seebeck tensor strongly depends on the transient carrier density along each axis. The inset of Fig. 4 shows the schematic drawing under the irradiation of a UV pulse laser. If we divided the laser into two components of being parallel and being perpendicular to the tilting direction, the transient carrier density along each axis is proportional to the illuminated power perpendicular to the tilting direction, leading to $(S_{[100]} - S_{[001]}) \propto (\cos\alpha - \sin\alpha)$. So, we can get

$$E_x \propto l(\cos\alpha - \sin\alpha)\sin(2\alpha)(dT/dz)/2, \tag{2}$$

where the maximum occurs at α = 20.9°. From Fig. 3, we can see that the calculated results (dashed lines in Fig. 4) from Eq. 2 are in very good agreement with the measured points, which indicates that the present model of the mechanism for laser-induced voltage becomes convincing. This result also suggests the importance of the anisotropic thermoelectricity.

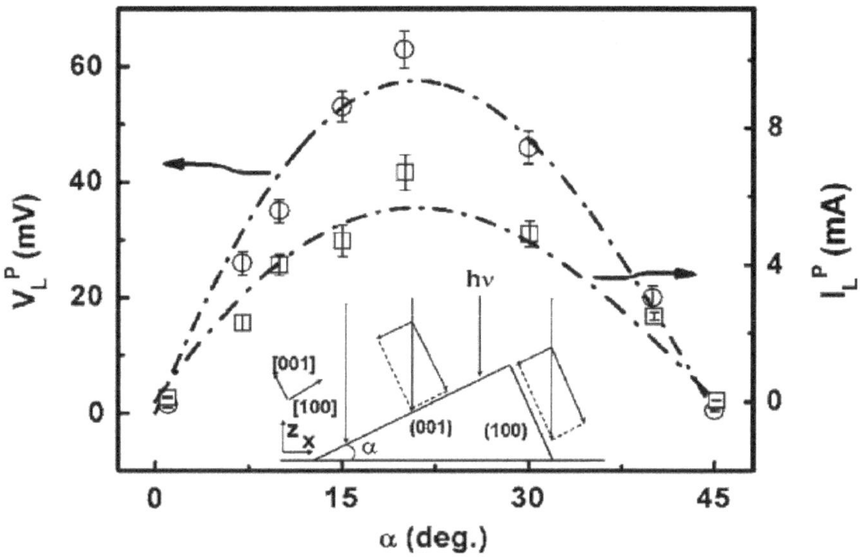

Fig. 4. Peak photovoltage (open circle) and peak photocurrent (open square) dependence of the tilting angle. The dashed lines show the fitting results using Eq. (2). [11]

To demonstrate the mechanism presented above, we also measured the irradiated area depended photovoltage, and the result showed that the measured voltage depended weakly on the length and linearly on the width of the irradiated area.[12] Figure 5(a) shows a diagram of the photoresponse measurement setup. The output of a Lambda Physik LEXTRA 200 XeCl excimer laser (308 nm wavelength, 20 ns pulse duration) was used as the laser source, with a single pulse energy density of 1.6 mJ/mm^2. The irradiated area was wl. The two electrodes were always kept in the dark to prevent any photovoltaic effect from the junctions. The photoresponse signal was monitored with a digital oscilloscope with 500 MHz resolution terminated into 50 Ω at ambient temperature.

Figure 6 displays the dependence of V_P on the irradiation area. For w = 3 mm and 10 mm, V_P is constant at ~ 0.05 V and ~ 0.15 V, respectively, independent of the value of l. However, V_P varies linearly with w for fixed l. Thus, from basic circuit theory, the total resistance in the l direction, can be expressed as

$$R_s = \frac{\rho l}{S} = \frac{\rho l}{wt} \tag{3}$$

Where ρ is the electrical resistivity along l direction and t is the thickness of the sample. Due to the configuration of the measurement setup (Fig. 5(b)), the circuit acts as a voltage divider, and the readout voltage V_T will be determined by

$$V_T = \frac{V_S R_T}{R_T + R_S} \approx \frac{V_S R_T}{R_S}, \tag{4}$$

where R_T = 50 Ω is the impedance at the input channel of the oscilloscope, and V_S is the actual signal generated by the sample, given by equations (2). Finally, according to the equation (2), one can obtain:

$$V_T \approx \frac{V_S R_T}{R_S} \propto \frac{R_T}{\frac{\rho l}{wt}} l(\cos\alpha - \sin\alpha)\sin(2\alpha)\frac{dT}{dz}\frac{1}{2} \propto w(\cos\alpha - \sin\alpha)\sin(2\alpha)\frac{dT}{dz}. \tag{5}$$

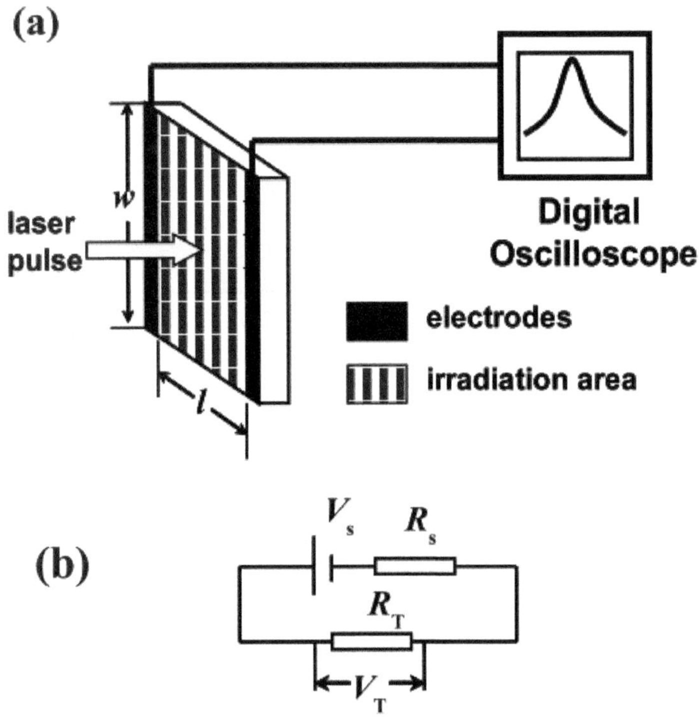

Fig. 5. (a) Schematic of the optical response measurement setup. l and w denote the length and width of the irradiated area, respectively. (b) Corresponding equivalent electrical scheme of the single crystal along the l direction. [12]

The suggested model for the source of the laser induced voltage in STO is fairly simplified, and only qualitatively explains the dependence of V_P on the irradiated width and length, as well as its dependence on the vicinal angles.

Picosecond photovoltaic effect has been observed in tilted Nb 0.7 wt %-$SrTiO_3$ (NSTO) single crystal under Nd:YAG pulse laser (a wavelength of 355 nm, a 25 ps duration, a 1 Hz repetition rate and an energy density of 1.07 mJ mm^{-2}) irradiation at ambient temperature without an applied bias.[23] The RT and FWHM are 828 and 670 ps, respectively.

In addition, ultrafast photoresponse were also observed in $LaSrAlO_4$ and MgO single crystals with their (001) planes tilted to the surface wafer at an angle of 10°.[24,25] When the samples were irradiated by 248 nm KrF laser pulses of 20 ns duration, open-circuit photovoltages were obtained. The RT and FWHM were ~10 ns and ~20 ns for tilted MgO and 6 ns and 19 ns for tilted $LaSrAlO_4$.

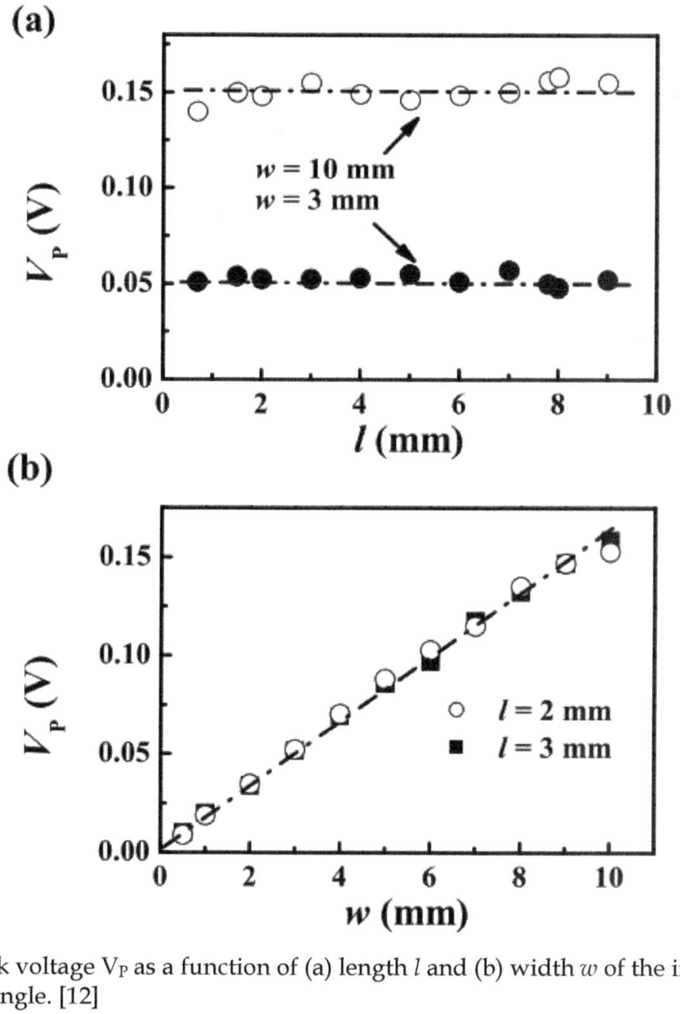

Fig. 6. The peak voltage V_P as a function of (a) length *l* and (b) width *w* of the irradiated area for 15° tilting angle. [12]

2.2 Thickness-dependent photovoltaic effects

The photovoltaic effects of tilted NSTO with different thicknesses were systematically investigated under the illuminations of ultraviolet pulsed lasers with the wavelength of 248 nm.[13] The as-supplied NSTO (001) (0.7 wt% Nb) single crystal wafer was investigated. The sample geometry is 5 × 5 mm^2 with a thickness of 0.5 mm and polished into several different thicknesses mechanically (d = 120 ~ 460 μm). The wafer orientation is 10° tilted from the exact (100) orientation which was confirmed by an XRD measurement with the θ~2θ scan. For the photovoltaic effect measurements, two Ag electrodes were applied and separated by 1 mm on the back of irradiated surface of the NSTO to avoid the photovoltaic effect due to contact barrier. Compex50 excimer pulsed laser with a laser spot of 5 × 13 mm², operating at a wavelength of 248 nm with 20 ns duration at a 2 Hz repetition rate, was used as the light source. The effective irradiation area is about 5 mm². The open-circuit photovoltaic signals were recorded using a digital oscilloscope terminated into 50 Ω.

Figure 7 shows the dependence of the peak photovoltaic signals (R_{VE}) on crystal thicknesses. With the decrease of thickness R_{VE} increased and reached to a maximum of 4.52 mV at 180 μm, which is ~2.63 times larger than that at 460 μm. However, R_{VE} decreased to 2.08 mV smaller than half maximum when the crystal is thinned from 180 μm to 120 μm. From the inset of Fig. 7, the R_{VE} linearly depended on the on-sample laser energy density E. Thus the thickness dependence of R_{VE} can be explained as following: there are many kinds of defects and traps (recombination centers) where the electrons and holes can locate in samples, owing to the doped 0.7 wt% Nb impurity, which is often found in single crystals and can be considered as carrier dilution effect. Only those carriers reaching to the electrodes in the back can be collected and give rise to photovoltaic signals. On one hand, with decrease of crystal thickness the amount of carriers collected by electrodes in the back increase due to lower loss, which is resulted from the shorter transport distances as well as less traps and recombination in thinner samples. Thus the photovoltaic signals for thinner samples are much larger than that for thicker samples. On the other hand, the decrease of thickness also leads to decrease of autologous carriers in NSTO single crystals, which limits the amount of carriers collected by electrodes. The competion between the above two factors results in a maximum photovoltaic signal.

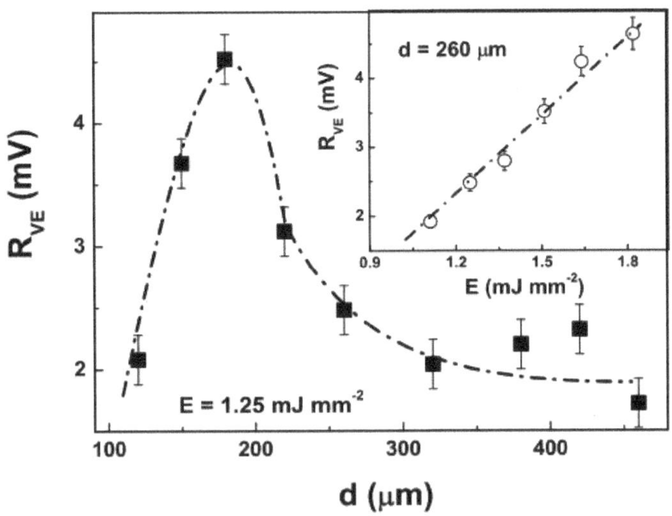

Fig. 7. Thickness and on-sample laser pulse energy density (inset) dependences of peak value responsivities of photovoltaic signals. [13]

Figure 8 shows the thickness dependence of the RT and FWHM of the photovoltaic signals. The response time is faster for the thinner sample. The differences of the RT and FWHM between the thickest and the thinnest crystals are about 4.49 ns and 9 ns, respectively. This dependence can be understood easily: the carriers induced by laser pulse were separated by lateral field in the NSTO single crystal, and traveled shorter distance to reach the electrodes for thinner samples so that faster photovoltaic responses can be observed.

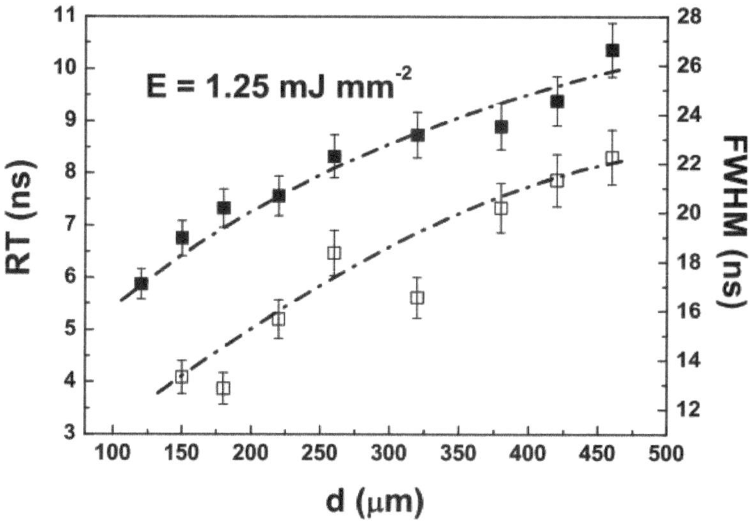

Fig. 8. Thickness dependence of RT (solid point) and FWHM (open point) of photovoltage. [13]

Experimental results show that the decrease in the crystal thickness is an effective method for improving the photovoltaic sensitivity in single crystals, suggesting the potential for optoelectronic detection applications.

2.3 Ultrafast photovoltaic effects in polar crystals

Silicon dioxide (SiO_2), which has good abrasion resistance, electrical insulation and high thermal stability, is widely used in electronics industry, refractory materials or glass production and so on.[26-28] Picosecond photovoltaic response has been observed in quartz single crystals under spectrally broadband laser irradiation from ultraviolet to near infrared without an applied bias at room temperature.

A mirror polished commercial quartz single crystal with 3 × 5 mm^2 in area, 0.5 mm in thickness (MTI Corporation Production, USA) and c-axis oriented (parallel to the normal of the illuminated surface) was used for the photoresponse studies. The samples have pure single crystal structures characterized by x-ray diffraction using Cu K_α radiation.

Figure 9 displays the temporal response of the open-circuit photovoltage responsivity R_{VE} equal to the photovoltage/on-sample-energy, and the primary photovoltaic signals are almost symmetrical. The inset of Fig. 9 displays the schematic circuit of the measurement. The Nd:YAG laser (pulse duration of 25 ps, repetition rate of 10 Hz) was used to irradiate the sample operating at the wavelengths of 1064, 532 and 355 nm. The photovoltaic signals were recorded using a sampling oscilloscope terminating into 50 Ω. The RT and FWHM are about 465 and 586 ps, 403 and 570 ps and 419 and 720 ps for 1064 nm, 532 nm and 355 nm, respectively.

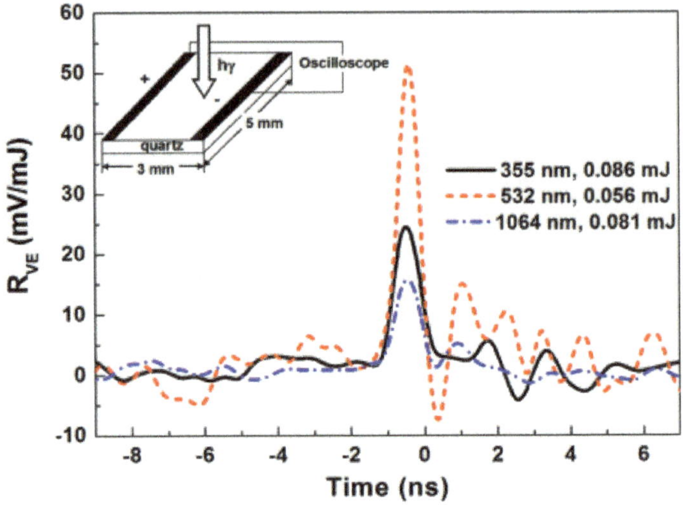

Fig. 9. Temporal response of a quartz single crystal under the irradiation of a Nd:YAG pulsed laser at room temperature without an applied bias. The inset displays the schematic circuit of the measurement. [14]

The peak amplitudes R^P_{VE} of R_{VE} as a function of the on-sample pulsed laser energy E_{in} are plotted in Fig. 10. It is found that the R^P_{VE} shows a high value when the sample was irradiated by a low energy pulse laser and drops gradually with the increase in E_{in}.

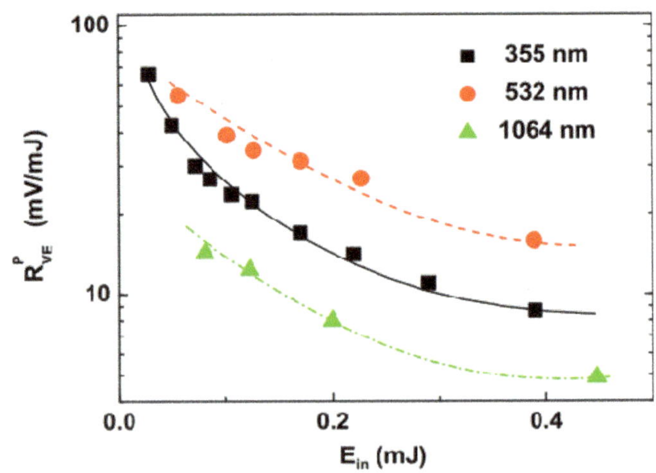

Fig. 10. Peak photovoltaic responsivity R^P_{VE} as a function of the on-sample laser pulse energy E_{in}. [14]

In addition, the samples were also illuminated directly by continuum solid state lasers (on-sample power: 20 mW) in the absence of an applied voltage at ambient temperature. The open-circuit photovoltaic signals were measured by a standard lock-in amplifier technique and recorded by a sampling oscilloscope terminating into 1 MΩ.

Figure 11 shows that the responsivity R_{VP}, equal to photovoltage/on-sample power, is about 0.476 μVmW^{-1}, 0.165 μVmW^{-1} and 0.188 μVmW^{-1} for 1064 nm, 808 nm and 532 nm, respectively, which resembles that under a bromine–tungsten lamp illumination (on-sample power: 1.08 mW). We also measured the picosecond laser-induced photovoltages in 10° tilting quartz single crystals and the signals were very close to those in the present nontilting samples, indicating that the Seebeck effect is ruled out.

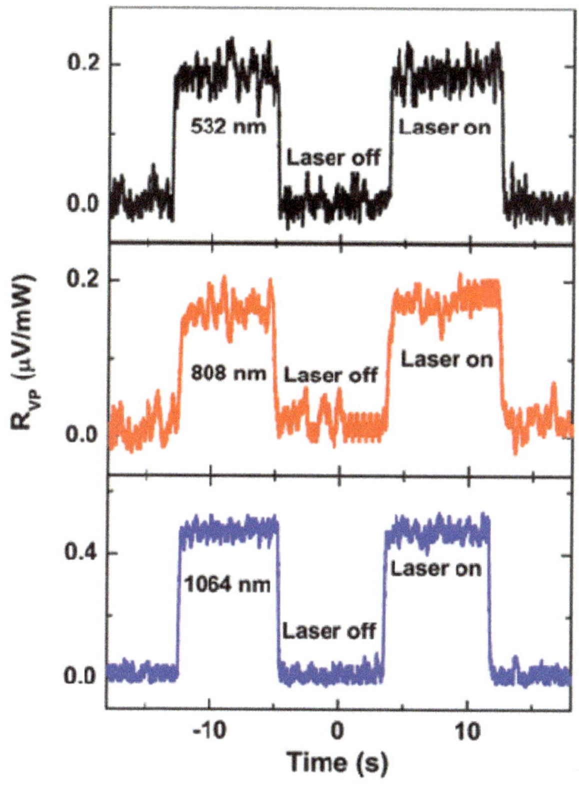

Fig. 11. Photovoltaic responsivity R_{VP} of a quartz single crystal under continuum laser illumination at room temperature without any bias. [14]

The present photovoltaic effect should relate to the trap centres existing in quartz single crystals. Based on metal–insulator contact theory,[29] the energy-band diagram for the interface between Ag and SiO_2 is shown in Fig. 12. The Fermi level (E_f) of Ag immediately after contact is higher than that of SiO_2. Thus, electrons will be transferred from the metal into SiO_2 to equalize the E_f. During the transmission, a portion of electrons are trapped by positively charged centres in the crystal and accumulated in traps. The contact potential difference (C.P.D.) in the equilibrium state is estimated as ~0.64 eV, which is determined by the difference between the metal and insulator work functions. Under illumination a number of trapped electrons are freed into the conduction band (C.B.) of SiO_2 and removed by the built-in electric field. Eventually, photovoltage occurred in the present system. In this case, the trap level below the C.B. was estimated to be ~1.165 eV (1064 nm photon energy). From Fig. 9, the transient photovoltaic responsivities under picosecond irradiation are about 0.375 μVmW^{-1} and 1.31 × 10^{-6} μVmW^{-1} at 0.081 mJ and 0.056 mJ for 1064 nm and 532 nm, respectively, are about six orders of magnitude smaller than that under continuum light illumination. This indicates that the above factor cannot be responsible for the photovoltage under picosecond laser irradiation. In short-pulse materials processing, where the transient power density is very high (10^7-10^8 Wcm^{-2} in our case), it is necessary to consider such physical events as internal electron-lattice coupling, thermal transport or lattice distortion, etc. The laser radiation is first absorbed by the electrons within the ultrashort pulse duration and subsequently transferred to the crystal lattice through electron–phonon interactions over a characteristic time. In addition, the crystal structures of α-quartz (trigonal form) and β-quartz (hexagonal) do not give rise to pyroelectricity due to the symmetry of quartz. In contrast, the β_2 tridymite quartz with internal strains is a candidate for pyroelectric material owing to its lower symmetry and intricate structure.[30] Here, the secondary and tertiary pyroelectric effects, which are produced by deformation and inhomogeneous temperature distributions within the crystal, were introduced to meet our present results.

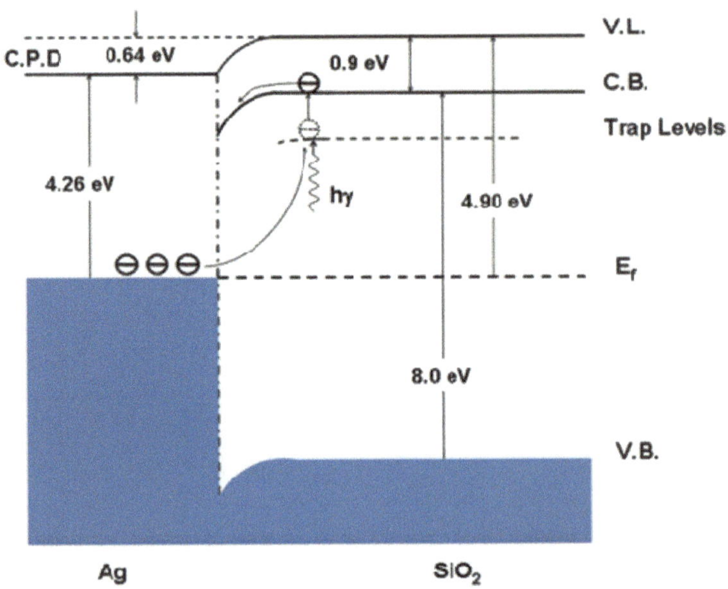

Fig. 12. The schematic band structure of the Ag-SiO_2 contact. C.P.D., V.L., C.B., V.B. and Ef denoted the contact potential difference, vacuum level, conduction band, valence band and Fermi level, respectively. [14]

From the heat flow equation $Q = \kappa \nabla T \approx \gamma E_{in}/\tau$, where Q is the heat current density, $\gamma \approx 1\%$ the absorption coefficient, $\kappa \approx 1.38$ $Wm^{-1}K^{-1}$ the thermal conductivity and τ the response time. With these values, the surface temperature can be estimated to be about 2×10^4 K when the sample is irradiated by a picosecond pulse laser. So the β_2 tridymite structure can be formed in crystals since the surface temperature is much higher than the transition temperature of 1143 K from β-quartz to β2 tridymite phases.[31] And also a non-uniform transient temperature distribution is induced on the crystal surface due to Gaussian intensity distribution of the laser pulse. Thus, non-uniform stress and strain distributions were created around the irradiated region as shown in Fig. 13.[32] Due to the pyroelectric effect of β_2 tridymite quartz, the positive and negative charge centres are separated slightly, leading to polarization (Fig. 13(c)), and eventually lateral photovoltage signals are observed.

Typical lateral ultrafast signals have also been observed in $LiNbO_3$ single crystals under Nd:YAG laser (pulse duration of 25 ps, repetition rate of 10 Hz) irradiation. The RT and FWHM are about 1.5 ns and 1~2 ns. Under the laser irradiation, photo-induced carriers are separated and assembled at the two electrodes by the spontaneous polarisation electric field.

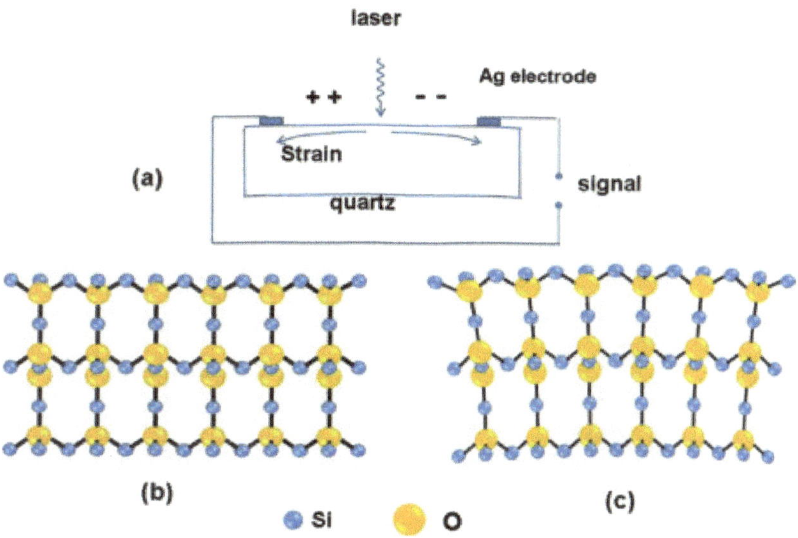

Fig. 13. (a) Diagram illustrating the sample structure. (b) and (c) are the β_2 tridymite structures before and after laser irradiation, respectively. [14]

3. Ultrafast photoresponse and its tunable effects in oxide films

3.1 Ultrafast photoresponse in oxide films

Recently, many researches have devoted to exploring photovoltaic properties and verifying new device concepts based on doped manganite thin films and heterostructures.[33-36] Technological interest has centered on solar cells, light-emitting diodes, and photoelectric detectors.[37,38] Previous researches showed that these materials represent good properties, such as ultrafast response, high-sensitivity, broad-spectrum and position-sensitivity, as optoelectronic device materials.

The $La_{1-x}Ca_xMnO_3$ (LCMO) thin films were deposited on several kinds of substrates (Si; $LaSrAlO_4$; MgO; NSTO) by facing-target sputtering technique.[39] The substrate temperature was kept at 680 °C and the oxygen partial pressure of 30 mTorr during deposition. The film thickness is uniform, controlled by sputtering time with the deposition rate (~0.03 nm/s). After the deposition, the vacuum chamber was immediately back-filled with 1 atm oxygen gas to improve the oxygen stoichiometry. Then, the samples were then cooled to room temperature with the substrate heater power cutting off.

In tilted LCMO films, picosecond ultrafast photoresponse were observed. The structure of the $La_{2/3}Ca_{1/3}MnO_3$/MgO sample characterized by XRD and transmission electron microscopy (TEM) shows that the $La_{2/3}Ca_{1/3}MnO_3$ film is a single phase and epitaxial growth. The sample of 5×5 mm^2 geometry has been used for the photoelectric properties measurement under the 355 nm Nd:YAG laser irradiation with 25 ps duration at a 2 Hz repetition. The waveforms were recorded by a sampling oscilloscope terminated into 50 Ω. Figure 14 displays the temporal waveform of the open-circuit photovoltage. The RT and FWHM are about 224 and 574 ps.

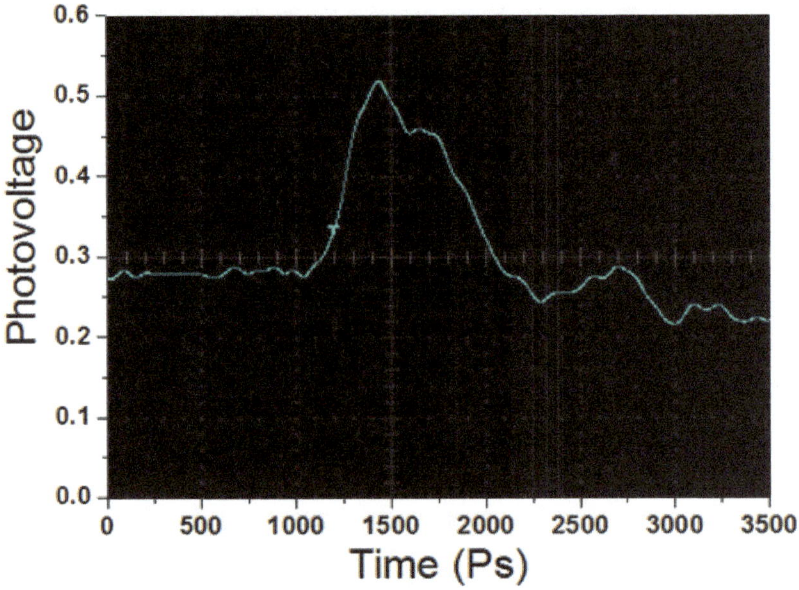

Fig. 14. The open-circuit photovoltaic waveform of $La_{2/3}Ca_{1/3}MnO_3$/MgO sample irradiated by a 355 nm laser pulse of 25 ps duration.

ZnO is a promising short-wavelength optoelectronic material because of its direct wide band gap (3.37 eV at 27 K), large exciton binding energy (60 meV) and high transparency (>80%) in the visible wavelength region. Over the past years a great deal of physical properties has been investigated in doping ZnO thin films. Lim et al. reported the fabrication of ZnO based LEDs using sputter deposited P-doped ZnO as the p-type layer.[40] Ye et al. used N–Al codoping for p-type doping and fabricated ZnO LEDs on Si with sputter deposition.[41] Myong et al. obtained highly conductive Al-doped ZnO thin films.[42] Ataev et al. reported a resistivity of 1.2×10^{-4} Ωcm for Ga-doped ZnO thin films

grown by chemical-vapor deposition.[43] The UV photovoltaic response of Ag-doped ZnO thin films deposited at different temperature has been studied.[44]
The films were prepared on fused quartz substrates by pulsed laser deposition (PLD). A KrF excimer laser (wavelength: 248 nm, pulse width: 30 ns, energy density: 1 J/cm^2) was used for ablation of a ZnO mosaic target (1/4 area of the target was uniformly covered with high-purity silver slices in the shape of a sector). In our experiment, the repetitive frequency of the laser was 4 Hz, the O_2 pressure was 5×10^{-4} Pa, and the temperature of the substrates varied from 350 to 550 °C. All the samples were cooled to room temperature under an O_2 pressure of 5×10^{-4} Pa in the chamber.

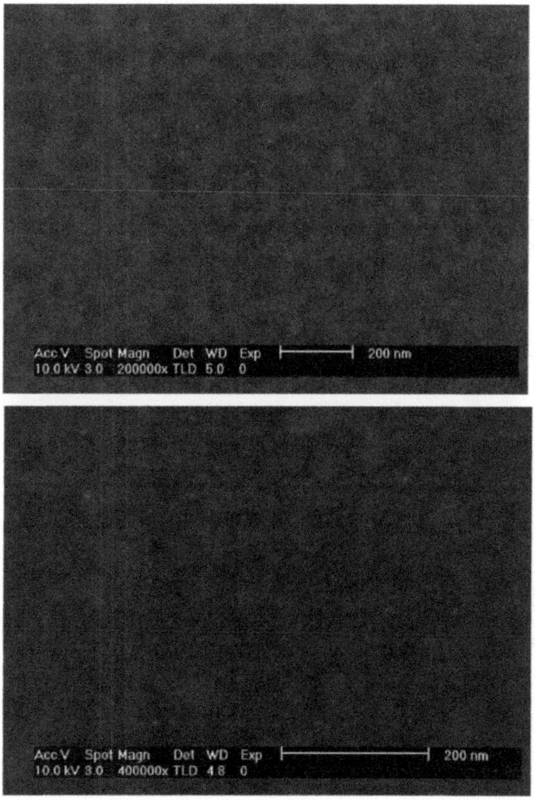

Fig. 15. SEM morphologies of Ag-doped ZnO films deposited at (a) 350 °C and (b) 550 °C. [44]

Figs. 15(a) and 15(b) show the SEM morphologies of Ag-doped ZnO films deposited at 350 and 550 °C, which is different from that at 450 °C reported in Ref. [45]. From the SEM image the Ag-doped ZnO films consist of grains separated by grain boundary (GB), and Ag would preferentially choose to sit in vicinity of grain boundaries due to its large ionic radius.
For the photovoltaic measurements a Nd:YAG laser (wavelength 266 nm) was used as the light source with a pulse energy of 1 mJ and a light spot of 6 mm in diameter. The photovoltaic signals were monitored with a 500 MHz oscilloscope terminated into 50 Ω.

Under the same condition, the 308 nm excimer laser was adopted to irradiate the samples. Fig. 16 presents typical open-circuit photovoltages transient of Ag-doped ZnO films deposited at different temperature (350 °C, 450 °C and 550 °C). Under pulsed 266 nm laser irradiation of 25 ps duration, the photovoltaic signals had FWHMs of 0.9, 0.8 and 1.0 ns, limited by the oscilloscope and peak photovoltages of 29, 72 and 28 mV for 350, 450 and 550 °C, respectively. Furthermore, Fig. 16(d) shows the temporal response of the Ag-doped ZnO film to a 20 ns 308 nm laser pulse. The peak photovoltage reaches ~29 mV. The FWHM and RT are ~20 and ~10 ns, respectively, which are limited by the excitation laser.

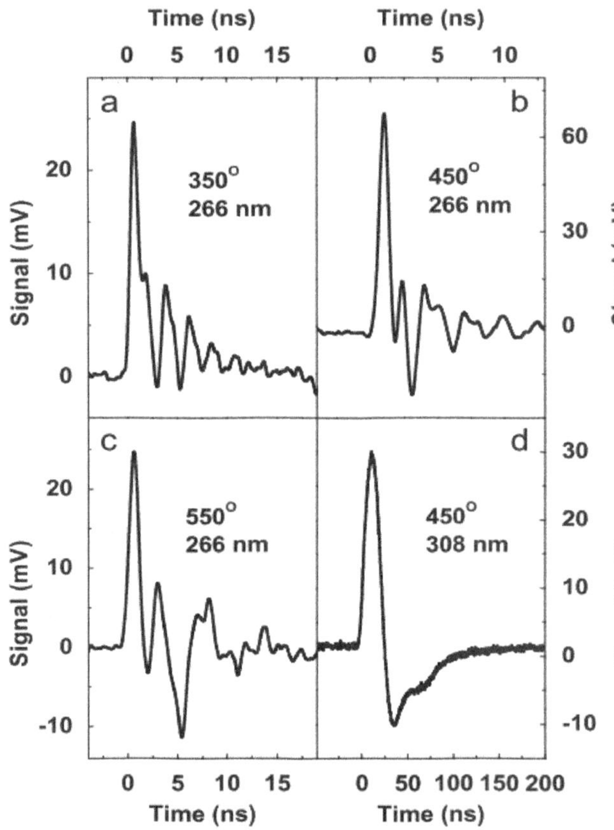

Fig. 16. Open-circuit photovoltages of Ag-doped ZnO films deposited at (a) 350 °C, (b) 450 °C and (c) 550 °C under the excitation of 266 nm laser. (d) Shows the photovoltage of the sample deposited at 450 °C under the excitation of 308 nm laser. [44]

We also studied the photovoltaic effect of ZnO films without Ag under the same condition and no voltaic signal appeared in our oscilloscope, indicating that Ag clusters in vicinity of grain boundaries play an important role in the present photovoltaic characteristics. Thus, a simplified model can be given on the origin of the photovoltaic signal. There is a chemical potential shift $\Delta\mu$ between the GB region and the grain, which might induce a depletion layer in the GB region and the build-in voltage V_b is given by $\Delta\mu$, $V_b = \Delta\mu$. Shu-Ting Kuo et al. calculated the electrostatic barrier of one-grain boundary of Ag-doped ZnO was approximately 2 V.[46] When the laser irridiates the sample, electron-hole pairs can be excited in the grains and GB regions since an energy gap of ZnO between occupied and empty electronic states is smaller than UV photon energies (4.0 eV for 308 nm and 4.7 eV for 266 nm). And then the nonequilibrium carriers are separated by the built-in electric field near the GB, eventually, leading to the appearance of an instant photovoltage.

Since the sample is polycrystalline, there should be an equal number of grains with photogenerated carriers shifting in one direction as in the opposite direction; hence there should be not net current flow and no photovoltage signal. This is not the case. In fact, it is uncertain why there is an overall preferred direction for the flow of the photo-generated current. This behavior may be due to the asymmetry of the lattice which induces an asymmetric moving of the excited carriers in a preferred direction. Further study on the nature of the photovoltaic properties of such a system is under way.

4. External field tunable effects

Perovskite-type manganites have attracted a great deal of interest because of the versatile electronic states that can be controlled by various kinds of external perturbations.[47-49] Photoexcitation offers an attractive method to vary the concentration of charge carriers without the added complication of a change in the chemical composition and the crystal structure.

The voltage tunable photodetecting properties have been studied in a $La_{0.6}Ca_{0.4}MnO_3$ film grown on 10° tilted LSAO (001) substrates under ultraviolet pulsed laser (248 nm; 20 ns) irradiation.[15] The photovoltaic effect of the $La_{0.6}Ca_{0.4}MnO_3$ thin film on 10° tilted LSAO substrate was measured under different bias at room temperature, and the open-circuit photovoltage transient V_R, across the input impedance of oscilloscope and recorded by oscilloscope, is displayed in Fig. 17(a). The RT and FWHM are about 12 and 23 ns, respectively. The V_R^b in Fig. 17(a) denotes the baseline recorded by the oscilloscope for laser-off state, which was caused by the external bias and the input impedance, and shifts from -0.69 to 0.69V symmetrically for V^b from -20 to 20 V. The photoinduced voltage V^p defined by $V^p = V_R^p - V_R^b$ is plotted in Fig. 17 (b) as a function of the applied bias V^b, and increases from 0.69 to 1.85 V with V^b from -20 to +20 V.

A 0.2 Ω resistance was connected in parallel with the sample (see the up inset of Fig. 18 (a)) for further studying the electric tunable photocurrent responses of the $La_{0.6}Ca_{0.4}MnO_3$ films. It is found that the baseline V_R^b recorded by the oscilloscope did not change due to the very small connected resistance (Fig. 18(a)), and *RT* is reduced to about 4 ns. The tail traces of the waveforms show some periodic oscillations persisting for tens ns which may be due to the signal reflection arising from an impedance mismatch in the circuit. The peak photocurrent I_{AB} was calculated by $I_{AB} \approx V_R^p/0.2\ \Omega$ where V_R^p is the peak voltage across the connected resistance 0.2 Ω, and increased monotonically to 0.74 A at $V^b = 20$ V (Fig. 18(b)), which is 1.82 times higher than 0.41 A at $V^b = -20$ V.

From the cross-sectional TEM image of $La_{0.6}Ca_{0.4}MnO_3$/LSAO (Fig. 19(a)) the film thickness was fairly uniform and about 120 nm. No appreciable interdiffusion occurred and no evidence of secondary phases was observed from the high resolution TEM (HRTEM) picture at the $La_{0.6}Ca_{0.4}MnO_3$/LSAO interface (Fig. 19(b)), which is agreement with the XRD investigations. Due to the 10° tilted substrate, the $La_{0.6}Ca_{0.4}MnO_3$/LSAO interface is not a plane and a terrace structure occurs, which is labeled in Fig. 19(b). This denotes that we have obtained exactly epitaxial the 10° tilted thin films, and the terrace structure is an important factor for the present photodetecting properties.

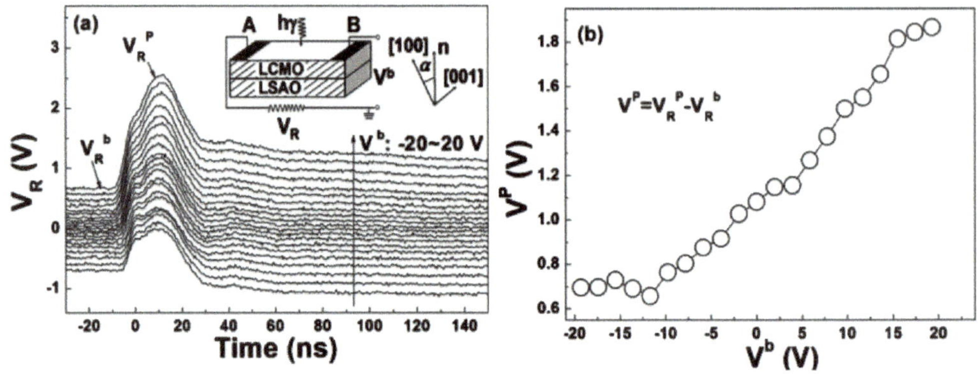

Fig. 17. (a) The photovoltaic pulses for $La_{0.6}Ca_{0.4}MnO_3$/LSAO under the illumination of a 248 nm laser at different biases recorded by an oscilloscope with an input impedance of 50. The inset displays the schematic circuit of the measurement. (b) The photoinduced voltage V^P defined by $V^P = V_R^P - V_R^b$ as a function of the applied bias V^b. The inset shows that V_R^P as a function of the applied bias V^b. [15]

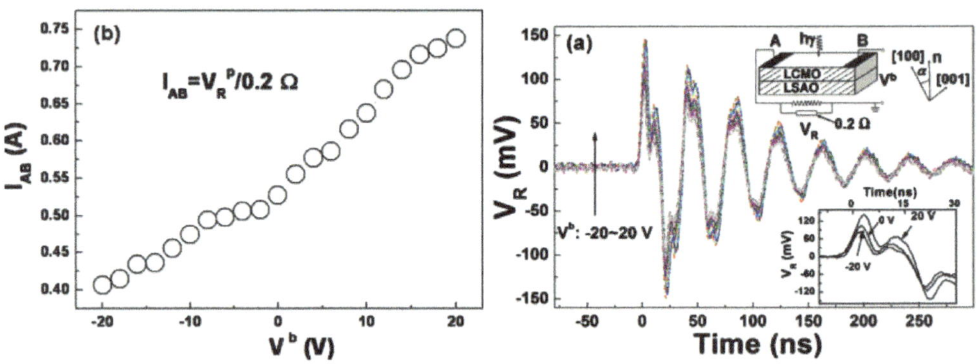

Fig. 18. (Color online) (a) The photovoltaic pulses and (b) peak photocurrent for $La_{0.6}Ca_{0.4}MnO_3$/LSAO under the illumination of a 248 nm laser at different biases recorded by an oscilloscope with a parallel resistance of 0.2 Ω. [15]

Because of the terrace structure in the interface of the $La_{0.6}Ca_{0.4}MnO_3$/LSAO, which is caused by the tilted angle of the substrate and clearly shown in the TEM image (Fig. 19(b)), we depicted schematic drawing under the irradiation of the pulsed laser as shown in Fig. 20. If the laser is divided into two components of being parallel and perpendicular to the tilting direction, the transient carrier density along each axis is proportional to the illuminated power perpendicular to the tilting direction. And the ratio between the irradiation intensities on (010) and (001) faces can be denoted as $I_{010}/I_{001}=\cos\alpha/\sin\alpha\approx5.67$ for α = 10°, so as the carries density. Since the photon energy of 248 nm wavelength (4.86 eV) is above the band gap of $La_{0.6}Ca_{0.4}MnO_3$ (~1.2 eV), electron-hole pairs are generated in the $La_{0.6}Ca_{0.4}MnO_3$ film, therefore, a gradient of carrier density was generated along the lateral orientation. In our case, the laser we used is a 248 nm KrF excite laser beam with an energy density of 0.31mJ/mm^2 in duration of 20 ns, so the amount of laser induced carriers should be comparable with or even much larger than that of the majority carriers in the $La_{0.6}Ca_{0.4}MnO_3$; on the other hand, there exists no built-in field which exists in the p-n junction to separate holes and electrons. Therefore, both the electrons and holes play an important in the photovoltaic, and a mechanism based on the difference between the mobilities of electrons and holes, such as Dember effect, was proposed to explain the photovoltaic effect in $La_{0.6}Ca_{0.4}MnO_3$/LSAO.

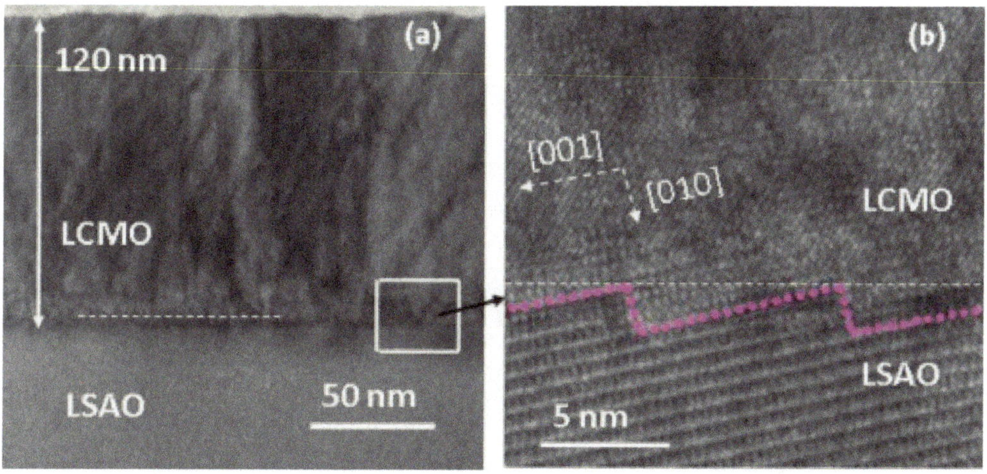

Fig. 19. (Color online) Cross-sectional TEM images of (a) LCMO/LSAO sample and (b) LSMO/LSAO interface. The dashed lines denote the interface position. [15]

Due to the applied bias and Dember effect (because of the difference of coefficient of diffusion between holes and electrons), an electric filed $E=E_b+E_d$ was generated where E_b and E_d denote the electric filed generated by applied bias and Dember effect, respectively. The drift current density j is presented as $j = j_p + j_n = n_p q v_p - n_n q v_n$, where j_p and j_n are the current density of the holes and electrons, n_p, v_p and n_n, v_n are the density, drift velocity of the holes and electrons, respectively, q is the unit of charge. Take electron as an example, we can obtain that $v_n(t) = v_{n0} - q/m_n^*Et$, where v_{n0} denotes the average velocity of the electrons after the scattering and should be 0 due to the random of the scattering, m_n^* is the effective mass of electron, E is the electric filed, and t is the average time between two scattering which can be denotes as relaxation time τ. Thus $v_n = -q/m_n^*\tau E$. Therefore, the drift carry density can be presented as $j = (\mu_p - \mu_n)E$, where $\mu_p = q/m_p^*\tau$ and $\mu_n = q/m_n^*\tau$ present the mobility of the hole and electron. When the applied bias is positive or negative, $E_b > 0$ or $E_b < 0$, therefore, the photovoltage is enhanced or reduced.

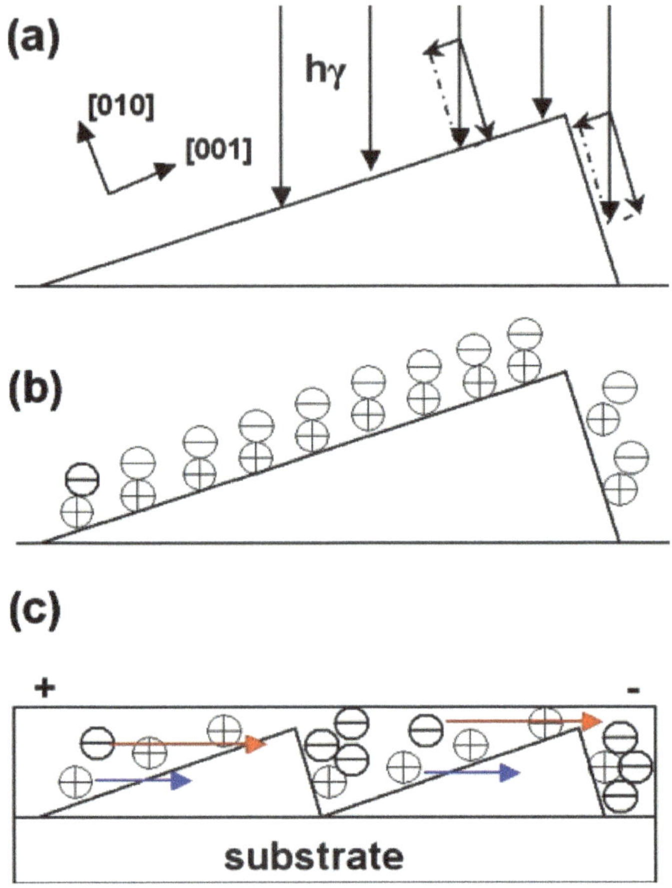

Fig. 20. (Color online) (a) The schematic drawing under the irradiation. (b) The difference of the carry destiny caused by the miscut angle. (c) The transient movement of the photogenerated carriers of the $La_{0.6}Ca_{0.4}MnO_3$. [15]

The enhanced magnetoresistance (MR) effect has been discovered under laser illumination in the $La_{2/3}Ca_{1/3}MnO_3$ film on the n-Si substrate at room temperature.[16]

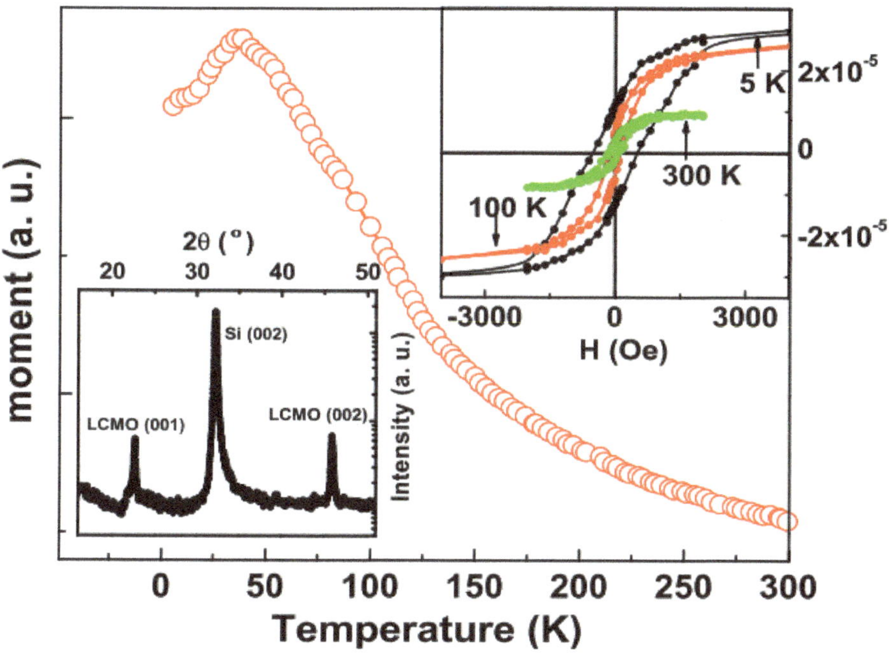

Fig. 21. The magnetic moment as a function of temperature of the junction. The left and right insets show the XRD pattern and magnetic hysteresis loop, respectively. [16]

XRD spectrum confirmed that the $La_{2/3}Ca_{1/3}MnO_3$ layers were oriented along the Si [001] direction as shown in the left inset of Fig. 21. The temperature dependence of magnetic moment was performed with SQUID. The magnetic hysteresis loop at room temperature shows a magnetic coercivity of 80 Oe in the right inset of Fig. 21, which reveals the existence of a few ferromagnetic phases near room temperature in the $La_{2/3}Ca_{1/3}MnO_3$ layer due to the phase separation scenario.[50] Accordingly, it is reasonable to infer the small MR in the junction at room temperature. From the TEM image of the heterojunction in Fig. 22, it is clear that the $La_{2/3}Ca_{1/3}MnO_3$ nano columns (LCMO2) are well crystallized with a single-phase perovskite structure. There is also an about 20 nm thick $La_{2/3}Ca_{1/3}MnO_3$ layer (LCMO1) on the surface of SiO_2, which could be due to interdiffusion. The lattice images can be clearly seen in the LCMO2 layer in Fig. 22(b). The atomically sharp interface between the SiO_2 and Si substrate was also observed in Fig. 22(c).

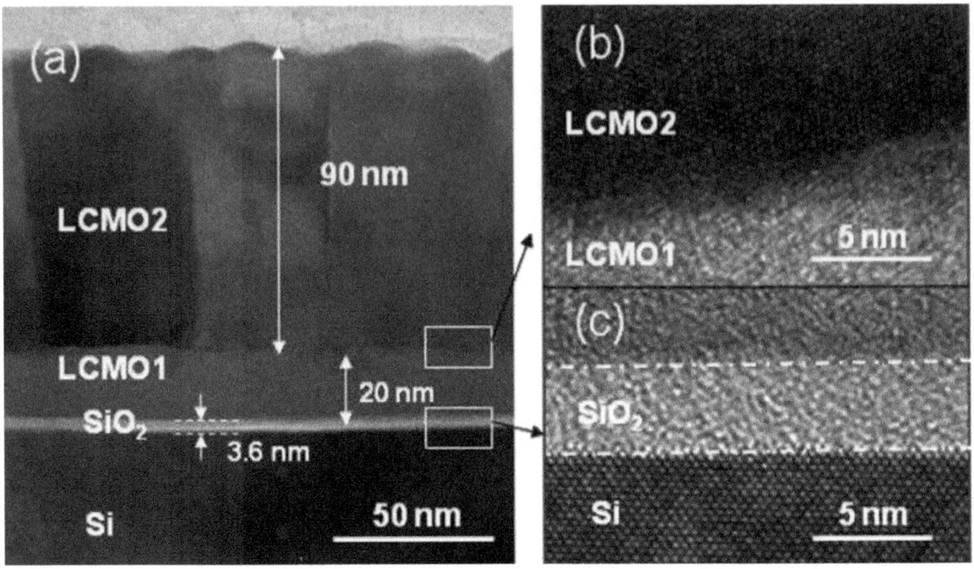

Fig. 22. A cross-section TEM image of the heterostructure. The SiO_2 layer can be clearly seen between the $La_{2/3}Ca_{1/3}MnO_3$ and Si substrate. [16]

The I-V characteristics of the junction were shown in Fig. 23 in different magnetic fields (H = 0, 1.81, 5.51 kOe) with and without laser irradiation. I+, V+, I- and V- are the four probes for the I-V measurement, and present current anode, voltage anode, current cathode and voltage cathode, respectively. The slope of the I-V curves becomes steeper and the junction current increases when irradiated by the laser in the reverse-bias case (bias current flows from Si substrate to $La_{2/3}Ca_{1/3}MnO_3$ film), which presents a decrease in junction resistance. This originates from the increased amount of the carriers due to light illumination. A great modification of resistance due to magnetic field was observed. At a reverse-bias current of -200 μA the output voltages change from -2.52 V to -2.46 V and -2.55 V in the selected fields of 1.81 kOe and 5.51 kOe, respectively. This effect is more striking under laser irradiation. For instance, the output voltages change from -1.88 to -1.84 and -2.10 V in 1.81 and 5.51 kOe under a bias of -200 μA when the junction is irradiated by laser.

Figure 24 reviews the dependence of MR on bias current in different magnetic fields, which were applied perpendicularly to the interface of the junction as shown in the inset of Fig. 24. The MR values become nearly constant under larger reverse bias current. When the junction was irradiated with a laser under reverse bias current, the MR values dramatically increase and are positive in larger magnetic field like 3.38 and 5.51 kOe. The ΔMR, defined as MR (laser on)—MR (laser off) for a fixed H, is larger under reverse current than under forward current as shown in the inset of Fig. 24.

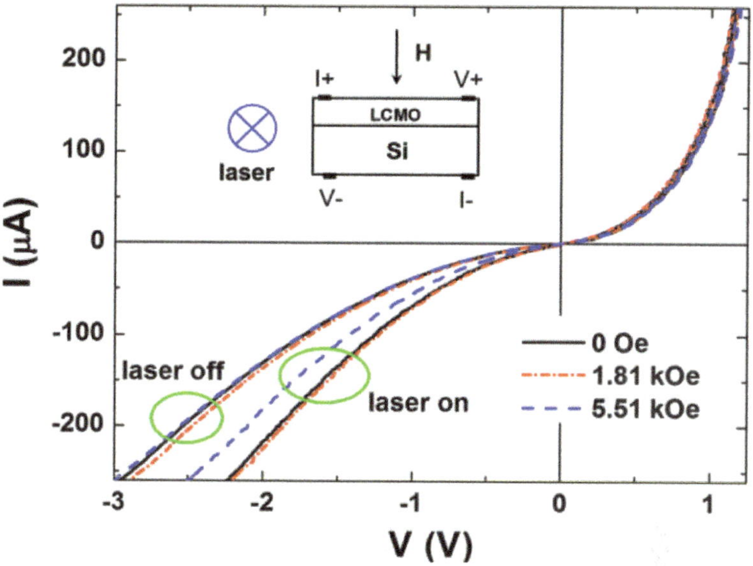

Fig. 23. I-V characteristics of the junction measured by tuning bias current in 0, 1.81 and 5.51 kOe at room temperature under 808 nm laser illumination on the interface. The inset shows the schematic measurement of I-V. The H was perpendicular to the junction interface and laser irradiation direction which was parallel. [16]

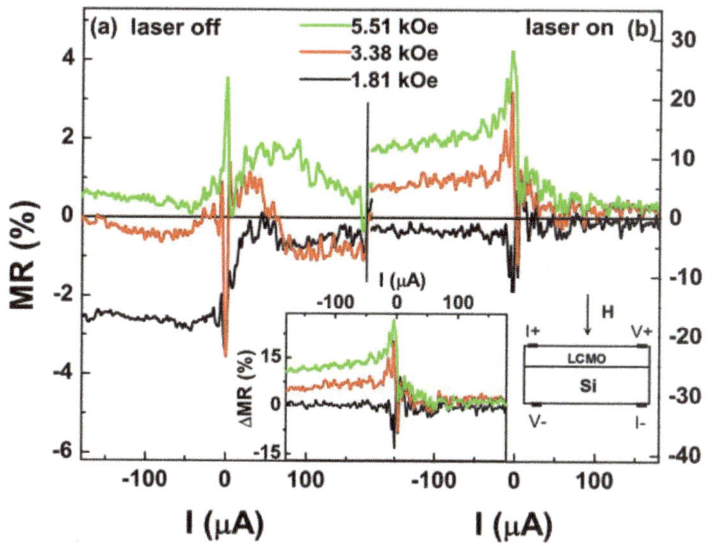

Fig. 24. The dependence of MR values of the junction on bias current in 1.81, 3.38 and 5.51 kOe at room temperature. The inset shows the dependence of ΔMR on bias current. [16]

Figure 25 shows the dependence of MR of the junction in magnetic field. The MR values increase and have a crossover from negative to positive values with increasing magnetic field H, which is different from the negative MR of the $La_{2/3}Ca_{1/3}MnO_3$ compound family. The MR values dramatically increase from 0.54% to 18% and 0.15% to 13% under -39 and -99 μA reverse-bias currents in 6.4 kOe, respectively, when the junction is irradiated by laser. This result demonstrates the possibility to gain great MR values in a broad temperature range by irradiating perovskite-type oxide heterojunction.

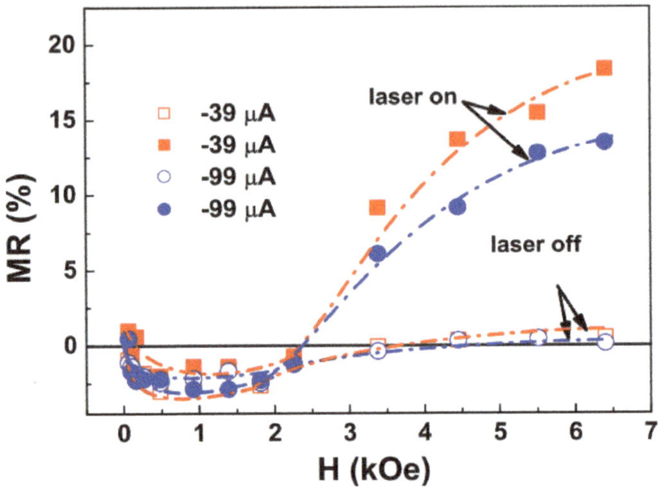

Fig. 25. The dependence of MR values on H under -39 and -99 μA with and without laser illumination in the junction at room temperature. [16]

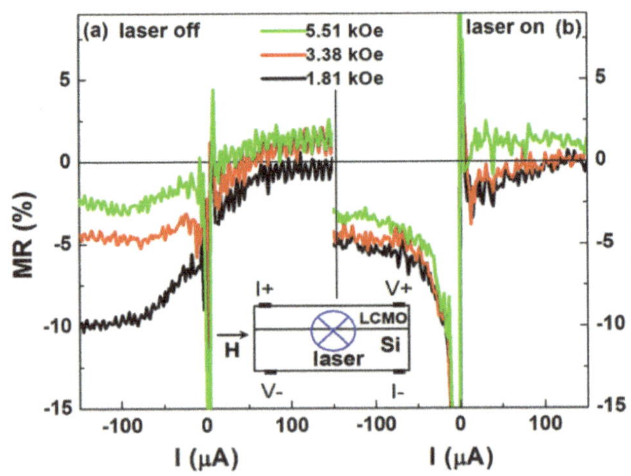

Fig. 26. The dependence of MR values of the junction on bias current in 1.81, 3.38 and 5.51 kOe fields at room temperature. The H was parallel to the junction interface and perpendicular to the direction of light which was parallel to the junction interface. [16]

In addition, when the H was applied parallel to the junction interface, the photo-induced enhanced MR effect has also been discovered at room temperature. As shown in Fig. 26, the MR ratios increase from -6.7%, -3.5% and -1.5% to -9.1%, -8.9% and -8.0% under -25 μA reverse- bias currents in 1.81 kOe, 3.38 kOe and 5.51 kOe, respectively, when the junction is irradiated by a 808 nm cw laser.

The high resolution TEM confirmed a native oxide (SiO_2) layer on the Si surface. Thus the junction can be treated as the $La_{2/3}Ca_{1/3}MnO_3/SiO_2/Si$ heterojunction. The photon energy (about 1.54 eV) of 808 nm laser is larger than the band gap of Si (about 1.12 eV) and $La_{2/3}Ca_{1/3}MnO_3$ (about 1 eV). When the laser directly irradiates onto the $La_{2/3}Ca_{1/3}MnO_3/$ SiO_2/Si interface, electrons in the valence bands (VB) of both $La_{2/3}Ca_{1/3}MnO_3$ and Si absorb photons and make a transition into the conduction bands (CB), generating electron-hole pairs. As indicated above, the ultrathin native SiO_2 oxide layer exhibits the large potential barrier, resulting in a large junction resistance.

When bias currents are applied, the carriers get energy to hop into the trap energy levels formed by defects in the SiO_2 insulator and are trapped in these levels leading to space charge [51]. It is easier for the charge carriers to tunnel through the SiO_2 space charge region than surmount the interfacial barrier. These tunnelling carriers form tunnelling current. The orbital motion and mobility reduction of the tunnelling charge carriers caused by the Lorentz force in magnetic field will depress tunnelling current and increase MR [52]. But the amplitude is relatively small. On the other hand, there was a ferromagnetic phase in $La_{2/3}Ca_{1/3}MnO_3$ at room temperature, which can lead to anisotropic MR (AMR) [53]. The intrinsic origin of AMR is the spin-orbit coupling and a larger probability of s-d scattering of electrons in the direction of magnetic field [54]. It is positive and grows with the applied magnetic field when the magnetic field is parallel to the current and negative when perpendicular [55, 56].

The present system under investigation consists of a photo-activated junction of manganite/oxide/n-Si and therefore the MR arising from this junction is tunnelling AMR (TAMR), because there is only one magnetic electrode. When the junction interface is illuminated by laser by side, great numbers of carriers are generated in both $La_{2/3}Ca_{1/3}MnO_3$ and Si side, which enhance tunnelling current and it can be seen from Fig. 23 that the I-V curves become steeper under laser illumination. When a negative bias current is applied, photogenerated electrons (with spin polarization) from $La_{2/3}Ca_{1/3}MnO_3$ contribute partly to tunnelling current. As the tunneling current is nearly parallel to applied magnetic field (the case in Fig. 23), TAMR is positive. As confirmed in Fig. 25, the TAMR will be enhanced when the magnetic field is perpendicular to tunnelling current. If a positive bias (from Si to $La_{2/3}Ca_{1/3}MnO_3$) is applied on the sample, there are no polarized carriers and no tunnelling current through the junction in the system. Thus the change of MR was small when positive currents were applied as shown in Fig. 24 and Fig. 26.

5. Conclusion

In conclusion, the characteristics of ultrafast photoresponses in oxide single crystals and films have been investigated. The RT and FWHM of the photovoltage signals were measured. Picosecond and nanosecond photoresponses were observed in tilted perovskite oxides (STO, NSTO and $LaSrAlO_4$), polar crystals (SiO_2 and $LiNbO_3$), tilted manganite films ($La_{1-x}Ca_xMnO_3/LaSrAlO_4$ and $La_{1-x}Ca_xMnO_3/MgO$) and Ag-doped ZnO films. Theoretical investigations about interesting experimental phenomena are presented. External field tunable effects for the photovoltaic effect were also systematically studied. Experimental results reveal the potential for optoelectronic detection applications.

6. Acknowledgement

This work was supported by the Program for NCET, NSFC, RFDP, Foresight Fund Program from China University of Petroleum (Beijing).

7. References

[1] M. Razeghi and A. Rogalski, J. Appl. Phys. 79, 7433 (1996).
[2] M. Topic, Appl. Phys. Lett. 78, 2387 (2001).
[3] J. W. Tomm, B. Ullrich, X. G. Qiu, Y. Segawa, A. Ohtomo, M. Kawasaki, and H. Koinuma, J. Appl. Phys. 87, 1844 (2000).
[4] F. Spaziani, M. C. Rossi, S. Salvatori, G. Conte, and P. Ascarelli, Appl. Phys. Lett. 82, 3785 (2003).
[5] W. Yang, R. D. Vispute, S. Choopun, R. P. Sharma, T. Venkatesan, and H. Shen, Appl. Phys. Lett. 78, 2787 (2001).
[6] M. Mikulics, M. Marso, P. Javorka, P. Kordoš, H. Lüth, M. Koèan, A. Rizzi, S. Wu, and R. Sobolewski, Appl. Phys. Lett. 86, 211110 (2005).
[7] W. R. Donaldson, A. M. Kadin, P. H. Ballentine and R. Sobolewski, Appl. Phys. Lett. 54 2470 (1989).
[8] M. Johnson, Appl. Phys. Lett. 59, 1371 (1991).
[9] F. A. Hegmann, R. A. Hughes and J. S. Preston, Appl. Phys. Lett. 64, 3172 (1994). Hegmann, D. Jacobs-Pekins, C. C. Wang, S. H. Moffat, R. A. Hughes, J. S. Preston, M. Currie, P. M. Fauchet, T. Y. Hsiang and R. Sobolewski, Appl. Phys. Lett. 67, 285 (1995).
[10] M. Lindgren, V. Trifonov, M. Zorin, M. Danerud, D. Winkler, B. S. Karasik, G. N. Gol'tsman and E. M. Gershenzon, Appl. Phys. Lett. 64, 3036 (1994).
[11] K. Zhao, K. J. Jin, Y. H. Huang, S. Q. Zhao, H. B. Lu, M. He, Z. H. Chen, Y. L. Zhou and G. Z. Yang, Appl. Phys. Lett. 89, 173507 (2006).
[12] K. Zhao, H. B. Lu and M. He, Eur. Phys. J. Appl. Phys. 41, 139–142 (2008).
[13] Z. J. Yue, K. Zhao, S. Q. Zhao, Z. Q. Lu, X. M. Li, H. Ni and A. J. Wang , Journal of Physics D: Appl. Phys. 43, 015104 (2010).
[14] H. Liu, K. Zhao, S. Q. Zhao, N. Zhou, H. Zhao, Z. Q. Lu and A. J. Wang, Journal of Physics D: Appl. Phys. 42, 075104 (2009).
[15] X. M. Li, K. Zhao, H. Ni, S. Q. Zhao, W. F. Xiang, Z. Q. Lu, Z. J. Yue, F. Wang, Y. C. Kong and H. K. Wong, Appl. Phys. Lett. 97, 044104 (2010).
[16] Z. J. Yue, K. Zhao, H. Ni, S. Q. Zhao, Y. C. Kong, H. K. Wong and A. J. Wang, Journal of Physics D: Appl. Phys. 44, 095103 (2011).
[17] S. H. Nam and H. G. Kim, J. Appl. Phys. 72, 2895 (1992).
[18] G. M. Rao and S. B. Krupanidhi, J. Appl. Phys. 75, 2604 (1994).
[19] M. H. Yeh, K. S. Liu and I. Nan, Jpn. J. Appl. Phys., Part 1 34, 2247 (1995).
[20] F. J. Walker, R. A. Mckee, H. W. Yen and D. E. Zelmon, Appl. Phys. Lett. 65, 1495 (1994).
[21] T. Hasegawa, S. Mouri, Y. Yamada and K. Tanaka, J. Phys. Soc. Jpn. 72, 41 (2003).
[22] H. Lengfellner, G. Kremb, A. Schnellbögl, J. Betz, K. F. Renk, and W. Prettl, Appl. Phys. Lett. 60, 501 (1992).

[23] N. Zhou, K. Zhao, H. Liu, Z. Q. Lu, H. Zhao, L. Tian, W. W. Liu and S. Q. Zhao, J. Appl. Phys. 105, 083110 (2009).
[24] Z. Q. Lu, K. Zhao, H. Zhao, S. Q. Zhao and Q. L. Zhou, Chinese Physics B, 18(10), 4521-4523 (2009).
[25] H. Zhao, K. Zhao, Z. Q. Lu and S. Q. Zhao http://www.paper.edu.cn/en/, (2009).
[26] T. Tomita, H. Utsunomiya, Y. Kamakura and K. Taniguchi, Appl. Phys. Lett. 71, 3664 (1997).
[27] D. Heh, E. M. Vogel and J. B. Bernstein, Appl. Phys. Lett. 82, 3242 (2003).
[28] C. R. Newman, R. J. Chesterfield, J. A. Merlo and C. D. Frisbie, Appl. Phys. Lett. 85, 422 (2004).
[29] W. J. Turner, Phys. Rev. 101, 1653 (1956).
[30] M. Fukuhara and A. Sampei, J. Mater. Sci.—Mater. Electron. 12, 131 (2001).
[31] R. B. Sosman, Trans. Brit. Ceram. Soc. 54, 665 (1955).
[32] T. Gorelik, M. Will, S. Nolte, A. Tuennermann and U. Glatzel Appl. Phys. A 76, 309 (2003).
[33] S. Jin, T. H. Tiefel, M. McCormack, R. A. Fastnacht, R. Ramesh, and L. H. Chen, Science 264, 413 (1994).
[34] H. Tanaka, J. Zhang, and T. Kawai, Phys. Rev. Lett. 88, 27204 (2002).
[35] R. Cauro, A. Gilabert, J. P. Contour, R. Lyonnet, M.-G. Medici, J. C. Grenet, C. Leighton, and I. K. Shuller, Phys. Rev. B 63, 174423 (2001).
[36] H. Katsu, H. Tanaka, and T. Kawai, Appl. Phys. Lett. 76, 3245 (2000).
[37] M. Rajeswari, C. H. Chen, A. Goyal, C. Kwon, M. C. Robson, R. Ramesh, T. Venkatesan, and S. Lakeou, Appl. Phys. Lett. 68, 3555 (1996).
[38] L. Méchin, J.-M. Routoure, B. Guillet, F. Yang, S. Flament, and D. Robbes, Appl. Phys. Lett. 87, 204103 (2005).
[39] X. T. Zeng and H. K. Wong, Appl. Phys. Lett. 66, 3371 (1995).
[40] J. H. Lim, C. K. Kang, K. K. Kim, I. K. Park, D. K. Hwang and S. J. Park, Adv. Mater. 18, 2720 (2006).
[41] Z. Ye, J. G. Lu, Y.Z. Zhang, Y. J. Zeng, L. L. Chen, F. Zhuge, G. D. Yuan, H. P. He, L. P. Zhu, J. Y. Huang and B. H. Zhao, Appl. Phys. Lett. 91, 113503 (2007).
[42] S. Y. Myong, S. J. Baik, C. H. Lee, W. Y. Cho and K. S. Lim, Jpn. J. Appl. Phys. Part 2 36, L1078 (1997).
[43] B. M. Ataev, A. M. Bagamadova, A.M. Djabrailov, V. V. Mamedo and R. A. Rabadanov, Thin Solid Films 260,19 (1995).
[44] W. W. Liu, S. Q. Zhao, K. Zhao, W. Sun, Y. L. Zhou, K. J. Jin, H. B. Lu, M. He and G. Z. Yang, Physica B 404, 1550-1552 (2009).
[45] S. Q. Zhao, Y. L. Zhou, K. Zhao, Z. Liu, P. Han, S. F. Wang, W. F. Xiang, Z. H. Chen, H. B. Lu, B. L. Cheng and G. Z. Yang, Physica B 373, 154 (2006).
[46] S. T. Kuo, W. H. Tuan, J. Shieh and S. F. Wang, J. Eur. Ceram. Soc 27, 4526 (2007).
[47] S. Jin, T. H. Tiefel, M. McCormack, R. A. Fastnacht, R. Ramesh, and L. H. Chen, Science 264, 413 (1994).
[48] A. Moreo, S. Yunoki, and E. Dagotto, Science 283, 2034 (1999).
[49] M. B. Salamon and M. Jaime, Rev. Mod. Phys. 73, 583 (2001).
[50] Q. L. Zhou, K. Zhao, K. J. Jin, D. Y. Guan, H. B. Lu, Z. H. Chen, G. Z. Yang, A. Li and H. K. Wong, Appl. Phys. Lett. 87, 172510 (2005).

[51] P. L. Lang, Y. G. Zhao, C. M. Xiong, P. Wang, J. Li and D. N. Zheng, J. Appl. Phys. 100, 053909 (2006).
[52] J. R. Sun, C. M. Xiong and B. G. Shen, Appl. Phys. Lett. 85, 4977 (2004).
[53] R. W. Li, H. B. Wang, X. W. Wang, X. Z. Yu, Y. Matsui, Z. H. Cheng, B. G. Shen, E. W. Plummer and J. D. Zhang, Proc. Natl Acad. Sci. 106, 14224 (2008).
[54] I. Genish, Y. Kats, L. Klein, J. W. Reiner and M. R. Beasley, J. Appl. Phys. 95 6681 (2004).
[55] C. Prados, D. V. Dimitrov, C. Y. Ni, A. Hernando and G. C. Hadjipanayis, Phys. Rev. B 56 14076 (1997).
[56] F. Lesmes, A. Salcedo, J. J. Freijo, D. Garcia, A. Hernando and C. Prados, Appl. Phys. Lett. 69 2596 (1996).

Laser Pulses Applications in Photovoltaic Effect

Kui-juan Jin, Chen Ge, Hui-bin Lu and Guo-zhen Yang
Beijing National Laboratory for Condensed Matter Physics
Institute of Physics, Chinese Academy of Sciences
China

1. Introduction

Laser pulses have been utilized in detecting photovoltaic properties and exploring functional characteristics in the field of solid-state physics (Jain & Landis, 1998; Sun et al., 2004; Lu et al., 2005; Yan et al., 2007; Jin et al., 2009; Lv et al., 2009). Experimental and theoretical investigations into photovoltaic effect under laser pulse are of particular interest for both physics and engineering.

Recent advances in seeking fast photovoltaic response material (Lu et al., 2005; Zhao et al., 2005; Zhao et al., 2006a; Zhao et al., 2006b; Zhao et al., 2006c; Huang et al., 2006), finding one-order-of-magnitude enhancement of lateral photovoltaic (LPV) effect in the perovskite oxide heterostructures compared with the substrates (Jin et al., 2007; Jin et al., 2009), and investigating the dependence of the photovoltaic effect on the thin films and substrates thickness (Qiu et al., 2007; Wen et al., 2009), are attributed to laser pulses applications in the photovoltaic effect. Several years ago, transient photoelectric effects were observed in $La_{0.7}Sr_{0.3}MnO_3$ (LSMO3)/Si heterostructure fabricated by pulsed laser deposition (Lu et al., 2005), which offers opportunities for designing new fast photodetectors. Moreover, the unusual LPV, ascribed to the Dember effect (Pankove, 1971) induced by large numbers of photo-generated carriers under laser pulses, was observed in the heterostructures of both $La_{0.9}Sr_{0.1}MnO_3/SrNb_{0.01}Ti_{0.99}O_3$ (LSMO1/SNTO) and LSMO3/Si (Jin et al., 2007; Jin et al., 2009). A one-order-of-magnitude enhancement of the LPV was found (Jin et al., 2009). Furthermore, the dependence of photovoltage on the thin films and substrates thickness was presented. The photovoltage becomes larger with the increase of the LSMO1 film thickness, while the film thickness is less than the depletion layer of the heterostructures. This is ascribed to the increase of the carrier amount of the LSMO1 layer and the enhancement of the built-in electric field in the space-charge region of the LSMO1/SNTO heterostructure (Qiu et al., 2007). Faster photoelectric response was observed in $LaAlO_{3-\delta}$/Si (LAO/Si) heterostructures, and the photoelectric sensitivity was greatly improved by decreasing the thickness of the Si substrates (Wen et al., 2009).

In order to reveal the underlying physical origin and the dynamic process of the photoelectric effect in the oxide heterostructure, time-dependent drift-diffusion model was employed (Liao et al., 2009a, 2009b, 2009c; Liao et al., 2010; Ge et al., 2010). The theoretical calculations showed that the modulation of Sr doping in $La_xSr_{1-x}MnO_3$ is an effective

method to accommodate the sensitivity and the speed of photovoltaic response (Liao et al., 2009a), that a smaller parallel resistance should result in faster photoelectric response (Liao et al., 2009b), and that the photodoping effect mainly occurs in the space-charge region of heterostructure (Liao et al., 2009c). By self-consistent calculations, it was demonstrated that with the increase in irradiated laser pulse energy, the Dember effect plays a more and more important role in LPV. A unified description for conventional LPV and Dember-effect-induced LPV can be obtained within the frame of drift-diffusion model (Liao et al., 2010). Two mechanisms for the great enhancement of LPV in the heterostructures were also presented by self-consistent calculations: the LPV of *p*-type material is larger than that of *n*-type material owing to the larger drift electric field induced in the *p*-type material than that in the *n*-type material; the built-in electric field at the interface between the thin film and substrate can enhance the LPV (Ge et al., 2010). Position-sensitive detectors (PSD) based on LPV are widely applied in various fields (Henry & Livingstone, 2001), and photo-Dember effect can produce THz in semiconductors under laser pulses (Dekorsy et al., 1993; Dekorsy et al., 1996 ; Gu et al., 2002 ; Liu et al., 2006; Tonouchi, 2007; Krotkus, 2010). Thus, such an understanding of the mechanisms for the enhancement of LPV should be helpful in further designing of the structures of PSD and new THz sources.
This chapter is dedicated to give a brief description of the intriguing experiment phenomena and basic theoretical understanding of photovoltaic effect under laser pulse. It will be organized as follows: the first section is an introduction; the second section focuses on the photovoltaic experiments; the third section addresses the theoretical investigations based on the drift-diffusion model, starting with a detailed description of the self-consistent numerical model; conclusions and perspectives are exhibited in the last section.

2. Photovoltaic experiments

Oxide heterostructures have been emerged as a leading topic in condensed-matter physics, since various fascinating phenomena, including high mobility (Ohtomo A. & Hwang H. Y., 2004) or even superconductivity (Reyren et al., 2007) between insulating oxides, were observed in these systems. In well-fabricated oxide heterostructures, a series of photovoltaic experimental results under laser pulses(Lu et al., 2005; Jin et al., 2007; Qiu et al., 2007; Jin et al., 2009; Wen et al., 2009) showed that desirable unusual effects can be observed, and outstanding functional properties can be obtained in these systems. It should provide avenues into understanding the fundamental properties of these systems and designing of new multifunctional devices.

2.1 Transient photoelectric effects

Many efforts have been devoted to explore functional properties under laser irradiation and design new device based on manganese oxide (Zhang et al., 2002; Sun et al., 2004; Lu et al., 2005; Yu & Wang, 2010). In the past few years, we have investigated the photoelectric effects under laser pulses in a variety of oxide heterostructures (Lu et al., 2005; Zhao et al., 2005; Zhao et al., 2006a; Zhao et al., 2006b; Zhao et al., 2006c; Huang et al., 2006; Wen et al., 2009). It was found that the photoelectrical process is in a picosecond or nanosecond order in the oxide heterostructures, which opens up a door for the applications of oxide heterostructures in the optoelectronic detection devices.
LSMO3 thin films, usually recognized as a *p*-type semiconductor, were deposited on *n*-Si substrate by a computer-controlled laser molecular-beam epitaxy (Laser-MBE) system (Yang

et al., 2001). The growing process of thin films was in situ monitored by reflection high-energy electron diffraction (RHEED). The photoelectric behaviours of LSMO3/Si were further investigated using a 1064 nm Nd:YAG laser (pulse width 25 ps) and measured by an oscilloscope of 130 ps rise time (Tektronix® TDS7254B) at ambient temperature. Figure 1 (Lu et al., 2005) shows a typical open-circuit photoelectric pulse when the LSMO3/Si was irradiated by 1064nm pulsed laser. The rise time is about 10 ns and the FWHM is about 12 μs. So as to reduce the influences of the measuring system and the capacitance of the heterostructure, a 0.2 Ω resistance is connected in parallel with the LSMO3/Si heterostructure. As shown in Fig. 2 (Lu et al., 2005), the rise time dramatically reduces to about 210 ps and the FWHM also reduces to about 650 ps. The photovoltaic process can be easily understood as follows. The Hall measurement showed that the hole concentration of LSMO3 is about 3×10^{18} cm^{-3}, while the electron density of Si is about 1×10^{16} cm^{-3}. When the LSMO3 was grown onto the Si substrate, electrons from *n*-type Si leaked out into the *p*-type LSMO3, while holes from *p*-type LSMO3 moved into the *n*-type Si due to the carrier diffusion. The diffusion induced a built-in electric field around the interface in the space charge region. Under the illumination of pulsed laser, the electron-hole pairs were generated, and then quickly separated by the built-in field. Eventually, the photovoltage occurred in the heterostructure. In addition, we observed that the rise time is about 9 ns in $SrTiO_{3-\delta}$/Si heterostructure (Zhao, et al., 2005), 23 ns in LSMO1/SNTO heterostructure (Huang et al., 2006) and 86 ps in $LaAlO_{3-\delta}$/Si (Wen et al., 2009).

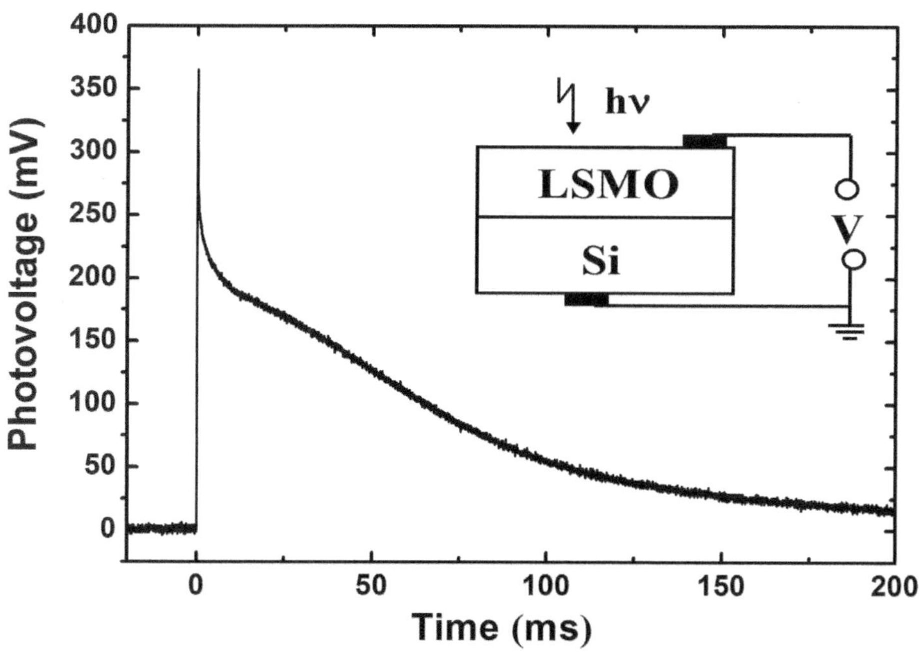

Fig. 1. Variation of the open-circuit photovoltage with time after excitation with a 1064 nm laser pulse on the LSMO3/Si. The schematic circuit of the sample measurement is shown in the inset.

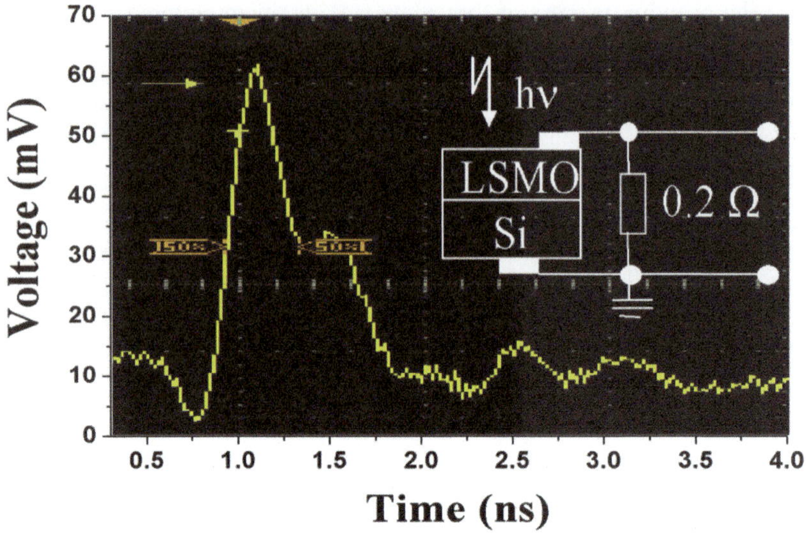

Fig. 2. Variation of the photovoltage with time under the same condition as in Fig. 1, but with a 0.2 Ω resistance connected in parallel across the heterostructure. Inset displays a schematic circuit of the sample measurement.

2.2 Enhancement of lateral photovoltage in oxide heterostructures

As stated above, all these works about ultrafast photoelectric effects focused on the photovoltaic effect across the oxides heterostructures. Furthermore, LPV experiments under the illumination of pulsed laser were also carried in order to further study the photovoltaic effect in oxides heterostructures (Jin et al., 2007; Jin et al., 2009). The well-established LPV theory (Wallmark, 1957; Lucovsky, 1960), that in nonuniformly irradiated *p-n* junctions electric potential near the irradiation center is higher than that far from the center on the *p*-type side while it is lower than that far away from the center on the *n*-type side, was challenged by the observed unusual LPV (Jin et al., 2007). In other words, the signs measured on both sides of *p-n* junction will be opposite according to the conventional LPV whereas they are same in the observed unusual LPV. Therefore, Dember effect (Pankove, 1971), induced by the different mobility of photogenerated electrons and holes, was introduced to qualitatively explain the unusual LPV in the oxide heterostructures (Jin et al., 2007).

LSMO1/SNTO and LSMO3/Si heterostructures were fabricated by growing a *p*-type LSMO1 (LSMO3) layer on a *n*-type SNTO (Si) substrate, with the computer-controlled laser-MBE technique. The schematic setup for LPV measurements is shown in the insets of Figure 3. A small area of 0.5 mm diameter on the *p*-LSMO surface was illuminated by a 308 nm XeCl excimer laser beam (pulse width of 20 ns, pulse energy of 0.15 mJ, and repetition rate of one pulse every 5 min to avoid the heating effect). We moved the samples in the lateral direction and recorded the LPV values, by a sampling oscilloscope of 500 MHz terminated into 1 MΩ at room temperature. Here, the photovoltage, which denotes the peak value of the LPV between the indium electrodes A (x = -3 mm) and B (x = 3 mm) on the LSMO1 (LSMO3) surface, depends on the laser spot position in the LSMO1/SNTO (LSMO3/Si)

heterostructure. In addition, the diameter of indium electrodes is about 1 mm. Particularly, the electrodes were always kept in dark to prevent any electrical contact effects.
Figure 3(a) (Jin et al., 2009) exhibits the measured LPV values in LSMO1/SNTO heterostructure. Figure 3(b) (Jin et al., 2009) depicts the experimental LPV values in LSMO3/Si heterostructure. From Fig. 3(a) and Fig. 3(b), it can be seen that the signs of LPV are same on both sides of heterostructures. This phenomenon can be explained as follows. As both electrons and holes are induced by photons and flow out on two sides of oxide heterostructures, the mobility difference of electrons and holes causes the same sign (positive or negative) of LPV on both sides of oxide heterostructures. It should be noted that only strong light can make the Dember-effect-induced LPV large enough to be observed. To compare the LPV in the heterostructures and that in substrates, the LPV measurements for *n*-SNTO and *n*-Si substrates were also carried out and the results are shown in Fig. 3(c) (Jin et al., 2009). Evidently, it can be concluded that a one-order-of-magnitude enhancement of the LPVs was observed, as compared with those of the substrates.

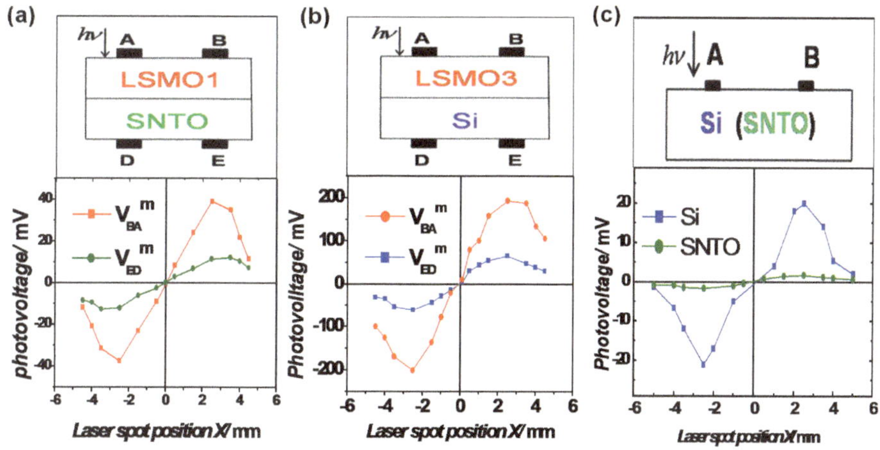

Fig. 3. The peak values of LPV V_{BA}^{m} and V_{ED}^{m} as a function of the position of the laser spot in the x direction in (a) the LSMO1/SNTO and in (b) LSMO3/Si heterostructures, the upper panel displays the schematic setup for the LPV measurement. A (-3 mm), B (3 mm), D (-3 mm), and E (3 mm) denote the electrodes. (c) The peak values of LPV, V_{BA}^{m}, for *n*-SNTO (open triangles) and Si (open squares) substrates.

In fact, it is well known that the optical generation of THz radiation can be produced in the conventional semiconductors using the photo-Dember effect under ultra-short pulse laser illumination (Dekorsy et al., 1993; Dekorsy et al., 1996; Gu et al., 2002; Liu et al., 2006; Tonouchi, 2007; Krotkus, 2010). Thus, the understanding of the underlying physical mechanisms for such an enhancement is fundamentally important and should be of great value for the further designing of the structures of potential applications in powerful THz sources. Mechanisms for the enhancement of LPV in oxide heterostructures will be discussed in detail later.

2.3 Dependence of photovoltage on the films and substrates thickness

For LPV effect, a one-order-of-magnitude enhancement of photovoltage can be obtained in oxide heterostructures compared with that of substrates. Then, to find out how to improve the sensitivity of photovoltaic effect across the heterostructures, surveys on the dependence of photovoltage on the films and substrates thickness were systematically carried out (Qiu et al., 2007; Wen et al., 2009).

A series of LSMO1/SNTO heterostructures with various films thickness were fabricated by Laser-MBE, for investigating the dependence of photovoltage on the films thickness. The growing process of LSMO1 layers was in situ monitored by RHEED. One period in the RHEED intensity oscillations marks the completion of one unit cell. Hence the thickness of LSMO1 ultrathin films was accurately controlled by the growth time which represented the number of the periods of the RHEED intensity oscillations.

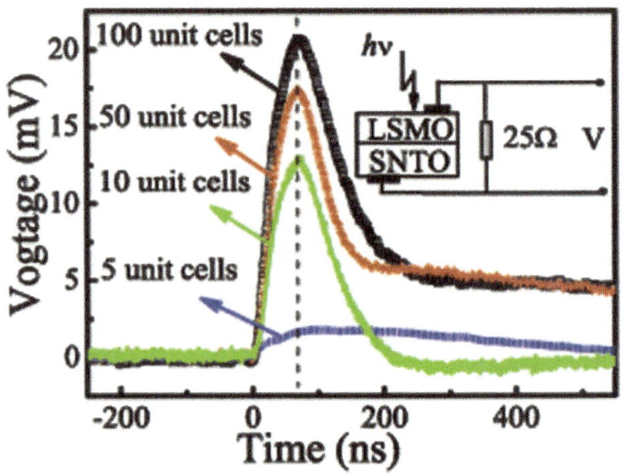

Fig. 4. The open-circuit photovoltage as a function of time to LSMO1/SNTO heterostructures. The inset presents the schematics of the measurement circuit.

Figure 4 shows that the maximum open-circuit photovoltage was about 2, 13, 17 and 21mV for the LSMO1/SNTO heterostructures with LSMO1 thicknesses of 20, 40, 200 and 400Å, respectively. The photovoltage becomes larger with the increase of the thickness of the LSMO1 layer. This result is ascribed to the increase of the carrier amount and the enhancement of the built-in electric field in the space-charge region of the LSMO1/SNTO heterostructure with the increase of the thickness of LSMO1 layer (Qiu et al., 2007). However, it is found that the speed of photovoltaic response is almost independent of the thickness of LSMO1 layer in the heterostructure. It should be useful to comprehend the photovoltaic characteristics in complex oxides materials.

Not only the thickness of films but also the thickness of the substrates can remarkably affect the photoelectric properties of oxide heterostructures. The 400 nm LAO thin films were epitaxially grown on *p*-type Si substrates (thickness 0.71 mm) by a computer controlled Laser-MBE system using two-step method (Wen et al., 2009). Then, three pieces of Si substrates will be polished mechanically until the thickness is 0.44, 0.19, and 0.10 mm, respectively. For convenience, the four samples of various substrate thicknesses will be denoted as F1 (0.71 mm) F2 (0.44 mm) F3 (0.19 mm) and F4 (0.10 mm) respectively. For the photovoltaic measurements, two indium electrodes were painted on the surfaces of LAO film and Si substrate. During the measurements, the electrodes were always kept in the dark to prevent the generation of any electrical contact photovoltage. All of the photoelectric measurements were carried out at room temperature.

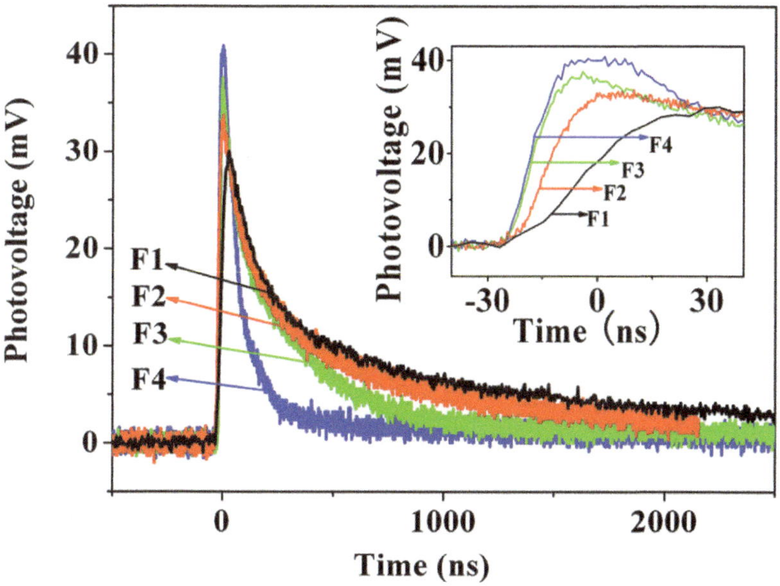

Fig. 5. The photovoltaic responses of LAO/Si heterostructures under the illumination of XeCl pulsed laser for samples F1, F2, F3, and F4. The inset is the rise time of photovoltaic response.

The influence of substrate thickness on photoelectrical effects of LAO/Si heterostructure was systematically investigated on samples F1, F2, F3, and F4 by a 308 nm XeCl excimer pulsed laser (pulse width 20 ns and energy density 0.1 J/cm^2) with a laser spot diameter of 2 mm. Figure 5 shows the dependence of the photovoltaic response on the substrate thickness. The photovoltaic signals were measured by an oscilloscope with 500 MHz bandwidth and an input impedance of 50 Ω. For the four samples F4, F3, F2, and F1, the 10%-90% rise times were 13.3, 13.9, 21.4, and 38.3 ns, respectively, and the FWHMs were 79.2, 134.7, 180.6, and 258.2 ns, respectively. The rise time of the photoelectric signal is faster for the heterostructure with thinner substrate. Meanwhile, the FWHM decreases with the decrease in substrate thickness. It can be understood in the following way: photogenerated carriers were separated by built-in electric field at the interface between LAO and Si, and those holes reached the surface of the substrate faster for thinner substrate so that faster photoelectric response can be observed. Smaller FWHM for the heterostructure with thinner substrate can be understood in the similar way. Overall, the decrease in the substrate thickness is an effective method for improving the photovoltaic sensitivity in oxide heterostructures.

3. Theoretical investigations

Photovoltaic experiments exhibit attractive properties and interesting phenomena in oxide heterostructures under the illumination of pulsed laser as mentioned above. Although several qualitative explanations were proposed, it is rather significant to further quantitatively clarify the physical origins and underlying mechanisms. It is therefore necessary to carry out self-consistent calculations based on the time-dependent drift-diffusion model.

3.1 Time-dependent drift-diffusion model

3.1.1 One-dimensional drift-diffusion model

Self-consistent calculation of drift-diffusion model, first proposed by Scharfetter and Gummel (Scharfetter & Gummel, 1969), is a powerful and effective method of describing carrier transport behaviour in semiconductor devices and is still widely used today(Grasser et al., 2003). In oxide heterostructures, some early theoretical works confirmed that numerical calculations based on drift-diffusion model could be employed to depict not only the steady transport behaviour (Zhou et al., 2005; Han et al., 2007; Hu et al., 2008; Hu et al., 2009) but also the transient transport behaviour (Liao et al., 2009a, 2009b, 2009c; Liao et al., 2010; Ge et al., 2010). One-dimensional time-dependent drift-diffusion model consists of the Poisson equation and carrier continuity equations as follows:

$$\frac{\partial^2 \phi(x,t)}{\partial x^2} = -\frac{e}{\varepsilon}[p(x,t) - n(x,t) + N] \tag{1}$$

$$\frac{\partial n(x,t)}{\partial x} = \frac{1}{e}\frac{\partial j_n(x,t)}{\partial x} + G(x,t) - R(x,t) \tag{2}$$

$$\frac{\partial p(x,t)}{\partial t} = -\frac{1}{e}\frac{\partial j_p(x,t)}{\partial x} + G(x,t) - R(x,t) \tag{3}$$

$$j_n(x,t) = -e\mu_n n(x,t)\frac{\partial \phi(x,t)}{\partial x} + kT\mu_n \frac{\partial n(x,t)}{\partial x} \tag{4}$$

$$j_p(x,t) = -e\mu_p p(x,t)\frac{\partial \phi(x,t)}{\partial x} - kT\mu_p \frac{\partial p(x,t)}{\partial x} \tag{5}$$

Where $\phi(x,t)$, $n(x,t)$ and $p(x,t)$ denote electrostatic potential, electron and hole densities; x and t is position and time coordinates; e, ε and N is electron charge, dielectric permittivity, net ionized impurity densities, respectively; $j_n(x,t)$ and $j_p(x,t)$ denote the electron and hole current densities respectively; μ_n and μ_p represent electron and hole mobility respectively. k and T represent Boltzmann constant and temperature taken as room temperature. The electron-hole pairs generation rate $G(x,t)$ can be expressed as $G(x,t) = I_0(t)\alpha\beta\exp(-\alpha x)$, where $I_0(t)$, α and β denote the density of incident photon flux, absorption coefficient and quantum efficiency. $R(x,t)$ denotes the recombination rate, here presented by the Shockley-Read-Hall (SRH) model (Shockley & Read, 1952). The interface condition for solving the Poisson equation Eq. (1) is taken as $\phi(x_{inter} - \delta, t) = \phi(x_{inter} + \delta, t)$, where x_{inter} denotes the position of the interface and δ is infinitesimal. At the interface of heterostructures, the Richardson thermionic emission current is employed as interface conditions (Yang et al., 1993) for solving Eq. (2) and (3). The initial values $\phi(x,0)$, $n(x,0)$ and $p(x,0)$ for self-consistent calculations were obtained by solving the Poisson equation coupled with the Boltzmann approximation (Selberherr, 1984).

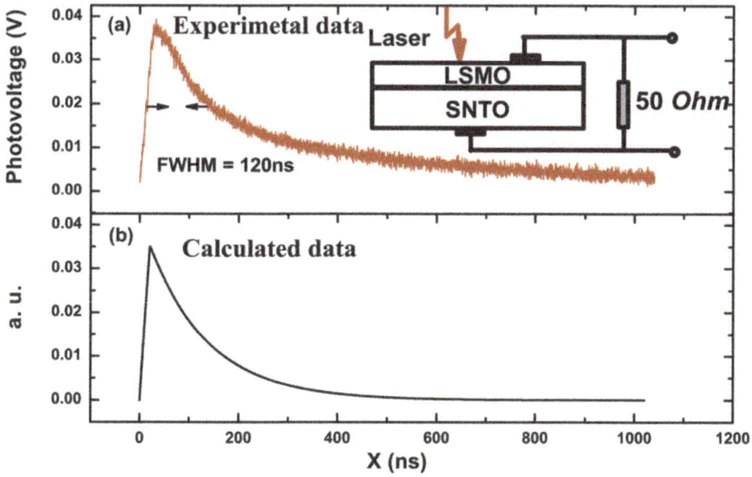

Fig. 6. (a) Experimental evolution of photovoltage in the LSMO1/SNTO heterostructure connected with 50 Ω resistance. The inset presents the illustration of heterostructure for voltaic measurement. (b) The calculated evolution of photovoltage.

The dependence of photovoltaic response on the time is shown in Fig. 6 (a) (Liao et al., 2009c) in LSMO1/SNTO heterostructure irradiated by a XeCl pulsed laser (wavelength of 308 nm, repetition rate of 1 Hz and duration of 20 ns). The inset in Fig. 6 presents the illustration of heterostructure for photovoltaic measurement. Figure 6 (b) shows the numerical result by self-consistently solving Eq. (1)-(5). The necessary parameters for numerical calculations are listed in Table 1. More details about the experiment and calculation can be found in Ref. Liao et al., 2009c. Apparently, the theoretical results are in good agreement with the experimental data in LSMO1/SNTO. Consequently, it can be concluded that the time-dependent drift-diffusion model can also well describe the transient transport of the oxide heterostructures.

On the basis of the good numerical simulations, the dependence of transient photovoltaic response on the various physical parameters was further analyzed (Liao et al., 2009a, 2009b, 2009c), which could be useful to the future experimental investigations on the photovoltaic properties of oxide heterostructures irradiated by pulsed laser. The theoretical caculations represented that the modulation of Sr doping in $La_xSr_{1-x}MnO_3$ is an effective method to accommodate the sensitivity and the speed of photovoltaic response (Liao et al., 2009a), that a smaller parallel resistance should result in faster photoelectric response and the increase of the carrier mobilities induced by applying higher energy photons can decrease the rise time but increase the peak value of the photovoltage (Liao et al., 2009b), and that the large variation of carrier concentration mainly locates in the space-charge region of the oxide heterostructure (Liao et al., 2009c).

	LSMO1(3)	SNTO	Si
Temperature (K)	300	300	300
Dielectric constant(ε_0)	10.0	300.0	11.9
Electron mobility ($cm^2/(V.s)$)	10.0	8.0	1450
Hole mobility ($cm^2/(V.s)$)	1.8	0.1	500
Band gap (eV)	0.8	2.8	1.12
Ionized impurity concentrations ($/cm^3$)	$4.0\times10^{19}(1.0\times10^{20})$	1.63×10^{20}	1.0×10^{17}
Photon absorption coefficient(cm^{-1})	1.5×10^5	1.2×10^5	4×10^5

Table 1. The parameters for self-consistent calculations. (Ge et al., 2010, and references therein)

3.1.2 Two-dimensional drift-diffusion model

As mentioned above, the further theoretical analysis of unusual LPV phenomenon and great enhancement of LPV is fundamentally important in terms of both physics and engineering. A quantitative approach was thereby introduced based on two-dimensional time-dependent drift-diffusion model (Liao et al., 2010; Ge et al., 2010). The equations are as follows:

$$\frac{\partial^2\phi(x,y,t)}{\partial x^2}+\frac{\partial^2\phi(x,y,t)}{\partial y^2}=-\frac{e}{\varepsilon}[p(x,y,t)-n(x,y,t)+N] \tag{6}$$

$$\frac{\partial n(x,y,t)}{\partial t}=\frac{1}{e}\nabla\cdot\vec{j}_n(x,y,t)+G(x,y,t)-R(x,y,t) \tag{7}$$

$$\frac{\partial p(x,y,t)}{\partial t}=-\frac{1}{e}\nabla\cdot\vec{j}_p(x,y,t)+G(x,y,t)-R(x,y,t) \tag{8}$$

$$\begin{aligned}\vec{j}_n(x,y,t)=&[-e\mu_n n(x,y,t)\frac{\partial\phi(x,y,t)}{\partial x}+eD_n\frac{\partial n(x,y,t)}{\partial x}]\cdot\boldsymbol{i}_x\\&+[-e\mu_n n(x,y,t)\frac{\partial\phi(x,y,t)}{\partial y}+eD_n\frac{\partial n(x,y,t)}{\partial y}]\cdot\boldsymbol{i}_y\end{aligned} \tag{9}$$

$$\begin{aligned}\vec{j}_p(x,y,t)=&[-e\mu_p p(x,y,t)\frac{\partial\phi(x,y,t)}{\partial x}-eD_p\frac{\partial p(x,y,t)}{\partial x}]\cdot\boldsymbol{i}_x\\&+[-e\mu_p p(x,y,t)\frac{\partial\phi(x,y,t)}{\partial y}-eD_p\frac{\partial p(x,y,t)}{\partial y}]\cdot\boldsymbol{i}_y\end{aligned} \tag{10}$$

where x and y denote the transverse and lateral axes, respectively, and i_x and i_y are the unit vectors along the x and y axes, respectively. Similar to the one-dimensional drift-diffusion model, all the symbols denote the same physical quantities. $G(x,y,t)$ is the generation rate of the photoinduced electron-hole pairs, expressed as $G(x,y,t)=I_0(y,t)\alpha\beta\exp(-\alpha x)$, where α and β denote the absorption coefficient and quantum efficiency respectively. $I_0(y,t)$ denotes the incident photon flux density. Furthermore, in our calculations, $I_0(y,t)$ is considered as a Gauss distribution function. Actually, Eqs. (6)-(10) are rather difficult to be solved. Owing to the large difference in the scale between the two dimensions, width and the thickness, a huge rounding error is caused in the numerical calculation of Poisson equations. For solving this problem, we decompose the Poisson equation Eq. (6) into two following independent equations:

$$\frac{\partial^2 \phi_x(x,y,t)}{\partial x^2} = Q_x(x,y,t) \tag{11}$$

$$\frac{\partial^2 \phi_y(x,y,t)}{\partial y^2} = Q_y(x,y,t) \tag{12}$$

where $Q_x(x,y,t)$ and $Q_y(x,y,t)$ represent the charge densities in transverse and lateral direction, respectively, $\phi_x(x,y,t)$ and $\phi_y(x,y,t)$ denote the electric potential distributions excited by the charge in transverse and lateral direction, respectively. The total potential distribution is $\phi(x,y,t) = \phi_x(x,y,t) + \phi_y(x,y,t)$. The two-dimensional continuity equations can also be decomposed into two groups of one-dimensional equations. These equations can be solved on the basis of the self-consistent calculations of transverse photovoltaic effect stated above (Liao et al., 2009a, 2009b, 2009c). More details of numerical calculations can be found in Ref. Liao et al., 2010.

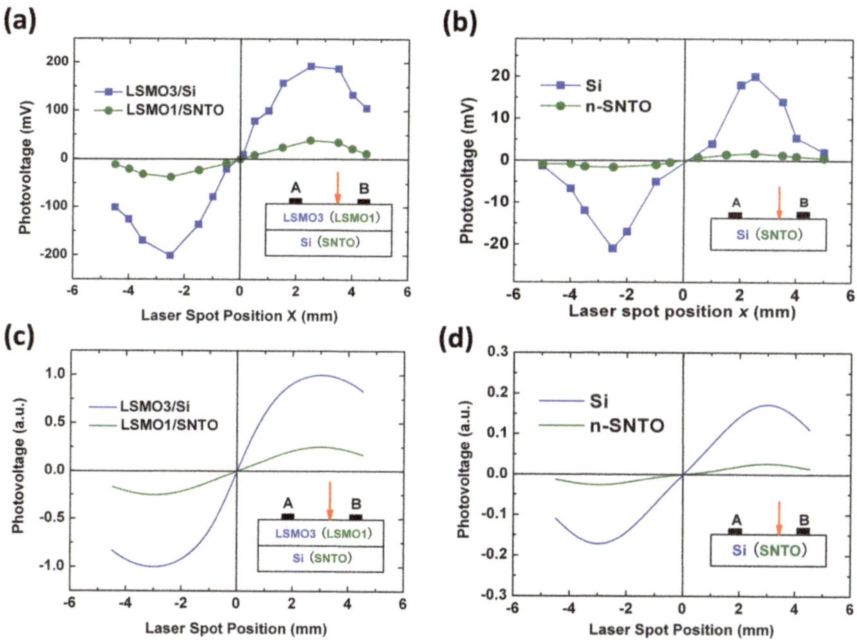

Fig. 7. The experimental (a) and theoretical (c) LPVs of the LSMO side in the LSMO1/SNTO denoted by the green curve and LSMO3/Si denoted by the blue curve; the experimental (b) and theoretical (d) LPVs of SNTO denoted by the green curve and Si denoted by the blue curve. The photovoltage denotes the peak value of LPV between the indium electrodes A (x = −3 mm) and B (x = 3 mm). The inset shows the schematic setup for LPV measurement.

Figure 7 (a) (Jin et al., 2007, 2009) exhibits the experimental LPVs of LSMO side in the LSMO1/SNTO (the green circular curve) and LSMO3/Si (the blue square curve) heterostructures. Figure 7 (b) (Jin et al., 2009) depicts the experimental LPVs of SNTO (the green circular curve) and Si (the blue square curve) substrates. Our calculated results for LPVs on the LSMO side in LSMO1/SNTO and LSMO3/Si heterostructures are shown in Fig. 7 (c) (Ge et al., 2010) as the green curve and blue one, respectively. The calculated LPVs on SNTO and Si substrates are displayed in Fig. 7 (d) (Ge et al., 2010) by the green curve and the blue curve, respectively. The necessary parameters for the calculations are listed in Table 1 (Ge et al., 2010). We use a.u. to show our calculated results to compare with the experimental data in the curves, due to the fact that our model purely describes the charge contribution in the heterostructure and the experimental data includes more complicated factors in the measuring circuit, contacts, and so on. It can be seen that the calculated results are in good accordance with the experimental data. Then a much clearer insight into the unusual LPV and one-order-enhancement of LPV can be obtained from the present calculations.

3.2 Theoretical discussion based on self-consistent calculations

3.2.1 Unified descriptions of the conventional and unusual LPV

So as to find out the reasons that the unusual LPV differs from the conventional LPV considerably, we carried out a throughout investigation of the dependence of LPV on the laser pulse energy.

As shown in Fig. 8 (a) (Liao et al., 2010), the calculated values of the electric potential on the LSMO1 side turn to be larger and larger with the increase in laser pulse energy. The trends of electric potential distribution on the two sides of the heterostructure are opposite to each other under the laser pulse energy lower than 0.015 mJ, while they turn to be the same with each other under the laser pulse energy higher than 0.015 mJ. As shown in Fig. 8 (d) (Liao et al., 2010), the corresponding calculated LPV on the SNTO side exhibits laterally modulated behaviours under the laser pulse energy of 0.015 mJ, while calculated LPV on the LSMO1 side becomes larger and larger with the increase in laser pulse energy as shown in Fig. 8 (c) (Liao et al., 2010). This laterally modulated LPV effect can be explained by the competition between Dember and conventional LPV processes.

Under critical laser pulse energy, the conventional LPV effect and the Dember effect are comparable to each other. Hence, neither of them can dominate the LPV all over the region. In the region near the irradiation center (-2.0 mm, 2.0 mm) where the carrier density is high due to the strong laser pulse irradiation, the Dember effect is stronger than the conventional LPV effect, as shown in Fig. 8. Therefore, in this region, the farther the position is away from the irradiation center, the smaller electric potentials are on both sides. While in the region far away from the irradiation center (2.0 mm, 7.5 mm) and (-7.5 mm, 2.0 mm) where the carrier density is low, the Dember potential is weak, as shown in Fig. 8. As a result, the conventional LPV effect is the main contributor to the LPV. Thus, our calculated results unified the description of the conventional LPV and the Dember effect into the drift-diffusion model.

The evolution and the competition process of the conventional and Dember LPV effect have been revealed theoretically. With the increase in irradiated laser pulse energy, the Dember effect plays a more and more important role in lateral photovoltaic effect. In conclusion, the laser pulse energy is the key factor in determining whether the conventional or the Dember LPV effect dominate the LPV. The corresponding experiments are highly expected.

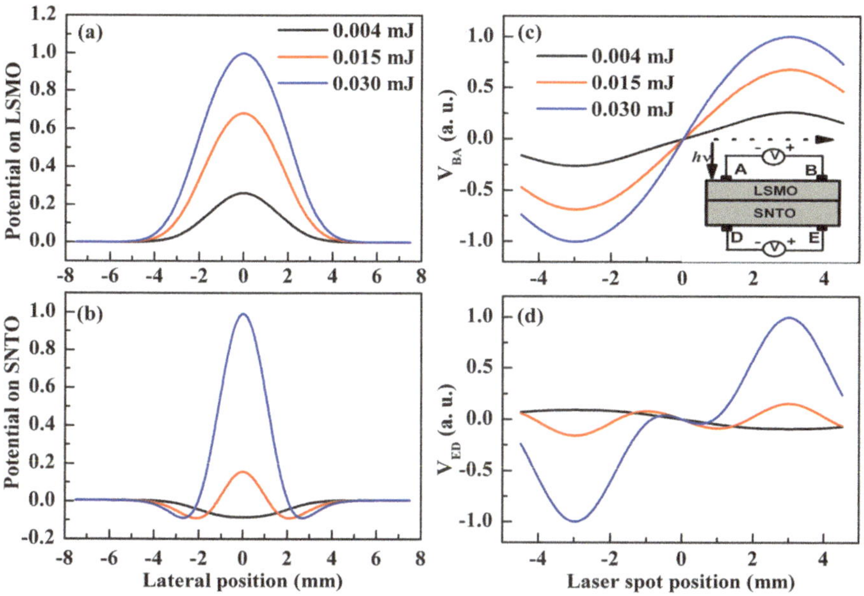

Fig. 8. The electric potential distributions near the irradiation center of the LSMO1/SNTO heterostructure with varied irradiation laser pulse energies (0.004, 0.015, 0.030 mJ) (a) on the LSMO1 side and (b) on the SNTO side. The calculated LPVs of the LSMO1/SNTO heterostructure with varied irradiation laser pulse energies (0.004, 0.015, and 0.030 mJ) (c) on the LSMO1 side and (d) on the SNTO side. The inset in Fig. 8 (c) shows the schematic setup for LPV measurement.

3.2.2 Mechanisms for the great enhancement of LPV

From the above calculations, the mechanisms for the one-order-of-magnitude enhancement of the Dember-effect-induced LPV can be revealed. There are mainly two physical origins. Firstly, we find that the Dember-effect-induced LPV of the *p* type material is larger than that of the *n* type material with the same carrier concentration. As illustrated in Fig. 9 (a) (Ge et al., 2010), we calculated the Dember-effect-induced LPVs of the *p* type material and *n* type material with the same carrier concentration of 1×10^{17} cm^{-3}. It can be seen that the LPV of the *p* type material is almost twice as large as that of the *n* type material. This can be totally ascribed to the difference between the mobilities of electrons and holes. The amounts of photo-generated electrons and holes are the same in both the *p* type material and *n* type material under the same condition. Consequently, the lateral diffusion current densities are the same. However, the main drift carriers for *p* type material are holes, while the main drift carriers for *n* type material are electrons. Assuming that the system is in the steady state that the drift current densities can just balance the diffusion current densities, the lateral drift current densities are of the same value for the *p* type material and *n* type material. Thus, the drift electric field of the *p* type material should be larger than that of the *n* type material, as the mobility of holes is smaller than that of electrons. Hence, it can be concluded that the Dember-effect-induced LPV of the *p* type material is larger than that of the *n* type material. Moreover, the Dember-effect-induced THz radiation from the surface of *p*-InAs wafer has been reported as a strong THz emitter (Tonouchi, 2007; Krotkus, 2010). We believe that the present study gives an insight into the physical origin of such an application.

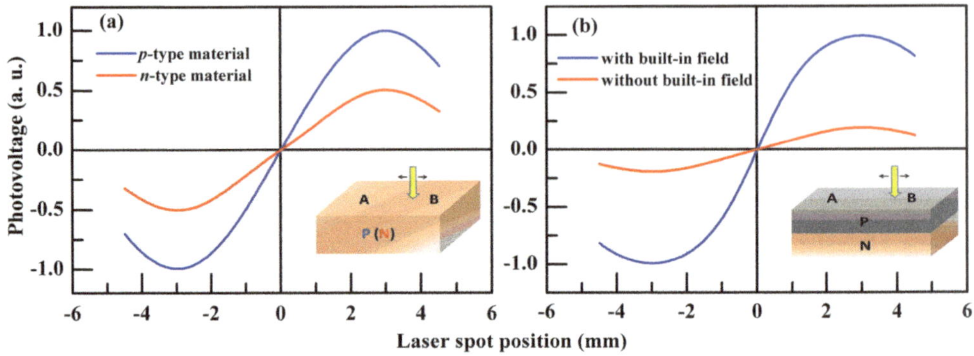

Fig. 9. (a) The calculated LPVs between the electrodes A (x = −3 mm) and B (x = 3 mm) in the same material with different doping type. The blue and red curves denote the LPVs in the *p*-type and *n*-type material, respectively. The inset exhibits the schematic setup. (b) The blue and red curves denote the calculated LPVs between the electrodes A (x = −3 mm) and B (x = 3 mm) in the same heterostructures with built-in field and without built-in field, respectively. The lower inset exhibits the schematic setup.

Secondly, from our calculations we find that the built-in electric field at the interface between the thin film and the substrate also plays an important role in the LPV effect. For revealing the influence of the built-in electric field, we assumed that the potential difference between the *p* type region and the *n* type region was zero and 0.52 Volt for the structure without and with the built-in electric field in our self-consistent calculations, respectively. Figure 9 (b) (Ge et al., 2010) shows the calculated LPVs for the heterostructure with and without the built-in electric field denoted by the blue curve and the red curve, respectively. From Fig. 9 (b), it can be estimated that the heterostructure with a small built-in field of 0.52 Volt can produce a five times larger LPV than that of the heterostructure without the built-in field. For the structure with the built-in electric field, the photo-generated electron-hole pairs can be separated by the built-in electric field. Thereby, the photo-generated holes are swept into the *p* type layer, and the potential of the irradiation region is raised relative to the situation without the built-in electric field. As a result, the Dember-effect-induced LPV for the structure with the built-in electric field is enhanced compared to the one without the built-in electric field. The combination of the above two mechanisms can well explain the one-order-of-magnitude enhancement of the LPV in the oxide heterostructures compared with that in the substrates.

The understanding of the mechanisms for the enhanced LPV in oxide heterostructures should be useful in further designing of the structures of potential applications in novel functional devices, for instance high sensitive PSDs and powerful THz sources.

4. Conclusion and outlook

In conclusion, we have sought to give a brief description of currently most important applications of laser pulses in photovoltaic effect. Fast transient photoelectric effects were observed in LSMO3/Si heterostructure fabricated by pulsed laser deposition technique, which opens up possibilities to design new photodetectors based on oxides. Furthermore, the unusual LPV was observed in the heterostructures of both LSMO1/SNTO and LSMO3/Si. Then, a one-order-of-magnitude enhancement of the LPV was found. Moreover, the dependence of photovoltage on the thin films and substrates thickness was elaborated. The photovoltage becomes larger with the increase of the LSMO1 film thickness, while the film thickness is less than the depletion layer of the heterostructures. This is attributed to the increase of the carrier amount of the LSMO1 layer and the enhancement of the built-in electric field in the space-charge region of the LSMO1/SNTO heterostructure. Faster photoelectric response was observed in LAO/Si heterostructure, and the photoelectric sensitivity was greatly improved by decreasing the thickness of the Si substrates.

To give much insight into the underlying physical origins of the fascinating phenomena under laser pulses, theoretical investigations were carried out on the basis of the time-dependent drift-diffusion model. The theoretical calculations of one-dimensional drift-diffusion model proved that the modulation of Sr doping in $La_xSr_{1-x}MnO_3$ is an effective method to accommodate the sensitivity and the speed of photovoltaic response, that a smaller parallel resistance should result in faster photoelectric response, and that the photodoping effect mainly occurs in the space-charge region. By self-consistent calculations, it can be seen that with the increase in irradiated laser pulse energy, the Dember effect plays a more and more important role in LPV. A unified description for conventional LPV and Dember-effect-induced LPV can be achieved within the frame of drift-diffusion model. The two mechanisms for the great enhancement of LPV in the heterostructures were also presented by self-consistent calculations: the LPV of *p*-type material is larger than that of *n*-type material owing to the larger drift electric field induced in the *p*-type material than that in the *n*-type material; the built-in electric field at the interface between the thin film and substrate can enhance the LPV. Such an understanding of the mechanisms for the enhancement of LPV should be helpful in further designing of the structures of PSD and new THz sources.

The comparison of the theoretical results presented in this chapter with various experimental data and their satisfactory agreement are evidences that the proposed model and the results obtained on its basis are sufficiently accurate and reliable. The methods and approaches presented can be used for the further development of new approaches of laser pulse applications in photovoltaic effect. Currently, there are still a large number of open questions in this filed. For instance, novel phenomenon could be observed under shorter and stronger laser pulse. In terms of theoretical studies, this may lead to the adoption of the more sophisticated simulation program in order to capture the essential physics beyond the drift-diffusion model. Hence, the further studies are highly expected in both experimental and theoretical aspects.

5. Acknowledgment

The authors acknowledge the financial support from the National Natural Science Foundation of China and the National Basic Research Program of China.

6. References

Dekorsy T. ; Auer H. ; Bakker H. J. ; Roskos H. G. & Kurz H. (1996). THz electromagnetic emission by coherent infrared-active phonons, *Phys. Rev. B*, Vol.53, pp. 4005-4014

Dekorsy T.; Pfeifer T.; Kütt W. & Kurz H. (1993). Subpicosecond carrier transport in GaAs surface-space-charge fields, *Phys. Rev. B*, Vol. 47, pp. 3842-3849

Ge C.; Jin K. J.; Lu H. B.; Wang C.; Zhao G. M.; Zhang L. L. & Yang G. Z. (2010). Mechanisms for the enhancement of the lateral photovoltage in perovskite heterostructures. *Solid State Commu.*, Vol.150, pp. 2114-2117

Grasser T.; Tang T. W.; Kosina H. Selberherr S. (2003). A Review of Hydrodynamic and Energy-Transport Models for Semiconductor Device Simulation, *Proc. IEEE*, Vol.91, pp. 251

Gu P.; Tani M.; Kono S.; Sakai K. & Zhang X.-C. (2002). Study of terahertz radiation from InAs and InSb, *J. Appl. Phys.*, Vol.91, pp. 5533-5537

Han P.; Jin K. J.; Lu H. B.; Zhou Q. L.; Zhou Y. L. & Yang G. Z. (2007). The mechanism study on transport properties in perovskite oxide p-n junction, *Appl. Phys. Lett.*, Vol.91, pp.182102

Henry J. & Livingstone J. (2001). Thin-film amorphous silicon position-sensitive detectors, *Adv. Mater.*, Vol.13, pp. 1-4

Hu C. L.; Jin K. J.; Han P.; Lu H. B.; Liao L. & Yang G. Z. (2008). The effect of phase separation on the temperature dependent magnetoresistance in perovskite oxide heterojunction, *Appl. Phys. Lett.*, Vol.93, pp.162106

Hu C. L.; Jin K. J.; Han P.; Lu H. B.; Liao L. & Yang G. Z. (2009). Theoretical study on the positive magnetoresistance in perovskite oxide p-n junctions, *Solid State Commun.*, Vol.149, pp.334

Huang Y. H.; Jin K. J.; Zhao K.; Lu H. B.; He M.; Chen Z. H.; Zhou Y. L.; Yang G. Z. & Ma X. L. (2006). Photoelectric Characteristic of $La_{0.9}Sr_{0.1}MnO_3/SrNb_{0.01}Ti_{0.99}O_3$ p-n Heterojunctions, *Chin. Phys. Lett.*, Vol.23, pp. 982

Jain R. K. & Landis G. A. (1998). Transient response of gallium arsenide and silicon solar cells under laser pulse, *Solid-State Electron.*, Vol.42, pp. 1981-1983

Jin K. J.; Lu H. B.; Zhao K.; Ge C.; He M. & Yang G. Z. (2009). Novel Multifunctional Properties Induced by Interface Effects in Perovskite Oxide Heterostructures. *Adv. Mater.*, Vol.21, pp. 4636-4640

Jin K. J.; Zhao K.; Lu H. B.; Liao L. & Yang G. Z. (2007). Dember effect induced hotovoltage in perovskite p-n heterojunctions, *Appl. Phys. Lett.*, Vol.91, pp. 081906

Krotkus A. (2010). Semiconductors for terahertz photonics applications, *J. Phys. D: Appl. Phys.*, Vol. 43, pp. 273001

Liao L.; Jin K. J.; Ge C.; Hu C. L.; Lu H. B. & Yang G. Z. (2010). A theoretical study on the dynamic process of the lateral photovoltage in perovskite oxide heterostructures. *Appl. Phys. Lett.*, Vol.96, pp. 062116

Liao L.; Jin K. J.; Han P.; Zhang L. L.; Lu H. B. & Ge C. (2009a). Acceptor Concentration Effects on Photovoltaic Response in the $La_{1-x}Sr_xMnO_3/SrNb_yTi_{1-y}O_3$ Heterojunction, *Chin. Phys. Lett.*, Vol.26, pp. 057301

Liao L.; Jin K. J.; Lu H. B.; Han P.; He M. & Yang G. Z. (2009b). Dynamic process of photovoltaic effect in $La_{0.9}Sr_{0.1}MnO_3/SrNb_{0.01}Ti_{0.99}O_3$ heterojunction, *Solid State Commu.*, Vol.149, pp. 915-918

Liao L.; Jin K. J.; Lu H. B.; Qiu J.; Han P. & Zhang L. L. (2009c). Theoretical study on the photodoping effects in $La_{1-x}Sr_xMnO_3/SrNb_yTi_{1-y}O_3$ p–n heterojunction, *Phys. Status Solidi A*, Vol.206, pp. 1655-1659

Liu K.; Xu J. Z.; Yuan T. & Zhang X.-C. (2006). Terahertz radiation from InAs induced by carrier diffusion and drift, *Phys. Rev. B*, Vol.73, pp. 155330

Lu H. B.; Jin K. J.; Huang Y. H.; He M.; Zhao K.; Cheng B. L.; Chen Z. H.; Zhou Y. L.; Dai S. Y. & Yang G. Z. (2005). Picosecond photoelectric characteristic in $La_{0.7}Sr_{0.3}MnO_3$/Si p-n junctions, *Appl. Phys. Lett.*, Vol.86, pp. 241915

Lucovsky G. (1960). Photoeffects in nonuniformly irradiated p-n junction, *J. Appl. Phys.*, Vol.31, pp. 1088

Lv X.; Huang Y.; Liu Z. B.; Tian J. G.; Wang Y.; Ma Y.; Liang J. J.; Fu S. P.; Wan X. J. & Chen Y. S. (2009). Photoconductivity of Bulk-Film-Based Graphene Sheets, *Small*, Vol. 5, pp. 1682

Ohtomo A. & Hwang H. Y. (2004). A high-mobility electron gas at the $LaAlO_3/SrTiO_3$ heterointerface, *Nature*, Vol.427, pp. 423

Pankove J. I. (1971). *Optical Processes in Semiconductors*, Prentice-Hall, Englewood Cliffs, New Jersey, U.S.

Qiu J.; Lu H. B.; Jin K. J.; He M. & Xing J. (2007). Manganite-layer thickness-dependent photovoltaic effect of $La_{0.9}Sr_{0.1}MnO_3/SrNb_{0.01}Ti_{0.99}O_3$ p-n heterojunction, *Physica B*, Vol.400, pp. 66-69

Reyren N.; Thiel S.; Caviglia A. D.; Kourkoutis L.; Hammerl G.; Richter C.; Schneider C. W.; Kopp T.; Rüetschi A. S.; Jaccard D.; Gabay M.; Muller D. A.; Triscone J. M. & Mannhart J. (2007). Superconducting Interfaces Between Insulating Oxides, *Science*, Vol.317, pp. 1196

Scharfetter D. L. and Gummel H. K. (1969). Large-signal analysis of a silicon read diode oscillator, *IEEE Trans. Electron Devices*, Vol.ED-16, pp. 64–77

Selberherr S. (1984). *Analysis and Simulation of Semiconductor Devices*, Springer-Verlag, Wien, New York

Shockley W. & Read W. T. (1952). Statistics of the Recombinations of Holes and Electrons, *Phys. Rev.*, Vol.87, pp. 835

Sun J. R.; Xiong C. M.; Shen B. G.; Wang P. Y. & Weng Y. X. (2004). Manganite-based heterojunction and its photovoltaic effects, *Appl. Phys. Lett.*, Vol.84, pp. 2611-2613

Sun J. R.; Xiong C. M.; Shen B. G.; Wang P. Y. & Weng Y. X. (2004). Manganite-based heterojunction and its photovoltaic effects, *Appl. Phys. Lett.*, Vol.84, pp. 2611

Tonouchi M. (2007). Cutting-edge terahertz technology, *Nat. Photonics*, Vol.1, pp. 97-105

Wallmark J. T. (1957). A new semiconductor photocell using lateral photoeffect, *Proc. IRE*, Vol.45, pp. 474

Wen J.; Jin K. J.; He M.; Lu H. B.; Yang F. & Yang G. Z. (2009). The substrate thickness dependence of the photovoltage in $LaAlO_{3-\delta}$/Si heterostructures, *Appl. Phys. Lett.*, Vol.94, pp. 061118

Yan Z. J.; Yuan X.; Xu Y. B.; Liu L. Q. & Zhang X. (2007). Photovoltaic effects in obliquely deposited oxygen-deficient manganite thin film, *Appl. Phys. Lett.*, Vol.91, pp. 104101

Yang G. Z.; Lu H. B.; Chen F.; Zhao T. & Chen Z. H. (2001). Laser molecular beam epitaxy and characterization of perovskite oxide thin films, *J. Cryst. Growth.*, Vol.227-228, pp. 929-935

Yang K.; East J. R. & Haddad G. I. (1993). Numerical modeling of abrupt heterojunctions using a thermionic-field emission boundary condition, *Solid State Electron.*, Vol.36, pp. 321

Yu C. Q. & Wang H. (2010). Light-induced bipolar-resistance effect based on metal-oxide-semiconductor structures of Ti-SiO_2-Si, *Adv. Mater.*, Vol.22, pp. 966–970

Zhang P. X.; Lee W. & Zhang G. Y. (2002). Time dependence of laser-induced thermoelectric voltages in $La_{1-x}Ca_xMnO_3$ and $YBa_2Cu_3O_{7-\delta}$ thin films, *Appl. Phys. Lett.*, Vol.81, pp. 4026

Zhao K.; Huang Y. H.; Zhou Q. L.; Jin K. J.; Lu H. B.; He M.; Cheng B. L.; Zhou Y. L.; Chen Z. H. & Yang G. Z. (2005). Ultraviolet photovoltage characteristics of $SrTiO_{3-\delta}$/Si heterojunction, *Appl. Phys. Lett.*, Vol.86, pp. 221917

Zhao K.; Jin K. J.; Huang Y. H.; Lu H. B.; He M.; Chen Z. H.; Zhou Y. L. & Yang G. Z. (2006a). Laser-induced ultrafast photovoltaic effect in $La_{0.67}Ca_{0.33}MnO_3$ films at room temperature, *Physica B*, Vol.373, pp. 72-75

Zhao K.; Jin K. J.; Huang Y. H.; Zhao S. Q.; Lu H. B.; He M.; Chen Z. H.; Zhou Y. L. & Yang G. Z. (2006b). Ultraviolet fast-response photoelectric effect in tilted orientation $SrTiO_3$ single crystals, *Appl. Phys. Lett.*, Vol.89, pp. 173507

Zhao K.; Jin K. J.; Lu H. B.; Huang Y. H.; Zhou Q. L.; He M.; Chen Z. H.; Zhou Y. L. & Yang G. Z. (2006c). Transient lateral photovoltaic effect in p-n heterojunctions of La0.7Sr0.3MnO3 and Si, *Appl. Phys. Lett.*, Vol.88, pp. 141914

Zhou Q. L.; Jin K. J.; Lu H. B.; Han P.; Chen Z. H.; Zhao K.; Zhou Y. L. & Yang G. Z. (2005). Transport property in $SrTiO_3$ p-n junction, *Europhys. Lett.*, Vol.71, pp. 283

15

Integrating Detectors for Photoacoustic Imaging

Hubert Grün[1], Thomas Berer[1], Karoline Felbermayer[1], Peter Burgholzer[1], Markus Holotta[2], Gerhard Zangerl[3], Robert Nuster[4] and Günther Paltauf[4]
[1]*RECENDT Research Center for NonDestructive Testing GmbH, Linz*
[2]*Department of Radiology, Innsbruck Medical University, Innsbruck*
[3]*Computational Science Center, University of Vienna, Vienna*
[4]*Department of Physics, Karl Franzens University, Graz*
Austria

1. Introduction

The aim of medical imaging is an unerring diagnosis of diseases. Up to now several well established imaging modalities like e.g. computed tomography (CT), magnetic resonance tomography (MRT), single photon emission computed tomography (SPECT), positron emission tomography (PET) or ultrasound imaging (US) are known. Each imaging modality exhibits advantages and shortcomings. Computed tomography images the absorption of X-ray quanta and is suitable for imaging bone structures, brain imaging, angiography (imaging of blood vessels) but involves ionising X-rays. The contrast mechanism in MRT is the relaxation time of excited protons and therefore this method images soft tissue and vessels (using a contrast agent with the drawback that it can trigger an allergic reaction of the human body) best. But MRT is an expensive technology; the huge magnetic field is not easy to shield and disqualifies some patients with old models of cardiac pacemakers and other metallic implants. A new imaging modality called magnetic particle imaging (MPI) - which is just topic of research – uses also high magnetic fields for imaging. In this case the fields generated by magnetic nanoparticles are imaged. Nuclear techniques like SPECT or PET involve a radionuclide for imaging functional processes like the metabolic rate - which is for instance higher in cancerous tissue than in healthy organs. The radionuclide is attached to a specific molecule and distributed in the body during the blood flow. The radioactive decay measured by adequate detectors shows the spatial distribution of the incorporated radioisotope which is higher in cancerous tissue compared to healthy tissue. Although these are important imaging modalities for cancer screening the radioactive substances which are incorporated in the body are one drawback apart from the high costs per examination. Nuclear imaging techniques only image functional processes but no anatomical structures for which reason other complimentary techniques (e.g. CT) are necessary. Ultrasound imaging displays the backscattering of ultrasonic waves on a boundary layer between different tissues or organs. Although US is a cheap and safe imaging modality, its contrast mechanism is only related to changes in acoustic properties. Since cancer arises from neoplastic cells, the properties of the cancer and the surrounding tissue are almost identical in terms of acoustic contrast during the first stages of cancer development. So there is no distinct boundary and therefore imaging contrast is low. Only at a progressed, but often too late stage the tumor can be imaged by US. To conclude, these contrast mechanisms and imaging modalities show the requirement for an additional imaging technique, since cancer screening and other applications are not satisfactorily covered by the established techniques.

Breast cancer screening is an important topic as many surveys document. Up to now X-ray mammography is the gold standard in breast imaging although several false-positive (there seems to be a cancerous lesion while in fact there isn't any cancer) and false-negative (an existing cancer is not detected) findings happen (Xu & Wang, 2006). Investigation of new imaging modalities using alternative contrast mechanisms is motivated by the potential of earlier diagnosis and reduced invasiveness. Recently several optical methods have been further developed, e.g. diffuse optical imaging (DOI) with the advantage of using non-ionizing light, but with the drawback of low spatial resolution due to strong optical scattering in tissue. Generally, all pure optical imaging techniques have limited spatial resolution at depths of a few millimetres (Boas, 2011). Since new medical imaging techniques for applications like early breast cancer detection need high spatial resolution, further ideas have to be developed. A new promising imaging modality that addresses the problems/needs mentioned above is photoacoustic imaging.

Photoacoustic imaging shows promise to overcome the challenge of imaging with high contrast - as known from pure optical imaging methods - and high spatial resolution in deeper tissue layers. Although the principle of photoacoustics was investigated in 1880 by Alexander Graham Bell (Bell, 1880), the technology to use optically generated acoustic waves for imaging became possible only with the introduction of short laser pulses with duration in the range of picoseconds to nanoseconds. Photoacoustic imaging uses short pulses of electromagnetic energy, such as short laser pulses, to illuminate the sample/tissue. The imaging modality is therefore different from that described in early publications about photoacoustics (Rosencwaig & Busse, 1980; Wong et al., 1979), where thermal waves generated by modulated electromagnetic radiation at relatively low frequencies interact with features of a sample.

In photoacoustic tomography the excitation beam is expanded to illuminate the whole sample at once. Best imaging quality is achieved when the excitation is as homogeneous as possible, which means an illumination from all sides around the sample/tissue is best suited. Depending on the wavelength the electromagnetic wave penetrates the sample/tissue in the range from several micrometers to a few centimetres (Xu & Wang, 2006). Inside the sample/tissue the electromagnetic energy is absorbed according to the local electromagnetic absorption properties, causing local heating in the range of millikelvin. This quantity is called specific absorption rate (SAR) [W/kg]. Due to this absorption thermoelastic expansion occurs, which launches broadband ultrasonic waves - the photoacoustic signal. Thus photoacoustic imaging is based on the conversion of the absorbed light into ultrasound waves and permits to image optical contrast with ultrasonic resolution. Photoacoustic signals contain frequencies in the range from several kilohertz up to several megahertz depending on the size of the objects: the smaller the structures the higher the frequencies. Short laser pulses absorbed in a semitransparent sample/tissue generate an initial pressure distribution

$$p_0(\mathbf{r}) = \Gamma W(\mathbf{r}) \tag{1}$$

proportional to the volumetric density $W(\mathbf{r})$ of the locally absorbed electromagnetic energy and the Grüneisenparameter Γ. Efficient generation of photoacoustic signals is only possible if the acoustic confinement and the thermal confinement are fulfilled (Xu & Wang, 2006). The former requires that the heating time (i.e. the pulse duration) of a certain absorbing object should be shorter than the time an acoustic wave needs to escape from that object. Thermal confinement is easier met, because it requires the heating time to be shorter than the characteristic time of, the relatively slow, heat diffusion from the absorbing object.

The excited photoacoustic signals that carry information about the optical absorption of the tissue sample can be detected outside the object either by conventional piezoelectric transducers (Xu & Wang, 2006), novel detectors like optical point detectors (Zhang et al., 2009; Berer et al., 2010) or integrating detectors which are described in detail in section 2. Collecting signals from a sufficient number of different positions around the sample allows tomographic image reconstruction in two and three dimensions. In section 3 the mathematics of integrating detectors is explained and principle ideas of image reconstruction are presented.

Many biological and medical applications arise from the advantage of photoacoustic techniques to provide optical contrast measured at ultrasonic resolution. Because of the high optical contrast between haemoglobin in the blood and the surrounding tissue, it can be used to visualize single blood vessels as well as the blood distribution in whole organs. The formation of new blood vessels (angiogenesis) can be an indication for cancer formation - hence imaging of angiogenesis is an important diagnostic method and could be done by photoacoustics. But photoacoustic imaging also shows promise for a number of other possible applications. For example brain imaging through an intact skull of a rhesus monkey was done by Xu et al. (Xu & Wang IEEE, 2006). Another group showed functional and structural imaging of a rat's cortex using photoacoustic tomography (Wang et al., 2003). Especially in preclinical and clinical studies non-invasive in-vivo methods are of interest. Small animal imaging is integral to a variety of preclinical imaging applications with researchers monitoring changes in organs and tissues, for example in response to changed physiological or environmental influences. Researchers want to see how an organ is working inside the body. Ex-vivo analysis isolates the organs, while in-vivo imaging, of course, provides a much better picture of what is happening inside the body. Photoacoustic tomography (using a small animal imager) could therefore provide solutions for such research topics (Boas, 2011). In section 4 first biological and medical experiments done with different types of integrating line detectors are presented, including images of insects (ant, flea) and an ex-vivo mouse model with myocardial infarction.

2. Integrating line detectors

For medical applications fast imaging techniques are necessary - real time imaging is the ultimate goal. Therefore many detectors have to be arranged around the sample to collect data simultaneously. By using many finite size conventional piezo elements it is possible to obtain a closed detection surface around the sample with the drawback of missing space for a homogeneous illumination from all sides. One possible approach to overcome this drawback was presented by Buehler et al (Buehler et al. 2010) were a bundle of glass fibers between the detectors was used. The detector size influences the achievable spatial resolution: the smaller the detecting element the better is the resolution. Due to physical limitations it is not possible to miniaturize detectors arbitrary also the signal to noise ratio (SNR) gets worse for a smaller detector volume.

Because of these reasons new concepts for detectors were developed and investigated. Haltmeier et al. (Haltmeier et al., 2004) and Burgholzer et al. (Burgholzer et al, 2005) introduced integrating area and line detectors. Their size is at least $\sqrt{8} \cdot D$ larger than the imaged object, whereas D is the diameter of a circle enclosing the sample and tangentialy touching the line detector. This way one can overcome the problem of miniaturizing the detector, which can be more easily done with optical approaches compared to piezo detectors. Moreover, integrating line detectors can be based on laser beams or optical fibers (as described in section 2.1) and therefore these detectors are no barrier for the illumination. The excitation laser beam can cross the detector without any disturbance.

An integrating detector integrates the pressure along two axes (area detector) or at least along one axis (line detector). An integrating area detector can be fabricated using a piezoelectrical film, e.g. made of PVDF (polyvinylidene fluoride), which offers better broadband sensitivity compared to the conventional piezo elements. Burgholzer et al. (Burgholzer et al., 2005) presented first measurements using a PVDF film integrating area detector. Unfortunately there are some drawbacks using PVDF. The high capacity in combination with the resistance of the electric circuit operates like a high pass filter and important frequencies get lost. In general illumination through PVDF films is also not possible although Nierderhauser et al. (Niederhauser et al., 2005) presented a transparent detector made of indium tin oxide (ITO). For an exact reconstruction one has to move the large area detector tangentially on the surface of a sphere enclosing the sample to collect data from all sides. Moving the detector this way is mechanically challenging. Dividing an integrating area detector into many integrating line detectors simplifies the movement of the detector and thus the setup. If many integrating line detectors are arranged in a circle around the sample only one rotation axis is needed as depicted in figure 1.

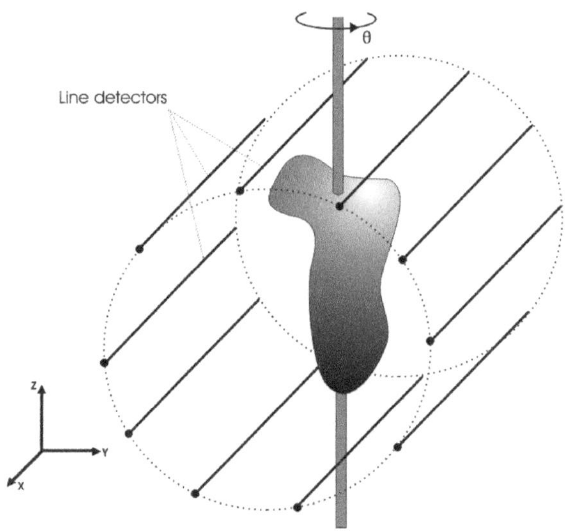

Fig. 1. Integrating line detectors arranged in a circle around the sample. In this mechanically simple setup only one rotation axis for three dimensional imaging is needed.

Such an integrating line detector integrates the pressure along one dimension - i.e. along a line. Therefore it measures the integral of a cylinder with the radius $c \cdot t$ where c is the speed of sound and t is the time of flight. As described mathematically in section 3 this integrating behaviour results in a projection along the direction of the line detector. From a set of data acquired with line detectors arranged around the sample, e.g. in a circle (see figure 1) a projection image is obtained, which is analog to images taken with standard X-ray mammography. For three dimensional imaging one has to rotate the sample (or the array of line detectors) to collect projection images of different angles. A subsequent inverse Radon transform reconstructs three dimensional images as explained in section 3. Some approaches of the implementation of optically integrating line detectors are presented in the following section.

2.1 Implementation of integrating line detectors

Since 2004 when Haltmeier et al. (Haltmeier et al., 2004) introduced the idea of integrating detectors several approaches of integrating line detectors have been investigated. A stripe of PVDF film was used as integrating line detector with some of the drawbacks mentioned before for piezoelectric sensors. As an alternative the use of optical interferometers as line detector has been proposed. In this approach a laser beam (or an optical fiber which guides a laser beam) being part of an interferometer is the sensitive element. The interferometer measures variations of the refractive index causing a variation of the change of the optical path length induced by the acoustic pressure (elasto-optic effect) which results in a change of the light intensity at the interferometer output. This section will give an overview about the implementation of different approaches of interferometric line detectors. The sensitive part of the interferometer - either the laser beam itself or the light guiding optical fiber - is immersed in a tank filled with water which operates as acoustic coupling medium for the wave propagation from the sample surface to the detector.

2.1.1 Free-beam integrating line detectors

Free-beam Mach-Zehnder interferometer

The simplest type of an optically integrating line detector is the Mach-Zehnder interferometer (MZI), which is a two-beam interferometer. Paltauf et al. implemented a free-beam MZI as integrating line detector and demonstrated first photoacoustic measurements (Paltauf et al., 2006). The schematic setup is depicted in figure 2. A continuous wave laser (633nm, HeNe-Laser) is split into two beams - one acting as reference beam and the other one as signal beam. Next to the signal beam the sample/tissue is located. Afterwards both beams are recombined and they interfere at the detecting photodiode with subsequent amplification. For stabilisation of the operating point (i.e. at quadrature condition) a piezo-electrically driven mirror in one path is used. Spatial resolution of an imaging setup is directly influenced by the size of the detector - in the case of a free-beam integrating line detector the diameter of the laser beam. A smaller diameter corresponds to better spatial resolution. By using lenses the laser beam is focused next to the sample/tissue to enhance spatial resolution. After the proof of principle (Paltauf et al., 2006) this kind of detector was further improved and the focus - and therefore the spatial resolution - was further enhanced. Paltauf et al. (Paltauf et al., 2009) characterized this integrating Mach-Zehnder line detector in detail. This characterization (shape of the signal from a point source, length of the line detector, directivity, and resolution) is also valid for all acoustically integrating line detectors.

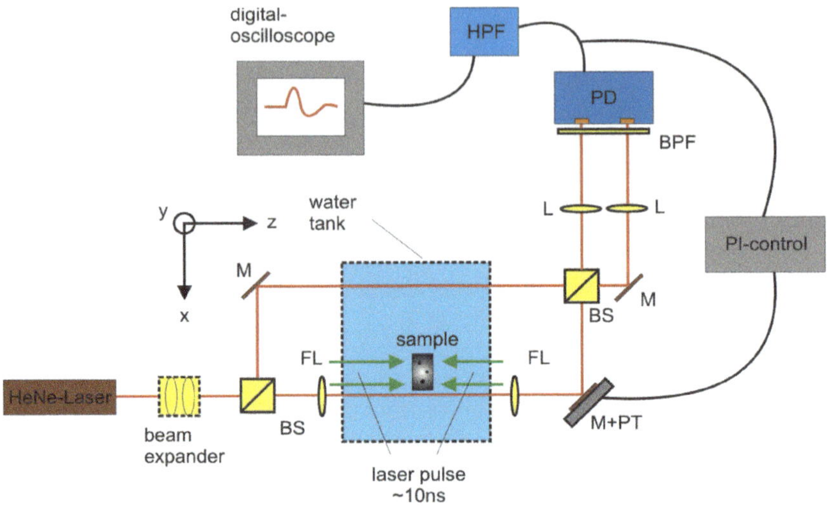

Fig. 2. Experimental setup for photoacoustic tomography using a Mach-Zehnder interferometer as integrating line detector. BS: beam splitter, M: mirror, L: lens, PD: photodetector, PT: piezo transducer, BPF: band pass filter, HPF: high pass filter, FL: focusing lens

Free-beam Fabry-Perot interferometer

Another approach for enhanced signal sensitivity is the use of a Fabry-Perot interferometer, which is a multi beam interferometer. A continuous laser beam is reflected several times between two semitransparent mirrors as depicted in figure 3. The signal enhancing mechanism of this device can be understood in a way that the incoming pressure wave does not only act once on the laser beam but several times. Depending on the reflectivity of the mirrors the light circulation in the cavity is different which influences the sensitivity. Higher reflectivity results in a better sensitivity. Burgholzer et al. (Burgholzer et al., 2006) built a setup using a free-beam Fabry-Perot line detector as depicted in figure 3. A continuous

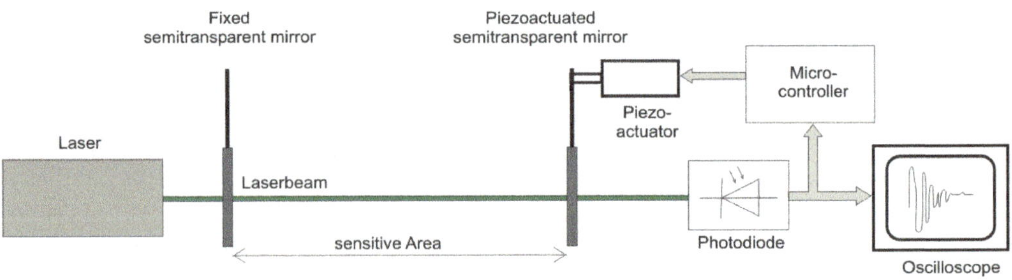

Fig. 3. Schematic of the used free-beam Fabry-Perot interferometer. A laser beam is reflected multiple times between two semitransparent mirrors and acts as integrating line detector.

wave laser (532nm) was coupled into the Fabry-Perot cavity using two semitransparent mirrors with a reflectivity of approximately 95%. The mirrors were realized by metalizing two microscope slides made of glass with thin layers of silver. On the opposite side of the Fabry-Perot cavity the modulated laser beam was detected by a photo diode with subsequent electrical amplification. For operation point stabilisation a piezo-electric actuator moved one of the semitransparent mirrors. In this case the laser beam was not focused and the diameter of the laser beam was constant along the line detector.

2.1.2 Fiber-based integrating line detectors

Both types of free-beam interferometers can also be realized in optical fibers. In this case the laser is coupled into and guided through an optical fiber. In general there are two types of optical fibers available that are applicable as sensor: glass optical fibers (GOF) and polymer optical fibers (POF). Glass optical fibers are standard components in telecommunication and are therefore cheap and reliable. Many components beside simple patch cords, e.g. beam splitters, attenuators, couplers and many more are available and easy to acquire. Recently also polymer optical fibers are used for short distance telecommunication. Due to high damping at the standard telecommunication wavelengths of 1310nm and 1550nm POFs are only used for the last few meters in the telecommunication chain (e.g. a network within a building). One advantage for standard applications is the low costs of POFs. For photoacoustic imaging polymer optical fibers are superior for realizing a detector compared to glass optical fibers as is described in detail in section 5. Depending on how many modes (different possible paths of light) can propagate through an optical fiber one can distinguish between single mode fibers or multimode fibers. The smaller a core diameter of an optical fiber the fewer modes can propagate through the wave guide. As described above the fiber diameter also influences the achievable spatial resolution - thus single mode fibers (with the smallest outer diameter) are more convenient than multi mode fibers. Varying the types of interferometer (Mach-Zehnder interferometer or Fabry-Perot interferometer), the material of the optical fibers (glass or polymer) and the geometry of the core (single mode or multi mode fiber) result in various possible setups for fiber-based line detectors as described in the next section.

Fiber-based Mach-Zehnder interferometer

First, single mode glass optical fibers were used for a Mach-Zehnder setup. A laser with adjustable wavelength and a center wavelength of 1550nm was coupled into a single mode glass fiber and split into two fibers. Analog to the free-beam Mach-Zehnder interferometer, one is the reference path and the other one is the signal fiber. The latter is placed in a tank filled with water next to the sample/tissue. At the sensitive region of the interferometer all outer layers of the fiber were removed, leaving only the core, cladding (both made of glass) and the inner surrounding polymer coating. The protection layers were removed to avoid damping of the pressure in the material. Both fibers - the reference path and the signal path - were combined afterwards and the interference of the laser beams was detected by a photodiode and a subsequent electrical amplifier (see figure 4). For operation point stabilization no mirrors had to be moved. Instead the wavelength of the laser was adjusted.

Alternatively to the single mode GOF a perfluorinated graded index polymer optical fiber was used as signal path (Grün et al., 2009). The rest of the setup was identical with the single mode GOF setup described above. By using a more sensitive polymer fiber, signal detection could be improved. At the used wavelength of 1550nm standard polymer optical fibers exhibit huge damping. Therefore a perfluorinated fiber - where the hydrogen molecules responsible for the damping are replaced by fluor molecules - was used. Unfortunately such perfluorinated fibers were only available as graded index fibers (a special case of multi mode fibers) with a core diameter of 50 microns.

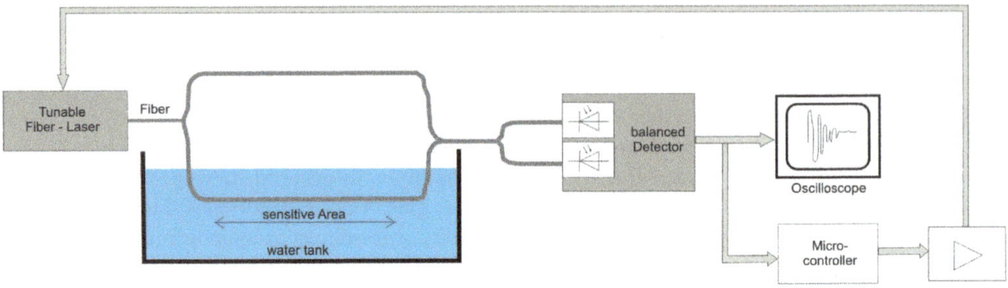

Fig. 4. Glass optical fibers (GOF) were used to realize a fiber-based Mach-Zehnder interferometer as optical line detector for photoacoustic imaging.

Fiber-based Fabry-Perot interferometer

As described earlier a Fabry-Perot interferometer is, because of the multiple reflections, more sensitive than a MZI. Therefore also fiber-based Fabry-Perot interferometers were realized. In glass optical fibers semitransparent mirrors are commonly realized by fiber Bragg gratings (FBG). A Bragg grating is a region in the fiber with a periodically changing refractive index (Dutton, 1989). Writing FBGs into GOFs is a standard procedure and these FBGs are commercial available. However, no stable Bragg gratings, which do not degrade in a seasonable amount of time, are available for polymer fibers. Alternatively the end faces of the fibers can be metalized by thin layers of metal (e.g. silver). This, however, comes at a cost of light absorption in the metallic layer. Fiber Bragg gratings are non-absorbing structures and therefore in general better than the metalized fiber end faces. A glass optical Fabry-Perot detector with Bragg gratings (Burgholzer et al., 2006) is depicted in figure 5. The laser beam is coupled into a single mode GOF. The detecting fiber has two FBGs with a reflectivity of approx. 99%. The sensitive area, i.e. the area between the gratings, is uncoated - as described above for the fiber-based MZI detector. The photodiode and the amplifier are the same as used for the MZI setup. Operation point stabilization is again done by adjusting the wavelength of the laser. A similar setup was also realized using a perfluorinated graded index POF - the same used for the MZI (Grün et al, 2010 SPIE). In this case, because of lacking the possibility of stable Bragg gratings, the end-faces of the fiber were metalized.

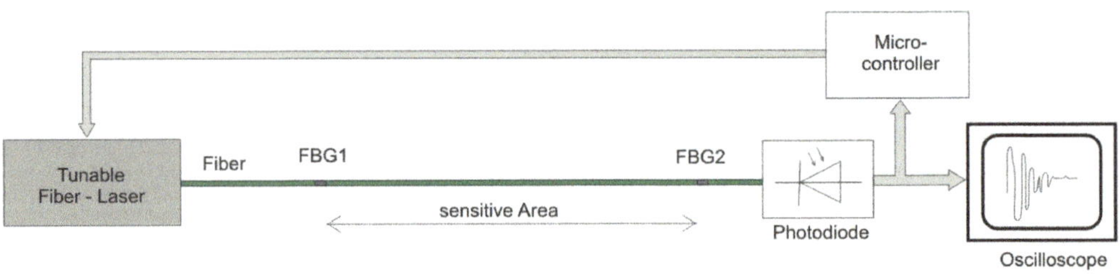

Fig. 5. Glass fiber-based Fabry-Perot line detector. Two fiber Bragg gratings (acting like semitransparent mirrors) form the cavity where the pressure transients are measured.

After these implementations of optically integrating line detectors a brief introduction in the mathematics of integrating detectors is given in the next section.

3. Mathematics of integrating detectors

Photoacoustic imaging aims for reconstructing a three dimensional image of electromagnetic absorption properties from pressure data outside the object. The conversion of pressure measurements into a three dimensional image requires a model that describes acoustic wave propagation. Mathematically, the propagation of sound waves in tissue is modelled as a Cauchy Problem for the wave equation with constant sound speed. Denoting the thermoacoustic pressure by p it satisfies the following Cauchy problem

$$\partial_t^2 p(\mathbf{x},t) = c^2 \Delta p(\mathbf{x},t) \qquad (\mathbf{x},t) \in R^3 \times (0,\infty) \tag{2}$$

$$p(\mathbf{x},0) = f(\mathbf{x}) \qquad \mathbf{x} \in R^3$$

$$\partial_t p(\mathbf{x},0) = 0 \qquad \mathbf{x} \in R^3$$

for the three dimensional wave equation, where $\mathbf{x}$ denotes a point in a three dimensional space R^3 and $t \geq 0$ denotes the time variable. Further, we refer to $f(\mathbf{x}) \in C_0^\infty(\Omega)$ as the initial density or initial pressure of the considered Cauchy problem.

Commonly, in photoacoustic imaging it is assumed that acoustic pressure is known pointwise on a surface nearby the object and several reconstruction formulas have been derived that assume pointwise pressure data. However, in practice, acoustic pressure is measured by piezoelectric transducers, which can only provide approximate pointwise (pointlike) measurements. For this reason, the formulas above yield images with a spatial resolution that is essentially limited by the size of the piezoelectric transducers (Xu et al., 2003). The size of a piezoelectric detector is typically a square with a side length of about 400μm.

A first approach that tried to overcome the problem of finite aperture size of piezoelectric transducers has been given by M. Xu and L. Wang in (Xu et al., 2003). Therein the spatial blurring caused by detectors is modelled as a convolution with a point spread function of ultrasonic transducers and is used to improve reconstructions by deconvolution.

A practical approach, that tries to overcome the limitations due to finite aperture size of detectors, was proposed by Burgholzer et al. (Burgholzer et al., 2005), where linear and planar integrating, or shortly, linear and planar detectors were introduced. Planar, respectively, linear detectors measure a quantity that is proportional to the integral of acoustic pressure over a planar area or a line segment. Planar and linear detectors reduce the problems, caused by finite aperture size of piezoelectric transducers, because they approximate over a line or a plane very accurately. In the following we will describe the mathematical reconstruction process in the case of integrating measurement data.

3.1 Planar detectors

Planar detectors have been realized in experiments in (Burgholzer et al., 2005). Therein a planar piezoelectric film is used to approximate pressure measurements over a plane E. For simplicity we will assume for the moment that E is the yz plane. Then the measurements collected by the plane E are given by

$$G_{planar}(t) = \int_E p(0,y,z,t)\,dydz \tag{3}$$

where p satisfies the wave equation. Haltmeier et al. (Haltmeier et al., 2004) derived a reconstruction algorithm for the measurement setup that is described and depicted in figure 6.

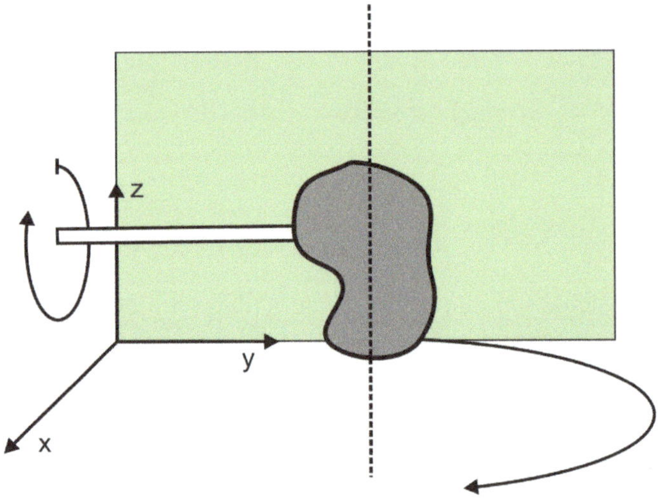

Fig. 6. The measurement procedure for planar detectors: After each measurement the object is rotated by a small angle increment and acoustic pressure is measured again. When the object has fully turned the planar detector is rotated around the dashed vertical axis and the whole measurement procedure is repeated. Once the planar detector has turned by 180 degree the process ends.

The reconstruction algorithm for planar detectors is based on a dimension reduction, that is based on the fact that the integrated pressure

$$p_{planar}(x,t) = \int_E p(x,y,z)\,dydz \tag{4}$$

satisfies the one dimensional wave equation

$$\partial_t^2 p_{planar}(x,t) - c^2 \partial_x^2 p_{planar}(x,t) = 0 \tag{5}$$

$$p_{planar}(x,0) = F(x) = \int_E f(x,y,z)\,dydz$$

$$\partial_t p_{planar}(x,0) = 0$$

By D'Alembert's formula the unique solution of the one dimensional wave equation is given by

$$p_{planar}(x,t) = \frac{1}{2}\left(F(x+ct) + F(x-ct)\right) \tag{6}$$

Thus, at the time t the planar detector E at $x = 0$ measures

$$G_{planar}(t) = p_{planar}(0,t) = \frac{1}{2}F(ct) \tag{7}$$

where we have used that $f(-ct)$ vanishes since f is only supported on one side of the detector. From the latter equation we see that a planar detector E measures integrals of the original f over planes parallel to E. Rotating the object at the detection planes as described in figure 6 yields the three dimensional Radon transform of the function f. Application of its inverse transform leads to a reconstruction of f.

3.2 Linear detectors

In the following we describe measurements that are collected by linear detectors as illustrated in figure 1. For convenience we will assume that the linear detectors are perpendicular to the yz plane E. The measurements that are collected by the circular arranged array of lines in figure 1 are given by

$$\mathrm{p}_{\mathrm{linear}}(\mathrm{y},\mathrm{z},\mathrm{t}) = \int_{\mathrm{L}} \mathrm{p}(\mathrm{x},\mathrm{y},\mathrm{z},\mathrm{t})\,\mathrm{dx} \tag{8}$$

where z,y lie on the circle $S_R \subset E$ (corresponds to the circle in figure 1). Equation (8) satisfies the two dimensional Cauchy problem.

$$\partial_t^2 p_{linear}(\mathbf{x},t) = c^2 \Delta p_{linear}(y,z,t) \qquad (y,z,t) \in R^2 \times (0,\infty) \tag{9}$$

$$\mathrm{p}_{\mathrm{linear}}(\mathrm{y},\mathrm{z},0) = \mathrm{F}(\mathrm{y},\mathrm{z}) = \int_{\mathrm{L}} \mathrm{f}(\mathrm{x},\mathrm{y},\mathrm{z})\,\mathrm{dx} \qquad \mathbf{x} \in \mathrm{R}^3$$

$$\partial_t p_{linear}(y,z,0) = 0 \qquad \mathbf{x} \in R^3$$

Unfortunately, the relation between **F** and the measurements G_{linear} is more complicated as for the one dimensional wave equation. However, it is possible to obtain an exact relationship (Haltmeier et al., 2007) between G_{linear} and **F** by means of integral transforms by

$$\mathrm{H}_\lambda\{\mathrm{F}_\lambda\}(\omega) = \frac{\left(2/\pi\right)^{1/2} \mathrm{C}\{\mathrm{G}_{\mathrm{linear},\lambda}\}(\omega)}{\omega \mathrm{J}_{|\lambda|}(\omega \mathrm{R})} \tag{10}$$

whenever $\omega J_{|\lambda|}(\omega R) \neq 0$.

In this equation $H_{|\lambda|}$ denotes the Hankel transform of order λ, C the cosine transform, ω the frequency variable of time and $F_\lambda, G_{linear,\lambda}$ are the Fourier coefficients of F and G_{linear} from their Fourier series expansion with respect to the angular variable.

From the latter equation F_λ can be computed by applying the inverse Hankel Transform. Accordingly F can be computed from its Fourier series expansion if all the F_λ are known. Note that F consists of integrals over lines that are perpendicular to the plane E. Furthermore, the computations above would also hold for an arbitrary plane E. Thus one can obtain a family of projections F_α of the original function f by rotating E around the z axis. For a fixed $z = z_0$ this family is nothing but the two dimensional Radon transform of the restriction of f to the plane $z = z_0$. Applying the inverse Radon transform in each plane yields f.

4. Biological and medical experiments with integrating line detectors

Different types of line detectors and biological samples have been used to show the applicability of these detectors in the field of biomedical research. Some examples are given in the following section.

Free-beam Mach-Zehnder interferometer for biological applications

The free-beam Mach-Zehnder line detector was used, among others, for imaging of a flea. The small insect was chosen in order to test the imaging resolution and accuracy. To achievean adequate spatial resolution the sensing laser beam - which is the signal path of the interferometer - was focused to a diameter of the beam waist of 40 μm. For fixation the flea was embedded in gelatine which shows good impedance matching with the surrounding water of the setup. For one projection image data were acquired at 90 detector positions arranged on a semi circle with a radius of 5 mm. 400 of such projection images

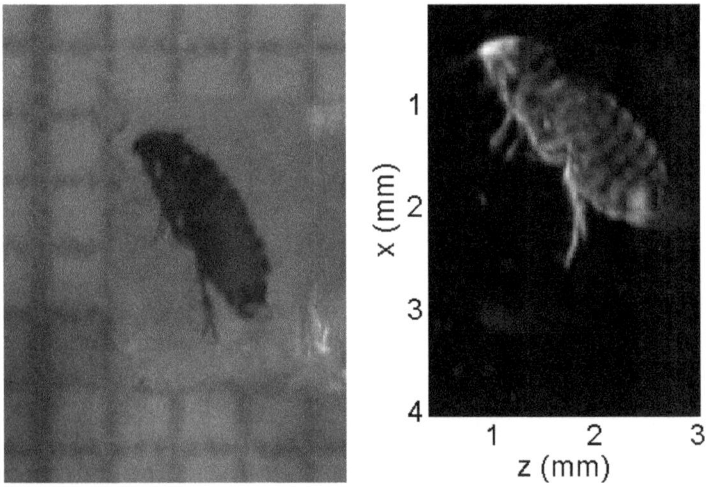

Fig. 7. Left: Photograph of a flea embedded in gelatine. Photograph is taken in front of a sheet of millimeter paper. The right picture shows a maximum amplitude projection (MAP) of the imaged flea in the xz-plane and was reconstructed using the universal back projection algorithm.

distributed over an angle of 360° were recorded. Laser pulses with a wavelength of 502 nm (Nd:YAG in combination with an OPO), and 8 ns pulse duration were used to excite photoacoustic signals in the flea. The radiant exposure was chosen with 10mJ/cm^2 which is below the allowed maximum value for biological tissue (20mJ/cm^2).
Looking at the reconstructed image (figure 7) of the flea one can clearly see the shape of the flea and the outer structures like the trotters. Also the striped structures at the back of the flea can be clearly identified.

Free-beam Mach-Zehnder interferometer for preclinical applications

Another study using the same type of setup was done a few months later. Nuster et al. (Nuster et al., 2010) and Holotta et al. (Holotta et al., 2011) presented first preclinical results on ex-vivo mouse hearts. They achieved a resolution in the range of 100 micrometers during imaging ex-vivo mouse hearts. Holotta et al. (Holotta et al., 2011) visualized ischemic areas within mouse hearts and compared it to already established imaging modalities like CT and MRT. In order to induce myocardial infarction, the left anterior descending artery has been ligated in-vivo in wild type mice. After varying survival periods the mice were sacrificed, the hearts were excised and immediately transferred into a 4% formaldehyde solution for conservation. Afterwards the hearts were embedded into 12% Agar and provided with a small stick to enable mounting on a translation stage.
Photoacoustic signals were induced by 5ns laser pulses (10Hz repetition frequency) at a wavelength of 750nm where best results were obtained. The radiant exposure was maintained below the maximum value for biological applications (20mJ/cm^2).
For data acquisition the sample was moved around the measurement beam in a 270° arc and turned on its own axis continuously. A total of 283 projection images were recorded and used for three dimensional image reconstruction as shown in figure 8.

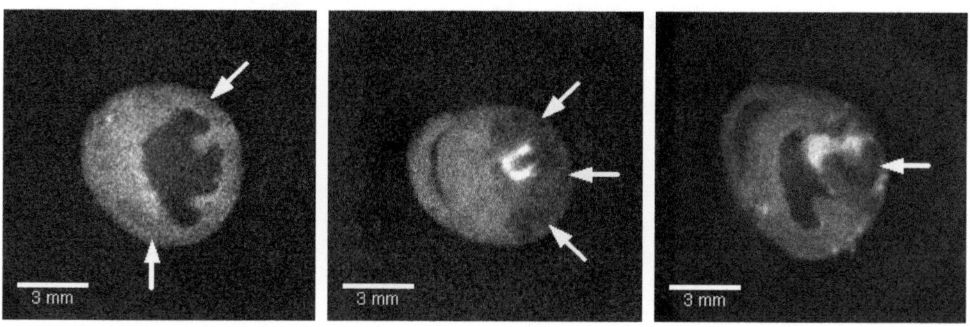

Fig. 8. Slices of mouse hearts - ischemic regions are marked by arrows. Left: mouse survived for 2 days. Middle: mouse survived for 3 days. Right: mouse survived for 4 days.

Due to the degradation of haemoglobin, the appearance of interstitial edema and various reparative processes (Friedrich, 2008; Kawasuji et al., 2000) the ischemic regions in the photoacoustic images (marked by arrows in figure 8) are characterized by lower absorption compared to the surrounding healthy muscle tissue in the myocardial wall. Agar - which is surrounding the ex vivo mouse heart - has a high water content and is, depending on its concentration, almost optically transparent. Therefore, it is hardly visible in the photoacoustic images shown in figure 8.

Fiber-based Fabry-Perot interferometer for biological applications

Grün et al. (Grün et al., 2010) presented measurements of an ant utilizing a Fabry-Perot interferometer made of a single mode glass fiber as depicted in figure 5. The reflectivity of the FBGs was 81% and the distance between the FBGs (i.e. the sensitive area) was 11,5cm. Data for one projection image were acquired at 200 line detector positions arranged in a circle (radius: 10mm) around the object. For three dimensional image reconstruction the sample was rotated and 25 projection images with an angle increment of 7.2 degrees were recorded. The ant was mounted on a nylon rod to allow rotation of the sample. In the lab where the tomography was carried out, unfortunately only an excitation wavelength of 532nm was available. At this wavelength most of the absorption takes place in the outer layers of the ant, i.e. no signals from inside the ant were excited. Thus, only the surface of the ant was reconstructed. For illumination laser pulses from a frequency-doubled Nd:YAG-Laser (532nm) with a pulse duration of 6ns and a repetition rate of 20 Hz were used. The pulse energy was adjusted to keep the radiant exposure below the allowed maximum value for biological tissue (20mJ/cm²).

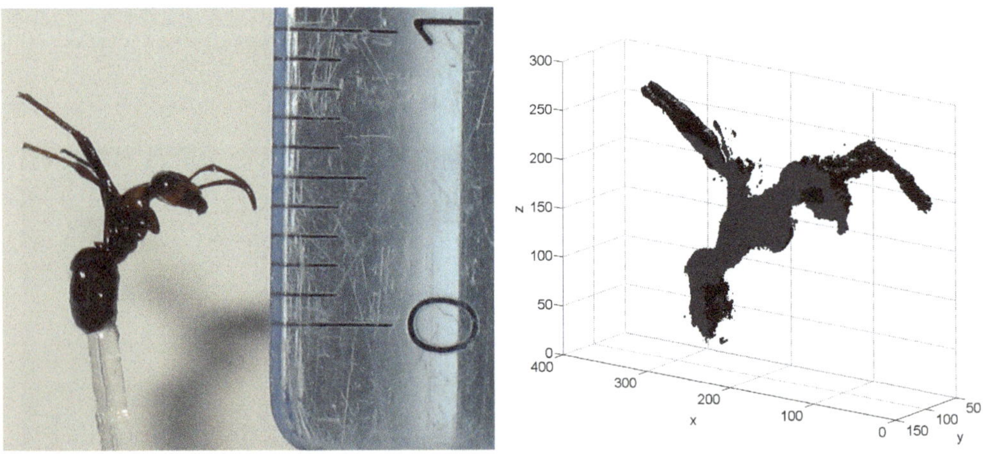

Fig. 9. On the left, a photograph of the ant - mounted on a nylon rod - can be seen. On the right a three dimensional reconstruction (using time reversal reconstruction) of the ant is displayed. 25 projection images, each consisting of 200 detector positions, were acquired for the whole tomographic image (Grün et al., 2010).

In figure 9 one can see the reconstruction of the photoacoustically imaged ant. The reconstruction image is sliced in the middle of the ant. For image reconstruction a time reversal algorithm was used. One can distinguish the head (caput) with one feeler, the body (mesosoma) with two superposed trotters and the metasoma.

5. Discussion

In this section some advantages and disadvantages of the different approaches of optical line detectors are discussed as well as general considerations about optical line detectors.

Optical line detectors are more convenient compared with other detector types in general. These detectors are transparent - free-beam interferometers as well as fiber-based detectors - and therefore are no optical barrier for the excitation laser pulses. The sample can be illuminated across the transparent detectors, which would not be able using conventional piezo elements.

Using free-beam interferometers as well as fiber-based detectors one overcomes the directivity effects of piezo detector elements and finite size plane detectors. Due to the omni directional sensitivity of optical line detectors they can be easily scanned around an object without loss of sensitivity and signal quality.

Spatial resolution is an important issue in medical imaging. In photoacoustic imaging the bandwidth of the detector has influence on the spatial resolution. For smaller objects the emitted ultrasonic frequencies are higher. At the same time extended objects emit rather low frequencies. Thus it is necessary to collect the data from several kilohertz up to frequencies in the range of tens of megahertz to reconstruct all structures in an object. Interferometers as detector show a high bandwidth and are therefore appropriate for high resolution photoacoustic imaging.

Another parameter that influences the spatial resolution is the diameter of the detecting beam in an optical line detector. Smaller diameters correspond to a better spatial resolution. With a free-beam Mach-Zehnder interferometer one can achieve good spatial resolution by focusing the beam to a small diameter. But there are physical conditions which cannot be overcome. Focusing the beam to a tiny diameter leads to a short focal depth. Objects outside of this focal range can only be imaged with worse spatial resolution - the high resolution is only achievable for small objects. This issue was described by Paltauf et al. (Paltauf et al., 2009). Fiber-based line detectors, by contrast, provide constant diameter along the whole fiber - and therefore constant spatial resolution along the whole detector. Thus fiber-based line detectors are applicable for imaging of bigger objects.

Comparing the fiber Mach-Zehnder and the fiber Fabry-Perot interferometer the latter has several advantages in applications. Within a MZI all optical paths are sensitive. This means the signal path but also the reference path is sensitive to ultrasonic vibrations. Thus, disturbances from the environment and movements of the detector lead to additional noise. Only disturbances acting on both paths of the interferometer are compensated by the interferometer itself. By contrast, the FPI is only sensitive between the two mirrors, i.e. between the fiber Bragg gratings for the fiber-based approach. Therefore, the movement of the detector around the object and environmental influences are not that critical. Another advantage of the FPI is the higher sensitivity due to its multi beam interference. This leads to a steeper transfer curve, which is the relation between intensity change at the interferometer output and the optical phase variation caused by the acoustic pressure.

Operation point stabilization in free-beam interferometers can be easily done by a mirror mounted on a piezo actuator, which slightly changes the length of the light path. This is a simple and well known approach which was realized in the free-beam MZI. This operation point stabilization also was realized in the free-beam FPI but resulted in severe troubles. A piezo actuator outside the water tank was connected to one of the microscope slides (which was metalized and therefore acted like a semitransparent mirror). Due to the big surface of the microscope slide it showed resistance while moving in the water. In combination with the long distance (as can be seen in figure 3) between the end of the microscope slide (with the metalized part) and the piezo actuator this construction began to resonate and stable operation point stabilization was achievable only for low frequencies. Therefore, the MZI detector is more applicable for the free-beam approach.

For fiber-based interferometers the operation point stabilization can be done by two different approaches. One possibility is to tune the wavelength of the laser for operation point stabilisation. A tuneable laser - where the wavelength is changed in the range of a view of tens of picometers - was used for the interferometers presented in this work. This approach works well but is also an expensive method, beside of the fact that such suitable tuneable lasers are only available at the telecommunication standard wavelength of 1550 nm. For some reasons shorter wavelengths for detection are be preferable. The damping in the used polymer fibers is much less for shorter wavelengths in the visible. A second reason is the higher phase shift of the detection light at the same pressure acting on the interferometer. This results in higher signal amplitudes and better SNR. Another approach for operation point stabilization, more equivalent to the free-beam operation point stabilization, is changing the optical path length. This can be done by electro-optical modulators which change the refractive index of the fiber or by piezo elements which change the length of the fiber. The latter can be realized by wrapping a fiber around an annular piezo element which changes the diameter or by stretching the fiber with a linear piezo stack changing its length. By using such an operating point stabilization one is independent of the detection wavelength. Furthermore, this is a cheap approach.

Another advantage of fiber-based detectors is that they offer the opportunity of arbitrarily shaped sensors. A line detector can be e.g. bent to an annular detector, as described for example by Berer et al. (Berer et al., 2009) and Grün et al. (Grün et al., 2011). This way new imaging opportunities arise. Only slight modifications of the detector are necessary as the detection concept is the same. Having biomedical applications in mind, optical line detectors have another advantage. Fiber-based detectors, especially those made of polymer, have excellent compatibility with organic material, thus giving them great potential for biomedical applications (Peters, 2011). This is important although the fiber-based detector is not in direct contact with the tissue when used as line detector for tomography. Free-beam line detectors of course, as they exhibit no material, are of course also compatible for organic materials. Therefore, they are also a good solution as detector for biomedical imaging.

Nuster et al. (Nuster et al, 2009) presented a comparison of the different types of optically integrating line detectors. At this stage the free-beam MZI was the most sensitive line detector, but the potential of more sensitive fiber-based line detectors could be clearly seen. In this study also the higher sensitivity of a FPI compared to a MZI was shown. For this reason the fiber-based detectors were further developed and many different approaches were implemented. In general, polymer optical fibers are more sensitive than glass optical fibers. This has two reasons. First, due to the lower Young's modulus of the polymer a pressure acting on a polymer fiber results in greater deformation compared to a glass fiber. As the strain-optic constants are similar the change in refractive index and thus the optical path length is greater and the same pressure transient generates a bigger signal in a polymer fiber than in a glass fiber. Second, the polymer is better impedance matched to the surrounding water. Thus most of the ultrasonic energy is transmitted into the fiber, while for glass optical fibers a large part is reflected. Using a single mode fiber, as described in section 2.1, results in the highest achievable spatial resolution when using fiber-based detectors. Summarizing these facts a single mode polymer fiber-based FPI with high finesse would be the most sensitive fiber-based line detector for photoacoustic imaging. Unfortunately at the time when the measurements, presented in section 4, were carried out only multimode polymer fibers were available. A first drawback of using this multimode fiber - in this case a graded index multimode fiber with a core diameter of 50 microns - was the worse spatial resolution compared with the single mode glass fiber-based detector. Another drawback was the multimode behaviour itself. When coupling light into the multimode fiber - independent of the interferometer type - always several modes are excited. The operation point, however, is stabilized on one of these excited modes. Whenever the fiber is moved - or whenever it relaxes between the points where it is attached to the detector holder - a different mode pattern is formed at the fibers end and the coupling into the successive single mode fiber changes. This results in unstable conditions for operation point stabilization. However, the proof of principle was shown and two dimensional image was demonstrated (Felbermayer et al., 2011). For time consuming three dimensional imaging, however, this method is up to now probably too unstable and cannot compete with the single mode glass fiber-based FPI. Another drawback of the used POF-FPI setup is related to the metalized fiber end faces instead of FBGs. Up to now it is not possible to write stable FBGs into polymer. For some reasons the FBGs written in POFs vanish or lose reflectivity after a period of days or weeks (Harbach, 2008). A FBG is a non absorbing structure in the fiber, the thin layers of silver, however, are absorbing structures and therefore reduce signal intensity. A polymer optical fiber-based Fabry-Perot line detector with fiber Bragg gratings acting as semitransparent mirrors seems be the goal for a future implementation.

The following points are not particularly specific to the topic of integrating detectors but are generally important for developing photoacoustic tomography devices. They have to be also discussed in relation to the detectors described in this chapter. As described earlier the smaller an object is the higher the emitted frequencies are. Therefore, frequencies ranging from several kilohertz up to some tens of MHz have to be detected. Unfortunately, in imaging of tissue in biological and medical application the effect of frequency depending attenuation is rather high. Higher frequencies are more damped than lower frequencies. This results in the loss of information about small structures in a certain depth. Thus, it is necessary to choose between a high bandwidth, high-frequency detector for microscopic imaging with high spatial resolution in small objects (or within a superficial region of few millimetres in a larger object) and a low frequency detector for tomographic imaging with lower spatial resolution in a bigger volume. Of course compensation of frequency depending attenuation improves the reconstructed image quality. But frequencies damped below the noise level of the detector cannot be retrieved (Bauer-Marschallinger et al., 2009; Burgholzer et al., 2009; Burgholzer et al., 2010).

For image reconstruction often constant speed of sound in the whole sample is assumed. Obviously this is just an approximation, as in biological samples the speed of sound is spatially varying. For example, between bones and surrounding tissue the difference in sound speed is considerable – but also in different types of tissue the sound velocity is slightly changing. Ignoring the spatial varying sound velocity can result in blurred images (Muratikov, 2004; Grün et al., 2008). However, compensation of the spatially varying sound velocity is not trivial, as one has to know the exact velocity distribution. This can be obtained by another imaging modality. Another approach is compensation by iterative image reconstruction algorithms.

For biological and medical imaging a short data acquisition time is sometimes crucial. Up to now only one optical line detector was moved around the sample – or in the case of the free-beam line detector the sample was moved around the laser beam. To speed up imaging parallel data acquisition is necessary. Instead of one line detector moving around the object many detectors could be arranged in an array. This way the data for one projection image can ultimately be obtained by one laser pulse. Only the detector array, or alternatively the object, has to be rotated to acquire many projection images for three dimensional image reconstruction. Parallelisation should be easier for fiber-based line detectors than for free-beam interferometers, although concepts for the latter one are available, such as the use of a CCD camera (Nuster et al., 2010).

6. Conclusion and outlook

Photoacoutic tomography is a new imaging method which is attractive for medicine and biology because it is capable to provide a three dimensional image of electromagnetic absorption properties of biological tissue – which is dependent of the used wavelength - without ionizing radiation. These properties are of considerable interest for medical diagnostics as they are related to the molecular composition of tissue and reveal its pathological condition. Furthermore photoacoustic imaging is a cheap imaging modality compared to well known techniques like CT or MRT. For this rather new imaging modality new detectors – like the presented integrating detectors - are developed. In conclusion one can state that the optical approach of integrating line detectors works quite well. Not only the proof of principle but also the applicability for biological and (pre-) clinical imaging using different types of optical integrating line detectors has been shown. Several advantages like the frequency range, the spatial resolution, or the insensitivity against electromagnetic disturbances from the environment justifies this development and further enhancements of sensitivity and spatial resolution.

Future work will focus on parallelization of many line detectors – independent of the kind of optical line detector that will be further developed - to speed up this technology and collect all data for one projection image by only one single laser pulse. For fiber optical detection another enhancement would be the use of single mode polymer fibers which are much more sensitive than glass optical fibers but do not exhibit stability problems as the multi mode optical fibers. Therefore improved signal-to-noise ratio is expected which further speeds up imaging as less signal averaging is required. Because of their smaller dimensions single mode fibers will further increase the spatial resolution compared to the multimode polymer fibers in use now. Inscribing fiber Bragg gratings into polymer optical fibers for Fabry-Perot detectors would reduce losses, which appear when using metalized fiber end faces. Last but not least excitation pulse lasers with a higher repetition rate in the range of kilohertz also would shorten imaging time and therefore push photoacoustic imaging towards real-time clinical imaging.

7. Acknowledgment

This work has been supported by the Austrian Science Fund (FWF), Project Nos. S10502-N20, S10503-N20, S10505-N20, S10506-N20, S10508-N20, L418-N20 and TRP102-N20; by the European Regional Development Fund (EFRE) in the framework of the EU-program Regio 13; and by the federal state of Upper Austria.

8. References

Bauer-Marschallinger, J.; Berer, T.; Grün, H.; Camacho-Gonzales, F.; Burgholzer, P. (2009). Experimental Determination of Frequency Dependent Acoustic Attenuation for Photoacoustic Imaging. *Proceedings of SPIE 2009 European Conferences on Biomedical Optics: Novel Optical Instrumentation for Biomedical Applications IV*, Munich, Germany, 2009

Bell, A.G. (1880). On the Production and Reproduction of Sound by Light: the Photophone. *American Journal of Science,* Vol. 20, pp. 305-324

Berer, T.; Grün, H.; Hofer, C.; Burgholzer, P. (2009). Photoacoustic Microscopy With Large Integrating Optical Annular Detectors, *Proceedings of SPIE 2009 European Conferences on Biomedical Optics: Novel Optical Instrumentation for Biomedical Applications IV*, Munich, Germany, 2009

Berer, T.; Hochreiner, A.; Zamiri, S.; Burgholzer, P. (2010). Remote photoacoustic imaging on solid material using a two-wave mixing interferometer. *Optics Letters,* Vol. 35, 4151

Boas, G. (2011). Small Animals, Big Achievements, *BioPhotonics*, pp. 18-20

Buehler, A.; Herzog, E.; Razansky, D.; Ntziachristos, V. (2010), Video rate optoacoustic tomography of mouse kidney perfusion. *Optics Letters,* Vol. 35, pp. 2475-2477

Burgholzer, P.; Hofer, C.; Paltauf, G.; Haltmeier, M.; Scherzer, O. (2005). Thermoacoustic Tomography With Integrating Area And Line Detectors. *IEEE Transactions on Ultrasonics, Ferroelectrics, and Frequency Control,* Vol. 52, pp. 1577-1583

Burgholzer, P.; Hofer, C.; Matt, G.J.; Paltauf, G.; Haltmeier, M.; Scherzer, O. (2006). Thermoacoustic tomography using fiber based Fabry–Perot interferometer as an integrating line detector, *Proceedings of SPIE 2006 Biomedical Optics: Photons Plus Ultrasound: Imaging and Sensing 2006,* San Jose, California, USA, January 21-26, 2006

Burgholzer, P.; Comacho-Gonzales, F.; Sponseiler, D.; Mayer, G.; Hendorfer G. (2009). Information Changes and Time Reversal For Diffusion-Related Periodic Fields. *Proceedings of SPIE 2009 Biomedical Optics: Photons Plus Ultrasound: Imaging and Sensing 2009,* San Francisco, California, USA, 2009

Burgholer, P.; Roitner, H.; Bauer-Marschallinger, J.; Paltauf, G. (2010). Image Reconstruction in Photoacoustic Tomography Using Integrating Line Detectors Accounting for Frequency-Dependent Attenuation. *Proceedings of SPIE 2010 Biomedical Optics: Photons Plus Ultrasound: Imaging and Sensing 2010,* San Francisco, California, USA, 2010

Dutton, H.J.R. (1998). *Understanding Optical Communications,* IBM, ISBN 0130201413, USA

Felbermayer, K.; Grün, H.; Berer, T.; Burgholzer, P. (2011). Operation point stabilization of fiber-based line detectors for photoacoustic imaging. *Proceedings of SPIE 2011 European Conferences on Biomedical Optics: Novel Optical Instrumentation for Biomedical Applications V,* Munich, Germany, 2011

Friedrich, M.G. (2008), "Tissue Characterization of Acute Myocardial Infarction and Myocarditis by Cardiac Magnetic Resonance," *Jacc-Cardiovascular Imaging. Vol. 1, pp. 652.*

Grün, H.; Nuster, R.; Paltauf, P.; Haltmeier, M.; Burgholzer, P. (2008). Photoacoustic Tomography of Heterogeneous Media using a Model-Based Time Reversal Method. *Proceedings of SPIE 2008 Biomedical Optics: Photons Plus Ultrasound: Imaging and Sensing 2008,* San Jose, California, USA, 2008

Grün, H.; Berer, T.; Nuster, R.; Paltauf, G.; Burgholzer, P. (2009). Fiber-Based Detectors for Photoacoustic Imaging, *Proceedings of SPIE 2009 European Conferences on Biomedical Optics: Novel Optical Instrumentation for Biomedical Applications IV,* Munich, Germany, 2009

Grün, H.; Berer, T.; Burgholzer, P.; Nuster, R.; Paltauf, G. (2010). Three-dimensional photoacoustic imaging using fiber-based line detectors. *Journal of Biomedical Optics,* Vol. 15, No 2

Grün, H.; Berer, T.; Pühringer, K.; Nuster, R.; Paltauf, G.; Burgholzer, P. (2010). Polymer Fiber Detectors for Photoacoustic Imaging. *Proceedings of SPIE 2010 Biomedical Optics: Photons Plus Ultrasound: Imaging and Sensing 2010,* San Francisco, California, USA, 2010

Grün, H.; Altmisdört, H. ; Berer, T.; Paltauf, G.; Zangerl, G. ; Haltmeier, M.; Burgholzer, P. (2009). Photoacoustic tomography with integrating fiber-based annular detectors, *Proceedings of SPIE 2011 Medical Imaging: Ultrasonic Imaging and Signal Processing,* Orlando, Florida, USA, 2011

Haltmeier, M.; Scherzer, O.; Burgholzer, P.; Paltauf, G. (2004). Thermoacoustic Computed Tomography With Large Planar Receivers. *Inverse Problems,* Vol. 20, pp. 1663-1673

Haltmeier, M.; Scherzer, O.; Burgholzer, P.; Nuster, R.; Paltauf, G. (2007). Thermoacoustic Tomography and the Circular Radon Transform: Exact Inversion Formula. *Mathematical Models and Methods in Applied Sciences,* Vol. 17, pp. 635-655

Harbach, N.G. (2008). Fiber Bragg Gratings in Polymer Optical Fibers. PHD-Thesis, Ecole Polytechnique Federale de Lausanne, Switzerland

Holotta, M.; Grossauer, H.; Kremser, C.; Torbica, P.; Volkl, J.; Degenhart, G.; Esterhammer, R.; Nuster, R.; Paltauf, G.; Jaschke, W. (2011), Photoacoustic tomography of ex vivo mouse hearts with myocardial infarction. *Journal of Biomedical Optics* Vol. 16, No 3

Kawasuji, M.; Ikeda, M.; Sakakibara, N.; Fujii, S.; Tomita, S.; Watanabe, Y. (2000), "Near-infrared monitoring of myocardial oxygenation during ischemic preconditioning," *Ann. Thorac. Surg*. Vol. 69, pp. 1806.

Muratikov, K.L. (2004). Laer photoacoustic imaging of inhomogeneous objects. *Technical Physics Letters*, Vol. 30, pp. 956-958

Niederhauser J.J.; Jaeger, M.; Hejazi, M.; Keppner, H.; Frenz, M. (2005). Transparent ITO coated PVDF transducer for optoacoustic depth profiling. *Optics Communications*, Vol. 253, pp. 401-406

Nuster, R.; Gratt, S.; Passler, K.; Grün, H.; Berer, T.; Burgholzer, P.; Paltauf, G. (2009). Comparison of Optical and Piezoelectric Integrating Line Detectors, *Proceedings of SPIE 2011 Biomedical Optics: Photons Plus Ultrasound: Imaging and Sensing 2009*, San Francisco, California, USA, 2009

Nuster, R.; Holotta, M.; Kremser, C.; Grossauer, H.; Burgholzer, P.; Paltauf, G. (2010). Photoacoustic microtomography using optical interferometric. *Journal of Biomedical Optics*, Vol. 15, No 2

Nuster, R.; Zangerl, G.; Haltmeier, M.; Paltauf, G. (2010). Full Field Detectiron in Photoacoustic Tomography. *Optics Express*, Vol. 18, pp. 6288-6299

Paltauf, G.; Nuster, R.; Haltmeier, M.; Burgholzer, P. (2006). Photoacoustic tomography using a Mach-Zehnder interferometer as acoustic line detector. *Applied Optics*, Vol. 46, pp. 3352-3358

Paltauf, G.; Nuster, R.; Burgholzer, P. (2009). Characterization of integrating ultrasound detectors for photoacoustic tomography. *Journal of Applied Physics*, Vol. 105

Peters, K. (2011). Polymer optical fiber sensors - a review. *Smart Materials and Structures*, Vol.20

Rosencwaig, A; Busse, G (1980). High resolution photoacoustic thermal wave microscopy. *Applied Physics Letters*, Vol. 36, pp. 725-727

Wang, X; Pang, Y; Ku, G; Xie, X; Stoica, G; Wang, L. (2003). Noninvasive laser-induced photoacoustic tomography for structural and functional in vivo imaging of the brain. *Nature Biology*, Vol. 21 (7)

Wong, Y.H.; Thomas, R.L.; Pouch, J.J. (1979). Subsurface structures of solids by scanning photoacoustic microscopy. *Applied Physics Letters*, Vol. 35, pp. 368-369

Xu, M.; Wang, L. (2003). Analytic explanation of spatial resolution related to bandwidth and detector aperture size in thermoacoustic or photoacoustic reconstruction. *Physical Review E*, Vol. 67

Xu, M; Wang, L. (2006). Rhesus Monkey Brain Imaging through Intact Skull with Thermoacoustic Tomography. *IEEE Transactions on Ultrasonics, Ferroelectrics, and Frequency Control*, Vol. 53 (3)

Xu, M; Wang, L. (2006). Photoacoustic imaging in biomedicine. *Review of Scientific Instruments*, Vol. 77

Zhang, E.Z.; Laufer, J.; Pedley, R.; Beard, P. (2009). In vivo high-resolution 3D photoacoustic imaging of superficial vascular anatomy. *Physics in Medicine and Biology*, Vol. 54, pp. 1035-1064

16

Optical Properties of Quantum-Confined Semiconductor Structures Driven by Strong Terahertz Fields

Tong-Yi Zhang and Wei Zhao
State Key Laboratory of Transient Optics and Photonics, Xi'an Institute of Optics and Precision Mechanics, Chinese Academy of Sciences
People's Republic of China

1. Introduction

The development of femtosecond pulse lasers is one of the key breakthroughs in the field of terahertz (THz) technology. THz radiation (also called T-rays) lies in the frequency gap between the infrared and microwaves, loosely referred to the frequencies from 300 GHz to 30 THz. THz radiation has long been used in astronomy and analytical science. Despite great scientific interest, however, the THz frequency range remains one of the least developed regions of the electromagnetic spectrum, due to relative lack of convenient radiation sources, detectors and transmission technology. Thanks to the development of femtosecond pulse lasers, THz research has come into the center stage during the last quarter century. Most milestones in the development of THz technology, such as THz time-domain spectroscopy and THz pulse imaging, rely on the generation of THz pulse by employing high-power femtosecond laser pulse.

In parallel to the development of THz pulsed sources, there has been a rapid expansion in developing continuous or quasi-continuous THz wave sources, such as THz free-electron lasers, photomixers, and quantum cascade lasers. High-power and frequency-tunable THz free-electron lasers have been successfully applied in scientific research into optical and transport properties in different semiconductor structures. The strong THz ac fields can coherently modify the optical properties of semiconductors. This modulation of optical properties is an essential ingredient for advanced optoelectronic devices.

This chapter reviews two topics about THz radiation and is organized as follows. In section 2, the generation of THz pulses by exploiting femtosecond laser pulses to excite photoconductive antennas, nonlinear optical crystals, and quantum-confined structures is introduced, with emphasis on the large-aperture photoconductive antennas. In section 3, the fundamental theory about optical absorption and formulations of semiconductor Bloch equations (SBEs) are presented, which have been used to investigate the optical response of semiconductor structures pumped by intense THz radiation and probed by an infrared pulse. The modulated optical properties in quantum-confined semiconductor structures driven by strong THz fields are discussed in section 4. Finally, a brief summary is given in the last section. The contents of the chapter are based on the work by the authors' research group at Xi'an Institute of Optics and Precision Mechanics, Chinese Academy of Sciences, but some relevant results of other groups are also introduced. We apologize to those whose work were not mentioned owing to space limitations.

2. THz pulse generation with ultrashort laser pulses

Two well-developed techniques of THz pulse generation are based on photoconductive antennas and optical rectification, which are the backbone of THz time-domain spectroscopy and THz pulse imaging. They rely on the excitation of different materials with ultrashort laser pulses. Quantum-confined semiconductor structures, as well as bulk semiconductors and semiconductor surfaces, emit transient THz radiation when excited with femtosecond laser pulses. And it has been developed into a new THz emission spectroscopic technique, which has been effectively used to investigate the coherent dynamics of a variety of materials and structures. In this section, THz pulse generation from photoconductive antennas, optical rectification, and quantum-confined structures is briefly introduced. The discussion is concerned principally with large-aperture photoconductive antennas because of their relative simplicity, both in fabrication and analysis, and their capabilities for high-power THz generation. A simple but unified description of their behavior is presented and the principal characteristics of these large-aperture photoconductive antennas are discussed.

2.1 Photoconductive antennas

A standard THz pulse generation scheme of photoconductive antenna is shown in Fig. 1. (Sakai & Tani, 2005) A photoconductive antenna consists of a coplanar transmission line and a dipole antenna with a small gap at the center. It is fabricated on a photoconductive substrate with short carrier lifetime, high mobility and high breakdown voltage, such as low temperature grown GaAs. A dc voltage is applied to the antenna, which generate an electric field in the semiconductor. The photoconductive gap is illuminated with femtosecond laser pulses (with photon energy greater than the bandgap of the material) to create electron-hole pairs. The free carriers accelerate in the bias field and form a short pulsed photocurrent, which radiate a transient electromagnetic pulse into free space. The photoconductive antenna behaves as a Hertzian dipole. Thus the radiated electric field $E(r,t)$ at a distance r (much greater than the wavelength of the radiation) and time t is described by the dipole antenna radiation formula

$$E(r,t) = \frac{l_e}{4\pi\varepsilon_0 c^2 r}\frac{\partial J(t)}{\partial t}\sin\theta, \tag{1}$$

where $J(t)$ is the current in the dipole, l_e is the effective length of the dipole, ε_0 is the dielectric constant of a vacuum, c is the velocity of light in a vacuum, and θ is the angle from the direction of the dipole. This source operates with comparatively low power, but the beam is stable and coherent with well known temporal characteristics. Hence it is commonly used for spectroscopy and imaging with high spectral resolution and excellent signal-to-noise ratio.

In order to generate high-power and wide beam THz pulse, large-aperture photoconductive antenna biased with a high dc voltage is used by pumping them with amplified femtosecond laser pulses. These photoconductive emitters are capable of relatively large average THz powers of 40 μW and bandwidths as high as 4 THz. Large-aperture photoconductive antenna are defined to have an optically illuminated area of dimensions much greater than the center wavelength of the emitted THz radiation. The characteristics of large-aperture

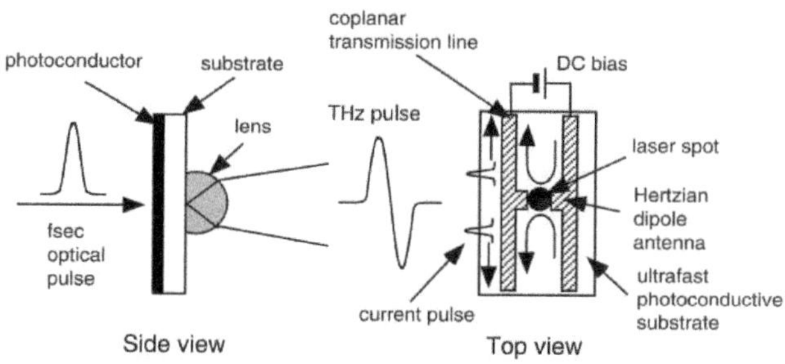

Fig. 1. Standard THz-pulse generation from a photoconductive antenna pumped with a femtosecond optical pulse. Propagation of current pulses along the coplanar transmission line are shown in the top view. (Sakai & Tani, 2005)

photoconductive antenna are well described by the current surge model. (Darrow et al., 1992; Zhang & Cao, 2004)

The current surge model assumes that THz electromagnetic field is radiated by a transient current generated at the surface of a photoconductive antenna. The surface current density $J_s(t)$ is determined by Ohm's law

$$J_s(t) = \sigma_s(t)\left[E_b + E_s(t)\right], \tag{2}$$

where $\sigma_s(t)$ is the time-dependent surface conductivity, E_b is the bias field, and $E_s(t)$ is the generated radiation field at the antenna surface. From the boundary conditions of Maxwell's equations and the finite size of the large-aperture antenna, the generated radiation field at antenna surface is related to the surface current density as

$$E_s(t) = -\frac{\eta_0}{1+\sqrt{\varepsilon}} J_s(t), \tag{3}$$

where $\eta_0 = 1/(\varepsilon_0 c) = 376.7\ \Omega$ is the impedance of free space and ε is the relative dielectric constant of the photoconductor. From Eq. (2) and Eq. (3), the radiation field at the antenna surface can be expressed through the bias field as

$$E_s(t) = -\frac{\eta_0\sigma_s(t)}{\eta_0\sigma_s(t) + (1+\sqrt{\varepsilon})} E_b. \tag{4}$$

The surface current can be expressed through the applied electric field from Eq. (2) and Eq. (4) as

$$J_s(t) = -\frac{(1+\sqrt{\varepsilon})\sigma_s(t)}{\eta_0\sigma_s(t) + (1+\sqrt{\varepsilon})} E_b. \tag{5}$$

The transient current at the antenna surface will radiate electromagnetic radiation. From Maxwell's equations, in the Coulomb gauge, the radiation field is given by

$$E_{rad}(r,t) = -\frac{1}{4\pi\varepsilon_0 c^2}\frac{\partial}{\partial t}\int \frac{J_s(r', t - |r-r'|/c)}{|r-r'|} ds', \tag{6}$$

where the surface current J_s in the emitting antenna is evaluated at a retarded time, r is the displacement from the antenna center, ds' is the increments of surface area at a displacement

r' from the antenna center, and the integration is taken over the surface of the illuminated region of the antenna. Eq. (6) shows that the far field is proportional to the time derivative of the surface current density. As a first order approximation, we substitute the integration by multiplying the area of the radiating surface to the surface current density. Thus the radiated field at far field is given by

$$E_{far}(t) = \frac{A}{4\pi\varepsilon_0 c^2 z} \frac{(1+\sqrt{\varepsilon})^2 \dot{\sigma}_s(t - z/c)}{\left[\eta_0 \sigma_s(t - z/c) + (1+\sqrt{\varepsilon})\right]^2}, \tag{7}$$

where A is the effective emitting area, z is the distance of observation point away from the center of the area, and the dot above the $\sigma_s(t)$ implies differential respective to time.
The surface conductivity induced by an optical beam of intensity $I_{opt}(t)$ is a convolution of $I_{opt}(t)$ with time-dependent mobility $\mu(t)$ and the carriers decay

$$\sigma_s(t) = \frac{q(1-R)}{hv} \int_{-\infty}^{t} \mu(t-t') I_{opt}(t') \exp\left(-\frac{t-t'}{\tau_c}\right) dt', \tag{8}$$

where q is the elementary charge, R is the optical reflectivity of the illuminated surface, hv is the photon energy, τ_c is the lifetime of carriers. From the equation of motion of an electron in a electric field

$$\frac{dv(t)}{dt} = \frac{qE}{m^*} - \frac{v(t)}{\tau_s}, \tag{9}$$

and with the definition of mobility we obtain the time-dependent mobility

$$\mu(t) = \frac{q\tau_s}{m^*}\left[1 - \exp\left(-\frac{t}{\tau_s}\right)\right] = \mu_s\left[1 - \exp\left(-\frac{t}{\tau_s}\right)\right], \tag{10}$$

where m^* is the effective mass of electrons in the photoconductor, τ_s is the relaxation time, and $\mu_s = q\tau_s/m^*$ is the static mobility. To analyze the effect of exciting optical pulse, we assume the pump pulse has a Gauss temporal profile as

$$I_{opt}(t) = \frac{F}{2\sqrt{\pi}\Delta t} \exp\left(-\frac{t^2}{\Delta t^2}\right), \tag{11}$$

where F is the total pump optical fluence, and $\sqrt{\ln 2}\Delta t$ is the full width at half maximum (FWHM) of the optical pulse. By inserting Eq. (10) and Eq. (11) into Eq. (8), we can deduce the time-dependent surface conductivity and its time differential as

$$\begin{aligned} \sigma_s(t) = \frac{q^2\tau_s}{m^*}\frac{(1-R)F}{4hv} & \left\{ \exp\left(-\frac{t}{\tau_c}\right) \exp\left(\frac{\Delta t}{2\tau_c}\right)^2 \left[1 + \operatorname{erf}\left(\frac{t}{\Delta t} - \frac{\Delta t}{2\tau_c}\right)\right] \right. \\ & - \exp\left(-t\left(\frac{1}{\tau_c} + \frac{1}{\tau_s}\right)\right) \exp\left(\frac{\Delta t}{2}\left(\frac{1}{\tau_c} + \frac{1}{\tau_s}\right)\right)^2 \\ & \left. \times \left[1 + \operatorname{erf}\left(\frac{t}{\Delta t} - \frac{\Delta t}{2}\left(\frac{1}{\tau_c} + \frac{1}{\tau_s}\right)\right)\right] \right\}, \end{aligned} \tag{12}$$

and

$$\dot{\sigma}_s(t) = \frac{q^2\tau_s}{m^*}\frac{(1-R)F}{4hv}\left\{\left(\frac{1}{\tau_c}+\frac{1}{\tau_s}\right)\exp\left(-t\left(\frac{1}{\tau_c}+\frac{1}{\tau_s}\right)\right)\exp\left(\frac{\Delta t}{2}\left(\frac{1}{\tau_c}+\frac{1}{\tau_s}\right)\right)^2\right.$$
$$\times\left[1+\operatorname{erf}\left(\frac{t}{\Delta t}-\frac{\Delta t}{2}\left(\frac{1}{\tau_c}+\frac{1}{\tau_s}\right)\right)\right]$$
$$\left.-\frac{1}{\tau_c}\exp\left(-\frac{t}{\tau_c}\right)\exp\left(\frac{\Delta t}{2\tau_c}\right)^2\left[1+\operatorname{erf}\left(\frac{t}{\Delta t}-\frac{\Delta t}{2\tau_c}\right)\right]\right\}. \tag{13}$$

Substituting Eq. (12) into Eq. (4), and Eq. (13) into Eq. (7), we obtain the explicit expressions for the surface field and the far field. The second term in the brace bracket in Eq. (12) and the first term in the brace bracket in Eq. (13) contain the effect of transient carrier mobility on the waveforms of the surface and radiated fields. The error function is the result of the convolution of optical intensity and the exponential decay. It reflects the two competing processes of carrier accumulating by photo-generation and carrier decaying by recombination. By defining a saturation fluence F_{sat}

$$F_{sat} = \frac{4hv(1+\sqrt{\varepsilon})}{q\mu_s\eta_0(1-R)}, \tag{14}$$

and denoting the parts in brace bracket in Eq. (12) and Eq. (13) as $\Sigma_1(t)$ and $\Sigma_2(t)$ respectively, we can express radiation field at surface and at far field in more compact forms, which the saturation property of the surface field at large F/F_{sat} is more obvious,

$$E_s(t) = -\frac{\frac{F}{F_{sat}}\Sigma_1(t)}{1+\frac{F}{F_{sat}}\Sigma_1(t)}E_b, \tag{15}$$

and

$$E_{far}(t) = \frac{A(1+\sqrt{\varepsilon})}{4\pi c}\frac{\frac{F}{F_{sat}}\Sigma_2(t-z/c)}{z\left[1+\frac{F}{F_{sat}}\Sigma_1(t-z/c)\right]^2}E_b. \tag{16}$$

Figure 2 shows the temporal shapes of the surface field and far field at different pumping fluence calculated by Eq. (15) and by Eq. (16), respectively. The normalized fluence F/F_{sat} varies from 1 to 10 at step 1. And the following typical values: $\tau_c = 1$ ps, $\tau_s = 0.5$ ps, $\Delta t = 0.1$ ps, are used. And the distance z is equal to 1 cm. It shows that the risetimes of the surface fields are about 500 fs. It indicates that the risetime of the surface field is determined by the carrier relaxation time. This is because it takes up a little time for the transient mobility to reach its quasi-equilibrium value. As the saturation property shown in Eq. (15), the spacing between the amplitudes of the surface fields is reducing as F/F_{sat} increases. With the fluence increasing, the peak of the radiation field increases and tend to saturation, the peak arrives earlier in time, the risetime of the main pulse decreases, and the pulse width decreases also. Figure 3 shows the influence of the relaxation time and the carrier lifetime on the amplitudes and the pulse width of the radiated far field at different pumping fluence, respectively. Longer relaxation time results in a smaller amplitude value. This is due to the carriers reach their static mobility more slowly, thus the average velocity or the conductivity is low. Longer lifetime of the carriers results in a slight larger amplitude of the radiation field. This is due to more

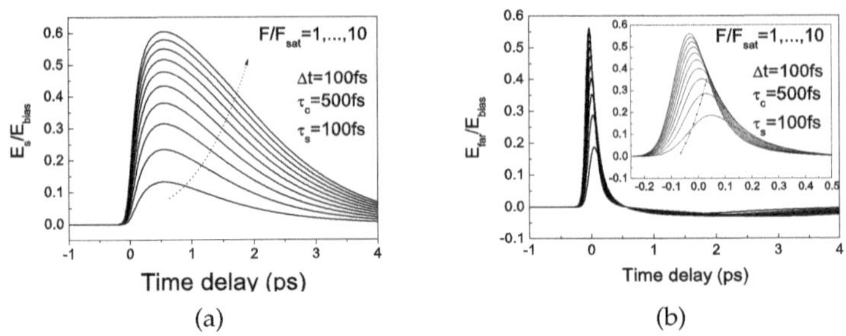

Fig. 2. Temporal shapes of the surface field (a) and far field (b) at different fluence, with the main pulses shown in the inset of (b). (Zhang & Cao, 2004)

accumulation of carriers and a larger current generated in the surface. But the influence of carrier lifetime is less prominent than the influence of relaxation time. The influence of the relaxation time on the pulse width of the radiated far field is much more prominent than the influence of the carrier lifetime. In the strong saturation regime ($F/F_{sat} = 10$), the influence of the relaxation time is reverse to that in the moderate saturation or in under-saturation regime ($F/F_{sat} = 1$ and $F/F_{sat} = 0.1$). The influence of the carrier lifetime on the pulse width is not monotonous. There are some fluctuations depended on the optical fluence.

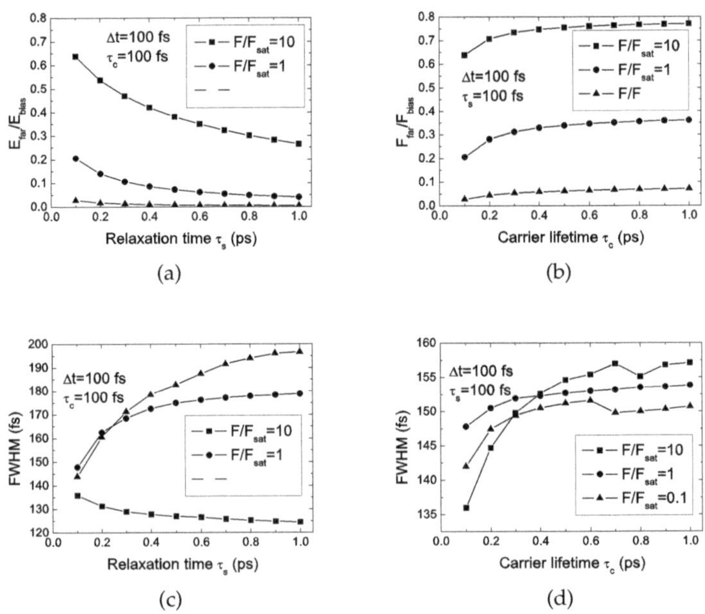

Fig. 3. Influence of the relaxation time and the carrier lifetime on the amplitudes (a) and (b), and on the pulse width (c) and (d) of the radiated far field at different pumping optical fluence. (Zhang & Cao, 2004)

2.2 Optical rectification

In optical rectification, the second-order nonlinear optical properties of materials excited by ultrafast laser pulses are used. These nonlinear effects arise when one illuminates a nonlinear optical crystal with high intensity laser. The ultrashort optical pulses contain various spectral components and the beatings between the different frequency components contained in a single ultrashort pulse will generate a broadband THz pulses. The nonlinear polarization created by a single optical pulse with the complex amplitude of the electric field $\mathcal{E}$ is given by

$$P(\omega) = \varepsilon_0 \chi^{(2)} \int_{-\infty}^{+\infty} \mathcal{E}_{opt}(\omega') \, \mathcal{E}_{opt}^{*}(\omega - \omega') \, d\omega', \tag{17}$$

where $\chi^{(2)}$ is the second-order susceptibility. The Fourier transformation of this equation gives second-order nonlinear polarization in the time domain

$$P(t) = \varepsilon_0 \chi^{(2)} \mathcal{E}_{opt}(t) \, \mathcal{E}_{opt}^{*}(t) \propto I_{opt}(t). \tag{18}$$

The first-order time derivative of this second-order nonlinear polarization gives the time-dependent current, thus the radiated THz field is proportional to the second-order time derivative of this nonlinear polarization. In this case, the optical beam itself is the origin of THz radiation, thus this process provides THz radiation only with very low efficiency, but has the advantage of very high bandwidths. Frequencies up to 41 THz have been accomplished. (Huber et al., 2000)

2.3 Quantum-confined semiconductor structures.

When pulsed lasers are used to excite coherently two energy levels in quantum well systems, beating phenomena can be observed. The beats involving spatial charge oscillations will emit electromagnetic radiation at the beat frequency. THz radiation emitted from a double coupled quantum well (Roskos et al., 1992) and a single quantum well (Planken et al., 1992) after optical excitation were observed. In the former, a coherent superposition of the symmetric and antisymmetric eigenstates of the coupled wells leads to the coherent tunneling of electrons between the two wells and therefore to a time-varying dipole moment. In the latter, a coherent superposition of the heavy hole (hh) and light hole (lh) excitons leads to charge oscillations in the well at the lh-hh splitting frequency. Both the oscillating charge emitted THz radiation into free space.

When quantum wells are embedded in a planar wavelength-scale optical microcavity, the light-matter interactions are largely modified owing to the confinement of the light field. In the strong coupling regime, hybrid exciton-photon modes (known as cavity polaritons) are formed. An excitation of such a system by a short laser pulse can trigger a complementary oscillation in the exciton and photon populations. The oscillation, called normal mode oscillation (NMO), corresponds to the intuitive time-domain picture of polaritons, where the energy is exchanged periodically between the excitons and photons. In GaAs-based semiconductor microcavities, the frequency of NMO lies in the THz regime. Then THz radiation is emitted from cavity-polaritons undergoing NMO.

For dc biased superlattices, resonant excitation of the lowest exciton connected to the optical transition between the first electron and hole miniband, electron-hole pairs are optically created in the miniband. The carriers perform "Bloch oscillation" both in real-space and momentum-space and emits THz radiation into free space. The Bloch frequency $\omega_B = eFd/\hbar$, where e is the elementary charge, $\hbar$ is the reduced Planck constant, F is the applied DC electric field, d is the period of the superlattice. Ultrafast experiments on high-quality samples unambiguously demonstrated that electrons in superlattice minibands perform at least a few cycle Bloch oscillations.

3. Formulation for theoretical analysis

The optical absorption coefficient $\alpha(\omega)$ of a dielectric medium is related to the optical dielectric function $\varepsilon(\omega)$ by

$$\alpha(\omega) = \frac{\omega}{n(\omega)c}\varepsilon''(\omega), \tag{19}$$

where $n(\omega) = \sqrt{\frac{1}{2}\left[\varepsilon'(\omega) + \sqrt{\varepsilon'^2(\omega) + \varepsilon''^2(\omega)}\right]}$ is the index of refraction, $\varepsilon'(\omega)$ and $\varepsilon''(\omega)$ are the real and imaginary parts of $\varepsilon(\omega)$, respectively. The optical dielectric function is related to the optical susceptibility $\chi(\omega)$ via $\varepsilon(\omega) = 1 + \chi(\omega)$. While the optical susceptibility relates the excitation optical field $\mathcal{E}(t)$ with its induced macroscopic polarization $P(t)$ in time-domain via

$$P(t) = \varepsilon_0 \int_{-\infty}^{t} dt' \chi(t;t')\mathcal{E}(t'). \tag{20}$$

For a system in equilibrium, the optical susceptibility depends only on the time difference. By performing a Fourier transform of eq.(20) with the time difference, $\chi(\omega)$ can be obtained from

$$\chi(\omega) = \frac{\tilde{P}(\omega)}{\varepsilon_0 \tilde{\mathcal{E}}(\omega)}, \tag{21}$$

where $\tilde{\mathcal{E}}(\omega)$ and $\tilde{P}(\omega)$ are the Fourier transform of the optical field $\mathcal{E}(t)$ and the polarization function $P(t)$. To obtain the macroscopic polarization, we use the powerful semiconductor Bloch equations (SBEs).

3.1 Semiconductor Bloch Equations and their extension in *k*-space

The SBEs are a set of coupled differential equations, which governs the coupled dynamics of electrons, holes and the optical polarization in the spectral vicinity of the semiconductor bandgap. They are the direct generalization of the equations for the free-carrier transitions, which can be put into the form of Bloch equations, and have become the standard model for semiconductor optics. The SBEs are of considerable complexity in derivation, and the detailed derivation can be found in Haug and Koch's textbook (Haug & Koch, 2004). They describes consistently the phenomena, such as, proper dependence on the light field, the inclusion of the valence conduction band continuum states, exciton effects, as well as band-filling dynamics. It successfully covers linear as well as nonlinear phenomena, such as pump-probe, four-wave-mixing, or photon echo experiments.

In k-space, the macroscopic interband polarization $\mathbf{P}(t)$ is the sum of the microscopic interband polarization $\mathbf{p_k}$, $\mathbf{P}(t) = 2d_{cv}\sum_{\mathbf{k}} \mathbf{p_k}(t)$, where d_{cv} is the interband dipole matrix element and the factor 2 is due to the spin degeneracy. In the two-band model, the microscopic interband polarization is defined as $\mathbf{p_k}(t) = \langle \beta_{-\mathbf{k}}\alpha_{\mathbf{k}}\rangle$, where $\alpha_{\mathbf{k}}$ ($\beta_{-\mathbf{k}}$) is the annihilation operator of an electron (a hole) with wavevector $\mathbf{k}$ ($-\mathbf{k}$) in the conduction (valence) band. Under Hartree-Fock approximation, from the Heisenberg equations of motion for the electron and hole operators, the well-known SBEs for microscopic polarization $p_{\mathbf{k}}(t)$ and carrier population $n^e_{\mathbf{k}} = \langle \alpha^\dagger_{\mathbf{k}} \alpha_{\mathbf{k}} \rangle$ and $n^h_{\mathbf{k}} = \left\langle \beta^\dagger_{-\mathbf{k}} \beta_{-\mathbf{k}} \right\rangle$ can be derived as

$$\frac{\partial}{\partial t} p_{\mathbf{k}} = -\frac{i}{\hbar}\left(\epsilon^e_{\mathbf{k}} + \epsilon^h_{\mathbf{k}}\right) p_{\mathbf{k}} + i\left(1 - n^e_{\mathbf{k}} - n^h_{\mathbf{k}}\right)\Omega_{R,\mathbf{k}} + \frac{\partial}{\partial t} p_{\mathbf{k}}|_{scatt}, \tag{22}$$

$$\frac{\partial}{\partial t} n^e_{\mathbf{k}} = -2\,\mathrm{Im}\left[\Omega_{R,\mathbf{k}} p^*_{\mathbf{k}}\right] + \frac{\partial}{\partial t} n^e_{\mathbf{k}}|_{scatt}, \tag{23}$$

$$\frac{\partial}{\partial t} n^h_{\mathbf{k}} = -2\,\mathrm{Im}\left[\Omega_{R,\mathbf{k}} p^*_{\mathbf{k}}\right] + \frac{\partial}{\partial t} n^h_{\mathbf{k}}|_{scatt}, \tag{24}$$

where $\epsilon^\lambda_{\mathbf{k}} = (E^\lambda_{\mathbf{k}} - \sum_{\mathbf{q}} V_{|\mathbf{k}-\mathbf{q}|} n^\lambda_{\mathbf{q}})/\hbar$ are the electron and hole energies renormalised by Coulomb interactions, $E^{(e,h)}_{\mathbf{k}}$ are the electron and hole band dispersions, $\Omega_{R,\mathbf{k}} = (d_{cv}\mathcal{E}(t) + \sum_{\mathbf{q}\neq\mathbf{k}} V_{|\mathbf{k}-\mathbf{q}|} p_{\mathbf{q}})/\hbar$ is the generalized Rabi frequency, $V_{\mathbf{q}}$ is the Fourier transform of Coulomb potential. The scattering terms represent the higher-order Coulomb correlations between carriers and carrier-phonon collisions.

In the presence of a driving field $\mathbf{F}$, the two-band SBEs are extended to the following forms (Meier et al., 1994; 1995; Zhang et al., 2007)

$$\frac{\partial}{\partial t} p_{\mathbf{k}} = -\frac{i}{\hbar}\left(\epsilon^e_{\mathbf{k}} + \epsilon^h_{\mathbf{k}}\right) p_{\mathbf{k}} + i\left(n^e_{\mathbf{k}} + n^h_{\mathbf{k}} - 1\right)\Omega_{R,\mathbf{k}} - \frac{e}{\hbar}\mathbf{F}\cdot\nabla_{\mathbf{k}} p_{\mathbf{k}} + \frac{\partial}{\partial t} p_{\mathbf{k}}|_{scatt}, \tag{25}$$

$$\frac{\partial}{\partial t} n^e_{\mathbf{k}} = -2\,\mathrm{Im}\left[\Omega_{R,\mathbf{k}} p^*_{\mathbf{k}}\right] - \frac{e}{\hbar}\mathbf{F}\cdot\nabla_{\mathbf{k}} n^e_{\mathbf{k}} + \frac{\partial}{\partial t} n^e_{\mathbf{k}}|_{scatt}, \tag{26}$$

$$\frac{\partial}{\partial t} n^h_{\mathbf{k}} = -2\,\mathrm{Im}\left[\Omega_{R,\mathbf{k}} p^*_{\mathbf{k}}\right] - \frac{e}{\hbar}\mathbf{F}\cdot\nabla_{\mathbf{k}} n^h_{\mathbf{k}} + \frac{\partial}{\partial t} n^h_{\mathbf{k}}|_{scatt}. \tag{27}$$

In quantum-confined semiconductor structures, the two-band SBEs have to be generalized to a multiband description of anisotropic semiconductors, (Binder et al., 1994; Meier et al., 1995) as

$$\frac{\partial}{\partial t} p_{ji,\mathbf{k}} = -\frac{i}{\hbar}\sum_{i'j'}\left(\epsilon^e_{ii',\mathbf{k}}\delta_{jj'} + \epsilon^h_{jj',-\mathbf{k}}\delta_{ii'}\right) p_{j'i',\mathbf{k}} - \frac{i}{\hbar}\mathcal{U}_{ij,\mathbf{k}}\left(1 - n^e_{i,\mathbf{k}} - n^h_{j,\mathbf{k}}\right) - \frac{e\mathbf{F}}{\hbar}\cdot\nabla_{\mathbf{k}} p_{ji,\mathbf{k}} + \frac{\partial p_{ji,\mathbf{k}}}{\partial t}|_{scatt} \tag{28}$$

$$\frac{\partial}{\partial t} n^e_{i,\mathbf{k}} = -\frac{i}{\hbar}\sum_{j'}\left(\mathcal{U}_{ij',\mathbf{k}} p^*_{j'i,\mathbf{k}} - \mathcal{U}^*_{ij',\mathbf{k}} p_{j'i,\mathbf{k}}\right) - \frac{e\mathbf{F}}{\hbar}\cdot\nabla_{\mathbf{k}} n^e_{i,\mathbf{k}} + \frac{\partial n^e_{i,\mathbf{k}}}{\partial t}|_{scatt} \tag{29}$$

$$\frac{\partial}{\partial t} n^h_{j,-\mathbf{k}} = -\frac{i}{\hbar}\sum_{j'}\left(\mathcal{U}_{i'j,\mathbf{k}} p^*_{ji',\mathbf{k}} - \mathcal{U}^*_{i'j,\mathbf{k}} p_{ji',\mathbf{k}}\right) - \frac{e\mathbf{F}}{\hbar}\cdot\nabla_{\mathbf{k}} n^h_{j,-\mathbf{k}} + \frac{\partial n^h_{j,-\mathbf{k}}}{\partial t}|_{scatt} \tag{30}$$

with

$$\mathcal{U}_{ij,\mathbf{k}} = \mu_{ij,\mathbf{k}}\mathcal{E}(t) - \sum_{i'j',\mathbf{k}'} V\begin{pmatrix}\mathbf{k}-\mathbf{k}'-\mathbf{k}\mathbf{k}' \\ i\,j'\,j\,i'\end{pmatrix} p_{j'i',\mathbf{k}}$$

$$\epsilon^e_{ii',\mathbf{k}} = E^e_{i,\mathbf{k}}\delta_{ii'} - \sum_{i'',\mathbf{k}''} V\begin{pmatrix}\mathbf{k}\mathbf{k}''\mathbf{k}\mathbf{k}'' \\ ii''i'i''\end{pmatrix} n^e_{i'',\mathbf{k}''}$$

$$\epsilon^h_{jj',\mathbf{k}} = E^h_{j,\mathbf{k}}\delta_{jj'} - \sum_{j'',\mathbf{k}''} V\begin{pmatrix}\mathbf{k}\mathbf{k}''\mathbf{k}\mathbf{k}'' \\ j'j''jj''\end{pmatrix} n^h_{j'',\mathbf{k}''}$$

where $E^{(e,h)}_{(i,j),\mathbf{k}}$ are the electron and hole band dispersions in subband i and j, respectively, $\epsilon^{e}_{ii',\mathbf{k}}$ and $\epsilon^{h}_{jj',\mathbf{k}}$ are the renormalized electron and hole band dispersions in subband i and j, respectively, $\mathcal{U}_{ij,\mathbf{k}}$ is the renormalized energies, $\mu_{ij,\mathbf{k}}$ is the optical dipole matrix element between conduction-band i and valence-band j states.

3.2 Semiconductor Bloch Equations in real-space

For some realistic one-dimensional structures, like V-groove and T-shaped quantum wires, it is very difficult to employ the SBEs in k-space to dealing the optical properties of these structures. This is due that the expansion in a problem-adapted basis leads to four-dimensional integrals for the Coulomb interaction, while an expansion in a basis where the Coulomb potential is not a problem will probably show a very slow convergence with basis size. For these structures, one has to take advantage of real-space method. The main idea is to solve directly a time-dependent inhomogeneousthe Schrödinger equation which describes the electron-hole dynamics in the presence of the Coulomb interaction, quantum confinements, and possible external magnetic and electric fields. The inhomogeneousthe Schrödinger equation takes the form of

$$i\hbar\frac{d\psi(\mathbf{r}_e,\mathbf{r}_h,t)}{dt} = \hat{H}\psi(\mathbf{r}_e,\mathbf{r}_h,t) - i\hbar\gamma\psi(\mathbf{r}_e,\mathbf{r}_h,t) - d_{cv}\delta(\mathbf{r}_e-\mathbf{r}_h)\mathcal{E}(t), \tag{31}$$

where $\psi(\mathbf{r}_e,\mathbf{r}_h,t)$ is the electron-hole-pair amplitude, $\hat{H}$ is the electron-hole Hamiltonian of the system considered, γ is a phenomenological dephasing constant. This equation can be considered essentially as SBEs in real space in low excitation limit, and many efficient numerical methods have been developed to solve it. (Glutsch, 2004; 1996) The macroscopic polarization in the semiconductor structure can be obtained from

$$P(t) = \frac{1}{\Omega}d^* \int d^3\mathbf{r}\psi(\mathbf{r},\mathbf{r},t), \tag{32}$$

where Ω is the normalization volume, and d^*_{cv} is the complete conjugate of the dipole matrix element d_{cv}. Some numerical results of us presented in the next section are obtained by this efficient real space approach.

4. Optical properties of semiconductor structures under external fields

An important tool in solid-state spectroscopy is the application of external fields, either static (dc) or/and dynamic (ac), electric and/or magnetic. These fields give rise to distinctive characteristic changes in the optical spectra yielding valuable information about the nature of the optical transitions. Moreover, the modulation of the optical properties of a semiconductor can be exploited for a number of advanced optoelectronic devices such as optical switches and modulators. In this section, the main characteristics in the absorption spectra of various semiconductor structures induced by external fields, particularly THz field, are reviewed.

4.1 Bulk semiconductors

The application of an external electric field will influence the relative motion of the electron-hole pair in semiconductors, which then gives rise change of the optical properties. In the case of a dc external electric field, the resulted change of optical absorption is called

Franz-Keldysh effect (FKE). (Franz, 1958; Keldysh, 1958) In Figs. 4(a) and 4(b), we show the calculated absorption spectrum without and with the excitonic effect considered. As shown in Fig. 4, pronounced oscillations above the bandgap in the absorption spectrum show up and a nonvanishing tail below the bandgap occurs. The amplitude of the oscillations decreases with increasing energy, while the oscillation period depends on the strength of the applied field. The exponential low energy absorption tail is caused by the field-induced tunneling of an electron from the valence band into the conduction band. As shown in Fig. 4(b), at high electric field, excitons are ionized, thus the results approach the free pair absorption as in Fig. 4(a).

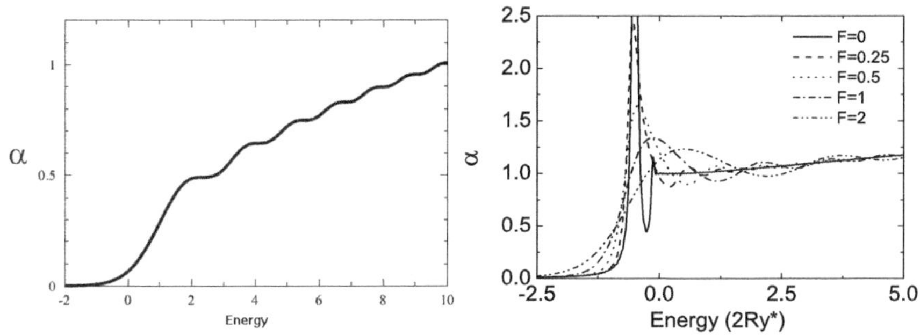

(a) (b)

Fig. 4. Franz-Keldysh effect without (a) and with excitonic effect (b). (Zhang & Zhao, 2008)

If the static electric field is replaced by a high-frequency electric field, the resulted effect is called dynamical FKE (DFKE), which was investigated by Yacoby (Yacoby, 1968) and recently by Johnsen and Jauho (Jauho & Johnsen, 1996; Johnsen & Jauho, 1998; 1999). The ac electric field modifys of the density of states both below and above the zero-field absorption threshold as shown in Fig. 5(a), and gives rise rich features in the excitonic absortion spectrum as shown in Fig. 5(b). These features are due to the replicas and THz sidebands of exciton states.

(a) (b)

Fig. 5. Density of states under THz field with different field strengthes (a) frequencies (b). (Zhang & Zhao, 2008)

Figure 6 (a) and 6(b) show the case of a THz field and a magnetic field applied to bulk semiconductor in Voigt configuration. (Zhang & Zhao, 2008) The magnetic field is $B = 4$ and $B = 2$, respectively, and the frequency of the THz field $\Omega = 2$. In both panels, the traces, from bottom to top, correspond to the THz field strength 0 to 3 at a step of 0.6. F, B, and $\Omega = 2$ are in units of excitonic energy. In the absence of the THz field (i.e. $F_0 = 0$), optical absorption spectrum exhibits a series of nearly equidistant magnetoexciton peaks, with the distance being the cyclotron frequency (B in our used units). These discrete magnetoexciton peaks stem from the Landau quantization of the orbits in the plane perpendicular to the magnetic field. The magnetoexciton resonance belonging to the lowest order Landau level shows Lorentzian lineshape, whereas the magnetoexciton resonances belonging to the higher order Landau levels show distinct Fano lineshapes. The Fano lineshapes originate from the coupling of discrete higher-order magnetoexcitons to the one-dimensional continua belonging to the lower-order Landau levels. In the presence of a strong driving THz field, the profile of the lowest magnetoexciton resonance evolves from the Lorentzian lineshape into the Fano lineshape. This change of the profile of the lowest magnetoexciton peak can be understood in the dressed states picture, i.e. the lowest magnetoexciton becomes degenerate with the continua of the lower THz-dressed magnetoexciton states. In the high frequency ($\Omega = 2$) region, the height of the lowest magnetoexcitonic resonance decreases and its width broadens. More remarkable, 2p replicas of the corresponding dark 2p magnetoexcitons of different order Landau levels show up in the continua between original magnetoexcitons. The two-THz-photon sidebands of the 1s and 2s magnetoexcitons, as well as of the 2p replica, can be identified below the lowest magnetoexciton peak and its 2p replica. Furthermore, even the two-THz-photon sideband of the 1s magnetoexciton belonging to the next neighboring higher order Landau levels can be identified. Because of the DFKE, the locations of the different magnetoexciton resonances, as well as the replicas and sidebands, blueshift to the higher energy (in addition to the diamagnetically shift).

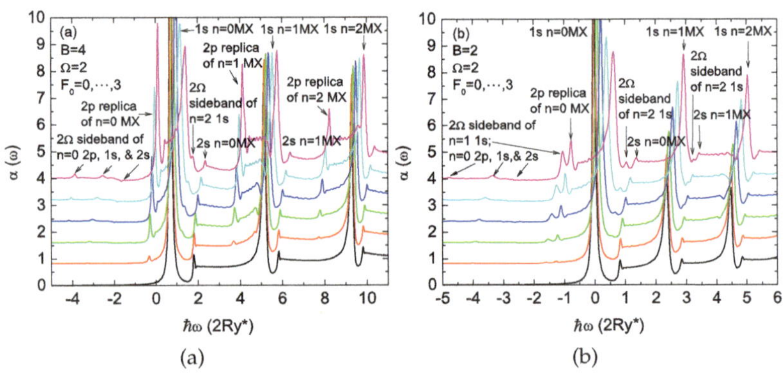

Fig. 6. (Color online) Magnetoexcitonic optical absorption spectrum of a THz-field-driven bulk GaAs in Voigt configuration. The magnetic field B=4 (a) and B= 2 (b), the frequency of the THz field $\Omega = 2$ for both case. The field strength increases from 0 to 3 from bottom to top at steps of 0.6. The index n labels the Landau levels, and MX refers to magnetoexciton. (Zhang & Zhao, 2008)

4.2 Quantum wells

Due to the quantum-confined motion of carriers in quantum wells, the excitonic and external field effects are more pronounced in semiconductor quantum wells. The spatial confinement in a quantum well prevents field ionization of the exciton up to very large field strengths. As a consequence, one can observe very large Stark shifts of exciton resonances in a field along the growth direction of the quantum well, quantum confined Stark effect (QCSE). (Miller et al., 1984; 1985)

The optical absorption of a quantum well under a growth-direction THz field was investigated by Maslov and Citrin both in SBEs approach (Maslov & Citrin, 2001) and coherent wave approach. Maslov & Citrin (2002) The main features are shown in Fig. 7. While a dc field leads to the Stark shift of excitonic peaks (QCSE) and gives rise to additional corresponding to transitions with different subband index for conduction and valence bands as in Fig. 7(a), a THz field leads to replicas around the main excitonic peak in addition to redshift and height decrease shown in Fig.7(b). Figure 7(c) shows the optical absorption with F_{dc}=50 kV/cm and different amplitudes of the THz field F_{ac} for a fixed value of the THz frequency. The excitonic peak decreases and multiple replicas appear as the THz field gets stronger. Figure 7(d) shows similar results as Fig. 7(c) but for a fixed THz field amplitude and different frequencies. For a 2 meV low-frequency THz field, the central peak is absent and only two replicas exist. If the THz frequency increases, a central peak appears and the replicas move away from the central peak. In addition to static and THz electric fields, if a magnetic field is simultaneously applied along the growth-direction of the quantum well, much richer features will show up in the absorption spectra. (Zhang & Cao, 2004)

The case in which the driving THz field polarized in the plane of quantum well was investigated by a number of authors, (Cerne et al., 1996; Jauho & Johnsen, 1996; Johnsen & Jauho, 1998; Nordstrom et al., 1998; Zhang et al., 2007) even in the presence of a magnetic field along the growth-direction. (Citrin & Hughes, 2000) It shows the following important features: (i) the main absorption edge moves to higher energies with increasing field strength, the bandedge blueshift by the ponderomotive energy, $E_{pon} = \frac{e^2F_0^2}{4m\Omega^2}$, i.e., the average kinetic energy of a particle of mass m and charge e in an electric field $F_0 \cos \Omega t$, and thus show above-band-gap transparency and increased absorption below the gap; (ii) there is a significant enhancement at the absorption edge; and (iii) the range of energies where the edge is modified increases with increasing field strengths. Figure 8 shows the experimental transmission of MQW near e1h1 exciton with $\hbar\omega_{THz} = 2.5$ meV (a) and $\hbar\omega_{THz} = 14$ meV (b) at $I_{THz} = 0, 1, 2, 4, 12$ (arbitrary units). Arrows connect calculated centers of experimental peaks and point in the direction of increasing I_{THz}. The experimental findings for the absorption are as follows (Nordstrom et al., 1998): the oscillator strength of the main absorption peak the 1s resonance is decreased by about 10-20% in the presence of the THz irradiation. There are two distinct regimes of experimental behavior as a function of THz frequency. When ω_{THz} is less than $\omega_{1s\rightarrow 2p}$, a redshift of the exciton for low THz intensities, i.e., an ac Stark shift. As THz intensity increases, this shift reaches a maximum and reverses, eventually becoming a net blueshift at the highest THz intensity; the DFKE blueshift dominates at high intensities as the ac Stark effect saturates (Fig. 8a). Conversely, if ω_{THz} is greater than $\omega_{1s\rightarrow 2p}$, a blueshift of the exciton which increases monotonically with increasing THz intensity; the ac Stark effect and DFKE act in concert, each contributing to the blueshift (Fig. 8b). In both cases, the exciton peak is broadened and suppressed, and the broadening increases with increased THz intensity.

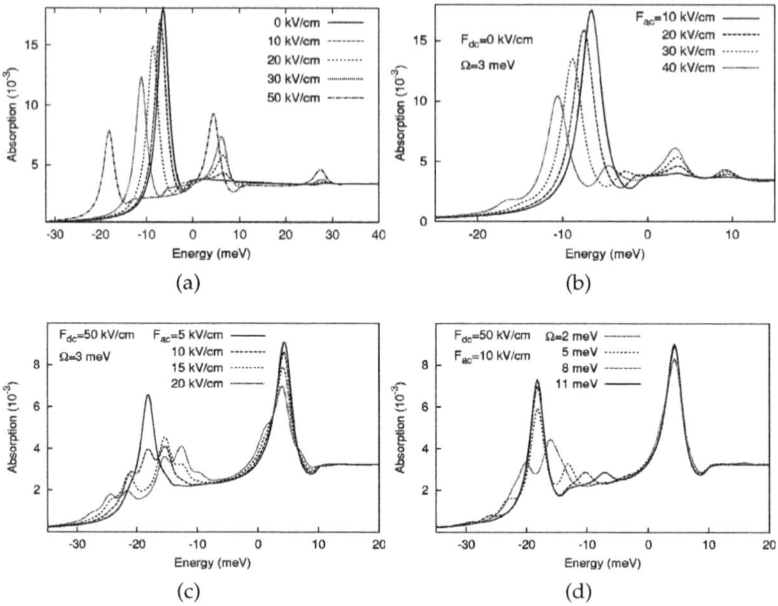

Fig. 7. Optical absorption including excitonic effects for (a) F_{dc}=0,10,20,30,50 kV/cm; (b) F_{dc}=0 kV/cm , $\hbar\Omega_{THz}$ =3 meV, and F_{ac}=10,20,30,40 kV/cm; (c) F_{dc}=50 kV/cm, $\hbar\Omega_{THz}$ =3 meV and F_{ac}=5,10,15,20 kV/cm; (d) F_{dc}=50 kV/cm, F_{ac}=10 kV/cm and $\hbar\Omega_{THz}$ =2,5,8,11 meV. (Maslov & Citrin, 2001)

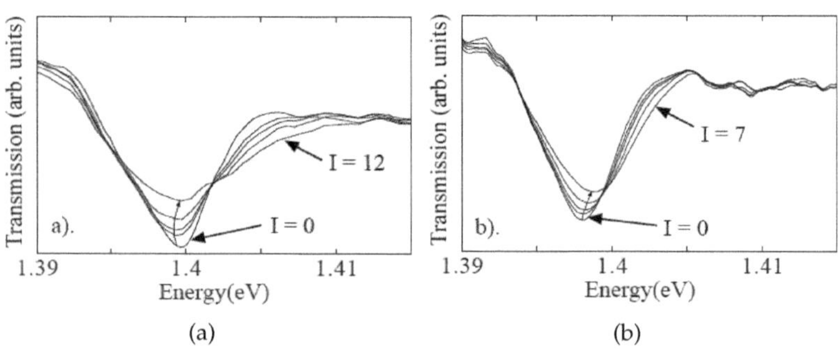

Fig. 8. Experimental transmission of MQW near e1h1 exciton. (a) $\omega_{THz} < \omega_{1s\rightarrow 2p}$, (b) $\omega_{THz} > \omega_{1s\rightarrow 2p}$. (Nordstrom et al., 1998)

The experimental features are firstly interpreted theoretically by Johnsen et al. (Nordstrom et al., 1998) by considering the interplay of DFKE and ac Stark effect in a quantum well exciton. Figiure 9 shows the calculated absorption of MQW with (a) $\hbar\omega_{THz}$ =2.5 meV<$\hbar\omega_{12}$ from top down at $\gamma = \frac{E_{pon}}{\hbar\omega_{THz}}$ =0, 0.03, 0.06, 0.1, 0.5, and 1.5. (b) $\hbar\omega_{THz}$ =14 meV>$\hbar\omega_{12}$ from top down at γ =0, 0.01, 0.02, 0.04, 0.08, and 0.15. They showed that a blue-shift of the 1s resonance if $\omega_{THz} > \omega_{1s\rightarrow 2p}$ and in a red-shift if $\omega_{THz} < \omega_{1s\rightarrow 2p}$. However as the intensity of the THz is increased the redshift saturates turns around and eventually becomes a net blue-shift.

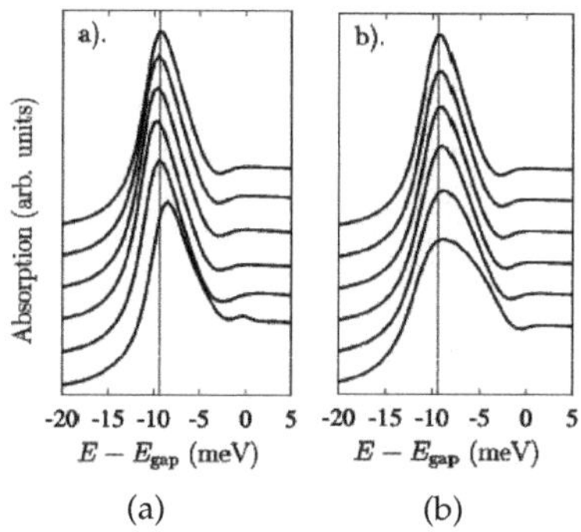

Fig. 9. Calculated absorption of MQW. (a)$\hbar\omega_{THz}$ =2.5 meV<$\hbar\omega_{12}$, (b) $\hbar\omega_{THz}$ =14 meV>$\hbar\omega_{12}$. (Nordstrom et al., 1998)

4.3 Quantum wires

Semiconductor quantum wires offer a playground for investigating the kinetics of electrons and holes in one-dimensional structures. A unique feature of the one-dimensional system is the inverse-square-root divergence of the joint density of states at the band edge. However, the Coulomb interactions between electrons and holes removes this divergence, and reduces the Sommerfield factor in contrast to the two-dimensional and three-dimensional cases. Consequently, the singular one-dimensional joint density of states does not show up at all in the linear absorption spectrum.

We have numerically analyzed the optical absorption spectra of a cylindrical quantum wire under a THz field along the axis of the wire. The cylindrical quantum wire was assumed 4200 nm in length with radius 4.2 nm. The potential at the lateral wall of the thin nanowire is assumed to be infinite, which implies the Bessel function of zero order for the envelope of the lateral ground state wave function. Considering the fundamental electron and hole subbands only, the quasi 1-D Coulomb interaction $V_{Coul}(z_e - z_h)$ is obtained from the average of the 3-D Coulomb interaction with the lateral ground state (Zhang & Zhao, 2006)

$$V_{Coul}(z_e - z_h) = \kappa \int_0^R \int_0^R \int_0^{2\pi} \int_0^{2\pi} \frac{\rho_e \rho_h J_0^2\left(\frac{\alpha_0 \rho_e}{R}\right) J_0^2\left(\frac{\alpha_0 \rho_h}{R}\right)}{\sqrt{(z_h - z_e)^2 + \rho_e^2 + \rho_h^2 - 2\rho_e\rho_h \cos(\theta_h - \theta_e)}} d\rho_e d\rho_h d\theta_e d\theta_h \qquad (33)$$

where $\kappa = e^2/[4\pi^3 \varepsilon R^4 J_1^4(\alpha_0)]$, ε the dielectric constant, R the radius of the nanowire, $\theta_{e(h)}$ and $\rho_{e(h)}$ are the polar angle and radial coordinate of electron (hole) respectively, J_0, J_1 the Bessel functions and α_0 is the zero of the J_0.

The optical absorption spectra is shown in Fig. 10(a). The frequency of the THz field is 5 THz, nearly resonant with the 1s exciton in the wire. With the field strength increasing, the main exciton peak splits, and the two peaks separate farther, and broaden, even become a band, and other peaks and oscillations emerge gradually. When field strength is lower (higher) than 9 kV/cm, the peak at high (low) energy side is dominant over the one at low (high) energy side. Furthermore, the 2s resonance and the band-edge blueshift due to dynamical FK effect. There appear two-photon replica in the continuum. With the THz field increasing, the two-photon replica redshift initially and turns into blueshift and broadens. This is because the ac-Stark effect and DFKE oppose each other. In very high field, the continuum becomes oscillating, and some other peaks corresponding to multi-photon replicas occur.

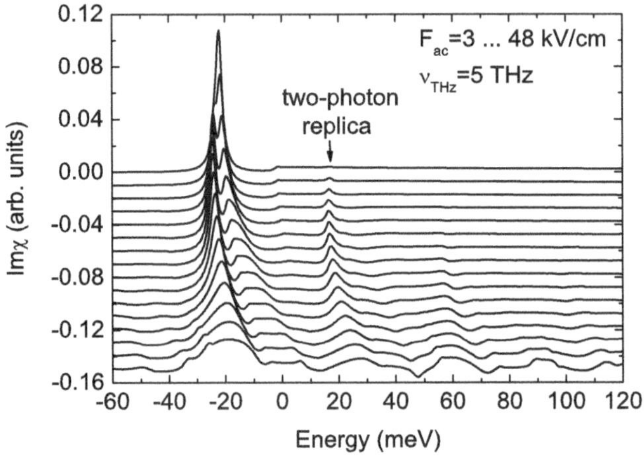

Fig. 10. (Color online) Optical absorption spectra of the cylindrical quantum wire under THz electric fields. The field strength F_{ac} varies from 3 to 48 kV/cm at the step of 3 kV/cm from top to bottom with frequency ν_{THz} of the THz field fixed at 5 THz. (Zhang & Zhao, 2006)

Figure 11 shows the spatiotemporal evolution of the polarization wave-packet for some different values of the THz field strength. The polarization wave-packet is distorted by the field and spreads to two directions following the alternating changing of the field. And there is a sharp drop for high field strength which indicates the ionization of the excitons immediately after they are generated.

The dependence of optical absorption spectra in the cylindrical quantum wire on the phase and frequency of the THz field are shown in Figs. 12(a) and 12(b). Figure 12(a) shows the influence of the frequency of the THz field on the main exciton peak. Note that the location of the dip in the splitted main 1s peak is one THz photon energy below the band-edge. Thus, this splitting is a single THz photon resonance induced coherent phenomenon. The separation and relative intensity are dependent on the frequency of THz field. Multi-photon replicas can be also identified. Figure 12(b) shows the dependence of the absorption spectra on the phase φ of the THz field at the time when the probe optical pulse reaches is peak value. The four panels show the optical absorption spectra at the frequency ν_{THz} =5 THz and the field strength F_{ac}=9 kV/cm with different phases of $\varphi = 0, \pi/2, \pi$, and $3\pi/2$, respectively. The peak emerging in the absorption continuum is the two-photon replica which is $2h\nu_{THz}$ above the 1s excitonic resonance. By changing this phase, the absorption can be enhanced or reduced at the location of the two-photon replica.

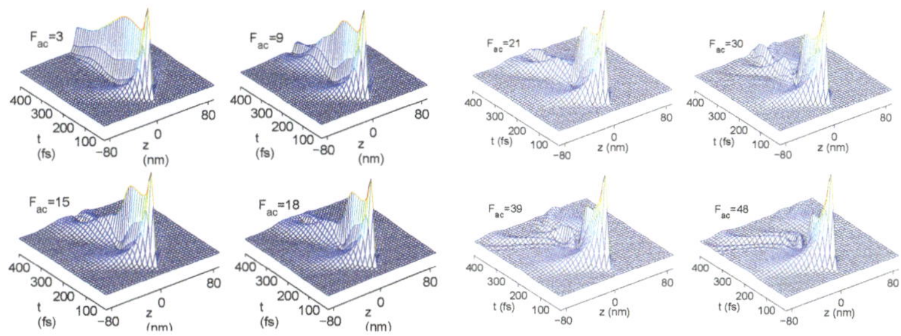

Fig. 11. (Color online) The spatiotemporal evolution of the polarization wave-packet when increasing the field strength with frequency ν_{THz}=5 THz. The eight panels correspond to different field strengths F_{ac}=3, 9, 15, 18, 21, 30, 39, and 48 kV/cm, respectively. (Zhang & Zhao, 2006)

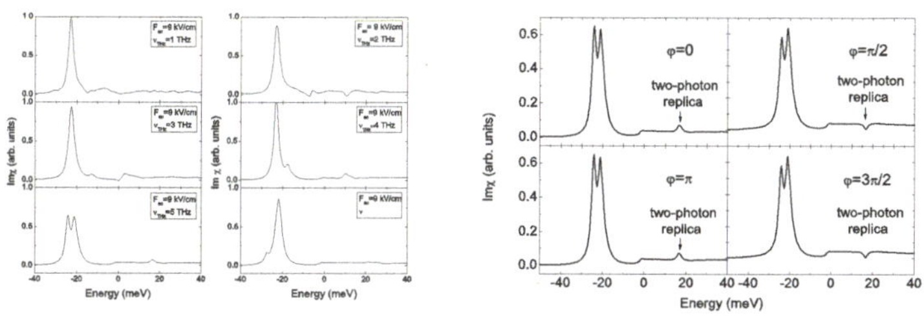

(a) (b)

Fig. 12. (a) Optical absorption spectra of the cylindrical quantum wire under THz electric fields at different frequency THz=1, 2, 3, 4, 5, and 6 THz with field strength fixed at 9 kV/cm, respectively. (b) Dependence of the optical absorption spectra of the cylindrical quantum wire on the phase under THz electric fields F_{ac}=9 kV/cm and ν_{THz}=5 THz. The four panels of b correspond to =0, $\pi/2$, π, and $3\pi/2$, respectively, with the phase of the THz wave at the time corresponding to the peak of the optical pulse. (Zhang & Zhao, 2006)

4.4 Quantum rings

Semiconductor quantum rings have been fabricated by self-organized Stranski-Krastanov growth. (Lorke et al., 2003) The not-simply-connected geometry makes them particularly interesting for studies in magnetic fields. Because the quantization of magnetic flux in the interior leads to a periodic change in the quantum mechanical properties of the electron system encircling the ring. And these periodic oscillations in the applied magnetic field are commonly called Aharonov-Bohm oscillations. This in turn affects all electronic and optical properties of rings.

Exciton absorption on a ring threaded by a magnetic flux was studied theoretically by Römer and Raikh, (Romer & Raikh, 2000) and by Hu et al. (Hu et al., 2001) Electronic structure of a quantum ring in a lateral electric field was studied theoretically by Llorens et al. (Llorens et al., 2001), and they show that the energy levels of large quantum rings exhibit a complex dispersion as a function of the electric field due to the break of axial symmetry by the field. The electric field effect on the Bohm-Aharonov oscillations is investigated by Barticevic.

(Barticevic et al., 2002) And Enhancement of the Aharonov-Bohm effect of excitons with a lateral static electric field is shown by Maslov and Citrin. (Maslov & Citrin, 2003)
We have investigated the effect of a lateral THz electric field on the excitonic AB effect using real space appraoch. (Zhang et al., 2005) Figure 13 shows the optical absorption in a nanoring with radius 50 nm under a lateral ac field and different magnetic flux Φ through the ring. The photon energy of the ac electric field in Figs. 13(a) and 13(b) is 0.256 meV and 10.22 meV, respectively, but with the same strength 5 kV/cm. We see that the shape of the exciton peak does not change noticeable with magnetic flux in the lower frequency ac field, but there is a slightly blue shift of the exciton peak with the flux increasing. However, in the higher frequency ac field, the exciton peak splits remarkably and a prominent peak appears bellow the main exciton peak, which is the replica of the dark exciton state just below the bandgap.

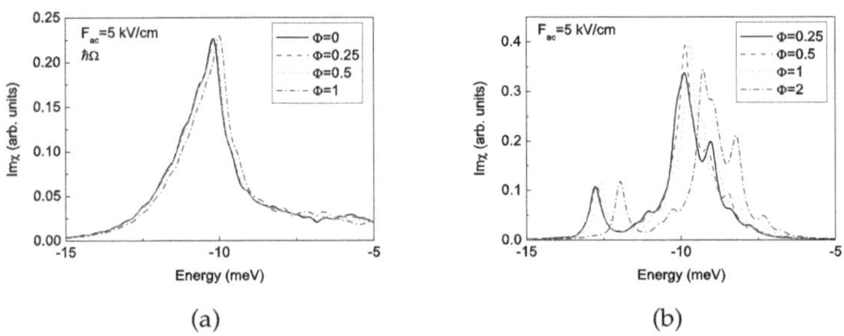

Fig. 13. (Color online) Optical absorption of the nanoring under a lateral THz field and a perpendicular magnetic field. (a) In the presence of a lateral THz field with $F_{ac} = 5$ kV/cm and $\hbar\Omega = 0.256$ meV, and different magnetic flux. (b) In the presence of the lateral THz field with $F = 5$ kV/cm and $\hbar\Omega = 10.22$ meV, and different magnetic flux. (Zhang et al., 2005)

4.5 Superlattices

Semiconductor superlattices provide insight into certain new areas of solid-state physics which cannot be accessed in conventional crystals, such as Bloch oscillation and dynamical localization. The observation of the Wannier-Stark ladder (WSL) in optical-absorption spectra of semiconductor superlattices ended a long-lasting debate about the formation of quasidiscrete energy levels when a strong dc electric field F is applied to an insulating crystal. In the absorption spectra a corresponding series of equidistant transitions with quasi-two-dimensional character show up. It is strikingly different from the FKE. Howevwer, the FKE was also expected to be obtained for low fields. The coexistence of both WSL and FKE in the same structure was demonstrated and a consistent explanation of both effects was given by Linder et al, (Linder, 1997) showing that the FKE basically consists of an intensity modulation of the WSL transitions.
Figure 14(a) shows calculated excitonic absorption spectra over a wide range of the electric field for a narrow-miniband superlattice. The narrow-miniband system is between the quasi-2D and the 3D wide-band limit of bulk material. The corresponding spectra without the Coulomb interaction shown in (b) for comparison. At high fields WS transitions can clearly be resolved as excitonic peaks in (a). For decreasing field, the energetic distance between the WS transitions falls below the value of the broadening parameter and individual peaks can

no longer be resolved. Gradually, the structures in the spectra transform to a pattern which is characteristic for FK oscillations in low fields.

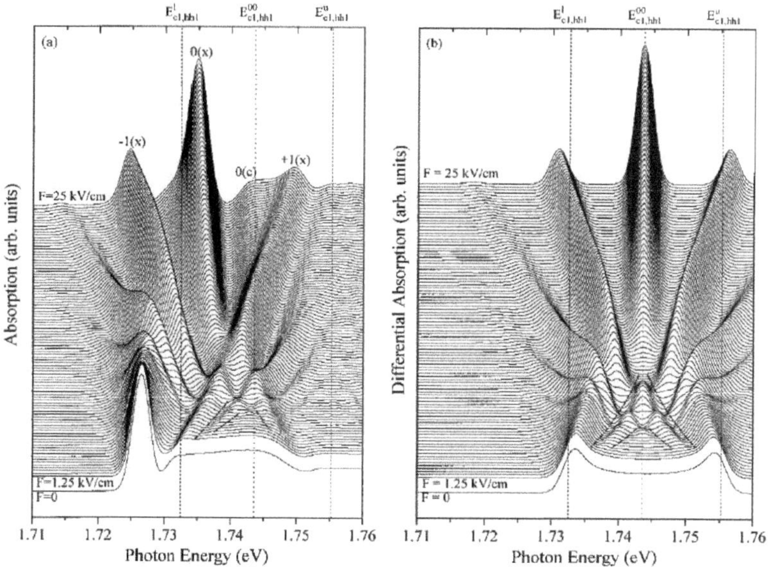

(a) (b)

Fig. 14. Excitonic absorption spectra over a wide range of the electric field for a narrow-miniband superlattice. (a) calculated with excitonic effect, (b) calculated without excitonic effect (Linder, 1997)

Figure 15 shows excitonic absorption spectra over a wide range of the electric field for a wide-miniband superlattice. In this case, the superlattice is closer to a 3D system. Figure 15(a) shows the experimental absorption spectra, and Fig. 15(b) and (c) shows the calculated results with and without excitonic effect considered. The theoretical results agree very well with the experimental results.

The optical absorption spectra of semiconductor superlattices under a combined dc-ac fields, $F_{dc} + F_{ac} \cos \upsilon_L t$, is also calculated based on SBEs. (Je et al., 2001; Zhang et al., 2003) Figure 16(a) shows the absorption spectra of a superlattice under a dc field ($eF_{dc}d$=20 meV) without applied ac field and those under an ac field ($f \equiv eF_{ac}d/\hbar\omega_L$=3.83 and $\hbar\omega_L$ =10 meV) without applied dc field. The dc field has sufficient strength to provide WSL's. The four dominant peaks correspond to WSL with $m = 0$, $m = -1$, $m = -2$, and $m = 1$, respectively. The Coulomb interactions result in smaller ladder spacing than the value of Bloch frequency. The WSL peak with $m = 1$ is largely broadened and much weaker than the WSL peak with $m = -1$, since they exist in the ionization-continuum of negative WSL states. The absorption spectra under the ac field have an exciton peak at 1.578 eV ($m' = 0$) and its second eigen state at 1.569 eV ($m' = -1$). Another peaks at 1.559 eV ($m' = -2$), 1.55 eV ($m' = -3$), and 1.54 eV ($m' = -4$) correspond to dynamical Stark ladders (DSL's) caused by the ac field. The ladder spacing estimated from these DSL's is almost $\hbar\omega_L$. Figure 16(b) displays the absorption spectra under the ac fields with different strengths f and the dc field of a fixed

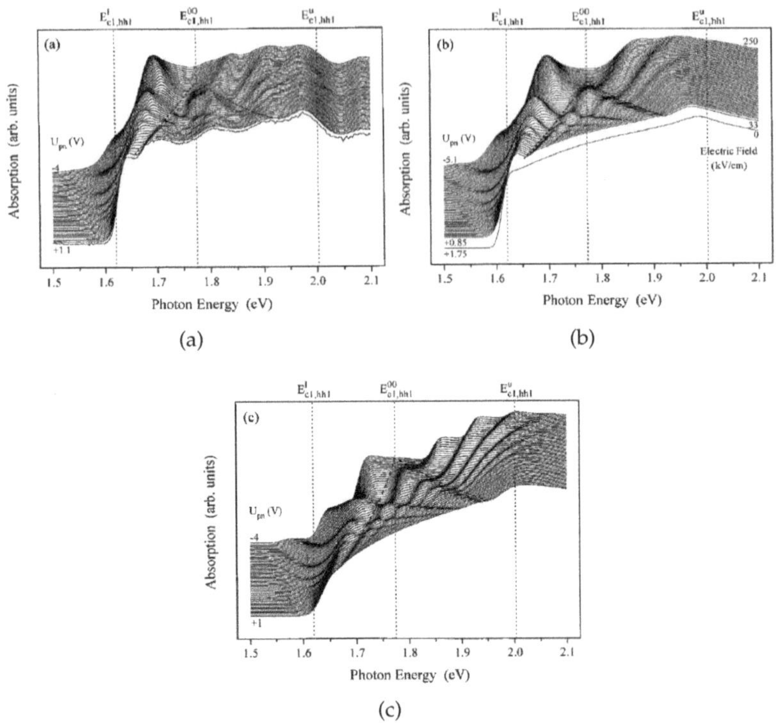

Fig. 15. Excitonic absorption spectra over a wide range of the electric field for a wide-miniband superlattice. (a) experimental,(b) calculated with excitonic effect, (c) calculated without excitonic effect. (Linder, 1997)

strength $eF_{dc}d = 20$ meV. A peak at 1.577 eV corresponds the WSL with the index $m = 0$. As the strength of the ac field increases, the DSL's appear with the same spacing energy $\hbar\omega_L$. The DFSL's between WSL's positions are also observed. Nearly all of the peaks, corresponding to the spectral positions at the formula $\epsilon_0 + m\hbar\omega_B + m\prime\hbar\omega_L$, exist. Almost all the peaks can be identified by (m, m').

4.6 Quantum-dot-superlattices

Recently, a structure consisting of a series of quantum dots interlaced along a nanowire has been grown. This quantum dot superlattice structure has characteristics of quantum dots, quantum wires and superlattices. As the band offset of different materials confines the carriers in one of the quantum dots, the electron states have zero-dimensional (0D) characteristics of a quantum dot. However, by tunneling between adjacent quantum dots the carriers can travel along the wire axis. Thus, a miniband electronic structure of an superlattice is formed. Due to strong lateral confinement, the carrier motion in the plane perpendicular to the wire axis is confined so that the system has characteristics of one-dimensional structures.

The effects of an external THz field applied along the quantum-dot-superlattice nanowire on the interband optical absorption spectra are shown in Fig. 17. Figure 17(a) shows

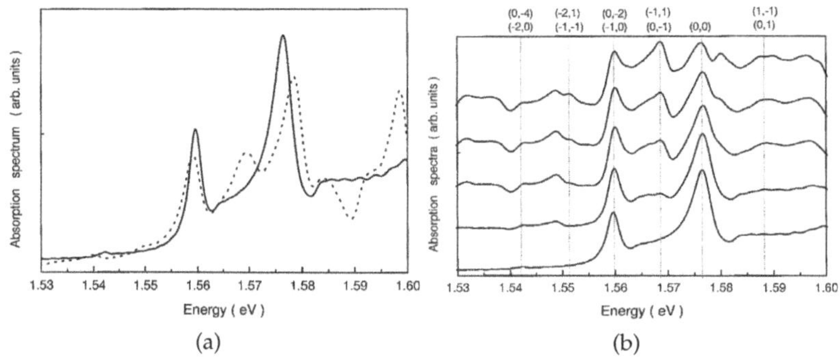

Fig. 16. (a) The absorption spectra for $eF_{dc}d$=20 meV and f=0 (solid line) and F_{dc}=0 and f=2.83 with $\hbar\omega_L$ =10 meV (dotted line). (b) Absorption spectra under the combined dc-ac fields when $eF_{dc}d$=20 meV and $\hbar\omega_L$ = 10 meV. The spectra are shown for the different strengths of the ac fields or f =0, 0.2, 0.4, 0.6, 0.8, and 1 from the bottom. Indices in the upper part of the figure is $(m, m\prime)$.(Je et al., 2001)

the absorption spectra for different frequencies of $\hbar\Omega$=5, 11, and 20 meV of the THz field, respectively, with the field strength fixed at F_{ac} = 30 kV/cm. The main exciton peak decreases, broadens and becomes asymmetric due to the applied THz field. It has small gain and small peaks appears in the spectra. These peaks are replicas and THz-sideband of the excitonic states dressed by THz photons. The excitonic absorption spectra for different field strengths F_{ac} with frequency fixed at $\hbar\Omega$=20 meV are shown in Fig. 17(b). The oscillator strength is completely transferred from the minibands to the exciton. Increasing field strength the main excitonic peak decreases, broadens and splits up. The asymmetric splitting resemble the well known Autler-Townes splitting of atoms in an ac driving field and is due to the coherent coupling of the lowest and higher exciton states. Furthermore, when the field strength increase, there exists one-photon and two-photon gain in the optical spectra.

Fig. 17. (Color online) Optical absorption in the QDSLW under a THz field F_{ac} along the wire. (a) Different frequencies $\hbar\Omega$=5, 11, and 20 meV, with field strength fixed at F_{ac}=30 kV/cm; (b) Different field strengths F_{ac}=20, 30, 50, and 60 kV/cm, with frequency fixed at $\hbar\Omega$=20 meV. (Zhang et al., 2005)

4.7 Microcavities

As semiconductor fabrication technology developed, it became possible to fabricate high-quality-factor (high-Q) semiconductor microcavities. In a system of quantum wells embedded within high-Q semiconductor microcavity, the strong exciton-photon coupling leads to the formation of two new eigenstates of the exciton-photon coupled system, called "microcavity exciton polariton" states. Thus results in two spectroscopically pronounced exciton polariton peaks in the absorption spectrum of the system. In a high-Q microcavity, most of the photons arising from the interband polarization associated with one exciton internal state are fed back into the system. This leads to a photon-mediated radiative renormalization of the optically allowed (i.e., s-like) exciton internal states. As shown in the subsection 4.2, a quantum well in the simultaneous presence of a THz field polarized in the growth direction and an incident optical field near an excitonic resonance results in substantial frequency mixing between the THz and optical fields, and leads to THz-sideband generation. In a high-Q optical microcavity, THz-sideband generation is strongly enhanced. Numerical estimate by Citrin (Citrin, 2002) showed that the values of the reflected intensity into the first THz sideband can be as large as 10% of the reflected intensity at the fundamental, and a factor of ~100 enhancement of doubly resonant THz-sideband generation in the microcavity compared with bare quantum well. In Fig. 18(a), the optical frequency dependence of the reflected field (solid) and the interband polarization (dashed) in the first THz sideband in a high-Q microcavity are shown. The frequency dependence of the reflected field in the fundamental from the same structure is shown in Fig. 18(b). The curves are normalized to the incident field amplitude.

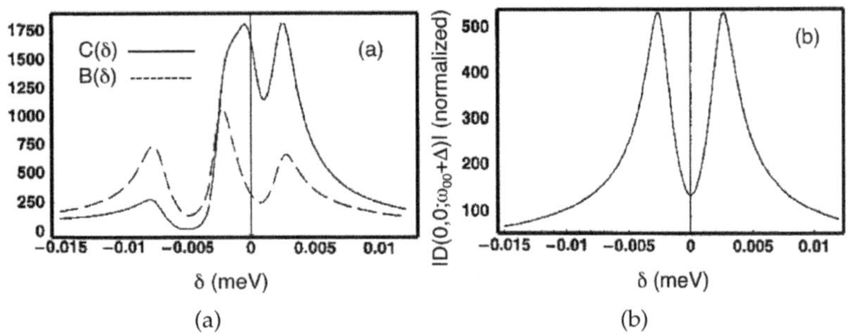

Fig. 18. (a) Optical frequency dependence of the reflected field (solid) and the interband polarization (dashed) in the first THz sideband in a high-quality-factor microcavity. (b) The frequency dependence of the reflected field in the fundamental from the same structure. The curves are normalized to the incident field amplitude. (Citrin, 2002)

Citrin (Citrin, 1999) also showed that ultrafast THz pulse can be used to control coherently the NMOs in microcavities. He showed that a THz pulse can introduce a phase shift, amplitude modulation, or both in the optical dynamics depending on whether the induced modulation is real, imaginary, or complex. Thus, by judiciously choosing the THz pulse, it is possible in principle to switch the coherent emission from a microcavity undergoing mode oscillations to emit at either norm mode following the incidence of the THz pulse. This is shown in Fig. 19. Fig. 19(a) shows the unmodulated signal. In (b), for $t < 0$, the emission occurs at the upper mode. Fig. 19(c) shows that the phase of the mode oscillations has been flipped at $t = 0$. And

Fig. 19(d) shows the case where the coherent emission for $t > 0$ is effectively quenched due to the modulation of the exciton linewidth by a THz pulse.

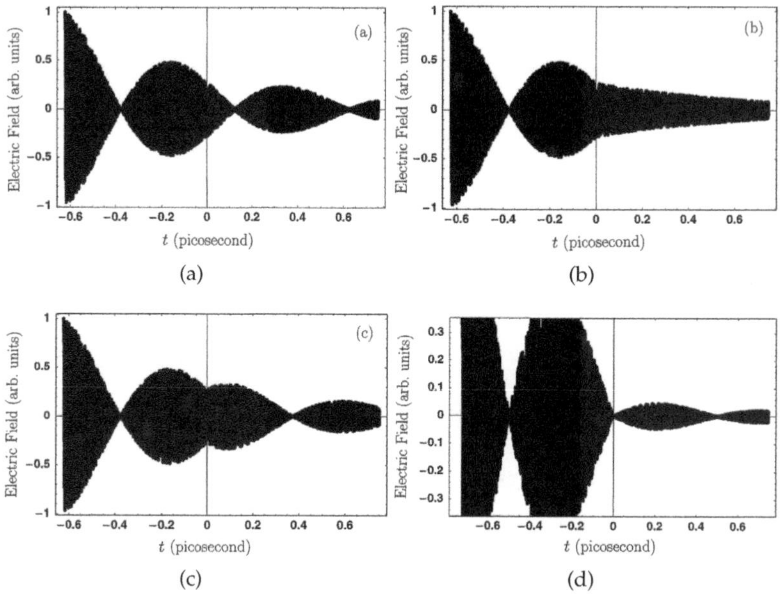

Fig. 19. Electric field in MC following ultrafast broad-band optical excitation. (a) The unmodulated case. (b) A THz pulse is used to shift the emission to the upper norm mode ε_+ at t=0. (c) Shows how a THz pulse can be used to invert the dynamics of the mode oscillations. (d) By introducing transient rapid dephasing when the excitation is in the exciton mode at time t=0, the coherent emission is quenched.

5. Conclusion

To summary, we have reviewed the generation of THz pulses by using femtosecond laser pulses to excite photoconductive antennas, nonlinear optical crystals, and quantum-confined semiconductor structures. THz radiation is emitted from these materials via transient photocurrent, optical rectification, and beat of a coherent superposition of different states, respectively. Then we present the fundamental formulism and the powerful extended SBEs used to calculate the macroscopic polarization and absorption coefficient in various semiconductor structures. The SBEs is presented in k-space and in real-space, respectively. We have applied these formulation to deal with the optical absorption in various different semiconductor structures driven by a strong THz electric field. The investigated objects involved bulk and quantum-confined semiconductor structures, such as quantum wells, quantum wires, quantum rings, superlattices, quantum-dot-superlattices, etc. Many fascinating phenomena, such as dynamical Franz-Keldysh effect, ac Stark effect, THz-sideband generation, replicas of resonant excitonic peaks, show up in the optical spectra of these structures in the simultaneous presence of a strong pumping THz field and a probing infrared pulse. These effects may be employed to develop advanced optoelectronic and THz devices.

6. Acknowledgments

This work is supported by the National Basic Research Program of China (973 Program) (2007CB310405) and the National Natural Science Foundation of China (10834015 and 60777017). The authors also gratefully acknowledge the support of K. C. Wong Education Foundation, Hong Kong.

7. References

Barticevic, Z.; Fuster, G. &Pacheco, M. (2002). Effect of an electric field on the Bohm-Aharonov oscillations in the electronic spectrum of a quantum ring. *Phys. Rev. B*, Vol.65(No.19): 193307-1-4

Binder, E.; Kuhn, T. &Mahler, G. (1994). Coherent intraband and interband dynamics in double quantum wells: Exciton and free-carrier effects. *Phys. Rev. B*, Vol.50(No.24): 18319-18329.

Černe, J.; Kono, J.; Sherwin, M. S.; Sundaram, M.; Gossard, A. C. &Bauer, G. E.W. (1996). Terahertz dynamics of excitons in GaAsyAlGaAs quantum wells. *Phys. Rev. Lett.*, Vol.77(No.6): 1131-1134.

Citrin, D. S. (1999). Terahertz sideband generation and coherent control in semiconductor microcavities. *Phys. Rev. Lett.*, Vol.82(No.15): 3172-3175.

Citrin, D. S. &Hughes, S. (2000). Franz-Keldysh effect on Landau levels and magnetoexcitons in quantum wells. *Phys. Rev. B*, Vol.61(No.8): R5105-R5108.

Citrin, D. S. (2002). Terahertz/optical mixing in symmetric semiconductor quantum wells embedded in optical microcavities. *J. Lightwave Technology*, Vol.20(No.12): 1983-1988.

Darrow, J. T.; Zhang, X. -C.; Auston, D. H. &Morses, J. D. (1992). Saturation properties of large-aperture photoconducting antennas. *IEEE J. Quantum Electron*, Vol.28(No.6): 1607-1616.

Franz, W. Z. (1958). Einfluβ eines elektrischen Feldes auf eine optische Absorptionskante. *Naturforsch*, Vol.13a 484-489.

Glutsch, S. (2004). *Excitons in Low-Dimensional Semiconductors*, Springer, Berlin.

Glutsch, S.; Chemla, D. S. &Bechstedt, F. (1996). Numerical calculation of the optical absorption in semiconductor quantum structures. *Phys. Rev. B*, Vol.54(No.16): 11592-11601.

Haug, H. &Koch, S. W. (2004). *Quantum Theory of the Optical and Electronic Properties of Semiconductors*, 4th ed., World Scientific, Singapore.

Hu, H.; Zhu, J.; Li, D. &Xiong, J. (2001). Aharonov-Bohm effect of excitons in nanorings. *Phys. Rev. B*, Vol.63(No.19): 195307-1-11.

Huber, R.; Brodschelm, A.; Tauser, F. &Leitenstorfer A. (2000). Generation and field-resolved detection of femtosecond electromagnetic pulses tunable up to 41 THz. *Appl. Phys. Lett.*, Vol.76(No.22): 3191-3193.

Jauho, A. -P. &Johnsen, K. (1996). Dynamical Franz-Keldysh effect. *Phys. Rev. Lett.*, Vol.76(No.24): 4576-4579.

Je, K.; Park, S. &Kim, Y. (2001). Optical absorption spectra and dynamical fractional Stark ladders in semiconductor superlattices. *Phys. Rev. B*, Vol.64(No.7): 075111-1-5.

Johnsen, K. &Jauho, A. -P. (1998). Linear optical absorption spectra of mesoscopic structures in intense THz fields: Free-particle properties. *Phys. Rev. B*, Vol.57(No.15): 8860-8872.

Johnsen, K. &Jauho, A. -P. (1999). Quasienergy spectroscopy of excitons. *Phys. Rev. Lett.*, Vol.83(No.6): 1207-1210.

Keldysh, L. V. (1958). The effect of a strong electric field on the optical absorption properties of insulating crystals. *Sov. Phys.-JETP*, Vol.34(No.7): 788-790. [*Zh. Eksp. Theor. Fiz.* Vol.34(No.):1138-1141.]

Linder, N. (1997). Excitons in superlattices: absorption asymmetry, dimensionality transition, and exciton localization *Phys. Rev. B*, Vol.55(No.20): 13664-13676.

Llorens, J. M.; Trallero-Giner, C.; García-Cristóbal, A. &Cantarero, A. (2001). Electronic structure of a quantum ring in a lateral electric field. *Phys. Rev. B*, Vol.64(No.3): 035309-

Lorke, A.; Garcia, J. M.; Blossey, R.; Luyken, R. J. &Petroff, P. M. (2001). Self-organized InGaAs quantum rings–fabrication and spectroscopy, *in* Kramer, B. (Ed.), *Adv. in Solid State Phys.*, Springer-Verlag, Berlin: Heidelberg, 125-137.

Maslov, A. V. &Citrin, D. S. (2001). Optical absorption of THz-field-driven and dc-biased quantum wells. *Phys. Rev. B*, Vol.64(No.15): 155309-1-10.

Maslov A. V. &Citrin, D. S. (2002). Numerical calculation of the terahertz field-induced changes in the optical absorption in quantum wells. *IEEE J. Selected Topics in Quantum Electronics*, Vol.8(No.3): 457-463.

Maslov, A. V. &Citrin, D. S. (2003). Enhancement of the Aharonov-Bohm effect of neutral excitons in semiconductor nanorings with an electric field. *Phys. Rev. B*, Vol.67(No.12): 121304-1-4.

Meier, T.; von Plessen, G.; Thomas, P. &Koch, S. (1994). Coherent electric-field effects in semiconductors. *Phys. Rev. Lett.*, Vol.73(No.6): 902-905.

Meier, T.; von Plessen, G.; Thomas, P. &Koch, S. (1995). Coherent effects induced by static and time-dependent electric fields in semiconductors. *Phys. Rev. B*, Vol.51(No.20): 14490-14497.

Meier, T.; Rossi, F.; Thomas, P. &Koch, S. (1995). Dynamic localization in anisotropic Coulomb systems: Field induced crossover of the exciton dimension. *Phys. Rev. Lett.*, Vol.75(No.13): 2558-2561.

Miller, D. A. B.; Chemla, D. S.; Damen, T. C.; Gossard, A. C.; Wiegmann, W.; Wood, T. H. &Burrus, C. A. (1984). Band-edge electroabsorption in quantum well structures: The quantum-confined Stark effect. *Phys. Rev. Lett.*, Vol.53(No.22): 2173-2176.

Miller, D. A. B.; Chemla, D. S.; Damen, T. C.; Gossard, A. C.; Wiegmann, W.; Wood, T. H. &Burrus, C. A. (1985). Electric field dependence of optical absorption near the band gap of quantum-well structures. *Phys. Rev. B*, Vol.32(No.2): 1043-1060.

Nordstrom, K. B.; Johnsen, K.; Allen, S. J.; Jauho, A.-P.; Birnir, B.; Kono, J.; Noda, T.; Akiyama, H. &Sakaki, H. (1998). Excitonic dynamical Franz-Keldysh effect. *Phys. Rev. Lett.*, Vol.81(No.2): 457-460.

Planken, P. C. M.; Nuss, M. C.; Brener, I.; Goossen, K. W.; Luo, M. S. C.; Chuang, S. L. &Pfeiffer L. (1992). Terahertz emission in single quantum wells after coherent optical excitation of light hole and heavy hole excitons. *Phys. Rev. Lett.*, Vol.69(No.26): 3800-3803.

Römer, R. A. &Raikh, M. E. (2000). Aharonov-Bohm effect for an exciton. *Phys. Rev. B*, Vol.62(No.11): 7045-7049.

Roskos, H. G.; Nuss, M. C.; Shah, J.; Leo, K.; Miller, D. A. B.; Fox, A. M.; Schmitt-Rink, S. &Köhler, K. (1992). Coherent submillimeter-wave emission from charge oscillations in a double-well potential. *Phys. Rev. Lett.*, Vol.68(No.14): 2216-2219.

Sakai, K. &Tani, M. (2005). Introduction to Terahertz Pulses, *in* Sakai, K.(Ed.), *Terahertz Optoelectronics*, Springer-Verlag, Berlin: Heidelberg, 6.

Yacoby, Y. (1968). High-frequency Franz-Keldysh effect. *Phys. Rev.*, Vol.169(No.3): 610-619.

Zhang, A.; Yang, L. &Dignam, M. M. (2003). Influence of excitonic effects on dynamic localization in semiconductor superlattices in combined dc and ac electric fields. *Phys. Rev. B*, Vol.67(No.20): 205318-1-7.

Zhang, T. Y. &Cao, J. C. (2004). Study of the surface and far fields of terahertz radiation generated by large-aperture photoconductive antennas. *Chin. Phys.*, Vol.13(No.10): 1742-1746.

Zhang, T. Y. &Cao, J. C. (2004). Optical absorption in terahertz-driven quantum wells under a magnetic field *J. Phys.: Condens. Matter*, Vol.16(No.49): 9093-9101.

Zhang, T. Y. &Cao, J. C. (2005). Optical absorption in semiconductor nanorings under a lateral terahertz electric field. *J. Appl. Phys.*, Vol.97(No.2): 024307-1-5.

Zhang, T. Y.; Zhao, W.; Cao, J. C. &Qasim, G. (2005). Interband optical-absorption spectra of a finite quantum dot superlattice in a cylindrical nanowire. *J. Appl. Phys.*, Vol.98(No.9): 094311-1-5.

Zhang, T. Y. &Zhao, W. (2006). Franz-Keldysh effect and dynamical Franz-Keldysh effect of cylindrical quantum wires. *Phys. Rev. B*, Vol.73(No.24): 245337-1-6.

Zhang, T. Y.; Zhao, W.; Liu, X. M. &Zhang, C. (2007). Nonlinear optical properties of semiconductor quantum wells under intense terahertz radiation. *Appl. Phys. Lett.*, Vol.91(No.4): 041909-1-3.

Zhang, T. Y. &Zhao, W. (2008). Excitonic optical absorption in semiconductors under intense terahertz radiation. *Chin. Phys. B* , Vol.17(No.11): 4285-4291.

Zhang, T. Y. &Zhao, W. (2008). Magnetoexcitonic optical absorption in semiconductors under strong magnetic fields and intense terahertz radiation in the Voigt configuration. *Europhys. Lett.*, Vol.82(No.6): 67001-1-6.

17

Photoemission Spectroscopy at Liquid Microbeams with a High Harmonics Table Top Radiation Source

Bernd Abel
University of Leipzig
Germany

1. Introduction

The use of ultrafast light pulses to visualize chemical dynamics has advanced our fundamental understanding of chemical and biochemical processes, by providing a means of monitoring the motion of atoms within a molecule in real time (Assmann et al. (2003)). However, closer to the chemist's heart is the evolution of the electron density in a molecule - as molecules transform - as a function of time. In fact, the chemist's dream is to trace orbitals in time during a chemical transformation (Kling et al. (2006)). The most attractive medium in this context is the liquid phase where chemistry happens (Siefermann et al. (2010)). Recent progress in electronic spectroscopy in the condensed phase (e.g., Link, Lugovoy, Siefermann, Liu, Faubel & Abel (2009) and Link, Voehringer-Martinez, Lugovoj, Liu, Siefermann, Faubel, Grubmueller, Gerber, Miller & Abel (2009)) has been driven by the development of new ultrashort pulsed light sources, as well as the development of new experimental techniques. Recent successful experiments have made use of sources of ultrafast extreme-UV as probes of ultrafast molecular and materials dynamics (Paul et al. (2001)). These new probes can in principle directly provide site-specific molecular dynamics or local order via photoelectron spectroscopy, providing the potential to directly observe chemical reactions in atomic-level detail as they occur without the need for specific chromophors. The new tools appear to provide a powerful new window into the microscopic world, particularly as the temporal resolution and energy range of these sources continues to improve. One recent area of research motivated by these goals and the quest for fundamental light matter interactions is the generation of ultrashort EUV (extreme ultraviolet) light pulses through the process of high-order harmonic generation (HHG). HHG pushes traditional nonlinear optics to an extreme, by coherently combining many laser photons together to generate coherent beams that span from the UV to the keV-region of the spectrum (Krausz & Ivanov (2009)). The pulses from HHG may be femtosecond to attosecond in duration. Thus, in the wavelength region, where these beams are bright and where they contain enough photons (<100 eV), they have been used to monitor a variety of processes in chemical and materials science and for applications such as holographic imaging and photoelectron emission spectroscopy (Krausz & Ivanov (2009), Kapteyn et al. (2007)).

As a liquid phase medium water has many unusual properties when compared with simple organic liquids (Head-Gordon & Johnson (2005)), and bulk water as well as single water molecules play a decisive role in many chemical and biological systems. The dynamical network of hydrogen bondend water in chemical and biological systems has been studied

with powerful spectroscopic techniques and theoretically in the past (Cowan et al. (2005), Fecko et al. (2003), Woutersen & Bakker (2006), Head-Gordon & Johnson (2005)).
Due to the technical and conceptual problems with (volatile) liquids in vacuum, liquid phase (high pressure) ESCA (electronic spectroscopy for chemical anylysis) is much less well established than XPS (x-ray photoelectron spectroscopy) at solid state surfaces (Siegbahn (1985), Lundholm et al. (1986)). Since Faubel et al. in Goettingen developed liquid beams in vacuum also volatile liquids like water could be investigated with photoelectron spectroscopy (Faubel & Kisters (1989), Faubel (2000)) Since then, photoelectron emission lines and the chemical shift in the static ESCA (Winter & Faubel (2006)), have also been particularly powerful observables for probing electron densities and molecular orbital energies in different intramolecular and intermolecular environments. Recently, we have combined powerful technologies such as photoelectron spectroscopy near volatile liquid interfaces in vacuum, ultrafast pump-probe spectroscopy, and table-top high-harmonics generation of soft X-ray radiation, which enabled us to perform liquid phase photoelectron spectroscopy with high time-resolution, i.e., adding the dimension of time to the liquid interface ESCA or liquid-phase XPS (Link, Lugovoy, Siefermann, Liu, Faubel & Abel (2009) Link, Voehringer-Martinez, Lugovoj, Liu, Siefermann, Faubel, Grubmueller, Gerber, Miller & Abel (2009)). The idea of ESCA combined with ultrafast EUV radiation has also been reported before (Drescher (2004)) but not for liquid phase matter. In the present chapter we will highlight and discuss, which insight this somewhat involved novel experimental approach can give us. As outlined above, it maps out a different kind of molecular information other than site specific chromophores or structural (diffraction) information - the orbital energy or electron density in a molecule in a special intramolecular and intermolecular environment. In the present contribution we aim to introduce into the new technology, highlight some recent results, and to show what is special and characteristic about time-resolved ESCA or XPS near liquid interfaces, in particular liquid water interfaces.

2. Liquid beams in vacuum and high harmonics generation

2.1 Liquid water beam technology

Liquid water jets in vacuum have been introduced by Faubel at the end of the 1980s (Faubel & Kisters (1989)). A major obstacle in handling common liquids in vacuum is their large vapour pressures, which make it virtually impossible to maintain the required vacuum without either freezing the sample or having it consumed by evaporation and spoiling the vacuum. A liquid jet in vacuum including the nozzle assembly is displayed in Fig. 1. While the left figure shows the nozzle assembly the right part shows the details of the micro beam relevant for an experiment.
The enlarged section in Fig. 1 shows a high speed image of a beam with a 2-3 mm continuous section of the beam that decomposes into droplets (classical Rayleigh limit). The liquid jet is typically driven by a HPLC-pump at 10 to 50 bar at flow speed of 20-50 m/s up to 100 m/s (Charvat et al. (2004)). The temperature of the beam (surface) due to evaporative cooling can be calculated and predicted. It is plotted against the distance from the nozzle orifice in Fig. 2 for liquid water, ethanol, and methanol, respectively. More details about the liquid jet technology are given in Ref. (Charvat et al. (2004))

2.2 High harmonics generation and their characteristics

Harmonic generation is a nonlinear optical process in which the frequency of laser light is converted into its integer multiples. Odd harmonics of very high orders are generated due to symmetry reasons from noble gas atoms exposed to intense (usually near-infrared) laser fields. The spectrum from the process of high harmonic generation (HHG) typically consists

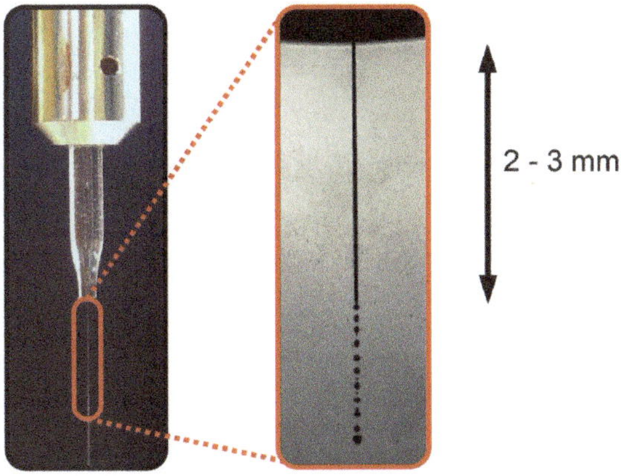

Fig. 1. Liquid beam (diameter = 15μm) and nozzle assembly. The enlarged section displays the continuous section of the liquid beam and the decay into droplets behind the Rayleigh limit.

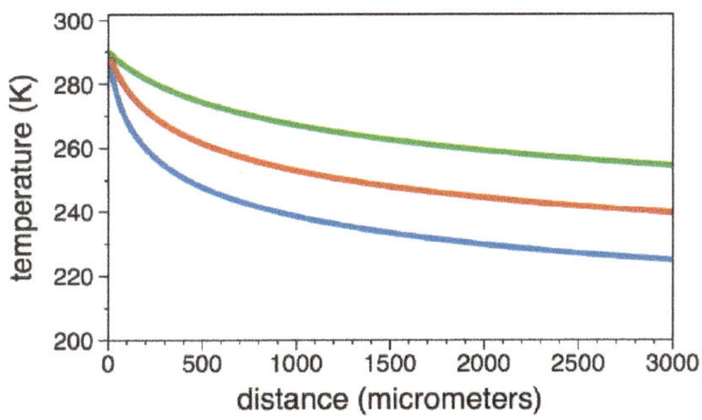

Fig. 2. Effect of evaporative cooling as a function of distance from the nozzle exit for a 17 μm nozzple at 0.3 ml/min flow and a flow speed of the jet of about 22 m/s. Color code: green: water; red: ethanol; blue: methanol.

of a plateau, where the harmonic intensity is nearly constant over many orders of magnitude followed by a sharp cutoff. The maximal harmonic photon energy E_c = hv is given by the well known cutoff law (Krause et al. (1992)),

$$E_c = I_p + 3.17U_p, \tag{1}$$

where I_p is the ionization potential of the target atom, and U_p the ponderomotive energy, respectively (Seres et al. (2004), Seres et al. (2006)). HHG has now been established as one of the best methods to produce ultrashort coherent light pulses covering a wavelength range from the vacuum ultraviolet to the soft x-ray region (Attwood (2000)). The development of HHG has opened new research areas such as attosecond science (Krausz & Ivanov

(2009)) and nonlinear optics in the extreme ultraviolet (xuv) region. Many features of HHG can be intuitively and sometimes even semi-quantitatively explained in terms of electron re-scattering trajectories, which are described by the semiclassical three-step model (Corkum (1993)). Since the first demonstration of high-harmonic order generation in 1987, (femtosecond) laser-driven high-harmonic generation sources have become increasingly important for the generation of coherent extreme-ultraviolet radiation and soft x-rays down to the water window (λ = 4.4 to 2.3 nm) and up to the keV-regime (Seres et al. (2006), Seres et al. (2005)). Compared to synchrotrons and x-ray free electron lasers (XFELs), these sources are small-scale, much less expensive, and highly versatile, and their resulting unique characteristic output can be tailored more or less according to the experimental requirements (Hentschel et al. (2001)). Some drawback in comparison to the devices above is certainly the limited photon flux - the big advantage is their reliable availability in the laboratory. Ultrashort pulses of table top systems of reasonable brightness and high coherence are used in various experimental setups and applications ranging from atomic and molecular spectroscopy in the gas phase, liquid phase and solid-state surfaces, and, due to its coherent nature (Jaegle (2006)), harmonic emission is increasingly used for coherent-diffractive imaging (Attwood (2000), Kapteyn et al. (2007)).
The setup used in Goettingen for ultrafast photoelectron spectroscopy near the liquid water interface (microjet surface) with UV or IR pump and HHG-probe is illustrated schematically in Fig. 3. The inset displays high harmonics on a multi-channel-plate detector behind the grating for Neon (a) and Argon (b) as a HHG medium. Note, the HHG radiation travels in vacuum only. It is dispersed and focused with a toroidal grating. The time-resolution of a single grating setup is on the order of a few hundred femtoseconds, depending upon the HHG spot and the illuminated grooves on the EUV-grating. It can be improved by employing dielectric mirrors or a dual grating setup (Nugent-Glandorf et al. (2001)). Further details of the setup are given in (Link, Lugovoy, Siefermann, Liu, Faubel & Abel (2009)).

3. Towards ultrafast angle-resolved photoelectron spectroscopy at liquid interfaces

In the ground state, the H_2O molecule exhibits the electron configuration

$$(1a_1)^2(2a_1)^2(1b_2)^2(3a_1)^2(1b_1)^2. \tag{2}$$

Here, $1b_1$, $3a_1$ and $1b_2$ are the three outer occupied molecular orbitals (MO) of water. The $1b_1$ orbital is the highest occupied molecular orbital (HOMO) and its nodal plane lies in the plane including all atoms. The $1b_1$ orbital has non-bonding character and hence is called a lone pair orbital. The $3a_1$ and $1b_2$ orbitals involve the O-H bond, whereby the $1b_2$ contributes the most. The $4a_1$ orbital is the lowest unoccupied molecular orbital (LUMO). In water aggregates, liquid water and ice, it accepts lone pair electrons from the $1b_1$ orbitals of neighboring water molecules - in line with the definition of Pauling of hydrogen bonding.
The direction of photoelectron emission in the gas phase and near interfaces (liquid or solid) is intrinsically non-trivial and more or less anisotropic in general (Banna et al. (1986)). The differential photoemission cross section for a single electron orbital in free atoms and molecules is given by

$$d\sigma/d\Omega = \sigma_0/4\pi(1+\beta P_2(cos\theta)). \tag{3}$$

σ_0 defines the total photoelectron emission cross for this orbital, and, β is called the anisotropy parameter. P_2 (cos θ) is the second Legrende polynom (Attwood (2000)).
Up to the present, the anisotropy parameters β and σ for the molecular valence orbitals of water have only been measured for isolated gas phase water molecules, but not for liquid water. It has instead been assumed that they do not differ significantly from the values

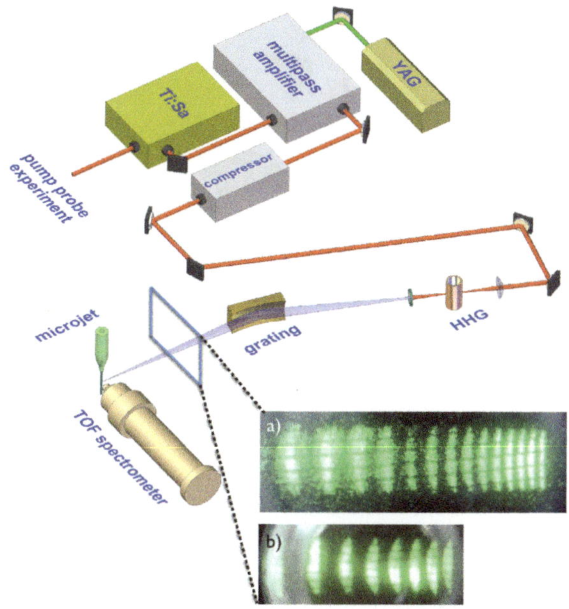

Fig. 3. Schematic view on the table top beam line experiment and the components for the photoelectron spectroscopy experiment. The essential units are depicted: Ti:Sa system, multipass amplifier, compressor, generation and selection of EUV light via HHG and toroidal grating. Inset: Generation of high harmonics of the fundamental 800 nm radiation in a neon or argon filled capillary. a) A sequence of high harmonics generated in neon, imaged by a XUV-sensitive CCD camera.The range of photon energies spans 40-70 eV b) A sequence of high harmonics generated in argon. The range of photon energies in this case is between 25 and 40eV. Note: it is only possible to generate the odd harmonics of the fundamental in both cases. Note: the EUV beam is in vacuum only up to the final chamber containing the liquid beam and the photoelectron spectrometer.

in the gase phase (Winter & Faubel (2006)). This lack of data is due to the challenges of an appropriate experimental setup. For the investigation of the angular distribution of photoelectron emission, measurements of photoelectrons at different angles between the photoelectron detecting direction and the polarization vector of the ionizing radiation have to be made. The adjustment of the photoelectron emission angle can be realized by rotating either the axis of the spectrometer or the polarization vector of the incident light. If the spectrometer is fixed in the setup, a suitable radiation source is needed, which can be changed in the polarization. Although, possible in principle at beamlines at a few stations polarization resolved experiments are rare - and not done for liquid water. In a table-top HHG experiment the EUV radiation polarization can be readily controlled via control of the fundamental radiation polarization (Link, Lugovoy, Siefermann, Liu, Faubel & Abel (2009)).
For the investigation of the angular distribution of emitted photoelectrons (at least) measurements at 3 different angles of polarization (HHG radiation) are necessary, if the spectrometer position is fixed. For this purpose, the polarization vector of the EUV light

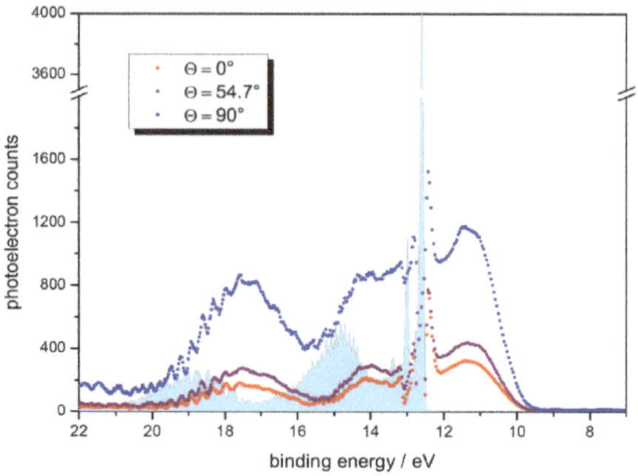

Fig. 4. Liquid beam photoelectron spectra at different angles. The photoelectron spectra are scaled to the $1b_1$ gas phase peak intensity. The light blue spectrum ist he spectrum of water in the gas phase measured with p-polarized EUV light.

will be tuned via the polarization of the fundmantal, since the photoelectron analyzer, a time-of-flight photoelectron spectrometer, is fixed perpendicular to the direction of light propagation and in the plane of the optical table. The $\lambda/2$ plate is positioned in the beam path of the fundamental radiation directed into the HHG chamber and in front of the iris that tailors the beam diameter of the 800 nm light. By rotating the optical axis of the $\lambda/2$ plate the electron emission is measured at the required angle. The initial polarization plane of the 800 nm fundamental light is parallel to that of the optical table, on which the HHG chamber is mounted. At p-polarization, the angle θ between the photoelectron detecting direction and the polarization vector of the ionizing radiation is $0°$, while $\theta = 90°$ for s-polarization. In addition, there is another relevant angle at $\theta = 54.7°$, also called the magic angle, which plays an important role in the characterization of the angular distribution of photoionization (Attwood (2000)). It must be noted that the focusing grating used here has different reflection properties for EUV radiation with various polarization vectors, which has been taken into account here.

The photoelectron emission spectra of liquid water for different angles of polarization of the high harmonics radiation are shown in Fig. 4. The spectra are normalized with respect to the $1b_1$ gas phase photoelectron emission peak. The pure gas phase spectrum (blue shaded) has been measured with a p-polarization of the EUV beam. These spectra translate into the anisotropy diagram in Fig. 5. For a HHG energy of 38.7 eV the anisotropy parameters β and σ of liquid in comparison to gas phase water for the differnt orbitals have been determined to be $1b_1(\beta = 0.8$ and $\sigma = 1)$, $3a_1$ $(\beta = 0.7$ and $\sigma = 0.88)$, and $1b_2$ $(\beta = 0.6$ and $\sigma = 0.94)$ for liquid water, and $1b_1(\beta = 1.4$ and $\sigma = 1)$, $3a_1$ $(\beta = 1.1$ and $\sigma = 0.88)$, and $1b_2$ $(\beta = 0.7$ and $\sigma = 0.94)$ for gas phase water. The latter are close to those measured by Banna et al. some time ago (Banna et al. (1986)). What is important and obvious here is the observation that the β -values of liquid and gas phase water are significantly different, while the σ-values are quite similar.

4. Hydrogen bonding in extreme states of water

If very short infrared radiation pulses tuned to a strong OH-stretch vibration absorption of water around 3 μm are used, it is possible to heat the water extremely, at a rate faster than the

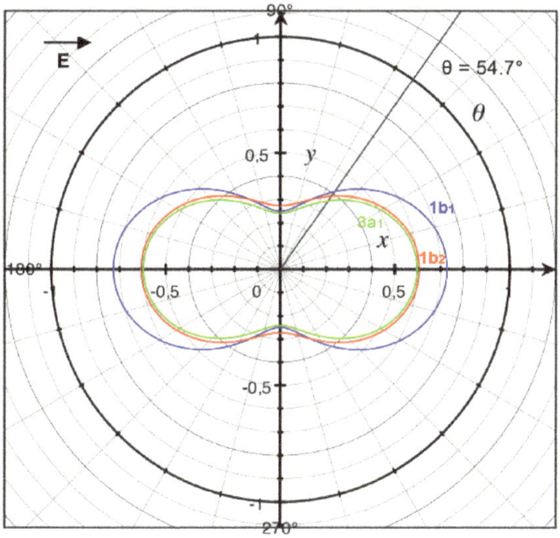

Fig. 5. Angular distribution of photoelectrons from the $1b_1$ (blue), $3a_1$ (green) and $1b_2$ (red) orbitals of liquid water. The ionizing photon energy is 38.7 eV. These curves are based on our experimentally obtained β and σ- values and assuming that the σ?- values are similar and do do not change significantly for water molecules in different aggregate states.

thermal expansion rate and to prepare extreme states of water (Link, Lugovoy, Siefermann, Liu, Faubel & Abel (2009), Link, Voehringer-Martinez, Lugovoj, Liu, Siefermann, Faubel, Grubmueller, Gerber, Miller & Abel (2009)). These states can have temperatures well above the boiling point and it has been shown that the water phase may be even heated significantly above the critical temperature (Debenedetti (1996)). We have recently investigated metastable, superheated and supercritical phases with time-resolved photoelectron spectroscopy with IR excitation and EUV probe. As shown in Ref. (Link, Voehringer-Martinez, Lugovoj, Liu, Siefermann, Faubel, Grubmueller, Gerber, Miller & Abel (2009)) the initial temperature can be estimated via the known absorption coefficients, penetrations depth of the IR beam into the water phase, and the laser focus diameter, after laser induced heating. If water is prepared at these extreme states, it is known that it literally explodes at the extreme conditions, however, at the femtosecond or picosecond time-scale it moves slowly in time and can thus be resolved with ultrafast spectroscopy. Different than with other spectroscopies time-resolved photoelectron spectra in the time-domain measure the chemical shift of the valence electrons of water in time, which provide interesting information about the question how the phase and the hydrogen-bonding network 'look like' and how they evolve in time. In a recent contribution we reported investigations on the molecular photoemission signature of the evolution of the phase and the hydrogen-bonding network of super-heated water below and above the critical point, as well as its timescales. Typical results are displayed in Fig. 6. In the plots the binding energy (being the difference of the photon energy and the kinetic energy of the photoelectrons measured in the experiment via a time-of-flight electron spectrometer, $E_b = E_{ph} - E_{kin}$) is plotted against the photoelectron emission counts resulting in a typical photoelectron spectrum in Fig. 6. In a typical valence electron photoemission spectrum of water the gas phase emission lines of the $1b_1$, $3a_1$, and $1b_2$ can be assigned easily (Winter et al.

(2004)). The corresponding lines of the orbitals in the liquid phase are shifted toward lower binding energies but they are significantly broader and overlapping. Since the corresponding $1b_1'$ orbital energy in the liquid phase is shifted most in the liquid phase it can be identified easily. A transition of water from the liquid phase to the gas phase is accompanied with a decrease in intensity in the most shifted $1b_1'$ emission line (color coded in red) of liquid water.

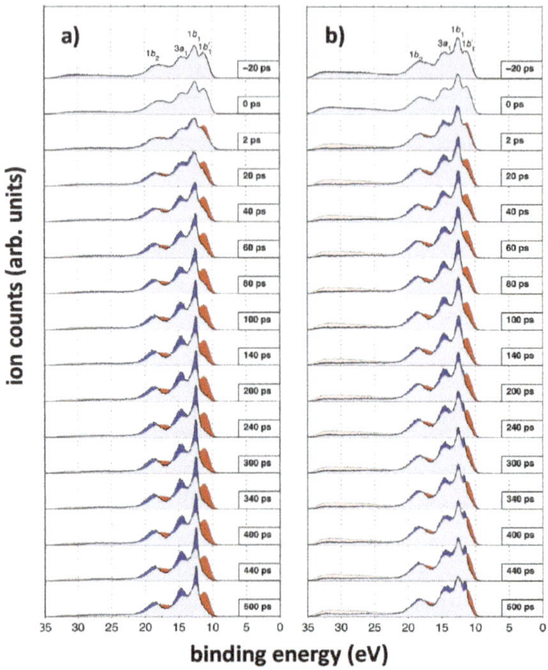

Fig. 6. Time-resolved photoelectron spectra of liquid (before excitation at neg. delay time) and metastable water (left: excitation at λ_{IR}=2650nm; right panel: λ_{IR}=2830nm:). For further details see the text and Refs. (Link, Lugovoy, Siefermann, Liu, Faubel & Abel (2009), Link, Voehringer-Martinez, Lugovoj, Liu, Siefermann, Faubel, Grubmueller, Gerber, Miller & Abel (2009)).

At the same time an increase in intensity of the gas phase lines ($1b_1$, $3a_1$, and $1b_2$) is observable (color coded in blue in Fig. 6). Different hot phases are correlated with the overall time-depedence of the signals and the dispersion of the liquid phase, which is observable in the binding energy region between the liquid phase $1b_1'$ peak and the coresponding gas phase peak. This is typically the range in which photoemission of more of less large clusters have been observed (see Refs in Link, Lugovoy, Siefermann, Liu, Faubel & Abel (2009) and Link, Voehringer-Martinez, Lugovoj, Liu, Siefermann, Faubel, Grubmueller, Gerber, Miller & Abel (2009)). In particular, the hotter phase in Fig. 6 (right panel) displays a higher degree of dispersion into smaller clusters. The overall chemical shift of the $1b_1'$ of the superheated phase towards larger binding energies is noticeable but small. In Ref. (Link, Voehringer-Martinez, Lugovoj, Liu, Siefermann, Faubel, Grubmueller, Gerber, Miller & Abel (2009)) we have shown that quantitative photoelectron emission detection and evaluation through a hot and moving interface is possible and that the observation depth is a few monolayers of water and alcohol molecules at the interface (Itikawa & Mason (2005)). The dynamics of superheated water and alcohols at the liquid-vacuum interface has been compared with large-scale molecular dynamics calculations that capture the dynamics of the solvents near their liquid and super heated interface. It is not unexpected that the dynamics of the phase evolution after laser heating is governed by the highly dynamical evolution of the network of hydrogen bonds (Link, Voehringer-Martinez, Lugovoj, Liu, Siefermann, Faubel, Grubmueller, Gerber, Miller & Abel (2009)).

5. Ultrafast concerted electron motion of a plasma at a liquid water surface

Various methods have been applied to investigate the parameters of laser-induced plasmas (see Ref. (Link, Lugovoy, Siefermann, Liu, Faubel & Abel (2009)) and refs therein). A plasma near a water interface constitutes a special case of coherent electron motion that can be observed with ultrafast photoelectron spectroscopy. Using optical near IR fs-pulses a surface plasma can be generated through multi photon absortion and ionization. If it is formed near an exploding liquid beam in vacuum the plasma first oscillates at the plasma frequency and then decays due to the overall decay of the hot water filament. High harmonics radiation can be employed to induce electron emission from the valence bands of dense water at the interface and gas phase water, which should be sensitive to the local plasma conditions. Thus we have anticipated that a plasma can be investigated with the new EUV photoelectron spectroscopy. In Ref. (Link, Lugovoy, Siefermann, Liu, Faubel & Abel (2009)) we report first experiments in which we have investigated the dynamics, i.e., the oscillation and decay of a plasma near a liquid water jet, as well as some of its characteristic parameters. IR pulses at an energy of of 20-30 μJ at 2650 nm were employed to trigger and induce the plasma. For pulse durations of 100 fs and a focus diameter of 80-100 μm a peak intensity of well above 10^{12} W/cm^2 is reached, which is sufficient to induce a plasma. The 25th harmonic of 800 nm radiation at 38.6 eV photon energy was chosen for the probe, i.e., the photoelectron emission from the valence bands of water. In the vicinity of a liquid beam in vacuum a significant but low vapor pressure was documented (Faubel et al. (1997)). In Fig. 7 (apparently) periodically changing 'binding energies' and photoelectron spectra as a function of time of a near liquid interface plasma, that displays distinct oscillations, is depicted (Link, Lugovoy, Siefermann, Liu, Faubel & Abel (2009)). The characteristics of the oscillating 'binding energies' suggest that the electrons stem from and belong to the water molecules, however, they are not tightly bound anymore to the water in the gas phase, but both are part of an ensemble of ionized cations and electrons oscillating coherently in time. The plot also shows photoelectrons with large kinetic energies close to the photon energy.

At early times the IR pump pulse initiates plasma formation near the liquid beam interface (a number of 22 photons at 2650 nm would be necessary to result in an ionization energy of 10 eV). In Ref. (Link, Lugovoy, Siefermann, Liu, Faubel & Abel (2009)) we have determined a period of oscillation of major photoemission lines of 1.5 ps corresponds to a plasma frequency of 0.67 THz and a free electron density close to of 5.6 x $10^{15} cm^{-3}$. This is a quite low level of electron density. It is worth noting that with this setup we have realized a pulsed ultrafast THz-Oscillator that should emit pulsed picosecond THz-radiation (not measured in (Link, Lugovoy, Siefermann, Liu, Faubel & Abel (2009))). At 2650 nm the resonance frequency of a laser driven plasma is 113 THz, which corresponds to a free electron density of 1.6 x 10^{20} cm^{-3}. This number is more than one order of magnitude lower than the critical resonance free electron density for 800 nm radiation. The time scale of the decay of the plasma in the few picoseconds regime is related to the transport of hot water molecules in the expanding (exploding) beam interface, which has been estimated from molecular dynamics simulations. This is again a nice example for the power of the pump-EUV-photoelectron

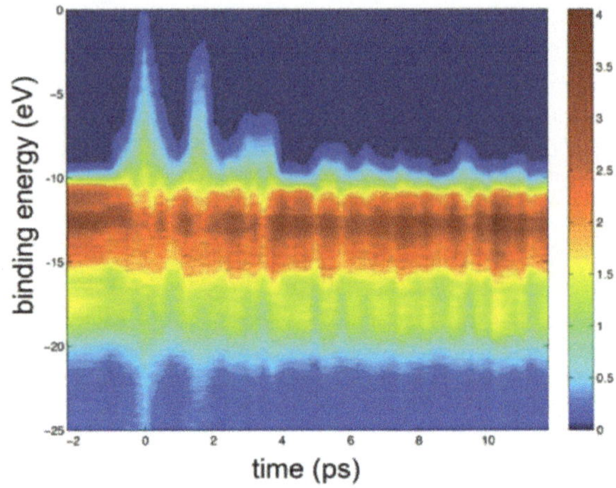

Fig. 7. Kinetic energies of photoelectrons as a function of time on top of the water photoelectron spectrum (spectra represent a vertical cut) in the case of an oscillating near liquid water surface plasma. An oscillation period of the fast electrons of 1.5 ps corresponding to a plasma frequency of 0.67 THz and a free electron density of 6×10^{15} cm^{-3} (for more details see the text).

emission probe technology near volatile liquid interfaces enabling real-time probing of laser-ionized plasmas with high temporal resolution, yielding information that is hardly accessible by other methods.

6. The solvated electron in water - news from a seemingly well known transient

Further development of liquid phase time-resolved photoelectron spectroscopy in our lab made it possible to directly measure the vertical binding energy (VBE) of one of the most important chemical transients in aqueous solution: the solvated electron in liquid water (Boag & Hart (1963)), which is also termed hydrated electron or $e^{-}(aq)$ (Siefermann et al. (2010)). In the most common picture, a hydrated electron is located in a non-spherical cavity formed by about 6 water molecules (Kevan (1981)). The VBE corresponds to the energy, which is required to completely remove the electron from this cavity, without changing the initial geometry of the system. The VBE is a crucial quantity for understanding the reactivity of hydrated electrons with DNA bases in DNA damaging processes - or many other electron transfer and attachment processes (Sanche (2009), Lu (2009)). Shortly after the report by Siefermann et al. in 2010 three more publications confirmed the binding energy of the bulk hydrated electron (Tang et al. (2010), Shreve et al. (2010), Luebcke et al. (2010)).

For a long time researchers tried to estimate the binding energy of hydrated electrons from data of large anionic water clusters (Neumark (2008)). In 2010, Siefermann et al. for the first time directly measured the vertical binding energies of $e^{-}(aq)$ (VBE = 3.3 ± 0.1 eV) and of electrons solvated/hydrated at the water surface $e^{-}(sf)$ (VBE = 1.6 ± 0.1 eV) using the experimental pump-probe setup employing UV pump- and EUV-probe pulses (see Fig. 3). The key feature of the experimental approach is the generation of solvated electrons by short pump pulses of 267 nm light (Chen & Bradforth (2008)) and recording photoelectron spectra at the same time employing time-delayed 38.7 eV (32 nm) high harmonic probe

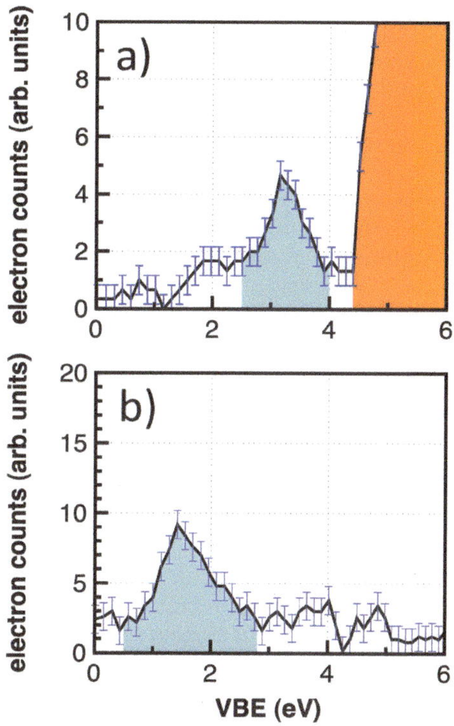

Fig. 8. Photoelectron spectrum and binding energy of the fully hydrated electron (a) and the surface hydrated electron (b) near the liquid water interface. While the first transient is prepared via a one photon excitation (low laser light intensity) of $K_4(Fe(CN)_6)$ in water the latter is prepared via two-photon excitation (high laser light intensity) of water at 270 nm. The emission bands from both hydrated electron binding motifs exist only if a 270 nm excitation pulse precedes the EUV-probe (both colored in blue). The brown emission band stems from the iron ion (Fe^{2+}) in the precursor complex. The noise level is indicated.

pulses. In order to access the properties of $e^-(sf)$, pure water was ionized via the two-photon absorption of 267nm light, which results in the formation of bulk ($e^-(aq)$) and surface solvated ($e^-(sf)$) electrons (Chen & Bradforth (2008)). When subsequently recording the photoelectron spectrum with the 38.7 eV probe pulse, the detection efficiency is highest for photoelectrons originating from the surface, and declines dramatically with every layer of water molecules (Ottosson et al. (2010)). The reason lies in the mean free path of photoelectrons, which is particularly short (Attwood (2000)) for the resulting photoelectron kinetic energies of about 35 eV (38.7 eV - VBE). This allowed us to directly measure properties of $e^-(sf)$, and it demonstrates that our experimental setup is particularly suited for the investigation of transient surface species, such as $e^-(sf)$.

In order to access the VBE of $e^-(aq)$ with our experimental setup, it is required to significantly increase the concentration of $e^-(aq)$, and at the same time suppress the formation of $e^-(sf)$. This 'inversion' type of situation is realized by employing $[Fe(CN)_6]^{4-}$ complexes in aqueous solution as precursors, which primarily yield $e^-(aq)$ after the absorption of one 267nm photon and which are literally repelled from the water surface (Siefermann et al. (2010)). Suppressing two-photon processes by lowering the 267 nm pump pulse intensity inhibits the formation of $e^{-}(sf)$ and allows for recording the signature of the fully hydrated electron $e^{-}(aq)$. Figure 8 presents the photoelectron spectra of $e^{-}(aq)$ (a) and $e^{-}(sf)$ (b). Evaluation of the relative signal heights reveals that signal intensities both species (binding motifs) are on the same order of magnitude. For different pump-probe time-delays between 7 ps and 100 ps, no significant changes in the photoelectron spectrum of $e^{-}(sf)$ were observed, which demonstrates that the lifetime of $e^{-}(sf)$ exceeds 100 ps, which was tentatively attributed to a dynamic barrier that stems from rearranging hydrogen bonds (Siefermann et al. (2010)). Follow-up experiments by a number of other groups employing UV light (267 - 213 nm) to probe the binding energy of the bulk hydrated electron yielded values of 3.27 eV (Tang et al. (2010)), 3.6 eV (Shreve et al. (2010)), and 3.4 eV (Luebcke et al. (2010)), which are close to the value initially reported in Ref. (Siefermann et al. (2010)).

7. Conclusions and outlook

In summary, this work demonstrates the potential for a new kind of ultrafast liquid phase spectroscopy, based on high-harmonics generation and liquid microbeam technology, that has unique advantages. First and foremost, the technique is very surface sensitive at wavelength in the extreme UV. This changes for shorter wavelengths. In particular, wavelength tuning should enable depth profiling from the liquid interface to the bulk - a very intruiging possibily. Time-resolution can be increased routinely nowadays (with commercial products) to a few femtoseconds enabling a new time window for molecular dynamics. This time window is reduced by recent efforts to produce stable single attosecond pulses in the extreme UV spectral range. Several groups demonstrated the generation of attosecond light pulses using high harmonic generation and investigated processes on the attosecond to few femtosecond timescale. One of our near future goals is it to improve the time-resolution of our experimental setup towards the few femtosecond timescale. The quest for higher and higher harmonics and shorter wavelength also enables new avenues to perform real ESCA experiments in the water window at around 3-4 nm. This is very close to the heart of a chemist, which enables us to watch chemical reactions though their change in electron density and orbital evolution as they happen in solution. Furthermore, the high harmonic probe should not be limited by 'dark states' present in other spectroscopies. High harmonics radiation can be employed as an *in situ* and local and site-specific probe of femtosecond and even attosecond nuclear and electron dynamics in molecules surrounded by a solvent or dense environment. Finally, the table top beamline setup is a very valuable tool to pre-study systems and optimal conditions for experiments at synchrotron beam limes such as the free electron lasers (XFEL, FLASH, LCLS) in Hamburg and Stanford, which will have superior performance but very limited beam time. To just repeat an experiment performed with somewhat limited laser driven table top systems with optimal conditions at a large scale x-ray laser facility is much more rewarding and promising than optimizing experiments on site. Therefore, experimental setups highlighted here will have a very bright future, even though large scale x-ray laser facilities are being build and ready to use in the near future.

8. Acknowledgement

This work has been generously supported by the Deutsche Forschungsgemeinschaft though the SPP1134 (*Investigation of Transient Structures of Photo-Induced Chemical Reactions in Solution with Ultrafast XUV-Photoelectron Spectroscopy*), the SFB 755 (*Nanoscale Photonic Imaging*), the Graduate School 782 (*Dynamics and spectroscopy of molecular aggregates, coils and networks*), the Fonds der Chemischen Industrie, as well as the Volkswagenstiftung.

Contributions of Dr. O. Link, Dr. K. Siefermann, Dr. Y. Liu, Dr. E. Lugovoy, and Dr. M. Faubel to the work and technology described here are gratefully acknowledged. The author also thanks M. Faubel, B. Winter, U. Buck, and E. Lugovoy for stimulating discussions on this topic.

9. References

Assmann, J., Kling, M. & Abel, B. (2003). Watching photo induced chemistry and molecular energy flow in solution in real-time, *Angewandte Chem Int. Ed.* 42: 2226–2246.

Attwood, D. (2000). *Soft X-Rays and Extreme Ultraviolet Radiation*, Cambridge University Press.

Banna, M. S., McQuaide, B. H., Malutzki, R. & Schmidt, V. J. (1986). The photoelectron spectrum of water in the 30-140 ev photon energy range, *J. Chem. Phys.* 84: 4739–4744.

Boag, J. W. & Hart, E. J. (1963). Absorption spectra in irradiated water and some solutions - absorption spectra of hydrated electron, *Nature* 197(486): 45.

Charvat, A., Lugovoy, E., Faubel, M. & Abel, B. (2004). New design for a time-of-flight mass spectrometer, *Rev.. Sci. Instr.* 75: 1209–1218.

Chen, X. & Bradforth, S. E. (2008). The ultrafast dynamics of photodetachment, *Annu. Rev. Phys. Chem.* 59: 203.

Corkum, P. B. (1993). Plasma perspective on strong-field multiphoton ionization, *Phys. Rev. Lett.* 71(13): 1994.

Cowan, M. L., Bruner, B. D., Huse, N., Dwyer, J. R., Chugh, B., Nibbering, E. T. & Elsaesser, T. (2005). Ultrafast memory loss and energy redistribution in the hydrogen bond network of liquid h2o, *Nature* 434: 199–202.

Debenedetti, P. G. (1996). *Metastable Liquids: Concepts and Principles*, Academic Press, Princeton, New Jersey.

Drescher, M. (2004). Time-resolved esca: a novel probe for chemical analysis, *Z.Phys. Chem.* 218: 1147–1168.

Faubel, M. (2000). *Photoionization and Photodetachment*, C. Y. Ng, World Scientific.

Faubel, M. & Kisters, T. (1989). Non-equilibrium molecular evaporation of carboxylic acid dimers, *Nature* 339(6225): 527.

Faubel, M., Steiner, B. & Toennies, J. P. (1997). Photoelectron spectroscopy of liquid water, some alcohols, and pure nonane in free micro jets, *J. Chem. Phys.* 106: 9013.

Fecko, C. J., Eaves, J. D., Loparo, J. J., Tokmakoff, A. & Geissler, P. L. (2003). Ultrafast hydrogen-bond dynamics in the infrared spectroscopy of water, *Science* 301: 1698–1702.

Head-Gordon, T. & Johnson, M. E. (2005). Tetrahedral structure or chains for liquid water, *PNAS* 103: 7973–7977.

Hentschel, M., Kienberger, R., Spielmann, C., Reider, G. A., Milosevic, N., Brabec, T., Corkum, P., Heinzmann, U., Drescher, M. & Krausz, F. (2001). Attosecond metrology, *Nature* 414: 509–513.

Itikawa, Y. & Mason, N. (2005). Cross sections for electron collisions with water molecules, *J. Phys. Chem. Ref. Data* 34: 1–22.

Jaegle, P. (2006). *Coherent Sources of XUV Radiation*, Springer-New York.

Kapteyn, H., Cohen, O., Christov, I. & Murnane, M. (2007). Harnessing attosecond science in the quest for coherent x-rays, *Science* 317: 775–778.

Kevan, L. (1981). Solvated electron structure in glassy matrices, *Acc. Chem. Res.* 14: 138. Kevan structure of hydrated electron.

Kling, M. F., Siedschlag, C., Verhoef, A. J., Khan, J. I., Schultze, M., Uphues, T., Ni, Y., Uiberacker, M., Drescher, M., Krausz, F. & Vrakking, M. J. J. (2006). Control of electron localization in molecular dissociation, *Science* 312(5771): 246.

Krause, J. L., Schafer, K. J. & Kulander, K. C. (1992). High-order harmonic generation from atoms and ions in the high intensity regime, *Phys. Rev. Lett.* 68: 3535.

Krausz, F. & Ivanov, M. (2009). Attosecond physics, *Rev. Mod. Phys.* 81(1): 163.

Link, O., Lugovoy, E., Siefermann, K. R., Liu, Y., Faubel, M. & Abel, B. (2009). Ultrafast electronic spectroscopy for chemical analysis near liquid water interfaces: concepts and applications, *Appl. Phys. A* 96(1): 117.

Link, O., Voehringer-Martinez, E., Lugovoj, E., Liu, Y., Siefermann, K., Faubel, M., Grubmueller, H., Gerber, R. B., Miller, Y. & Abel, B. (2009). Ultrafast phase transition in metastable water near liquid interfaces, *Faraday Discuss.* 141: 67.

Lu, Q.-B. (2009). Correlation between cosmic rays and ozone depletion, *Phys. Rev. Lett.* 102(11): 118501.

Luebcke, A., Buchner, F., Heine, N., Hertel, I. & Schultz, T. (2010). Time-resolved photoelectron spectroscopy of solvated electrons in aqueous nai solution, *PCCP* 12: 14692.

Lundholm, M., Siegbahn, H., Holmberg, S. & Arbman, M. (1986). Core electron spectroscopy of water solutions, *J. Electron. Spectrosc. Relat. Phenom.* 40(2): 163.

Neumark, D. (2008). Spectroscopy and dynamics of excess electrons in clusters, *Mol. Phys.* 106: 2183.

Nugent-Glandorf, L., Scheer, M., Samuels, D. A., Mulhisen, A. M., Grant, E. R., Yang, X., Bierbaum, V. M. & Leone, S. R. (2001). *Phys. Rev. Lett.* 87: 193002.

Ottosson, N., Faubel, M., Bradforth, S. E., Jungwirth, P. & Winter, B. (2010). Photoelectron spectroscopy of liquid water and aqueous solution: Electron effective attenuation lengths and emission-angle anisotropy, *J. Electron. Spectrosc. Relat. Phenom.* 177(2-3): 60.

Paul, P. M., Toma, E. S., Breger, P., Mullot, G., Auge, F., Balcou, P., Muller, H. G. & Agostini, P. (2001). Observation of a train of attosecond pulses from high harmonic generation, *Science* 292: 1689.

Sanche, L. (2009). Beyond radical thinking, *Nature* 461: 358–359.

Seres, E., Seres, J., Krausz, F. & Spielmann, C. (2004). *Phys. Rev. Lett.* 92: 163002.

Seres, J., Seres, E., Verhoef, A. J., Tempea, G., Streli, C., Wobrauschek, P., Yakovlev, V., Scrinzi, A., Spielmann, C. & Krausz, F. (2005). Source of coherent kiloelectronvolt x-rays, *Nature* 433: 596.

Seres, J., Wobrauschek, P., Streli, C., Yakovlev, V. S., Seres, E., Krausz, F. & Spielmann, C. (2006). *New J. Phys.* 8: 251.

Shreve, A. T., Yen, T. A. & Neumark, D. M. (2010). Photoelectron spectroscopy of hydrated electrons, *Chem. Phys. Lett.* 493: 216.

Siefermann, K. R., Liu, Y., Lugovoy, E., Link, O., Faubel, M., Buck, U., Winter, B. & Abel, B. (2010). Binding energies, lifetimes and implications of bulk and interface solvated electrons in water, *Nat. Chem.* 2(4): 274.

Siegbahn, H. (1985). Electron spectroscopy for chemical analysis of liquids and solutions, *J. Phys. Chem.* 89: 897.

Tang, Y., Shen, H., Sekiguchi, K., Kurahashi, N., Mizuno, T., Suzuki, Y.-I. & Suzuki, T. (2010). Direct measurement of vertical binding energy of a hydrated electron, *PCCP* 12: 3653.

Winter, B. & Faubel, M. (2006). Photoemission from liquid aqueous solutions, *Chem. Rev.* 106: 1176.

Winter, B., Weber, R., Widdra, W., Dittmar, M., Faubel, M. & Hertel, I. V. (2004). Full valence band photoemission from liquid water using euv synchrotron radiation, *J. Phys. Chem. A* 108(14): 2625.

Woutersen, S. & Bakker, H. J. (2006). Ultrafast vibrational and structural dynamics of the proton in liquid water, *Phys. Rev. Lett.* 96: 138305.

Permissions

The contributors of this book come from diverse backgrounds, making this book a truly international effort. This book will bring forth new frontiers with its revolutionizing research information and detailed analysis of the nascent developments around the world.

We would like to thank Anatoli V. Andreev, for lending his expertise to make the book truly unique. He has played a crucial role in the development of this book. Without his invaluable contribution this book wouldn't have been possible. He has made vital efforts to compile up to date information on the varied aspects of this subject to make this book a valuable addition to the collection of many professionals and students.

This book was conceptualized with the vision of imparting up-to-date information and advanced data in this field. To ensure the same, a matchless editorial board was set up. Every individual on the board went through rigorous rounds of assessment to prove their worth. After which they invested a large part of their time researching and compiling the most relevant data for our readers. Conferences and sessions were held from time to time between the editorial board and the contributing authors to present the data in the most comprehensible form. The editorial team has worked tirelessly to provide valuable and valid information to help people across the globe.

Every chapter published in this book has been scrutinized by our experts. Their significance has been extensively debated. The topics covered herein carry significant findings which will fuel the growth of the discipline. They may even be implemented as practical applications or may be referred to as a beginning point for another development.

The editorial board has been involved in producing this book since its inception. They have spent rigorous hours researching and exploring the diverse topics which have resulted in the successful publishing of this book. They have passed on their knowledge of decades through this book. To expedite this challenging task, the publisher supported the team at every step. A small team of assistant editors was also appointed to further simplify the editing procedure and attain best results for the readers.

Our editorial team has been hand-picked from every corner of the world. Their multi-ethnicity adds dynamic inputs to the discussions which result in innovative outcomes. These outcomes are then further discussed with the researchers and contributors who give their valuable feedback and opinion regarding the same. The feedback is then collaborated with the researches and they are edited in a comprehensive manner to aid the understanding of the subject.

Apart from the editorial board, the designing team has also invested a significant amount of their time in understanding the subject and creating the most relevant covers. They scrutinized every image to scout for the most suitable representation of the subject and create an appropriate cover for the book.

The publishing team has been involved in this book since its early stages. They were actively engaged in every process, be it collecting the data, connecting with the contributors or procuring relevant information. The team has been an ardent support to the editorial, designing and production team. Their endless efforts to recruit the best for this project, has resulted in the accomplishment of this book. They are a veteran in the field of academics and their pool of knowledge is as vast as their experience in printing. Their expertise and guidance has proved useful at every step. Their uncompromising quality standards have made this book an exceptional effort. Their encouragement from time to time has been an inspiration for everyone.

The publisher and the editorial board hope that this book will prove to be a valuable piece of knowledge for researchers, students, practitioners and scholars across the globe.

List of Contributors

Boeglin Christine
Institut de Physique et de Chimie de Strasbourg, Université de Strasbourg, France

Eisuke Miura
National Institute of Advanced Industrial Science and Technology (AIST), Tsukuba central 2, 1-1-4 Umezono, Tsukuba, Ibaraki, Japan

Vadim Belyaev and Anatoly Matafonov
Central Research Institute of Machine Building, Russian Federation

Serguei Y. Kalmykov and Bradley A. Shadwick
Department of Physics and Astronomy, University of Nebraska - Lincoln, Lincoln, USA

Arnaud Beck and Erik Lefebvre
CEA, DAM, DIF, Arpajon F-91297, France

Elena Kuznetsova and S. F. Yelin
Department of Physics, University of Connecticut, Storrs, Connecticut, United States of America
ITAMP, Harvard-Smithsonian Center for Astrophysics, Cambridge, Massachusetts, United States of America

Robin Côté
Department of Physics, University of Connecticut, Storrs, Connecticut, United States of America

T. Brudevoll and A. K. Storebo
FFI (Norwegian Defence Research Establishment), Kjeller

O. Skaaring
Kongsberg Defence and Areospace, Kjeller

C. N. Kirkemo
Department of Physics, University of Oslo, Oslo

O. C. Norum and O. Olsen
Department of Physics, Norwegian University of Science and Technology, Trondheim

M. Breivik
Department of Electronics and Telecommunications, Norwegian University of Science and Technology, Trondheim, Norway

R. Cabrera-Trujillo, J. Jiménez-Mier and A. M. Juárez
Universidad Nacional Autónoma de México, México

László Nánai and Szabolcs Beke
University of Szeged, Department of General and Environmental Physics Laser Technology Laboratory, Hungary

Koji Sugioka
RIKEN - The Institute of Physical and Chemical Research, Wako, Saitama, Japan

Enikoe Seres and Christian Spielmann
Institute of Optics and Quantumelectronics, Friedrich Schiller University Jena, Germany
Helmholtz Institute Jena, Germany

A.V. Andreev, S.Yu. Stremoukhov and O.A.Shoutova
Physics Department, M.V. Lomonosov Moscow State University, Russia

Philip Chan
Institute for Quantum Information Science, and Department of Electrical and Computer Engineering, University of Calgary, Canada

Itzel Lucio-Martínez, Xiaofan Mo and Wolfgang Tittel
Institute for Quantum Information Science, and Department of Physics and Astronomy, University of Calgary, Canada

Alessia Allevi
Dipartimento di Fisica e Matematica - Università degli Studi dell'Insubria, U.d.R. Como - Consorzio Nazionale Interuniversitario per le Scienze fisiche della Materia, Italy

Maria Bondani
Istituto di Fotonica e Nanotecnologie - Consiglio Nazionale delle Ricerche, U.d.R. Como - Consorzio Nazionale Interuniversitario per le Scienze fisiche della Materia, Italy

Hao Ni and Kun Zhao
Laboratory of Optic Sensing and Detecting Technology, China University of Petroleum, Beijing, People's Republic of China

Kui-juan Jin, Chen Ge, Hui-bin Lu and Guo-zhen Yang
Beijing National Laboratory for Condensed Matter Physics, Institute of Physics, Chinese Academy of Sciences, China

Hubert Grün, Thomas Berer, Karoline Felbermayer and Peter Burgholzer
RECENDT Research Center for NonDestructive Testing GmbH, Linz, Austria

Markus Holotta
Department of Radiology, Innsbruck Medical University, Innsbruck, Austria

Gerhard Zangerl
Computational Science Center, University of Vienna, Vienna, Austria

Robert Nuster and Günther Paltauf
Department of Physics, Karl Franzens University, Graz, Austria

Tong-Yi Zhang and Wei Zhao
State Key Laboratory of Transient Optics and Photonics, Xi'an Institute of Optics and Precision Mechanics, Chinese Academy of Sciences, People's Republic of China

Bernd Abel
University of Leipzig, Germany